Artificial Intelligence, Big Data and Data Science in Statistics

Ansgar Steland • Kwok-Leung Tsui
Editors

Artificial Intelligence, Big Data and Data Science in Statistics

Challenges and Solutions in Environmetrics, the Natural Sciences and Technology

Springer

Editors
Ansgar Steland
Institute of Statistics and AI Center
RWTH Aachen University
Aachen, Germany

Kwok-Leung Tsui
Grado Department of Industrial
and Systems Engineering
Virginia Polytechnic Institute and State
University
Blacksburg, VA, USA

ISBN 978-3-031-07154-6 ISBN 978-3-031-07155-3 (eBook)
https://doi.org/10.1007/978-3-031-07155-3

This Springer imprint is published by the registered company Springer Nature Switzerland AG
The registered company address is: Gewerbestrasse 11, 6330 Cham, Switzerland

Preface

The change to data-centrism in many fields, the need to extract information and knowledge from big data, and the increasing success of machine learning (ML) and artificial intelligence (AI) have created both opportunities and challenges to the field of statistics. These developments have, to some extent, led to the creation of data science, partially regarded as a new discipline, related to statistics and computer science. The intersections among ML/AI, data science, and statistics are much larger than people expect, particularly on theory, models, practical methods, and problems under investigation. All communities can learn a lot from each other.

The impressive successes of ML and AI methods, especially deep learners and convolutional networks, in many practical problems might seem to devalue statistical approaches. Quite a few researchers as well as practitioners regard machine learning as being more focused on problem solving and benchmark data sets than statistics. But, on the other hand, ML solutions are often tailored to a specific problem and thus can be difficult to generalize and implement for a wide range of applications.

Further, there is wide range of problems related to data for which statistics provides more appropriate or even optimal solutions and allows specific interpretable models. Stochastic models often provide mathematical descriptions of physical processes rather than relying on black boxes. Indeed, lack of model interpretability, potential bias, causality, and stability, and why and when deep learners may work are common questions for the ML approaches. Statistical thinking and approaches are good alternatives to rectify these problems, in terms of both theories, models, and practical methods. A further issue where statistics is indispensable is the question whether a given data set satisfies proper sampling designs, as studied by statistical sampling theory, and the sound statistical preprocessing, handling, and cleaning of data. Both topics are important to evaluate given data, to ensure high data quality, and to clarify what can be learnt from a certain data set. On the other hand, the flexibility of many ML and AI methods may yield superior results when reliable first-class data from well-selected variables are not available and one has to rely on noisy and surrogate data.

Focusing on environmental science, natural science, and technology, this book contributes to the discussions of various issues and general interplay among statistics, data science, machine learning, and artificial intelligence. The chapters cover theoretical studies of machine learning methods, expositions of general methodologies for sound statistical analyses of data, as well as novel approaches for modeling and analyzing data in specific areas and problems. In terms of applications, the chapters deal with data as arising in industrial quality control, autonomous driving, transportation and traffic, chip manufacturing, photovoltaics, football, transmission of infectious diseases, Covid-19, and public health.

The idea for this volume came from the meetings of the Section on Environmetrics, Natural Science and Technology of Deutsche Statistische Gesellschaft of the last few years, and most authors have presented research at the annual conferences Statistische Woche. All chapters of this volume have been peer reviewed, and the editors are grateful to those colleagues who helped in the evaluation process as anonymous reviewers. Nevertheless, the authors of each chapters are solely responsible for their work.

Aachen, Germany Ansgar Steland
Blacksburg, VA, USA Kwok-Leung Tsui
November 2021

Contents

Part II Challenges and Solutions in Applications

Part I
Methodologies and Theoretical Studies

One-Round Cross-Validation and Uncertainty Determination for Randomized Neural Networks with Applications to Mobile Sensors

Ansgar Steland and Bart E. Pieters

Abstract Randomized artificial neural networks such as extreme learning machines provide an attractive and efficient method for supervised learning under limited computing resources and for green machine learning. This especially applies when equipping mobile devices (sensors) with weak artificial intelligence. Results are discussed about supervised learning with such networks and regression methods in terms of consistency and bounds for the generalization and prediction error. Especially, some recent results are reviewed addressing learning with data sampled by moving sensors leading to non-stationary and dependent samples. As randomized networks lead to random out-of-sample performance measures, we study a cross-validation approach to handle the randomness and make use of it to improve out-of-sample performance. Additionally, a computationally efficient approach to determine the resulting uncertainty in terms of a confidence interval for the mean out-of-sample prediction error is discussed based on two-stage estimation. The approach is applied to a prediction problem arising in vehicle integrated photovoltaics.

Keywords Cross-validation · Extreme learning · Model comparison · Neural network · Photovoltaics · Uncertainty interval

1 Introduction

Artificial neural networks are an attractive class of models for supervised learning tasks arising in data science such as nonlinear regression and predictive analytics. There is a growing interest which is mainly driven by the development of highly

A. Steland (✉)
Institute of Statistics and AI Center, RWTH Aachen University, Aachen, Germany
e-mail: steland@stochastik.rwth-aachen.de

B. E. Pieters
Forschungszentrum Jülich, Institut für Energie- und Klimaforschung, Jülich, Germany
e-mail: b.pieters@fz-juelich.de

A. Steland, K.-L. Tsui (eds.), *Artificial Intelligence, Big Data and Data Science in Statistics*, https://doi.org/10.1007/978-3-031-07155-3_1

efficient and fast algorithms for training and an improved understanding of multi-layer deep neural network architectures, especially in terms of how to design and fit them for concrete machine learning tasks. Supervised learning tasks are pervasive and allow to equip devices and technical systems with weak artificial intelligence by processing data collected by sensors. This quickly results in big data sets difficult to handle by, for example, cars or smartphones, which have limited computing resources but can highly benefit form autonomous learning abilities. For example, one can attach solar panels to cars and trucks and use past and current data as well as planned routes and geographical data to predict the energy production of the solar panels and optimize their usage or storage during driving.

Extreme learning machines [10, 28], are widely used in applications due to their extremely fast learning compared to full optimization of feedforward neural networks. For this reason, they have been chosen as a benchmark classifier in the recently updated MNIST data set of handwritten characters, see [6] for details. They also performed very well in empirical comparison studies investigating 179 different classifiers for a large number of publicly available data sets [7]. Extreme learning machines select the parameters of all layers except the last (output) layer randomly. The weights of the last layer are optimized by minimizing a (regularized) least squares error criterion. Since the output layer uses a linear activation function, this step means that the random but data-dependent features generated by the preceding layers are linearly combined to explain the target values (responses). The optimization of the parameters of the output layer collapses to a linear least squares problem, which can be solved explicitly and does not require iterative minimization algorithms. Consequently, extreme learning machines optimize only a part of the parameters and choose the remaining ones randomly. Therefore, they belong to the class of randomized networks, see [20] for a broader review including random kernel machines using random Fourier features and reservoir computing based on randomized recurrent networks.

In this paper, cross-validation and two-stage estimation methodologies are proposed to handle the uncertainty resulting from the randomization of feedforward neural networks in a statistical sound way. The basic idea studied here is to apply a simple cross-validation scheme to evaluate network realizations to pick one with good out-of-sample generalization. The approach can be easily combined with model selection, especially the choice of the number of hidden neurons. Since the usual error criteria minimized by training algorithms are known to have many local minima, the random choice of the starting point used for optimization also leads to some degree of randomness of the trained network. This especially applies when using early stopping techniques to achieve better generalization abilities. Hence, the approach can also be used when optimizing all parameters of a neural net. Further, the basic idea can easily be applied to other machine learners, but in our presentation we focus on (randomized) feedforward networks resp. extreme learning machines.

Cross-validation is a well-established and widely used statistical approach [27], and has been extensively studied for nonlinear regression and prediction, see, e.g., [1] for a review and [23] for results addressing kernel smoothers for dependent data streams with possible changes. In its simplest form, one splits the data in a training (or learning) sample and a validation sample. The performance of a method estimated (trained) from the learning sample is then evaluated by applying it to the validation sample and tuned by choosing hyperparameters of the method. The final fit is evaluated using a test sample. In principle, there are well-established results when this works. Cross-validation for comparing regression procedures has been studied by Yang [29] for i.i.d. training samples. It is, however, worth mentioning that these results are usually not applicable to machine learning methods such as artificial neural networks, since this would require to train and apply the methods in such a way that consistency as statistical estimators is ensured. But this is frequently not met in practice.

As a second statistical tool, we discuss the construction of an uncertainty interval for the mean sample prediction error in the validation data set. Two-stage estimation is a well-established statistical approach leading to small required sample sizes, thus keeping the computational costs at a low level.

As a challenging data science problem we discuss the application to a prediction problem arising in photovoltaics (PV) and analyze data from a pilot study to illustrate the proposal. In recent years, especially with the electrification of transport, there is an increasing interest in Vehicle Integrated Photovoltaics (VIPV) applications [3, 13]. In this context there is a keen interest in photovoltaic (PV) yield prediction for such applications. Yield prediction is complicated by the ever changing orientation and location of the vehicle, with influences of buildings and other objects. Many vehicles have specific routes that often repeat (e.g., home/work commuting). Artificial neural networks may provide a powerful means to improve the PV yield prediction on those specific routes.

The organization of the paper is as follows. Section 2 reviews artificial feedforward neural networks, fully optimized and randomized ones, and discusses some recent theoretical results on fast training algorithms and learning guarantees in terms of consistency results and bounds for generalization errors. It draws to some extent on the expositions in [8, 10, 24]. The proposed cross-validation approach to deal with the randomness of the out-of-sample performance of randomized networks is presented in Sect. 3. The application to vehicle integrated photovoltaics including a data analysis is provided in Sect. 4.

2 Classical Neural Networks and Extreme Learning Machines

2.1 Hidden Layer Feedforward Networks

Suppose we are given input variables $z = (z_1, \dots, z_q)^\top$ and a vector $y = (y_1, \dots, y_d)^\top$ of output variables, which are related by an assumed functional relationship disturbed by a mean zero random noise ϵ,

$$y = f(z; \theta) + \epsilon,$$

where function $f(\bullet;\boldsymbol{\theta})$ is a parameterized nonlinear regression function. Artificial neural networks form a commonly used class of regression functions and correspond to specific forms of $f(\bullet;\boldsymbol{\theta})$. In what follows, we briefly review single hidden layer feedforward networks with linear output layer and their extension to multilayer (deep learning) networks, which have become quite popular, and discuss some recent progress in their understanding. Extreme learning machines, as introduced by Huang et al. [10], also called neural networks with random weights as first described by Schmidt et al. [21], belong to this class of artificial neural networks.

A single hidden layer feedforward network with q input nodes, p hidden neurons, d output neurons and activation function g computes the output of the jth neuron of the hidden layer for an input vector $z_t \in \mathbb{R}^q$ by

$$x_{tj} = g\left(b_j + \boldsymbol{w}_j^\top z_t\right), \qquad j = 1, \dots, p. \tag{1}$$

$\boldsymbol{w}_j \in \mathbb{R}^q$ are weights connecting the input nodes and the hidden units, and $b_j \in \mathbb{R}$ are bias terms, $j = 1, \dots, p$. The output of the jth node of output layer is then computed as

$$o_{tj} = \beta_0 + \sum_{j=1}^{p} \beta_j x_{tj} = \boldsymbol{\beta}_j^\top \boldsymbol{x}_t,$$

where $\boldsymbol{\beta}_j = (\beta_0^{(j)}, \dots, \beta_p^{(j)})^\top$ are weights and $\boldsymbol{x}_t = (1, x_{t1}, \dots, x_{tp})^\top$. There are various proposals for the activation function. Among the most popular ones are the sigmoid function $g(u) = 1/(1 + e^{-u})$, the rectified linear unit (ReLU) function $g(u) = \max(0, u)$ or the leaky ReLU, $g(u) = \delta x \mathbf{1}(x < 0) + x\mathbf{1}(x \geq 0)$, for some small $\delta > 0$, which allows for a small gradient when the neuron is not active. The d output neurons we have d weighting vectors resulting in a $d \times p$ parameter matrix $\boldsymbol{\beta} = (\boldsymbol{\beta}_1, \dots, \boldsymbol{\beta}_p)^\top$ and a net output $\boldsymbol{o}_t = \boldsymbol{\beta}\boldsymbol{x}_t$.

Although it is well-known that a single hidden layer network suffices to approximate rich classes of functions with arbitrary accuracy, see [9, ch. 16] for an accessible treatment, multilayer (deep) learning networks are quite popular and successful for specific problems, especially in imaging. Such a multilayered neural network is formally defined in terms of the successive processing of the input data through $r \in \mathbb{N}$ hidden layers. For r hidden layers with squashing functions $g_1, \dots, g_r$ and n_k neurons in the kth layer, the jth output of the kth layer is computed recursively via the equations

$$x_{tj}^{(1)} = g_1(b_1 + \boldsymbol{w}_j^{(1)\top} z_t), \qquad j = 1, \dots, n_1,$$

and

$$x_{tj}^{(k)} = g_k(b_{jk} + \boldsymbol{w}_j^{(k)\top} \boldsymbol{x}_t^{(k-1)}), \qquad j = 1, \dots, n_k,\ k = 2, \dots, r,$$

where $\boldsymbol{x}_t^{(k)} = (x_{t1}^{(k)}, \dots, x_{tn_k}^{(k)})^\top$ and $\boldsymbol{w}_j^{(k)}$ is a n_k-vector of connection weights, $k = 1, \dots, r$. Let $\boldsymbol{W}^{(k)} = (\boldsymbol{w}_1^{(k)}, \dots, \boldsymbol{w}_{n_k}^{(k)})^\top \in \mathbb{R}^{n_k \times n_{k-1}}$ be the matrix of weights connecting the neurons of the kth layer with the previous layer $k-1$ resp. the input layer if $k = 1$, and $\boldsymbol{b}_k = (b_{1k}, \dots, b_{n_k,k}) \in \mathbb{R}^{n_k}$ the bias terms. We may write

$$\boldsymbol{x}_t^{(k)} = \boldsymbol{g}_k^{(\boldsymbol{W}^{(k)}, \boldsymbol{b}_k)}(\boldsymbol{x}_t^{(k-1)}) = g_k(\boldsymbol{b}_k + \boldsymbol{W}^{(k)} \boldsymbol{x}_t^{(k-1)}),$$

where $\boldsymbol{g}_k^{(\boldsymbol{W}^{(k)}, \boldsymbol{b}_k)}(\bullet)$ is a vector functions consisting of the n_k real-valued functions $g_{k\ell}^{(\boldsymbol{W}^{(k)}, \boldsymbol{b}_k)}(\bullet)$, $\ell = 1, \dots, n_k$, and, for a real-valued function f with domain $\mathbb{R}$ and a vector $\boldsymbol{u} = (u_1, \dots, u_m) \in \mathbb{R}^m$, $f(\boldsymbol{u})$ is defined as the vector with entries $f(u_\ell)$, $\ell = 1, \dots, m$. To summarize, the output $\boldsymbol{x}_t$ of the rth hidden layer for an input $\boldsymbol{z}_t$ is given in terms of the composition operator $\circ$ by

$$\boldsymbol{x}_t = \boldsymbol{g}_r^{(\boldsymbol{W}^{(r)}, \boldsymbol{b}_r)} \circ \cdots \circ \boldsymbol{g}_1^{(\boldsymbol{W}^{(1)}, \boldsymbol{b}_1)}(\boldsymbol{z}_t).$$

2.2 *Training Neural Networks and Extreme Learning Machines*

Neural networks are trained given a learning or training sample $(\widetilde{\boldsymbol{y}}_t, \widetilde{\boldsymbol{z}}_t), t = 1, \dots, n$ of size n. We denote the training in this way, in order to reserve the symbols $\boldsymbol{x}_t, \boldsymbol{z}_t$, etc. for the validation sample addressed by the proposed statistical tools, see Sect. 3. A common approach dating back to the early days of machine learning is to minimize the least squares criterion corresponding to the quadratic loss function $\ell(\boldsymbol{u}) = \|\boldsymbol{u}\|_2^2 = \boldsymbol{u}^\top \boldsymbol{u}, \boldsymbol{u} \in \mathbb{R}^d$,

$$\boldsymbol{\theta} \mapsto L_n(\boldsymbol{\theta}) = \sum_{t=1}^n \|\widetilde{\boldsymbol{y}}_t - f(\widetilde{\boldsymbol{z}}_t; \boldsymbol{\theta})\|_2^2 = \sum_{t=1}^n \|\widetilde{\boldsymbol{y}}_t - \boldsymbol{\beta}^\top \boldsymbol{x}_t(\widetilde{\boldsymbol{z}}_t; \boldsymbol{\theta}')\|_2^2,$$

where $\boldsymbol{\theta} = (\boldsymbol{\theta}', \boldsymbol{\beta}) = (\boldsymbol{W}_1, \dots, \boldsymbol{W}_r, b_1, \dots, b_r, \boldsymbol{\beta})$ denotes the full set of parameters. This needs to be done by numerical optimization, usually a gradient-descent algorithm, where the special structure of feedforward nets allows simplified computations of the gradient by means of backpropagation, see [14] for a review. The optimizers mainly differ in how they choose the (gradient) direction and the learning rate (step size) in each step. Besides well-known and widely used classical optimizers such as BFGS or conjugate gradient methods, stochastic gradient descent, where the algorithm cycles through the data and selects at each step the gradient evaluated at a single observations, and ADAM, see [12] and, for a proof of its local convergence [4], are the most popular methods to train neural networks. Their efficiency in practice has certainly contributed to the success of deep learning networks. Nevertheless, the optimized artificial neural network and hence its performance in validation samples are to some extent random, since all algorithms require an initial starting value, which is chosen randomly. The best mathematical guarantee we can have is convergence at some fast rate to a local minimum, since the shape of the least squares criterion is generally known to be wiggly and characterized by many local extrema, often with almost negligible curvature.

Extreme learning machines resp. neural networks with random weights make use of the following observation: If the number of neurons of the last hidden layer is equal to the number of observations, n, such that the output matrix of that hidden layer, $\boldsymbol{X}_n$, is a $n \times n$ matrix, one can always find weights $\boldsymbol{\beta}$ with $\boldsymbol{X}_n\boldsymbol{\beta} = \widetilde{\boldsymbol{Y}}_n$, where $\widetilde{\boldsymbol{Y}}_n = (\widetilde{y}_1, \dots, \widetilde{y}_n)^\top$ is the $n \times d$ data matrix of the responses, *whatever* the values of the weights $\boldsymbol{\theta}'$ used to connect the remaining hidden layers among each other and the inputs with the first hidden layer. In this situation, we can perfectly explain the target values, i.e., the training data is interpolated. Here, the weights $\boldsymbol{\theta}'$ can also be random numbers. What happens, if $n_T \ll n$? Then it is no longer possible to interpolate the training data and instead it makes sense to minimize the least squares criterion as a function of the weights $\boldsymbol{\beta}$ of the output layer *given* randomly selected weights $\boldsymbol{\theta}'$. This is equivalent to fitting a multiple regression model by least squares for the covariates $\boldsymbol{x}_t(\widetilde{z}_t, \boldsymbol{\theta}')$ calculated for the randomly chosen $\boldsymbol{\theta}'$. Basically, we draw randomly a set of regressors depending on the inputs and use these regressors, which span a subspace of $\mathbb{R}^n$, to explain the responses by projecting them onto that subspace. Since one no longer fully optimizes the least squares criterion with respect to all unknowns, but instead only optimizes the output layer, the computational speed up is substantial. The reason is that the latter optimization only requires to solve the normal equations $\boldsymbol{X}_n^\top \boldsymbol{X}_n\boldsymbol{\beta} = \boldsymbol{Y}_n$ efficiently, i.e., a set of linear equations. To improve the generalization abilities it has been proposed to apply ridge regression at the output layer, also called Tikhonov regularization, which leads to the linear equations $(\boldsymbol{X}_n^\top \boldsymbol{X}_n + \lambda \boldsymbol{I})\boldsymbol{\beta} = \boldsymbol{Y}_n$ for some regularization (ridge) parameter $\lambda > 0$.

2.3 Approximation and Generalization Bounds

Let us briefly review the general approximation abilities of such machine learners [8]. When optimizing all parameters $\boldsymbol{\theta} = (\boldsymbol{b}, \boldsymbol{W}, \boldsymbol{\beta})$ of a single hidden layer feedforward net, $f_N(z) = \sum_{j=1}^{N} \beta_j \boldsymbol{x}_j(\boldsymbol{b} + \boldsymbol{W}z)$, $z \in \mathbb{R}^q$, it is known that the optimal achievable approximation error in the L_2-norm is independent of the dimension q of the inputs and is of the order $O(1/N^{1/2})$, see [2]. For *fixed* $\boldsymbol{b}$, $\boldsymbol{W}$ and when optimizing only $\boldsymbol{\beta}$, it has been shown that the approximation error uniformly achievable over a class of smooth functions is lower bounded by $C/(qN^{1/q})$, where the constant C does not depend on N [2]. If, however, the weights $(\boldsymbol{b}, \boldsymbol{W}) \sim \mu$ are set randomly [11] proved that the *expected* L_2 error is of the order $O(1/N^{1/2})$. This result asserts that there *exists* some distribution μ such that the expected error is $O(1/N^{1/2})$. It does not contradict the worst case order of the L_2-norm of the error $C/(qN^{1/q})$ for fixed weights, as it makes a statement about the average.

Generalization bounds for least squares estimation based on i.i.d. training samples resembling the results in [9, ch. 3] have been obtained by Liu et al. [17]. In their result μ is a uniform distribution and the mean approximation error is considered, where the expectation is taken with respect to the data distribution (as for fully optimized feedforward networks) and with respect to μ as well. This means, the averaged performance is studied here as well. The generalization bound is essentially also of the optimal form $O(n^{-2r/(2r+q)})$, up to a logarithmic factor, where r measures the smoothness of the true regression function and n is the number of samples.

Such learning results are more informative for applications than pure approximation results, since they consider the relevant case that the artificial neural network is optimized from data using the empirical least squares criterion. But one may formulate two critiques: These results consider the framework of learning from i.i.d. samples, which is too restrictive for complex data sets. Before discussing this issue in greater detail, let us pose a second critique: The classical learning guarantees and generalization bounds address, mathematically speaking, bounds for (a functional of) the empirical generalization error which are uniform over a certain class of regression functions f (i.e., networks). Such uniform bounds over a function class $\mathcal{F}$ are based on bounds for a fixed function and break the $\sup_{f \in \mathcal{F}}$ by imposing appropriate assumptions on the complexity of the class $\mathcal{F}$. Here, measures such as the Rademacher complexity, the Vapnik–Chervonenkis (VC) dimension or entropy measures provide the most satisfying and useful results, see, e.g., the monograph [18, 22] for recent results for deep learners.

But in applications, for a fixed problem and data set, the true function f is fixed, whether or not being a member of some nice class $\mathcal{F}$, and, therefore, the validity of learning guarantees and generalization bounds (anyway how these are defined) matters only for a single function. It has also been criticized by Neyshabur et al. [19] that such bounds often do not explain the phenomenon that over-parameterized nets improve in terms of the test error when increasing the size of the net. The authors establish generalizations for a two-layer network, which depend on two Frobenius matrix norms: Firstly, on the Frobenius norm of the weights of the top layer, $\boldsymbol{\beta}$, and, secondly, on the Frobenius norm of $\boldsymbol{W}_{tr} - \boldsymbol{W}_0$, where $\boldsymbol{W}_{tr}$ denotes the trained weights and $\boldsymbol{W}_0$ the randomly chosen initialization weights. The intuitive explanation of the authors is quite close to the heuristics behind extreme learning machines: If the number of hidden neurons gets larger and finally infinity, the hidden layer provides all possible (nonlinear) features, it mainly remains to pick and combined the right ones to explain the response and tuning the weights of the hidden layers is of less importance.

Let us proceed with a discussion of the critique that the i.i.d. training framework is too restrictive for data science problems. A major issue is that many complex big data sets used in machine learning are collected over time and may also have a spatial structure. In [24] extreme learning machines and multivariate regression have been studied for a non-stationary spatial-temporal noise model having in mind data collected by moving objects (cars, drones, smartphones carried by pedestrians, etc.), which especially covers many multivariate autoregressive moving average (ARMA) time series models. Since only the weights, $\boldsymbol{\beta}$, of the output layer are optimized, consistency of the least squares estimator, $\widehat{\boldsymbol{\beta}}_n$, is of interest as well as consistency of the related prediction $\widetilde{\boldsymbol{X}}_n\widehat{\boldsymbol{\beta}}_n$ for the truth $\widetilde{\boldsymbol{X}}_n\boldsymbol{\beta}$. In [24] it has been shown that, under quite mild regularity conditions given therein,

$$\|\widehat{\boldsymbol{\beta}}_n - \boldsymbol{\beta}\|_2^2 = O_P(p/n), \qquad \mathbb{E}\|\widehat{\boldsymbol{\beta}}_n - \boldsymbol{\beta}\|_2^2 = O(p/n),$$

where p denotes the number of hidden neurons of the (last) hidden layer. This means, even for dependent noise each parameter can be estimated with the rate $1/\sqrt{n}$. Having in mind applications and the typical goal of prediction when fitting artificial neural networks, bounds for the sample prediction error are even more interesting. The sample mean-square prediction error (MSPE) is defined by

$$\widehat{MSPE}_n = \frac{1}{n}\sum_{t=1}^{n}\left(\widetilde{\boldsymbol{x}}_t^\top\boldsymbol{\beta} - \widetilde{\boldsymbol{x}}_t^\top\widehat{\boldsymbol{\beta}}_n\right)^2$$

and measures the accuracy of the predicted targets in terms of the empirical 2-vector norm with respect to the training sample. Here, and in what follows, we assume univariate targets ($d = 1$). As shown in [24], under certain conditions it holds, given the (random) weights $\boldsymbol{b}$, $\boldsymbol{W}$,

$$\widehat{MSPE}_n = O_P(p/n).$$

For the ridge estimator similar learning guarantees have been established generalizing and complementing result from [5, 15] and [16], which are restricted to i.i.d. sampling. Under regularity conditions given there, one can show that if the regularization parameter satisfies

$$\lambda_n = o_{\mathbb{P}}(n/\sqrt{p}),$$

then the above statements on consistency of the estimated parameters of the last layer and in terms of consistency of the sample prediction error still remain true, see [24]. It is worth mentioning that this result allows the regularization parameter to be random. If $\lambda_n/n \to \lambda^0$ for some constant $\lambda^0 \geq 0$, then the estimator is biased.

3 Comparing and Cross-Validating Randomized Networks

Simply training an artificial neural network to a training sample $(\widetilde{y}_t, \widetilde{z}_t)$, $t = 1, \ldots, n$, should only be the first step. Comparing model specifications also taking into account additional criteria not covered by the training algorithm is generally advisable. We start with a discussion of a possible formal approach for such comparisons and evaluations.

As argued above, any artificial neural network fitted to a training sample is random given the training sample, especially randomized nets such as extreme learning machines. As a consequence, any evaluation of the out-of-sample performance is random as well. This implies that calculated performance measures quantifying the generalization ability are (nonnegative) random variables instead of fixed numbers. A simulation experiment discussed below demonstrates this effect.

Therefore, we propose and elaborate on a cross-validating approach using a validation sample $(\boldsymbol{y}_t, z_t)$, $t = 1, \ldots, n_V$, tailored for randomized networks, in order to make use of this uncertainty to improve the behavior of the final predictions. Lastly, a method is discussed to quantify the mean sample prediction error with minimal computational costs.

3.1 Model Comparison and Evaluation

Suppose we are given two model specifications in terms of $(\boldsymbol{b}_1, \boldsymbol{W}_1, \widehat{\boldsymbol{\beta}}_1)$ and $(\boldsymbol{b}_2, \boldsymbol{W}_2, \widehat{\boldsymbol{\beta}}_2)$, where $\boldsymbol{b}_i$ and $\boldsymbol{W}_i$ are the (random) biases and connection weights and $\widehat{\boldsymbol{\beta}}_i$ the optimized weights of the output layer. Model comparison is often conducted by looking at the optimized values of the chosen training criterion. But the comparison or evaluation can be based on a different measure than used for training, of course, and another choice often makes sense to take into account additional objectives. Among those objectives are data fidelity, sensitivity with respect to input variables, prediction accuracy and robustness, amongst others. Incorporating such criteria in the objective function minimized to optimize the weights of the output layer is possible, but the computational costs can increase dramatically compared with least squares and ridge regression.

Instead, one may simply select a specification (and thus a fit and prediction model) among a (small) set of candidate models, which has better behavior in terms of a selected criterion without conducting full optimizing the output layer.

Assume we have picked a criteria function C_n defined on $\mathbb{R}^d \times \mathbb{R}^{n \times p} \times \mathbb{R}^{p+1} \to \mathbb{R}$ and calculate

$$C_{ni} = C_n(\boldsymbol{Y}_n; \widetilde{\boldsymbol{Z}}_n^i, \widehat{\boldsymbol{\beta}}_i), \qquad i = 1, 2,$$

where $\widetilde{\boldsymbol{Z}}_n^i = (\widetilde{z}_1^i, \ldots, \widetilde{z}_n^i)^\top$ are the $n \times q_i$ matrices of the n observations taken from q_i input variables and $\widehat{\boldsymbol{\beta}}_i$ is the vector of optimized output layer weights, $i = 1, 2$. Observe that the comparison can be based on different input matrices of different dimensions. In this way, one may compare models using a different number of input variables. Especially, by putting $\widetilde{z}_t^{(1)} = (\widetilde{z}_{t1}, \ldots, \widetilde{z}_{tq})^\top$ and $\widetilde{z}_t^{(2)} = (0, \ldots, 0, \widetilde{z}_{t,q_1+1}, \ldots, \widetilde{z}_{tq})^\top$ one can analyze whether or not the first q_1 inputs are relevant. In this case, $\boldsymbol{W}_2$ is set to the last $q - q_1$ columns of $\boldsymbol{W}_1$ and $\boldsymbol{b}_2$ to the corresponding entries of $\boldsymbol{b}_1$. A reasonable decision function is to decide in favor of model 2, if and only if the improvement expressed as a percentage is large enough, i.e., if

$$C_{n2} < C_{n1} f$$

for some $0 < f < 1$.

Least Squares Data Fidelity The choice

$$C_n(\boldsymbol{Y}_n; \widetilde{\boldsymbol{Z}}_n^i; \widehat{\boldsymbol{\beta}}_i) = \frac{1}{n} \sum_{t=1}^{n} (Y_t - g(\boldsymbol{b}_i + \boldsymbol{W}_i \widetilde{\boldsymbol{z}}_t^i)^\top \widehat{\boldsymbol{\beta}}_i)^2$$

corresponds to the least squares training criterion and measures the achieved data fidelity of the fit.

Robustness Let $\rho : \mathbb{R} \to [0, \infty)$ be a (non-decreasing, bounded, ...) function and put

$$C_n(\boldsymbol{Y}_n; \widetilde{\boldsymbol{Z}}_n^i; \widehat{\boldsymbol{\beta}}_i) = \frac{1}{n} \sum_{t=1}^{n} \rho(Y_t - gt(\boldsymbol{b}_i + \boldsymbol{W}_i \widetilde{\boldsymbol{z}}_t^i)^\top \widehat{\boldsymbol{\beta}}_i).$$

Here, the (loss) function ρ is used to evaluate the residuals. A common choice corresponding to robust M estimation is Huber's ρ-function $\rho(u) = u^2/2\mathbf{1}(|u| \leq K) + K(|u| - a/2)\mathbf{1}(|u| > K)$ for some constant $K > 0$. For small $|u|$, the loss is quadratic and linear for larger values. In this way, the sensitivity to outliers is reduced.

Mean-Square Prediction Error Estimating the prediction error arising when predicting the true but unknown (optimal) mean responses by the optimized net outputs leads to the choice

$$C_n(\boldsymbol{Y}_n; \widetilde{\boldsymbol{Z}}_n^i; \widehat{\boldsymbol{\beta}}_i) = \frac{1}{n} \sum_{t=1}^{n} \left(\boldsymbol{\beta}^\top \boldsymbol{x}_t - \widehat{\boldsymbol{\beta}}_i^\top \widetilde{\boldsymbol{x}}_t^i \right)^2,$$

where $\widetilde{\boldsymbol{x}}_t^i = g(\boldsymbol{b}_i + \boldsymbol{W}_i \widetilde{\boldsymbol{z}}_t^i)$.

The following assumption ensures that, asymptotically, the sample-based criterion function converge to constants.

Assumption A $C_{ni}, i = 1, 2$, converge in probability to constants $c_i, i = 1, 2$, i.e.,

$$C_{ni} \xrightarrow{P} c_i, \tag{2}$$

as $n \to \infty$, for $i = 1, 2$.

Assumption A is rather weak, especially, because no rate of convergence is required. The question arises to which extent a model needs to improve upon a competitor.

Definition 1 Let us call model 2 (C_n, f)-preferable, if the constants from Assumption A fulfill the requirement $c_2 < fc_1$ for some $f \in (0, 1]$.

The following result follows almost automatically from Assumption A and tells us that the rule will select the right model with probability one in large samples.

Theorem 1 *Suppose that Assumption A holds true and let $f \in (0, 1]$. If model 2 is (C_n, f)-preferable, then the decision rule (2) selects the correct model with probability approaching zero, i.e.,*

$$P(C_{n2} > C_{n1} f) \to 0, \qquad n \to \infty.$$

The approach discussed above is mainly designed as an additional step when training a model from the learning sample by comparing a couple of model specifications in terms of the input variables and additional criteria such as in-sample prediction error and robustness, for a fixed choice of the (random) model parameters of the hidden layer(s). Their random choice, however, introduces uncertainty and, furthermore, the prediction accuracy should be quantified with new fresh data samples.

3.2 A Simulation Experiment

Before proceeding, let us discuss the results of a small simulation experiment conducted to illustrate the effect of randomly selecting part of the network parameters. A single hidden layer feedforward net with h neurons, 5 inputs and 1 output was examined for standard normal inputs, $z_t \sim \mathcal{N}(\mathbf{0}, I)$, and a univariate output modeled as $y_t = \boldsymbol{x}_t(z_t, \boldsymbol{b}, \boldsymbol{W})^\top \boldsymbol{\beta}_0 + \epsilon_t$, $t = 1, \ldots, n$, where the errors ϵ_t are i.i.d. standard normal. The training sample size was set to $n = 1000$ and the validation sample size to $n_V = 100$. Fixing realizations of the training and test data and a randomly chosen true coefficient vector $\boldsymbol{\beta}_0$, an extreme learning machine with random weights $\boldsymbol{b}, \boldsymbol{W}$ following a uniform distribution on $[-1, 1]$ was fitted to the training sample and then evaluated in the validation sample by calculating the associated sample mean prediction error, see below for a formula. This simulation step was repeated $1,000$ times to obtain for each network topology (given by h) an estimate of the distribution of the conditional mean prediction error. Figure 1 shows characteristics of the simulated distribution as a function of the number, h, of hidden neurons. One can observe that the support of the distribution opens the door for picking a realization of the weights leading to superior out-of-sample performance compared to single-shot fitting of a neural network with random weights.

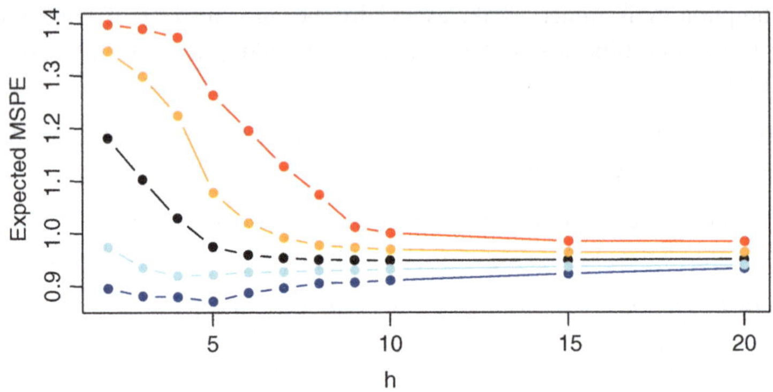

Fig. 1 Simulated distributions of the expected sample prediction error in a validation sample of size 100, for $h = 2, \ldots, 10, 15, 20$ hidden neurons. The curves (simulated points are joined by lines) represent from top to bottom the maximum, 95%-quantile, mean, 25%-quantile and minimum of the simulated distribution based on 1000 runs

3.3 One-Round Cross-Validation for Randomized Networks

In what follows, we assume that a validation sample (Y_i, z_i), $i = 1, \ldots, n_V$, of size n_V is available for evaluation of a fitted neural network. This corresponds to a one-round cross-validation. In principle, this could be generalized to a k-fold cross-validation scheme, but to keep the presentation simple and clean, we confine ourselves to the setting of a training of size n and a validation sample of size n_V as in the experiment reported above. In such a setting, one often works with ratios n/n_V around 80/20, whereas k-fold cross-validation would split the available data in k equal parts (folds). We elaborate on a single hidden layer network and leave the simple, notational changes for deep learning networks to the reader.

Suppose that $Z_j = Z(\boldsymbol{\eta}_j)$ is a nonnegative measure for the prediction accuracy calculated for a random draw $\boldsymbol{\eta}_j$ of a subvector $\boldsymbol{\eta}$ of the full parameter $\boldsymbol{\theta}$ of a neural network. In case of a hidden layer net we have $\boldsymbol{\eta} = (\boldsymbol{b}, \boldsymbol{W})$ and $\boldsymbol{\theta} = (\boldsymbol{\eta}, \boldsymbol{\beta})$. In view of the randomness of $\boldsymbol{\eta}_j$, Z_j is a random variable attaining values in $[0, \infty)$. For J i.i.d. draws given the training and validation samples, we obtain an i.i.d. sample $Z_1, \ldots, Z_J$, namely the J evaluations in the validation sample, when conditioning on the training and the validation samples.

The sample mean-square prediction validation error for the validation data set of size n_V, given in terms of the response n_v-vector $\boldsymbol{Y}_{n_V}$ and inputs $\boldsymbol{Z} = (z_1, \ldots z_{n_V})^\top$, a $n_V \times q$ matrix, is calculated as

$$\widehat{MSPE}_{n_V} = \frac{1}{n_V} \|\boldsymbol{Y}_{n_V} - \boldsymbol{X}_{n_V}(\boldsymbol{b}, \boldsymbol{W}, \boldsymbol{Z})\widehat{\boldsymbol{\beta}}_n\|_2^2 = \frac{1}{n_V} \sum_{i=1}^{n_V} (Y_i - g(\boldsymbol{b} + \boldsymbol{W} z_i)^\top \widehat{\boldsymbol{\beta}}_n)^2.$$

The predictions $\widehat{Y}_{n_V}$ for the validation data set (Y_i, z_i), $i = 1, \ldots, n_V$, are, therefore, computed using the output matrix of the hidden layer when fed with the validation inputs, i.e., using the $n_V \times p$ output matrix

$$X_{n_V}(b, W, Z) = \begin{bmatrix} g(b + Wz_1)^\top \\ \vdots \\ g(b + Wz_{n_V})^\top \end{bmatrix}$$

and, therefore, take the form $\widehat{Y}_{n_V} = X_{n_V}(b, W, Z)\widehat{\beta}_n$. Here $\widehat{\beta}_n$ denotes the least squares estimator of the output layer calculated from the training sample $(\widetilde{Y}_i, \widetilde{z}_i)$, $i = 1, \ldots, n$, of size n, and b, W are the random network parameters connecting the input and the hidden layer, i.e., using the $n \times p$ output matrix of the hidden layer

$$\widetilde{X}_n(b, W, \widetilde{Z}) = \begin{bmatrix} g(b + W\widetilde{z}_1)^\top \\ \vdots \\ g(b + W\widetilde{z}_n)^\top \end{bmatrix},$$

such that

$$\widehat{\beta}_n = (\widetilde{X}_n(b, W, \widetilde{Z})^\top \widetilde{X}_n(b, W, \widetilde{Z}))^{-1} \widetilde{X}_n(b, W, \widetilde{Z})^\top \widetilde{Y}_n.$$

This process is now iterated J times, i.e., we draw J sets of random parameters $(b(j), W(j))$, $j = 1, \ldots, J$, and calculate for each draw $(b(j), W(j))$ the associated estimate of the sample mean prediction error in the validation sample,

$$Z_j = \|Y_{n_V}(j) - X_{n_V}(b(j), W(j), Z)\widehat{\beta}_n\|_2^2. \tag{3}$$

These estimates are averaged to obtain an estimator of the conditional mean $\mathbb{E}(\widehat{MSPE}_{n_V} | (\widehat{Y}_n, \widehat{X}_n), (Y_{n_V}, X_{n_V}))$, where the expectation is with respect to the distribution of the randomized network weights. Further, we have simulated a sample of realizations of the $\widehat{MSPE}_{n_V}$ and may select the network leading to the best performance in the fixed validation sample, i.e., take $j^* \in \{1, \ldots, J\}$ with

$$Z_{j^*} = \min_{1 \le j \le J} Z_j$$

and thus use the network with the specification $(b(j^*), W(j^*))$ of the randomized parameters. This algorithm is summarized below. General results on the consistency of this approach including bounds for the estimated mean sample mean prediction error and model selection consistency are subject of ongoing research [25].

Observe that the above algorithm covers the case of model selection, classically understood as the selection of the number of hidden neurons of the net, as well as the selection of the network topology. This is so because the distribution G may take into account certain topologies, e.g., a convolutional layer which linearly processes all fixed-length subvectors of the inputs z by a linear filter with randomly drawn coefficients, followed by a maxpolling layer, which is then further processed. Similarly, G could be defined such that for a fully connected layer the connection weights of each neuron have ℓ_0-norm s_0, so that each neuron processes only s_0 of the outputs of the previous layer resp. of the inputs. In this way, one can try (randomly) different network topologies in a systematic way.

Algorithm

1. Draw $(\boldsymbol{b}(j), \boldsymbol{W}(j)) \overset{i.i.d.}{\sim} G,\ j = 1, \dots, J$.
2. For $j = 1, \dots, J$ do
3. Estimate $\boldsymbol{\beta}$ from the training sample using the output matrix $\widetilde{\boldsymbol{X}}_n(j) = \widetilde{\boldsymbol{X}}_n(\boldsymbol{b}(j), \boldsymbol{W}(j), \widetilde{\boldsymbol{Z}}_n)$ giving $\widehat{\boldsymbol{\beta}}_n(j)$.
4. Compute the predictions of the validation sample $\widehat{Y}_{nv}(j) = \boldsymbol{X}_{nv}(j)\widehat{\boldsymbol{\beta}}_n(j)$ with $\boldsymbol{X}_{nv}(j) = \boldsymbol{X}_{nv}(\boldsymbol{b}(j), \boldsymbol{W}(j), \boldsymbol{Z})$.
5. Compute the square prediction errors $Z_j = \frac{1}{nv}\|\boldsymbol{Y}_{nv} - \widehat{\boldsymbol{Y}}_{nv}(j)\|_2^2$.
6. Estimate the mean-square validation prediction error by

$$\widehat{MSPE}_{nv,J} = \frac{1}{J}\sum_{j=1}^{J} Z_j$$

7. Select the network by computing $j^* \in \{1, \dots, J\}$ with

$$Z_{j^*} = \min_{1 \le j \le J} Z_j$$

3.4 An Uncertainty Interval for the Mean Sample Prediction Error with Minimal Computational Costs

In order to deal with the uncertainty of the sample MSPE and to minimize the required computational costs, one can calculate a fixed-width confidence interval for the expected sample MSPE in the validation sample, $\mu = \mathbb{E}_{(\boldsymbol{b},\boldsymbol{W})}\left(\widehat{MSPE}_{nv}\right)$, corresponding to the black points in Fig. 1. For a fixed uncertainty $d > 0$, specified in advance as the half-length of an interval around the estimator $\widehat{MSPE}_{nv,J}$, one wants to determine J from data, such that the resulting interval has confidence $1-\alpha$,

$\alpha > 0$ small. This means, we want to determine the smallest J, such that the fixed-width interval

$$\left[\widehat{MSPE}_{nv,J} - d, \widehat{MSPE}_{nv,J} + d\right]$$

has coverage probability $1 - \alpha$. The problem to construct fixed-width uncertainty intervals has been studied for general parameters by Steland and Chang [26] for the classical asymptotic regime $d \to 0$ as well as the novel high-confidence regime $1 - \alpha \to 1$. A solution, $\widehat{J}_{opt}$, which is consistent for the theoretically optimal solution, and first as well as second order efficient, is as follows: One fixes a minimal number of draws, $\bar{J}_0$, and calculates

$$J_0 = \max\left\{\bar{J}_0, \left\lfloor \frac{\Phi^{-1}(1-\alpha/2)\widehat{\sigma}}{d} \right\rfloor + 1\right\}.$$

Here Φ^{-1} is the quantile function of the standard normal distribution function. $\widehat{\sigma}$ is the sample standard deviation of Z_j's of a small number of initial runs, which can be as small as 3 according to the simulation studies in [26]. Next, perform J_0 simulation runs and calculate $\widehat{\sigma}^2_{J_0} = \frac{1}{J_0}\sum_{j=1}^{J_0}(Z_j - \overline{Z})^2$. Lastly, one calculates the final number of runs given by

$$\widehat{J}_{opt} = \max\left\{J_0, \left\lfloor \frac{\widehat{\sigma}^2_{J_0}\Phi^{-1}(1-\alpha/2)^2}{d^2} \right\rfloor\right\}.$$

If $\widehat{J}_{opt} > J_0$, one conducts the required additional $\widehat{J}_{opt} - J_0$ draws of the random parameters of the neural net, determines the associated sample prediction errors $Z_1, \ldots, Z_{J_{\widehat{J}_{opt}}}$ according to (3), and eventually computes the interval $\overline{Z}_{\widehat{J}_{opt}} \pm d$.

4 Application to Vehicle Integrated Photovoltaics and Data Analysis

An interesting specific problem arising in VIPV is the prediction of the yield due to the integrated solar panels. Compared to panels mounted at the rooftop of a truck or car, panels mounted at the sides pose additional problems, since their energy yield depends on the orientation of the vehicle. The question arises to which extent one can predict their contribution to the total yield by the irradiance measured at the rooftop. The basic idea to explain the measurements of a sensor (or PV module) facing left (or right) in terms of a sensor facing up is that it is easier to derive, in advance, expected irradiance maps for horizontally aligned sensors. To a planned route one can then assign an expected irradiance trajectory for the sensor facing up. A prediction model then allows us to forecast the contribution of further sensors. In this way, one can predict the VIPV yield.

4.1 Vehicle Mounted Data Logger

Several sensors and a data logger were mounted on a vehicle. The sensors include 4 irradiance sensors, one acoustic wind sensor, a Global Positioning System (GPS), and a magnetometer.

The sensors are specifically chosen to provide relevant data for VIPV yield. For this purpose we obviously want to monitor the irradiance. The irradiance depends strongly on the orientation of the PV module. Thus, we use 4 irradiance sensors facing in different directions (top, left, right and back), and we log the vehicle orientation. While the vehicle is moving we can use GPS data to provide a good indicator for the vehicle orientation (assuming the vehicle is moving in forward direction). However, as vehicles are also often parked, we in addition use a magnetic sensor to provide information on the vehicle orientation.

Another important factor for yield is the module temperature, as PV modules are less efficient at higher temperatures. The module temperature itself depends on several environmental factors; wind, ambient temperature, and irradiance. In [13] it was shown that the head wind from driving provides a significant positive impact on PV yield as the additional wind cools the PV modules.

The data logger was developed around a Raspberry Pi single board computer. The Raspberry Pi is equipped with a GPS module and a magnetometer. Note that the magnetometer is used as GPS only provides information on the orientation of the vehicle while the vehicle is moving (assuming the vehicle is mover forward). However, most vehicle spend a large amount of time parked. The remaining wind and irradiance sensors are connected with two RS485 interfaces, one for the wind sensor and one for the four irradiance sensors. The logged sensor data is written to a USB thumb drive. The setup is powered from the 12 V car battery and is enclosed in a weather proof box mounted on a rooftop rack.

For the irradiance sensors we used four calibrated silicon sensors from Ingenieurbüro Mencke & Tegtmeyer GmbH of type SiRS485TC-T-MB. As the sensors are silicon reference cells the measured irradiance is of particular relevance for PV applications as the spectral range of the sensors matched that of typical PV modules. The four irradiance sensors are mounted on the same rooftop rack, facing up, left, right and backwards.

The wind sensor is an FT205 acoustic wind sensor from FT Technologies. The sensor measures both wind speed and direction (2D). The sensor also reports the acoustic air temperature, i.e., the air temperature derived from the temperature dependent speed of sound in air.

The data used in this paper was collected during several test drives of the system. We plan to use several car mounted data logging systems in the coming years on several cars with different use profiles.

4.2 Data Analysis

As a preliminary study, we analyzed a small pilot sample collected during three test drives. In view of the limited data available for this analysis, we can only get a first impression whether the information describing the position of the car, namely where it is located and in which direction it drives, can be exploited to predict measurements of a sensor facing left, right or backwards from measurements from the sensor facing up.

The available data was split in a training sample with $n = 3472$ data points and a test sample with 3669 observations. The validation sample was selected as observations 1000–2250 from the test sample, since such PV data is highly heterogenous, as irradiance differs substantially depending on the time of day and weather. For the present data set, the first part of the test sample was inappropriate.

In our nonlinear model it is assume that the tth voltage measurement of the sensor facing left, s_{2t}, is related to the sensor facing up, s_{1t}, via the equation

$$s_{2t} = s_{1t}(1 + f(a_t, x_t, y_t)), \qquad t = 1, \dots, n.$$

Here a_t denotes the angle (direction) of the car and (x_t, y_t) is the car's location at time t, expressed in terms of geographical coordinates (longitude and latitude). f is an unknown (nonlinear) function. A baseline (null) model would be to assume that f is equal to some constant value f_0. It is, however, clear that under idealized noiseless conditions, f is a function of angle and geographical location. For example, at a certain location the car's side but not the roof may be shadowed by a building. Of course, a more refined model needs to take into account time of day and season, but estimating such models requires sufficiently big data set over much longer time span than available for the present illustrative data analysis.

The function f can be modeled and estimated by a nonlinear regression approach,

$$y_t = f(a_t, x_t, y_t) + \epsilon_t, \qquad t = 1, \dots, n,$$

for mean zero random noise terms ϵ_t, using the targets (responses)

$$y_t = \frac{s_{2t} - s_{1t}}{s_{1t}}$$

and the input variables (regressors) $z_t = (\alpha_t, x_t, y_t)$. Having a prediction $\widehat{y}_t$ the corresponding forecast of s_{2t} is then calculated as $\widehat{s}_{2t} = s_{1t}(1 + \widehat{y}_t)$.

We compared two model specifications. Firstly, a linear specification, i.e., a classical multiple linear regression model, given by

$$y_t = b_0 + b_1 a_t + b_2 x_t + b_3 y_t + \epsilon_t,$$

for regression coefficients $b_0, \ldots, b_3 \in \mathbb{R}$. The second model is a single hidden layer feedforward network with $p = 4$ hidden units and a logistic squasher $1/(1 + \exp(-x))$,

$$y_t = f(z_t; \boldsymbol{\theta}) + \epsilon_t = \boldsymbol{x}_t(z_t; \boldsymbol{\eta})^\top \boldsymbol{\beta} + \epsilon_t,$$

where $\boldsymbol{\theta} = (\boldsymbol{\eta}, \boldsymbol{\beta})$ with $\boldsymbol{\eta} = (\boldsymbol{b}, \boldsymbol{W})$, and $\boldsymbol{\beta} \in \mathbb{R}^p$ represents the weights of the linear output layer. We also experimented with networks with a different number of hidden units, but the specification found and reported below uses $p = 4$ hidden layers. It is worth mentioning that fitting successfully neural networks requires to norm the input variables to the interval $[-1, 1]$. The neural net was trained as an extreme learning machine using ridge regression with ridge regularization parameter $\lambda = 0.2$ and random weights $\boldsymbol{\eta}$ with i.i.d. entries following a uniform distribution on the interval $[-1, 1]$. In order to get more robust results, the most extreme 5% of the observations of the training sample were omitted. Following the proposed cross-validation method, a realization of $\eta^* = (\boldsymbol{b}^*, \boldsymbol{W}^*)$ was chosen which yields the best prediction accuracy in the validation sample.

Table 1 provides the sample mean prediction error in the validation sample for all three prediction methods, the baseline null model, multiple linear regression and artificial neural network. In addition, for each model the statistic $\widehat{MSPE}_{n_V} = \frac{1}{n_V} \sum_{i=1}^n \widehat{e}_i$, where $\widehat{e}_i$ denotes the prediction error for the ith datapoint, e.g., $\widehat{e}_i = Y_i - \boldsymbol{x}_i(\boldsymbol{b}^*, \boldsymbol{W}^*, z_i)^\top \widehat{\boldsymbol{\beta}}_n$ for the neural network, was decomposed by computing the components

$$\widehat{MSPE}_{n_V,0.1} = \frac{1}{n_V} \sum_{\substack{i=1 \\ |\widehat{e}_i| \le q_{0.1}}}^{n_V} \widehat{e}_i^2,$$

$$\widehat{MSPE}_{n_V,0.1:0.9} = \frac{1}{n_V} \sum_{\substack{i=1 \\ q_{0.1} < |\widehat{e}_i| < q_{0.9}}}^{n_V} \widehat{e}_i^2,$$

$$\widehat{MSPE}_{n_V,0.9} = \frac{1}{n_V} \sum_{\substack{i=1 \\ |\widehat{e}_i| > q_{0.9}}}^{n_V} \widehat{e}_i^2,$$

Table 1 Prediction accuracy in the validation sample

Method	$\widehat{MSPE}_{n_V}$	$\widehat{MSPE}_{n_V,0.1}$	$\widehat{MSPE}_{n_V,0.1:0.9}$	$\widehat{MSPE}_{n_V,0.9}$
Null model	100,555.8	11,635.59	6,908.67	82,011.5
Linear regression	67,967.4	12,740.26	5,547.23	49,679.9
ELM neural network	74,502.3	4,065.44	2,133.83	68,303.0

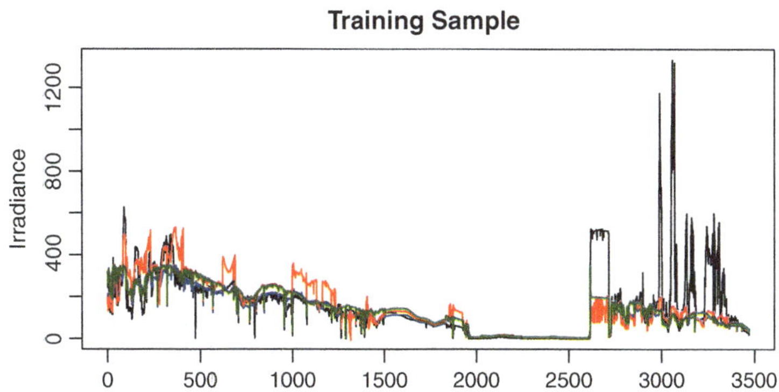

Fig. 2 Observed irradiance at sensor 2 and predictions for the training sample: null model (green), linear regression (red), extreme learning machine (blue)

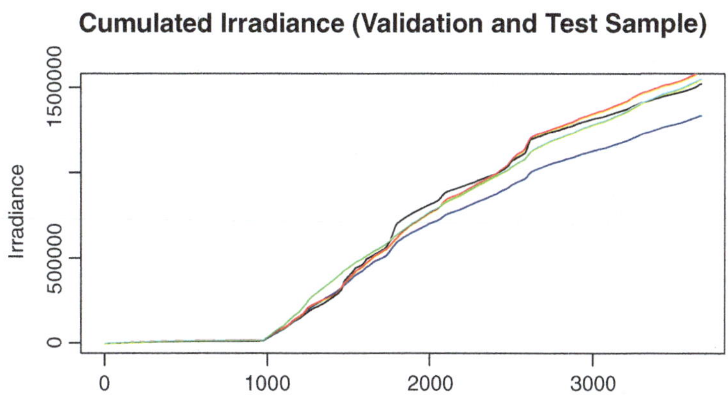

Fig. 3 Observed cumulated measurements and cumulated predictions for the validation and test sample: the neural net underestimates power generation in the test sample as several extrema are not properly predicted

where q_p denotes the p-quantile of the empirical distribution of the prediction errors $\widehat{e}_i, i = 1, \dots, n_V$. In this way, one can analyze how well a method works in the tails compared with the central 90% of the data. $\widehat{MSPE}_{n_V,0.1}$ measures the prediction error when the method overestimates and $\widehat{MSPE}_{n_V,0.9}$ if it underestimates. One can observe that the neural net predictions surprisingly well in the central part and also when it overestimates, but the prediction errors are large when it underestimates.

Figure 2 shows the predictions of the three prediction methods in the training sample, whereas Fig. 3 depicts the results for the validation and test sample. The predictions of the nonlinear neural network are in most cases closer to the observed data points, except for some extreme measurements, which are not nicely captured by the neural net. In Fig. 3 the cumulated measurements and their predictions, respectively, are plotted. Since the sensors provide data sampled at a fixed sampling

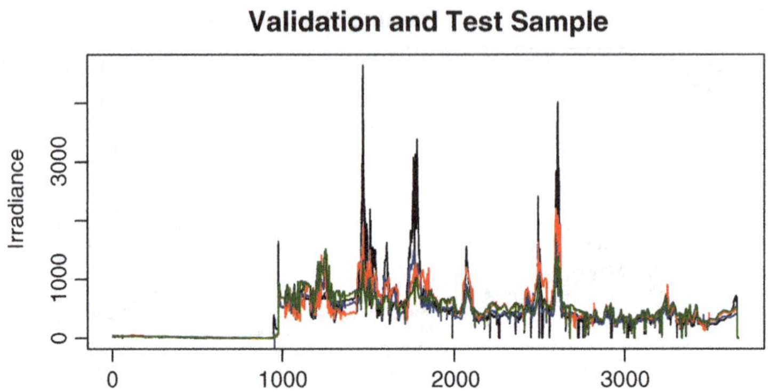

Fig. 4 Observed irradiance and predictions for the validation and test sample: the neural net yields better predictions in most cases, but underestimates several extrema

rate without gaps, the cumulated values can be regarded as proxies for the (total) energy yield. Because the neural network is not able to capture some extremes, it underestimates the yield.

However, the data set used in this pilot study is too small to draw conclusions, especially about the question to which extent artificial neural networks outperform linear methods for the problem of interest. It is also not clear whether the observed properties of the prediction errors are artifacts or will still be present when larger data sets are analyzed (Fig. 4).

Appendix: Proof of Theorem 1

If model 2 is (C_n, f)-preferable, then the probability of a false decision is given by

$$P(C_{n2} > C_{n1} f) = P(C_{n2} - \delta_2 > C_{n1} f - c_1 f - \delta_2 + c_1 f).$$

Consequently,

$$\begin{aligned} P(C_{n2} > C_{n1} f) &= P([C_{n2} - c_2] + [C_{n1} - c_1] f > c_1 f - c_2) \\ &\le P([C_{n2} - c_2] > (c_1 f - c_2)/2) + P([C_{n1} - c_1] > (c_1 f - c_2)/(2f)) \\ &\to 0, \end{aligned}$$

as $n \to \infty$, since $c_1 f - c_2 > 0$. From these simple bounds it is clear that a convergence rate for the criterion automatically yields a convergence rate for the error probability to select the wrong model.

Acknowledgments This work has been financially supported by the German Federal Ministry for Economic Affairs and Energy (Bundesministerium für Wirtschaft und Energie, BMWi), within the project "Street – Einsatz von hocheffizienten Solarzellen in elektrisch betriebenen Nutzfahrzeugen," grant no. 0324275A. The authors are responsible for the content. They gratefully acknowledge preparatory work of M.Sc. Nils Brügge.

References

1. Arlot, S., Celisse, A.: A survey of cross-validation procedures for model selection. Statist. Surv. **4**, 40–79 (2010). https://doi.org/10.1214/09-SS054
2. Barron, A.R.: Universal approximation bounds for superposition of a sigmoidal function. IEEE Trans. Inf. Theory **39**(3), 930–945 (1993)
3. Birnie, D.P.: Analysis of energy capture by vehicle solar roofs in conjunction with workplace plug-in charging. Sol. Energy **125**, 219–226 (2016)
4. Bock, S., Weiss, M.: A proof of local convergence for the Adam optimizer. In: IEEE International Joint Conference on Neural Networks (IJCANN). IEEE, New York (2019). https://opus4.kobv.de/opus4-oth-regensburg/frontdoor/deliver/index/docId/50/file/IJCNN2019.pdf
5. Bühlmann, P., van de Geer, S.: Statistics for high-dimensional data. In: Springer Series in Statistics. Springer, Heidelberg (2011)
6. Cohen, G., Afshar, S., Tapson, J., Schaik, A.: EMNIST: an extension of MNIST to handwritten letters, arXiv 1702.05373
7. Fernández-Delgadom M., Cernadas, E., Barro, S., Amorim, D.: Do we need hundreds of classifiers to solve real world classification problems? J. Mach. Learn. Res. **15**(90), 3133–3181 (2014)
8. Gaban, A.N., Tyukin, I.Y., Prokhorov, D.V., Sofeikov, K.I.: Approximation with random bases: Pro et contra. Inf. Sci. **364–365**, 129–145 (2016)
9. Györfi, L., Kohler, M., Krzyzak, A., Walk, H.: A Distribution-Free Theory of Nonparametric Regression. In: Springer Series in Statistics. Springer, New York (2002)
10. Huang, G.-B., Zhu, Q.-Y., Sieq, C.-K.: Extreme learning machine: a new learning scheme of feedforward neural networks. IEEE Int. Joint Conf. Neural Netw., 985–990 (2004)
11. Igelnik, B., Pao, Y.H.: Stochastic choice of basis functions in additive function approximation and the functional-link net. IEEE Trans. Neural Netw. **6**(6), 1320–1329 (1995)
12. Kingma, D.P., Ba, J.: Adam: a method for stochastic optimization (2017). https://arxiv.org/abs/1412.6980. Cited 27 October 2020
13. Kühnel, M., Hanke, B., Geissendörfer, S., von Maydell, K., Agert, C.: Energy forecast for mobile photovoltaic systems with focus on trucks for cooling applications. Prog. Photovolt. Res. Appl. **25**(7), 525–532 (2017)
14. LeCun Y.A., Bottou L., Orr G.B., Müller, K.R.: Efficient BackProp. In: Montavon, G., Orr, G.B., Müller, K.R. (eds.) Neural Networks: Tricks of the Trade. Lecture Notes in Computer Science, vol. 7700. Springer, Berlin (2012). https://doi.org/10.1007/978-3-642-35289-8_3
15. Lita da Silva, J.: Some strong consistency results in stochastic regression. J. Multivariate Anal. **129**, 220–226 (2014)
16. Liu, H., Yu, B.: Asymptotic properties of Lasso+mLS and Lasso+Ridge in sparse high-dimensional linear regression. Electron. J. Stat. **7**, 3124–3169 (2013)
17. Liu, X., Lin, S., Fang, J., Xu, Z.: Is extreme learning machine feasible? A theoretical assessment (Part I). IEEE Trans. Neural Netw. Learn. Syst. **26**, 7–20 (2015)
18. Mohri, M., Rostamizadeh, A., Talwalkar, A.: Foundations of Machine Learning. 2nd ed. MIT Press, Cambridge (2018)
19. Neyshabur, B., Li, Z., Bhojanapalli, S., LeCun, Y., Srebro, N.: The role of over-parametrization in generalization of neural networks. In: International Conference on Learning Representations (2019). https://openreview.net/forum?id=BygfghAcYX

20. Scardapane, S., Wang D.: Randomness in neural networks: an overview. Wiley Interdiscip. Rev. Data Min. Knowl. Disc. **7**(2), 1–42 (2017)
21. Schmidt, W.F., Kraaijveld, M.A., Duin, R.P.W.: Feedforward neural networks with random weights. In: Proceedings of 11th IAPR International Conference on Pattern Recognition, Conference B: Pattern Recognition Methodology and Systems, vol. 2, pp. 1–4 (1992)
22. Schmidt-Hieber, J.: Nonparametric regression using deep neural networks with ReLU activation function. Ann. Statist. **48**(4), 1875–1897 (2020). https://doi.org/10.1214/19-AOS1875
23. Steland, A.: Sequential data-adaptive bandwidth selection by vross-validation for nonparametric prediction. Commun. Stat. Simul. Comput. **41**(7), 1195–1219 (2012). https://doi.org/10.1080/03610918.2012.625853
24. Steland, A.: Extreme learning and regression for objects moving in non-stationary spatial environments. Preprint on arXiv (2020). https://arxiv.org/abs/2005.11115
25. Steland, A.: On consistency of cross-validation for randomized (deep) learners. Under preparation (2022)
26. Steland, A., Chang, Y.T.: High-confident nonparametric fixed-width uncertainty intervals and applications to projected high-dimensional data and common mean estimation. Seq. Anal. **40**(1), 97–124 (2020)
27. Stone, M.: Cross-validatory choice and assessment of statistical predictions (with discussion). J. Roy. Statist. Soc. Ser. B **8**, 111–147 (1974)
28. Tang, J., Deng, C., Huang, G.-B.: Extreme learning machine for multilayer perceptron. IEEE Trans. Neural Netw. Learn. Syst. **27**(4), 809–821 (2016)
29. Yang, Y.: Consistency of cross validation for comparing regression procedures. Ann. Statist. **35**, 2450–2473 (2007)

Scale Invariant and Robust Pattern Identification in Univariate Time Series, with Application to Growth Trend Detection in Music Streaming Data

Nermina Mumic, Oliver Leodolter, Alexander Schwaiger, and Peter Filzmoser

Abstract A method is proposed to identify a pre-defined pattern in univariate time series. The pattern could describe an expected trend, for example, the development of a "hit" in music streaming data, with a rapid increase of the number of streams, to a peak, and a slow decay. With this application in mind, the method is scale invariant in the time domain as well as for the values of the time series (e.g., number of streams). Moreover, it is suitable also for irregularly spaced time series, and robust against short-term seasonal movements, as well as to noisy and spiky time series. Simulation studies compare this proposal with a method for identifying breaks in a time series. If the number of breaks for this method is pre-defined, the windows with the simulated patterns can be well identified with both procedures. The new proposal can additionally filter out those time series which contain the pre-defined pattern. This method is applied to a big data base of digital music streaming data for the purpose of "hit" detection.

Keywords Pattern identification · Time series · Robustness

1 Introduction

With the increased digitalization, various industrial sectors from economy, technology to healthcare have to cope with huge amounts of collected data. Often only specific movements or patterns in the data are relevant for the user. Therefore, pattern detection in a series of observations becomes a striking issue in many disciplines. In the case of the music industry, agents and music labels are interested

N. Mumic
Legitary GmbH, Vienna, Austria
e-mail: nermina.mumic@legitary.com

O. Leodolter · A. Schwaiger · P. Filzmoser (✉)
Institute of Statistics and Mathematical Methods in Economics, TU Wien, Vienna, Austria
e-mail: peter.filzmoser@tuwien.ac.at

A. Steland, K.-L. Tsui (eds.), *Artificial Intelligence, Big Data and Data Science in Statistics*, https://doi.org/10.1007/978-3-031-07155-3_2

in finding and predicting music hits. Those might be characterized by a rapid increase in the number of streams or downloads to a peak and a slow decay. This is connected to several issues: the corresponding time series can show strong serial correlation, and they can be very noisy or contain spikes, representing exceptional events. In those cases, long term trends are contaminated with short-term movements. Thus, pattern detection algorithms should not focus on those seasonal patterns but rather be capable of identifying the long-run trend behind. Another challenge is the scale invariance, as the shape expressed by a "hit" can be the same in different songs, but the number of daily streams or downloads could be very different. For this reason, the invariance in both time (x) and measurement (y) scale is a crucial point in the development of an algorithm for pattern identification. Furthermore, it is of interest to precisely identify the position of a pre-defined pattern, with the starting and ending point. Such a procedure could be applied to a big data base of music streaming data in order to filter out songs containing pre-defined patterns. In a next step one could analyze external information to learn more about the reason for the occurrence of the pattern.

Current literature for pattern detection in time series suggests computing appropriate similarity measures between time series. Lin et al. [10] distinguish between supervised methods (classification, nearest neighbor, decision trees, support vector machines, artificial neural networks), semi-supervised learning (labeled training data is used to predict unlabeled test data), and unsupervised learning (hierarchical clustering, k-means). They also suggest approximations of the time series via some representation—like Symbolic Aggregate Approximation (SAX), Discrete Fourier Transform (DFT) or Discrete Wavelet Transform (DWF), to name a few, and compute distance measures based on extracted features.

Similarity measures between time series can also be defined based on their shape or their structure. Representatives of the shape based approach are Euclidean Distance or Dynamic Time Warping (DTW), which are suitable for short time series [2, 11]. DTW works unsupervised and established as standard approach in time series clustering. DTW fixed the issue of the Euclidean Distance not being invariant in the time space (x) but still lacks with invariance in the measurement(y). The structure based approach is grounded on extracting features of the time series, like information of the covariance structure or ARMA coefficients, see [5]. This approach is suitable for long time series but lacks in detecting long term patterns. Furthermore, it is not capable of searching specific patterns as they may vary in scale and covariance structure.

Deep Learning (DL) is currently a popular tool for data mining, as it works completely unsupervised without any bigger modeling requirements of the user [6]. To achieve decent results, a sufficiently big training set is required, which is mostly not available. Furthermore, DL is not capable of locating the position of the pattern and shows issues in dealing with spikes and strong seasonalities [15].

In this paper we present an approach for long-run pattern detection that performs well on noisy univariate time series and which is robust towards outliers and exceptional events. The wanted pattern must be defined in terms of x and y coordinates, which makes the algorithm applicable to search for any arbitrarily defined pattern and locates its exact position. Furthermore, no training data set is required like in the case of DL applications. In Sect. 2 we introduce our methodology of pattern detection. First we define the sliding windows approach and explain how x and y scale invariance are achieved within each window. Then we robustly regress the observations on the values of a pre-defined pattern. The resulting regression coefficients and the robustly estimated long-run variance of the resulting residuals inform about the likelihood of the presence of the pattern in the window. In the last step we combine this information and define candidates for the windows containing the wanted pattern, using appropriate cutoff values. In Sect. 3 we evaluate the algorithm by means of a simulated time series by varying their underlying pattern, length, magnitude, signal-to-noise ratio, and their correlation structure. In Sect. 4 we present real data examples, where we use a data set consisting of more than 160 million revenue records, representing the daily sales of music titles over the period of about 3 years. The final Sect. 5 summarizes and concludes.

2 Methodology

Depending on the application, it is possible to decompose a time series $y = \{y_t \mid t = 1, \ldots, T\}$ into

$$y_t = p_t + s_t + \varepsilon_t, \tag{1}$$

where p_t denotes the trend component or "pattern" which determines the general long-run direction of the time series, s_t the seasonal component capturing repeating patterns of a fixed frequency, and ε_t the stochastic error or residual component. In this section we want to develop an approach to identify an arbitrarily pre-defined, deterministic long-run pattern p_t with variable magnitude and duration, and to detect the exact position of the pattern. This method is applicable for very noisy, univariate time series, containing additional overlaying trends and seasonal components s_t. We assume that the duration of the pattern p_t is longer than the seasonal component s_t in order not to mix it up with pure seasonal movement. We also assume that the observed time series is in general longer than the pattern, thus it starts before the pattern appears, and it ends after the pattern.

The proposed approach is based on a sliding window, where a multitude of subsequences (windows) of varying length is extracted from a univariate time series. Then those subsequences are normalized to achieve scale invariance for the values of the time series. Within those windows, features are calculated, and this information is used for identifying the patterns and their exact position.

2.1 Sliding Windows

Given a time series $y = \{y_t \mid t = 1, \ldots, T\}$, a window w is defined by a starting index $s^w \geq 1$ and an ending index $e^w \leq T$. We denote the corresponding subsequence of y as $y^w = \{y_t \mid t = s^w, s^w + 1, \ldots, e^w - 1, e^w\}$. In order to detect a pattern independently of its position and length, we need a set W of windows where the windows $w \in W$ vary in the starting positions s^w, as well as in lengths $l^w := e^w - s^w + 1$. The finest possible set of suitable windows for a time series y is $W^f = \{w \mid s^w < e^w, 1 \leq s^w \leq T, 1 \leq e^w \leq T\}$. In most cases, using this entire set is unnecessarily computational intense. For this reason we define a reasonable subset by sliding windows of different lengths $l \in L$, with step lengths v_l, through the time series y. This results in the set

$$W = \Big\{ w : s^w \in \{1, 1 + v_l, 1 + 2v_l, 1 + (\lfloor T/v_l \rfloor - 1)v_l\}, e^w = s^w + l - 1 \Big\}, \quad (2)$$

where $\lfloor a \rfloor$ denotes truncation to the next integer $\leq a$.

With a suitably defined set of windows W we can continue defining appropriate features for the subsequences $\{y^w \mid w \in W\}$ and use unsupervised clustering methods to group windows with similar features. Galeano and Pena [5] suggest clustering according to similarity measures, like the autocorrelation function. This would only be a reasonable choice in the case of stationary time series without long-run trend patterns. Thus, we present alternative features that are capable of identifying long-run patterns.

2.2 Scale Invariance

The method should detect pre-defined patterns regardless of their length or the magnitude. We will further refer to this property as x and y scale invariance.

Invariance along the x axis is achieved by taking subsequences y^w of the time series y of varying lengths l, as described in Sect. 2.1. This ensures capturing the same pattern in varying length (x scale) among different time series.

To achieve comparability across both windows y^w within a single time series and those of different time series, it is necessary to normalize the subsequences y^w, $w \in W$. We will further denote the standardized time series as $\tilde{y}^w$. Here we want to scale such that the first observation of the series $\tilde{y}^w$ has a value of approximately 1, thus $\tilde{y}_1^w \approx 1$. However, simply dividing all observations of y^w by the first value y_1^w could be too sensitive to this value, and in order to gain more robustness against outliers, we propose as a scaling factor the median of the first $\left\lfloor \frac{1}{20} \cdot l^w \right\rfloor$ observations, where

l^w is the length of the series y^w, and thus

$$\tilde{y_t}^w := y_t^w / \operatorname{med}\left(y_s^w, s = 1, \ldots, \left\lfloor \frac{1}{20} \cdot l^w \right\rfloor\right), \qquad t = s^w, s^w + 1, \ldots, e^w. \tag{3}$$

2.3 Pattern Identification

Assume that we are looking for a pattern of the form

$$p = \left\{(x_1^p, y_1^p), (x_2^p, y_2^p), \ldots, (x_m^p, y_m^p)\right\},$$

where $x_1^p < x_2^p < \ldots < x_m^p$ correspond to positions on the time axis, and m is the number of time points of y^w. An example for constructing a pattern is to consider coordinates $\{(0, 0), (0.5, 1), (1, 0)\}$, which define a triangular in the range of a normed time window. If we assume equidistant time information in y^w, the pattern points $x_1^p, \ldots, x_m^p$ are simply set to $1, \ldots, m$, where $m = l^w$. The first half of the values $y_1^p, \ldots, y_m^p$ are linear interpolations between 0 and 1, and the second half are linear interpolations between 1 and 0, respectively. In case of non-equidistant time series, the linear interpolation scheme has to be adjusted accordingly.

Then we consider the linear model

$$\tilde{y_t}^w = \beta_0^w + \beta_x^w x_t^p + \beta_p^w y_t^p + \epsilon_t^w, \qquad t = s^w, s^w + 1, \ldots, e^w, \tag{4}$$

where, without loss of generality, s^w and e^w are reparameterized to 1 and m, respectively. The intercept term β_0^w allows for a vertical shift of the pattern, the slope parameter β_x^w enables to compress or stretch the time axis, and the slope parameter β_p^w provides information about the pattern fit. We will refer to this parameter as pattern coefficient in the following. The term ϵ_t^w is the error term in the model. Since the underlying time series can be noisy, the regression parameters in this model are estimated with the MM-regression estimator [16], which is highly robust against outliers.

2.4 Model Diagnostics

The estimated pattern coefficient $\hat{\beta}_p^w$ in model (4) provides information about the presence of the pattern in the time series. If the pattern corresponds to the long-run trend contained in the series y_t^w, the residuals ϵ_t^w should not contain a long-run trend. To evaluate the extent of uncaptured seasonality and long-run trends in ϵ_t^w, we propose the use of the long-run variance

$$\omega^2(\epsilon_t^w) = \gamma_{\epsilon_t}(0) + 2\sum_{i=1}^{\infty} \gamma_{\epsilon_t}(i) , \tag{5}$$

where $\sigma^2(\epsilon_t^w) = \gamma_{\epsilon_t}(0)$ is the variance, and $\gamma_{\epsilon_t}(i)$ is the autocorrelation to lag i.

Instead of computing the long-run variance on the raw residuals ϵ_t^w, we take trimmed and smoothed residuals $\tilde{\varepsilon}_t^w$ in order to reduce the effect of possible outliers. The residuals are smoothed with a running median to downweight potential spike effects. In our implementation, the smoothed residuals are obtained as

$$\epsilon_s^w(t) := \text{med}\left(\epsilon_{t-3}^w, \epsilon_{t-2}^w, \ldots, \epsilon_{t+2}^w, \epsilon_{t+3}^w\right) . \tag{6}$$

These smoothed residuals are then trimmed with a trimming factor $\alpha = 0.1$, and thus we obtain

$$\tilde{\epsilon}^\omega(t) = \epsilon_s^\omega(t) , \qquad t = \max\left(1, \lceil l^\omega \cdot 0.1 \rceil\right), \ldots, \min\left(l^\omega, \lfloor 0.9 \cdot l^\omega \rfloor\right) .$$

This is the input information to obtain the estimated long-run variance $\hat{\omega}^2\,(\tilde{\epsilon}^w)$. For this purpose we use the Andrews quadratic spectral kernel HAC estimator [1]. The R function `lrvar` from the package `sandwich` [17, 18] allows to compute this estimator.

Both the estimated pattern coefficient $\hat{\beta}_p^w$ and the trimmed smoothed residuals ϵ_s^w are used to evaluate the results from the regression fits for different windows $w \in W$. However, they first need to be made comparable for the different considered windows by centering and scaling appropriately. We assume that for the majority of the windows, the residuals will not contain any long-run structure, because there is either no structure present in the window or it was already considered by the coefficient β_p^w. However, residuals from windows that cover only a part of the pattern will contain remaining structure. These will lead to outliers in the distribution of the residuals. Thus we can assume that the distribution of the estimated long-run variances based on the different windows is bimodal, and the mode for the minority will refer to the interesting structure. We estimate the modes by a robust clustering procedure, here in terms of trimmed k-means clustering, implemented in the R package `tclust` [4]. Clustering is thus performed in the univariate space, based on the estimated long-run variances that result from all considered windows. Since the distribution of the long-run variances is skewed, we first log-transform (the square-root of) these values. The estimated mode of the smaller cluster is $\mathcal{L}(\hat{\omega}(\epsilon)_t^w) = M_1(\log(\hat{\omega}(\tilde{\epsilon}_t^w)))$. We also scale the (square-root of the) long-run variance, by using the robust scale estimator Q_n [12], which is based on pairwise absolute differences. Thus, the resulting centered and scaled version to compare the (square-root of the) long-run variance among different windows is

$$\tilde{\omega}(\tilde{\epsilon}_t^{\omega}) := \frac{\log(\hat{\omega}(\tilde{\epsilon}_t^{w})) - \mathcal{L}(\hat{\omega}(\tilde{\epsilon}_t^{w}))}{\hat{\sigma}\left(\hat{\omega}(\tilde{\epsilon}_t^{w})\right)}. \tag{7}$$

Also the pattern coefficient is scaled as

$$\tilde{\beta}_p^{w} = \frac{\hat{\beta}_p^{w}}{\hat{\sigma}\,(\hat{\beta})} \qquad \forall w \in W, \tag{8}$$

where $\hat{\beta} = \{\hat{\beta}_p^w \mid \tilde{\omega}(\tilde{\epsilon}_t^{w}) < 2\}$, and $\hat{\sigma}(\hat{\beta}) = Q_n(\hat{\beta})$. Thus, here we only focus on pattern coefficients that originate from windows resulting in sufficiently low long-run variance.

If a window $w \in W$ contains the defined pattern, the corresponding pattern coefficient is supposed to be clearly greater than zero. In the ideal case, the pattern already covers most of the long-run structure, such that $\tilde{\omega}(\tilde{\epsilon}_t^{\omega})$ is small. In this case, such a window is a candidate for further investigation. In the next section we will develop a strategy to decide whether such a candidate refers to a window with a pre-specified pattern or not.

The complete algorithm is summarized as pseudo-code in Algorithm 1.

3 Simulation Experiments

The aim of this section is to derive appropriate tuning parameters for the method, to illustrate and evaluate the proposed methodology, and to compare it to an alternative proposal. We will first explain the simulation design, present different variations of the proposed methodology, and then present the results.

3.1 Simulation Design

We will simulate time series based on five different types of patterns, denoted as $p^0, \ldots, p^4$. These patterns are denoted as Type 0, Type 1, etc. Within each pattern there will be 12 different variations of the time series structure. A simulated time series with values y_t, $t = 1, \ldots, T$, will be split along the time axis into three distinct time intervals o_1, o_2, o_3, defined as

$$o_1 = \left\lfloor \hat{k}_1 \cdot m \right\rfloor \quad o_2 = \left\lfloor \hat{k}_2 \cdot m \right\rfloor \quad o_3 = \left\lfloor \hat{k}_3 \cdot m \right\rfloor, \qquad m \sim U(800, 1200),$$

Input
Time series $(y_t)_{t=1,\ldots,T}$
Output
Found pattern type
for (each pattern Type 1, Type 2 ,Type 3 ,Type 4) **do**
Create set of suitable windows, where s^{ω} denotes the start index and e^{ω} the end index

$$W := \left\{ \omega : s^{\omega} \in \left\{ 1, 1+v_l, 1+2v_l, 1+\left(\left\lfloor \frac{T}{v_l} \right\rfloor - 1 \right) v_l \right\}, e^{\omega} = s^{\omega} + l - 1 \right\}$$

for (each subsequence y^{ω}, $y^{\omega} \in \{y^{\omega} | \omega \in W\}$) **do**
Scale y^{ω} and denote the scaled series with $\tilde{y}_t^{\omega}$
Solve the linear model where (x^P, y^P) depends on the Type

$$\tilde{y}_t^{\omega} = \beta_0^{\omega} + \beta_x^{\omega} x_t^P + \beta_p^{\omega} y_t^P + \epsilon_t^{\omega}, \qquad t = s^{\omega}, s^{\omega}+1, \ldots, e^{\omega}$$

Estimate the longrun variance $\hat{\omega}$ of the trimmed and smoothed residuals $\tilde{\epsilon}_t^{\omega}$
Scale the estimated pattern coefficient $\hat{\beta}_p^{\omega}$ and denote it with $\tilde{\beta}_p^{\omega}$
end for
if more than 5% of the windows fullfill $\hat{\omega}(\tilde{\epsilon}_t^{\omega}) < 2$ and $\tilde{\beta}_p^{\omega} > c$ **then**
Type is eligible
return window whose longrun variance is in the lower third of all windows and has the highest pattern coefficient out of those
else
return Type
end for
if Types were found **then**
Out of all eligible Types choose the one where most windows fullfill $\hat{\omega}(\tilde{\epsilon}_t^{\omega}) < 2$ and $\tilde{\beta}_p^{\omega} > c$
return Type
else
return No Type

Algorithm 1: Pseudo-code of the pattern detection algorithm

with

$$\hat{k}_1 = \frac{k_1}{\sum_{i=1}^4 k_i} \qquad \hat{k}_2 = \frac{k_2 + k_3}{\sum_{i=1}^4 k_i} \qquad \hat{k}_3 = \frac{k_4}{\sum_{i=1}^4 k_i}$$

and $k_i \sim U(0.1, 0.9)$ for $i = 1, \ldots, 4$. Thus, the sum of the three intervals $o_1 + o_2 + o_3$ equals T (up to truncating effects).

The time series values are generated according to

$$y_t = z + \epsilon_t\,, \qquad t = 1, \ldots, o_1\,, \tag{9}$$

$$y_t = p^j_{t-o_1} + \epsilon_t\,, \qquad t = o_1 + 1, \ldots, o_1 + o_2\,, \tag{10}$$

$$y_t = z + \epsilon_t\,, \qquad t = o_1 + o_2, \ldots, o_1 + o_2 + o_3\,, \tag{11}$$

where p^j (for $j = 0, 1, \ldots, 4$) is the pattern type, ϵ is the error term, and z is the starting level shift of the time series. The term z is generated according to $z \sim U(900, 1100)$. The pattern types $p^0, \ldots, p^4$ are generated with the following coordinate pairs (x, y):

Type 0: $S_0 := \{(0, 0), (1, 0)\}$
Type 1: $S_1 := \{(0, 0), (0.1, 1), (1, 0)\}$
Type 2: $S_2 := \{(0, 0), (0.2, 1), (1, 0)\}$
Type 3: $S_3 := \{(0, 0), (0.5, 1), (1, 0)\}$
Type 4: $S_4 := \{(0, 0), (0.9, 1), (1, 0)\}$

For each pattern we linearly interpolate between the given coordinate pairs. For that purpose we use $b \cdot o_2$ equidistant points in the interval $[0, 1]$, with $b \sim U(0.5, 0.9)$. Afterwards, the resulting y-coordinate values are multiplied by the factor h, resulting from $h = z \cdot P$, where $P \sim U(0.5, 0.9)$ defines the proportion to the starting level shift of the time series. Thus, for pattern type p^j, $j = 0, \ldots, 4$, $t = 1, \ldots, o_2$, and x_t equidistant $\in [0, 1]$ we have:

$$p^j = \left\{(t, y^p_t) : \; y^p_t = y_{i-1} + \frac{y_i - y_{i-1}}{x_i - x_{i-1}}(x_t - x_{i-1})h\,, (x_i, y_i) \in S_j,\; i = 2, \ldots, |S_j|\right\}. \tag{12}$$

Figure 1 shows examples of the resulting pattern time series. Type 0 contains no pattern, and subsequently this type will refer to random noise. Type 1 and 2 present a rapid increase at the beginning (which is stronger for Type 1), followed by a slower decay. Type 3 results in a symmetric pattern around the center of the pattern series, and Type 4 is the mirrored picture of Type 1.

The next step is to define the error component used in (9)–(11). We will use a setting which is the sum of an AR(1) process with randomly generated spikes. Define $a \in \{0.1, 0.5, 0.9\}$ and $r \in \{0.01, 0.05, 0.1, 0.2\}$. For $t = 1, \ldots, T$ we simulate:

$$\eta_t = a \cdot \eta_{t-1} + \xi_t\,, \qquad \eta_0 \sim N(0, \sigma^2) \tag{13}$$

$$\sigma^2 := \frac{1}{1.645} r \cdot z \cdot \sqrt{1 - a^2}\,, \qquad \xi_t \sim N(0, \sigma^2) \tag{14}$$

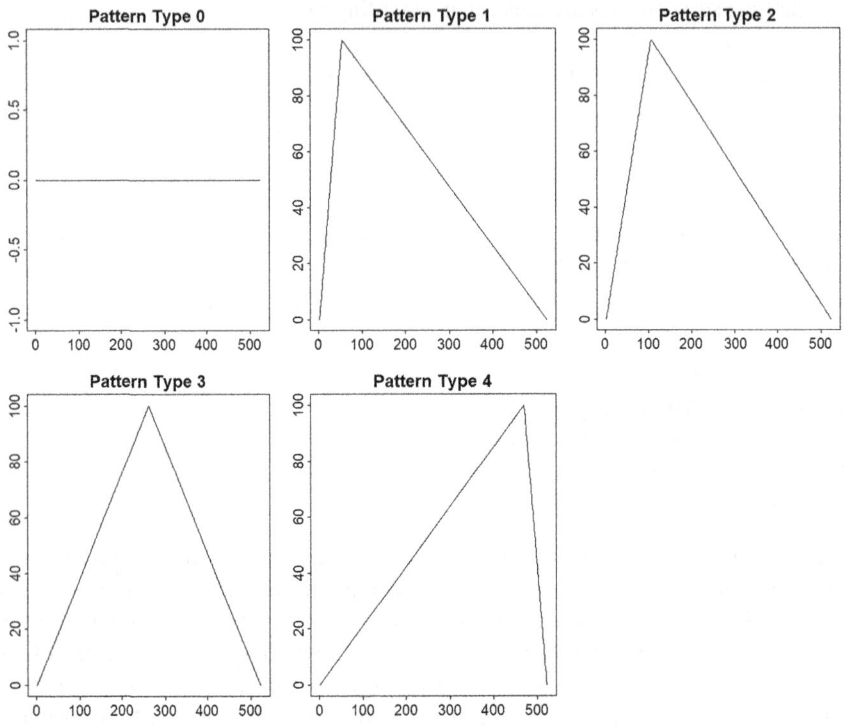

Fig. 1 Pattern types which used for the simulation study

$$\tau_t = \chi_t \cdot 6 \cdot \frac{1}{1.645} r \cdot z, \qquad \chi_t \sim \text{Bernoulli}(0.6) \tag{15}$$

$$\epsilon_t = \eta_t + \tau_t. \tag{16}$$

Here, 1.645 is the quantile 0.95 of the standard normal distribution. The choice of σ^2 has the effect that higher values of r and lower values of a increase the probability of a bigger shock in the AR(1) process (13). The magnitude of a determines the impact of a shock at time point t_i to future values at t_j, for $i < j \leq T$; bigger values of a lead to a stronger effect. The term τ_t generates outliers, which are getting stronger with increasing values of r. For every pattern we will generate values with all 12 combinations of a and r. The components p^j and ϵ are added according to (10).

Figure 2 presents examples of the simulated pattern for Type 1 to Type 4, with parameters a and r indicated on top of the plots.

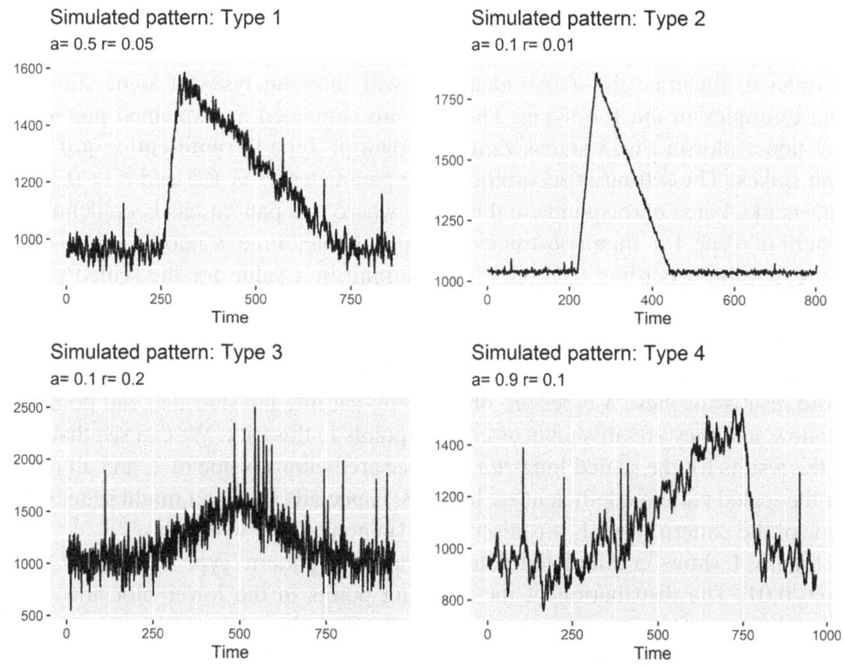

Fig. 2 Examples of simulated time series pattern types, with different values for a and r

3.2 *Analyses of the Simulated Data*

For analyzing the simulated time series with the methodology proposed in Sect. 2 we first need to define candidate windows $w \in W$, and in every window we will search for a pattern of Type 1 to Type 4. Ideally, the methodology should identify the correct pattern type and the correct window where the pattern has been simulated. The starting points s^w of the windows depend on the step length v_l and on the window length l^w. For our simulation we will use $v_l = 30$ and $l^\omega \in L = \{350, 400, 450, 500, 550, 600, 650\}$. The number of considered windows will be taken as $N_l = \lceil (T - l^w)/30 \rceil$, where the bracket refers to the next biggest integer. Then the starting points are defined as $s^w = 1 + k \cdot 30$, and the ending points as $e^w = l^w + k \cdot 30$, for $k = 0, \ldots, N_l$.

The remaining steps are as outlined before: The time series in a particular window is scaled as in (3). For the considered search pattern, the corresponding time series values are obtained by interpolation as defined in (12), where we use length l^w instead of o_2. Then, regression according to model (4) is carried out, and our particular interest is in the results for the scaled pattern coefficient $\tilde{\beta}_p^w$ and in $\tilde{\omega}(\tilde{\epsilon}_t^\omega)$, which we denote simply as scaled long-run variance.

3.3 Example Analyses of Simulated Data

In order to illustrate the above ideas, we will show analyses of some simulated data examples in the following. The data are simulated as explained previously. The upper plot in Fig. 3 shows simulated data of Type 0, representing just noise with spikes. The setting is according to the parameters $a = 0.5$ and $r = 0.1$. The dark marked area corresponds to the range where this pattern has been defined. A pattern of Type 1 is then constructed in a pre-defined time window, and it is used for regression according to model (4), resulting in a value for the scaled pattern coefficient $\tilde{\beta}_p^w$ and centered and scaled long-run variance $\tilde{\omega}(\tilde{\epsilon}_t^w)$. Thus, this leads to one red dot in the lower plot of Fig. 3. Similarly, other types of patterns are used in the regression model, resulting in three more points with different color in this plot. More results for these 4 types are obtained by varying the start end end point of a window, and they are all shown as colored points in this plot. We can see that most of the results for the scaled long-run variance are below a value of 2, and all results for the scaled pattern coefficient are below 3. Especially the latter might suggest that none of the pattern types 1–4 is discovered (in none of the windows).

Figure 4 shows a simulated data example for pattern Type 2, $a = 0.1$ and $r = 0.01$. The distribution of the resulting points in the lower plot now looks quite different from the previous results. Particularly, many of the points refer to a scaled pattern coefficient bigger than 3, which indicates the presence of the particular search pattern in the considered search window. Most of these points are for search pattern Type 2, especially if we select those which have a low scaled

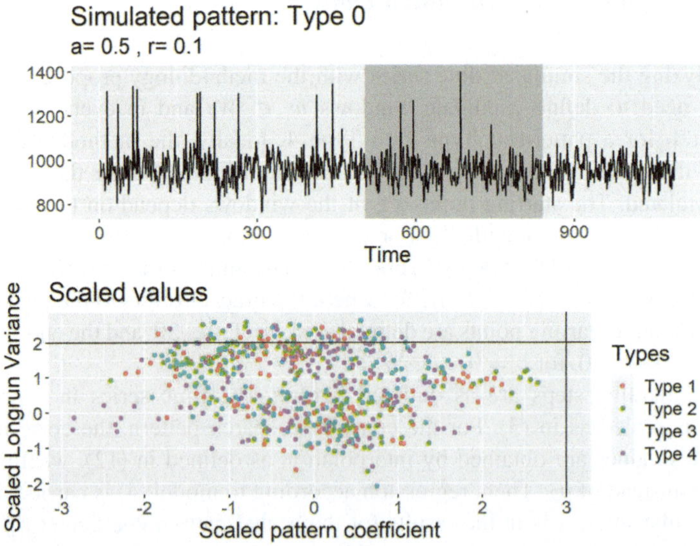

Fig. 3 Simulated time series from pattern Type 0, with $a = 0.5$ and $r = 0.1$ (top), and results from the evaluation (bottom)

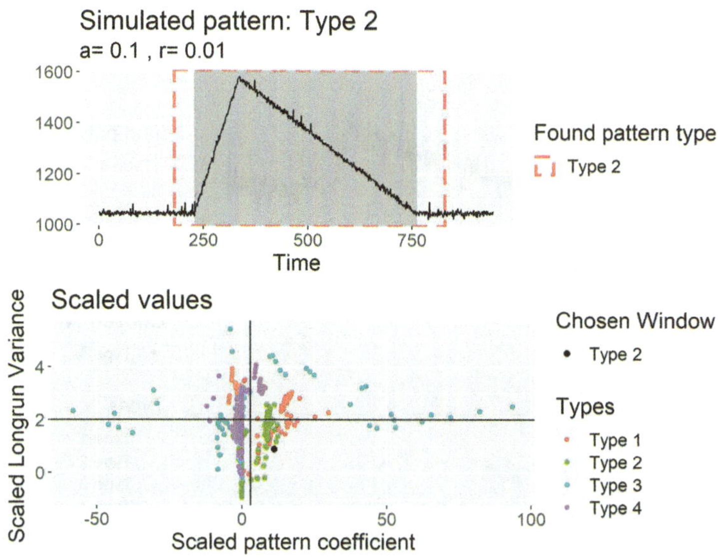

Fig. 4 Simulated time series from pattern Type 2, with $a = 0.1$ and $r = 0.01$ (top), and results from the evaluation (bottom)

long-run variance (say, smaller than 2), indicating no essential remaining structure. Thus, we focus on those points in the lower right quadrant of the plot, defined by the horizontal and vertical line. As a representative, we select the black indicated point, which is from a Type 2 search pattern. This point refers to the search window shown with red dashed lines in the upper plot. Indeed, this window is close to the window where the Type 2 pattern has been constructed.

A final simulated data example is shown in Fig. 5, originating from a pattern of Type 4 with $a = 0.5$ and $r = 0.2$. The points in the interesting lower right quadrant of the lower plot exclusively refer to search patterns of the correct Type 4, and the indicated black dot originates from a search window which is almost identical with the window of the simulated pattern.

3.4 Selection of the Pattern Type and Window

From the simulation example analyses before one could already see that a careful selection of the results for different windows and pattern types needs to be done. Before we had taken a particular point in the lower right quadrant of the bottom plot. Here we propose three different options:

- **Method 1:** We select those "initial" candidates for which we obtain $\hat{\omega}(\tilde{\epsilon}_t^w) < 2$ and $\tilde{\beta}_p^w > c$, for a cutoff value $c > 0$. In order to avoid artificial solutions, at

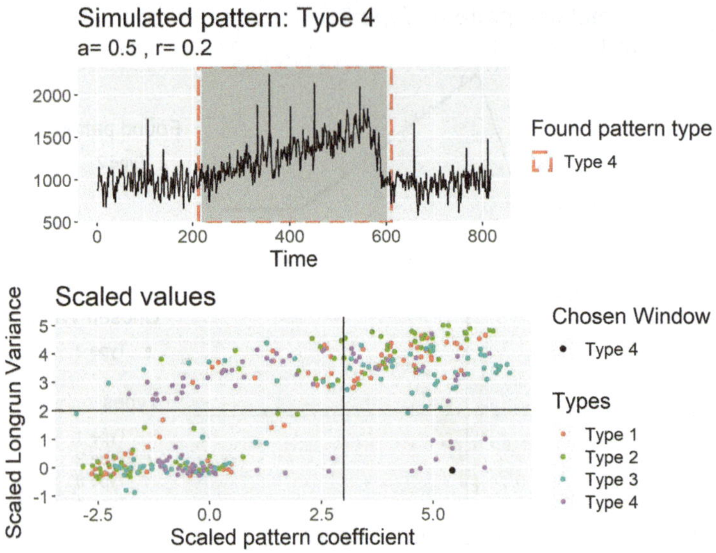

Fig. 5 Simulated time series from pattern Type 4, with $a = 0.5$ and $r = 0.2$ (top), and results from the evaluation (bottom)

least 5% of the windows of every type need to lead to results that fulfill these requirements. If this is not the case for a type, no result is reported for this type. Thus, it can be possible that "no window found" will be reported. For obtaining a unique solution, we ask for a compromise between low long-run variance and high pattern coefficient. Thus, we reduce the candidates per type to those which are in the lower third of the long-run variance. Out of those we take that solution with the biggest pattern coefficient. If there are solutions of more than one type, we take that which has the highest number of initial candidates.

- **Method 2:** This method is similar to Method 1, but we replace in Eq. (8) the condition $\hat{\beta} = \{\hat{\beta}_p^w \mid \tilde{\omega}(\tilde{\epsilon}_t^w) < 2\}$ by the new condition $\hat{\beta} = \left\{\hat{\beta}_p^w \mid \tilde{\omega}(\tilde{\epsilon}_t^w) < 2, \hat{\beta}_p^w > 0\right\}$. This makes sure that the pattern coefficients from windows which do not contain the pattern are not used for scaling. Windows resulting in negative pattern coefficients cannot include the complete pattern, because there should not be a negative relationship with the response.
- **Method 3:** Again similar to Method 1, but instead of model (4) we omit the term x_t^p and use the model

$$\tilde{y}_t{}^w = \beta_0^w + \beta_p^w y_t^p + \epsilon_t^w, \qquad t = s^w, s^w + 1, \ldots, e^w. \tag{17}$$

Thus, the slope parameter for the linear trend in the time axis is not considered, which might have affected the fit of the different pattern types.

We took two steps to fix the method and the tuning parameter c. These steps were based on 100 simulation replications for each parameter setting.

1. We vary c in $\{1, 1.2, 1.4, \ldots, 4\}$ and compute AUC (area under the curve) values from ROC (receiver operating characteristic) curves [3]. ROC curves present sensitivity and specificity, and thus provide a picture about correctly identifying the pattern type, and incorrectly specifying the wrong type. AUC values are in the range $[0, 1]$, the closer to one, the better the algorithm. A value of 0.5 corresponds to a random assignment [8]. According to the results we selected Method 1 as the overall best method. Some example results from the simulation are shown in the Appendix (Figs. 9 and 10).
2. Further simulations are conducted to select the cutoff value c. Depending on the value c, we count how often the four different pattern types are identified with Method 1 for a time series containing a particular pattern type. It turns out that a cutoff value $c = 3$ is a good choice. Some results from the simulation are provided in the Appendix (Figs. 11 and 12).

3.5 *Simulation Results*

We compare our method with the method BFAST [13, 14], which allows to decompose a time series into trend, season and remainder components. Trend and seasonal component are estimated iteratively, and also the number and position of breaks in the seasonal and trend component are estimated iteratively. Thus, BFAST is not able to identify pre-determined search patterns, but it flexibly detects breaks in the time series.

In order to allow for a fair comparison with our method, we fix the maximum number of breaks for the BFAST method with 3, which would correspond to what we would like to see with our search patterns. Then we can compare the time points of the first and last break with the time points of our simulated pattern windows. This is done by the RMSE (root mean squared error) over 100 simulation replications,

$$\frac{1}{100}\sum_{i=1}^{100}\sqrt{\frac{(s_i^w-\hat{s}_i^w)^2+(e_i^w-\hat{e}_i^w)^2}{2}},$$

where (s^w, e^w) is the true starting and ending position of the windows and $\left(\hat{s}^w, \hat{e}^w\right)$ the estimated positions (by BFAST or our method). Further, we compute

$$\alpha_s = \frac{|\hat{s}^w - s^w|}{T_w} \qquad \text{and} \qquad \alpha_e := \frac{|\hat{e}^w - e^w|}{T_w},$$

which provide information about the precision of estimating starting and ending point separately, relative to the window length T_w. Also for these measures we report

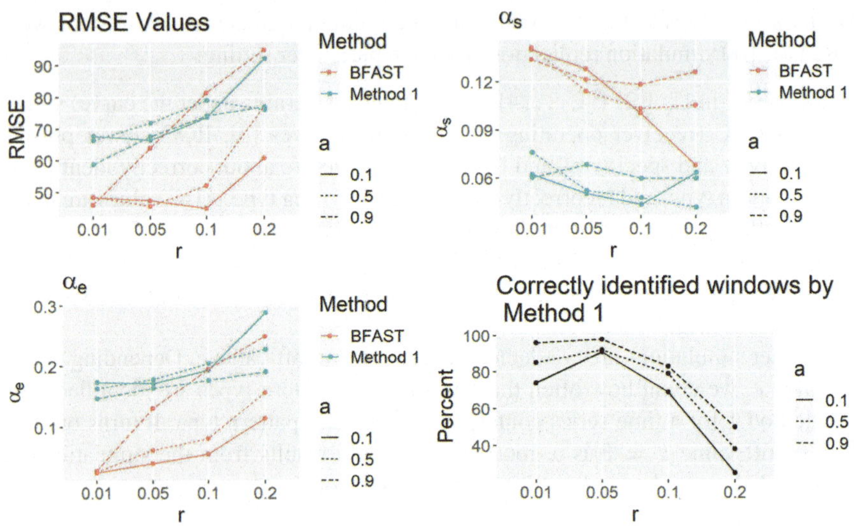

Fig. 6 Simulation results for simulated pattern Type 1

averages over the 100 simulations, for all combinations of $r = 0.01, 0.05, 0.1, 0.2$ and $a = 0.1, 0.5, 0.9$, and for each pattern type.

Figure 6 presents the results for these evaluation measures for simulated patterns from Type 1. For our Method 1 we only report the results when searching for the correct pattern Type 1. We can see that BFAST has more difficulties to identify the starting point of the window, but has better abilities to find the ending point. Overall, the RMSE is lower for most parameter settings. The bottom right plot shows that especially for lower values of r, our method is very successful with identifying the correct pattern type, and thus most of the simulation results have been used in the other plots. Further simulation results for the other pattern types are shown in the Appendix (Figs. 13, 14, and 15). Overall, BFAST and our method show similar performance.

In a final simulation we do not provide any information on the number of breaks for BFAST. Thus, we simply count over 100 simulations the total number of breaks identified by BFAST. Also for our method we count this total number, where one identified window counts for 3 breaks. Thus in the results shown in Fig. 7 we would expect to see around 300 counts. The symmetric Type 3 pattern is more difficult for Method 1 for higher values of r, and BFAST shows clearer deviations from the target 300 for higher values of r in all pattern types.

When simulating data without any pattern (Type 0), the resulting total number of breaks over 100 simulations is less than 10 for Method 1, for any combination of a and r. BFAST leads to around 40 identified breaks if $a = 0.5$, and to almost 400 for $a = 0.9$, for any choice of r.

As a summary it turned out that BFAST has a similar performance in identifying the breaks where the patterns start and end, specifically if the number of maximum breaks is pre-defined. If this is not done, BFAST has severe difficulties for higher values of a, specifically if only noise is present. Our proposed method has the additional advantage that not only the position of the windows are identified but it also aims to identify those windows which contain a pre-specified search pattern.

4 Examples from the Digital Music Industry

The streaming industry is one of the world's rapidly growing high-volume markets. According to the Global Music Report of 2019 of the International Federation of the Phonographic Industry (IFPI), the streaming industry's growth rate continues to rise by 20–60% since 2010 [9]. Covid-19 may accelerate the existing trend of music consumption shifting towards streaming [7].

Thus, it is of special interest for music right holders to understand, what drives the growth of streaming figures and detect underlying patterns and trends automatically, taking exploding data amounts into regard. In particular, we are interested in long-run trends in streaming counts of music titles, as they allow for a classification

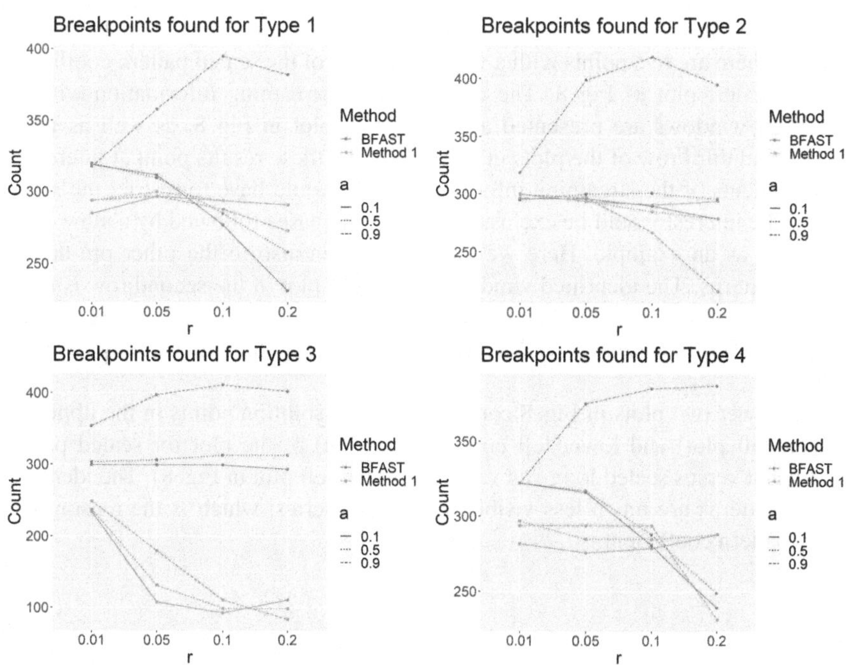

Fig. 7 Total number of identified breaks over 100 simulations for BFAST and Method 1

of organic and non-organic growth trends. For example, steep growth and decline of streaming figures might indicate non-organic consumption behavior like those arising from streaming bots, while continuously slow growth or decline gives an indication of organic behavior.

In this section we illustrate this use case of identifying growth patterns and classifying organically trending music titles. Thus, we examine real streaming accounting data provided by Rebeat Digital GmbH (https://rebeat.com), an Austria based music distribution company. We pick 100 music titles with the biggest numbers of total streams from their catalogue and analyze their streaming counts for the DSP (Digital Service Provider) Spotify (https://www.spotify.com/at/). Each title is represented by a univariate time series of streaming counts in a monthly frequency.

We use the same four search pattern types as defined in the previous section, and apply the algorithm for each search pattern to the individual music streaming time series. In about half of the streams, the algorithm did not identify an eligible window for these search patterns. For the other half of the songs we can find the result in the upper left plot of Fig. 8, which represents the resulting scaled pattern coefficients and scaled long-run variances. Thus, every point in this plot corresponds to the identified window of one music title, and the color informs about the window type. Since these are eligible solutions according to *Method 1*, all these points are arranged in the lower right quadrant of the plot, see vertical line at 3 and horizontal line at 2.

As we cannot present all the solutions as time series, we focus on some specific results. There are five points with a very high value of the scaled pattern coefficient, see upper left plot of Fig. 8. The corresponding streaming information with the identified windows are presented as upper right plot in Fig. 8, as well as in the second and third row of the plots in this figure. All these results point at interesting developments of the streaming information. One would have to restrict the search type if the interest would be exclusively in fast increases followed by a slow decay (Type 1), as an example. Here we have been open also to the other pre-defined search patterns. The identified window in the right plot of the second row is rather surprising, and it is a signal that the maximum window length used in the algorithm could have been chosen higher. Nevertheless, also this pattern reveals an interesting behavior.

The lower two plots in Fig. 8 correspond to the solution points in the upper left corner (left plot) and lower left corner (right plot) of the plot for scaled pattern coefficient versus scaled long-run variance (upper left plot in Fig. 8). The identified search patterns are much less visible in the time series, which is the reason for a lower pattern coefficient.

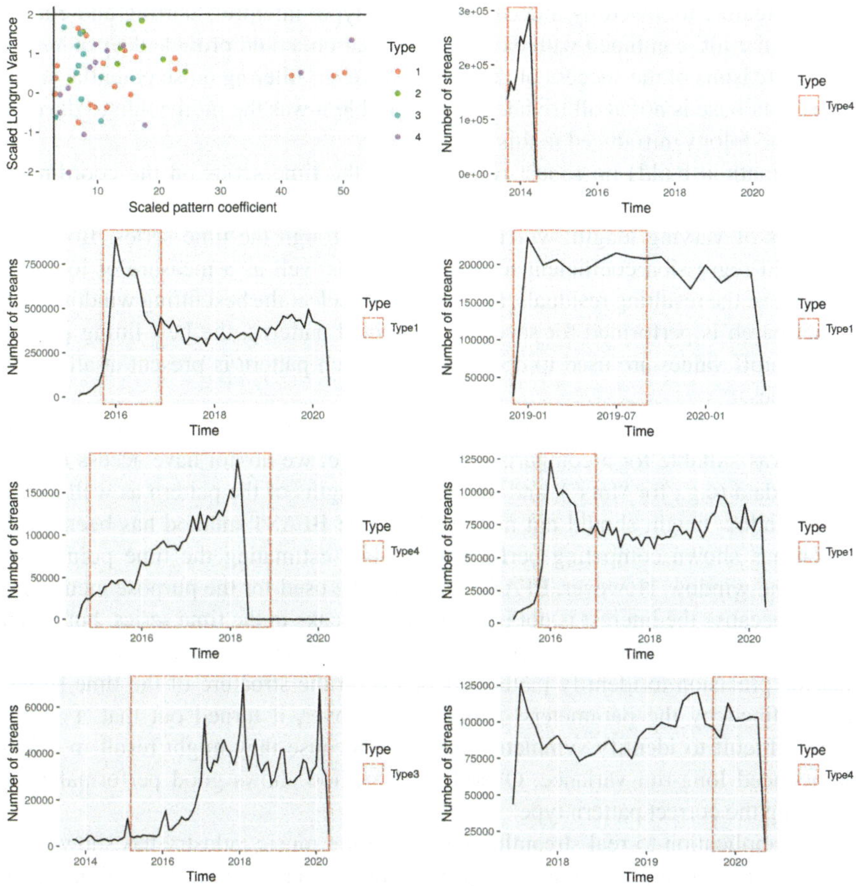

Fig. 8 Results from the application of the algorithm to 100 music streaming time series: for about half of the songs, a window has been identified, and the corresponding type, scaled pattern coefficient and scaled long-run variance for these songs is should in the upper left plot. The time series and identified windows corresponding to the 5 rightmost points are shown in the upper right plot, and in second and third row; the last row shows the results corresponding to the top left point (left plot) and to the bottom left point (right plot)

5 Summary and Conclusions

Bigger music labels have access to huge amounts of streaming information for various kinds of songs. It is of general interest to identify songs with specific streaming patterns, such as "hits" which might be characterized by a steep increase in the streams, followed by a decay. The song type, interpret, starting and ending point of the hit, combined with external information, could provide deeper insight into the reasons of the success of the song. However, filtering those potential songs in the data base is not at all trivial, and this problem was the motivation to develop the methodology introduced in this paper.

The method builds on robust regression of the time series on the coordinates which define the search pattern. Regression is performed for the values in time windows of varying length, which are moved through the time series. Both, the estimated regression coefficient for the pattern, as well as a measure of long-run variance of the resulting residuals, are the basis to select the best fitting window, and, if the search is performed for several pre-defined patterns, the best fitting pattern type. Cutoff values are used to decide if the search pattern is present at all in the time series.

Since the problem setting is rather specific, it was not obvious to find a method which was suitable for a comparison. For instance, we do not have access to pre-labeled data (e.g., for "hits"), and the window lengths of the pattern as well as the peak window height should not matter. Here, the BFAST method has been used, which has shown competing performance when estimating the time points that define the window. However, BFAST could not be used for the purpose mentioned above, because the interest is not in identifying breaks in the time series, but rather to find time series containing a pre-specified pattern. The simulations have shown that the precision to identify patterns depends on the structure of the time series, here defined by the parameters a and r. Moreover, it turned out that it seems more difficult to identify symmetric patterns, because they might result in a less pronounced long-run variance. Overall, the method shows good performance in detecting the correct pattern type.

The application to real streaming data from the music industry has shown that the algorithm can indeed identify songs with interesting patterns. We used those top 100 songs from a data base with the biggest numbers of streams in the considered time period, and in about half of the songs we have identified patterns. However, if one would be interested in a specific search pattern, much less songs would be identified containing such a pattern. One can also assume that songs with much smaller numbers of streams would not be identified with the search patterns, because the corresponding time series would rather show noise without specific patterns.

The computation time of the algorithm depends on various parameters, in particular, on the number of search windows. In our application, the pattern search took a few seconds per song on a standard computer. The algorithm would thus be appropriate for a search in bigger data bases. One also has to be aware that it might not be necessary to run the algorithm on a daily basis. In our application we used monthly aggregated data, which seems to be appropriate for the purpose.

Note that the method can also be applied to irregularly spaced time series. The algorithm would be appropriate also for any other application where a specific pattern in a time series is to be identified, and if computation time is important, one could parallelize the search, or limit the number of search windows and search patterns.

6 Appendix

See Figs. 9, 10, 11, 12, 13, 14, and 15.

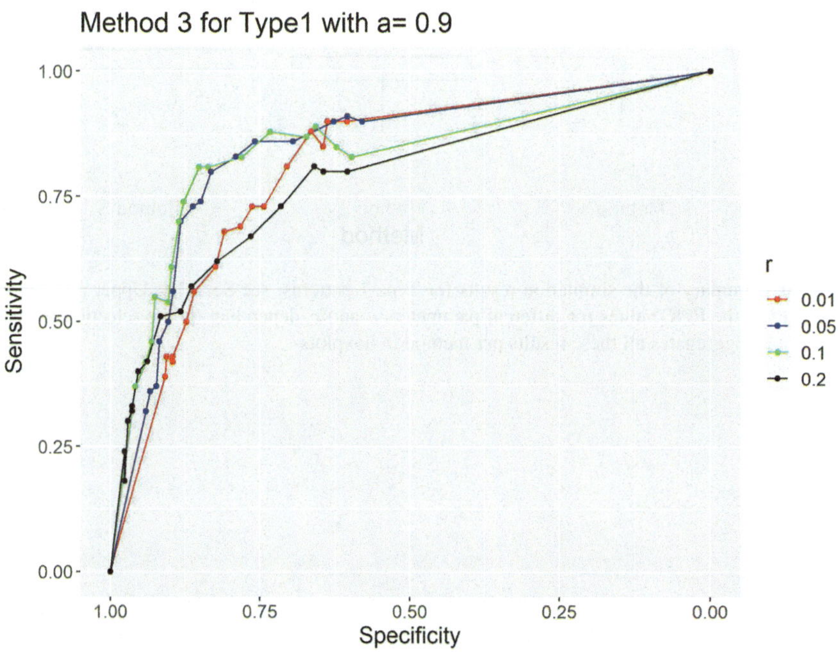

Fig. 9 Example of a ROC curve from a simulation of pattern Type 1 with $a = 0.9$, see Sect. 3.4: we vary c and search for the four different pattern types by using Method 3

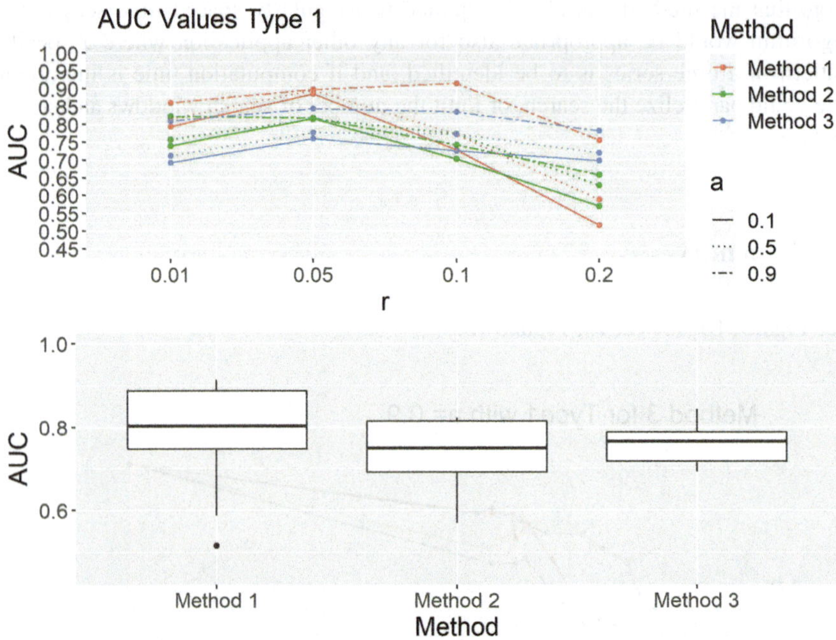

Fig. 10 Summary of the simulation results for Type 1 patterns, see Sect. 3.4. Upper plot reports averages of the ROC values for different parameters a and r, depending on the selection method. Lower plot aggregates all these results per method in boxplots

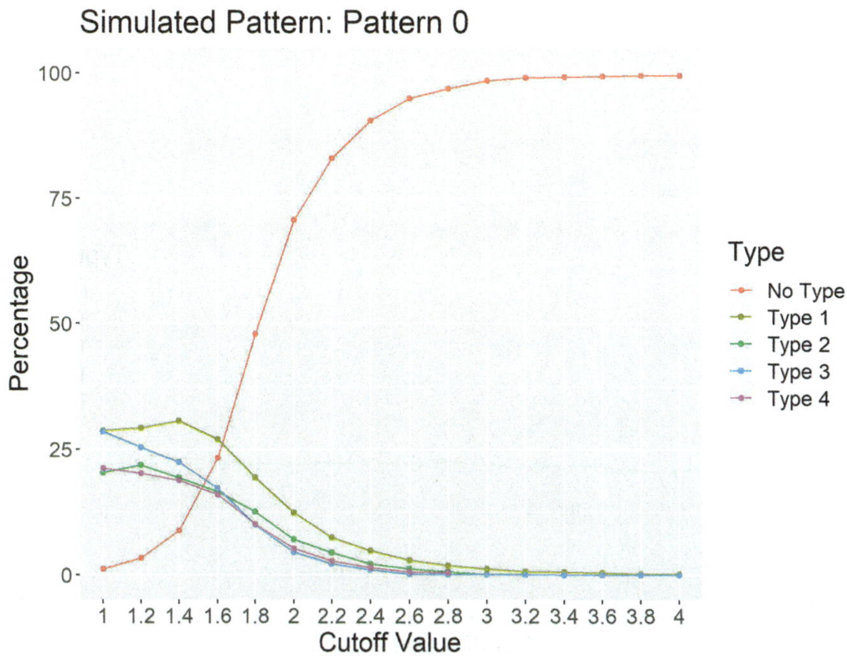

Fig. 11 We simulate 1200 time series (100 replications for the different values of a and r) containing no structure (Type 0), and count the percentage of identifying the different types of patterns, depending on the cutoff value c for Method 1. A cutoff $c = 3$ leads to a small proportion of wrong and to a high proportion of correct assignments

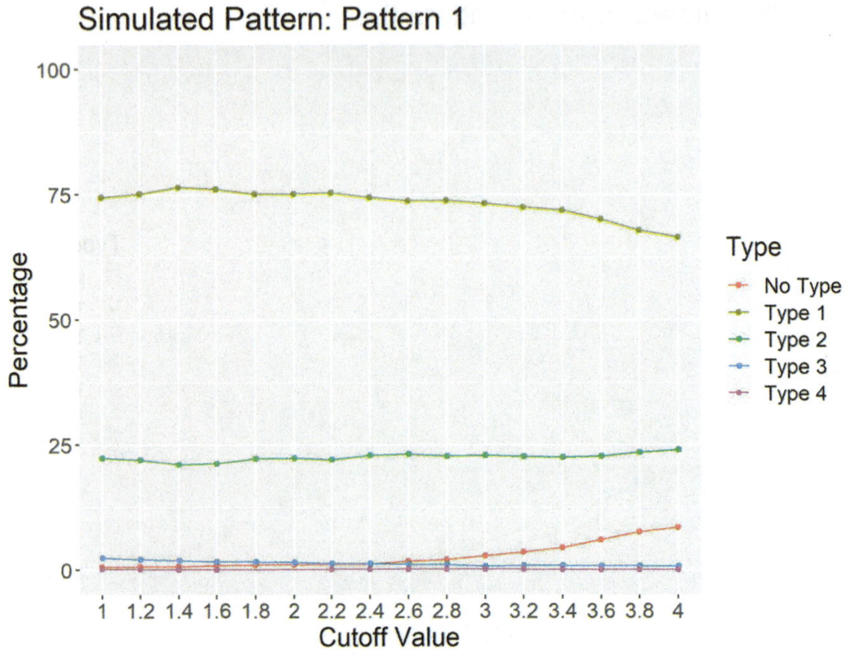

Fig. 12 We simulate 1200 time series (100 replications for the different values of a and r) containing pattern Type 1, and count the percentage of identifying the different types of patterns, depending on the cutoff value c for Method 1. A cutoff $c = 3$ leads to a small proportion of wrong and to a high proportion of correct assignments

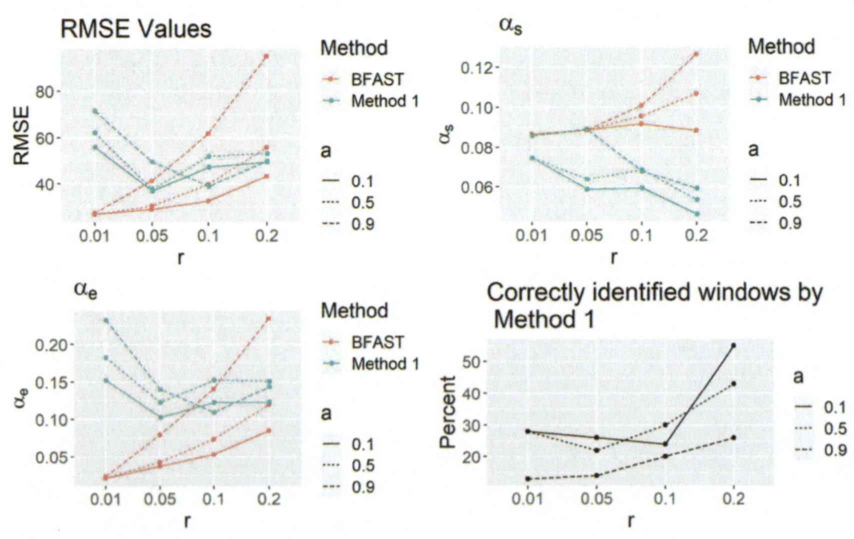

Fig. 13 Simulation results for simulated pattern Type 2, see Sect. 3.5

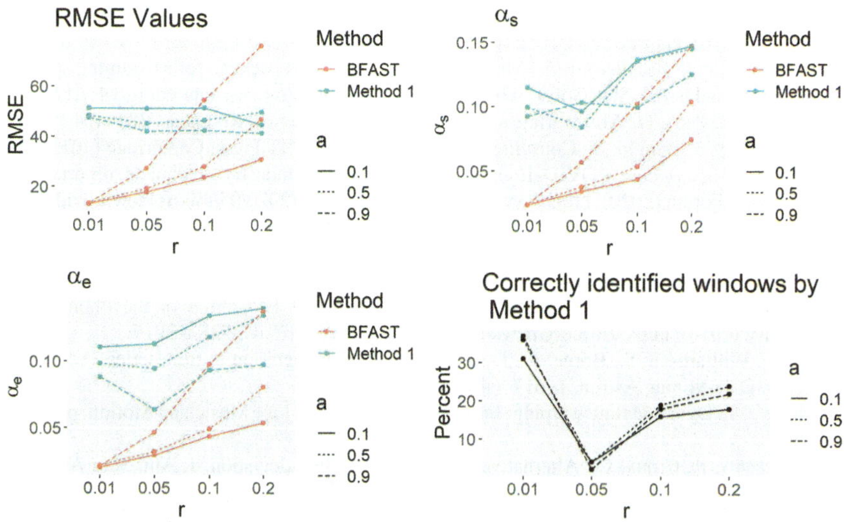

Fig. 14 Simulation results for simulated pattern Type 3, see Sect. 3.5

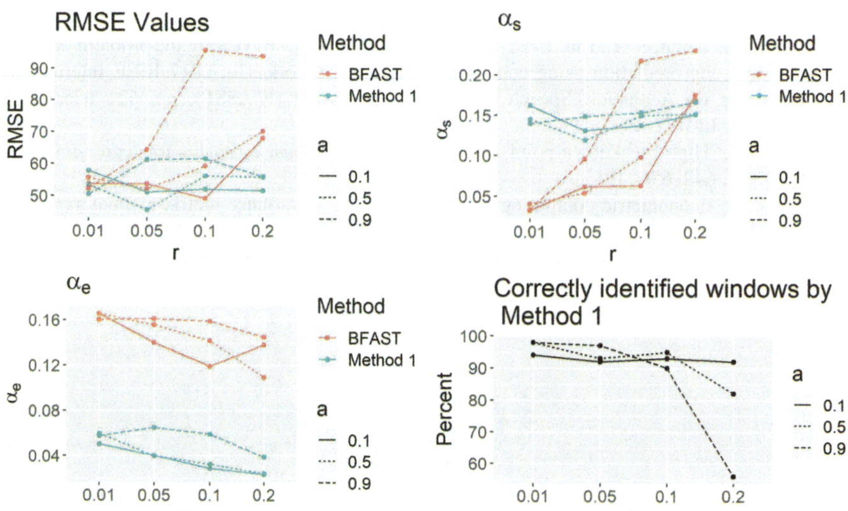

Fig. 15 Simulation results for simulated pattern Type 4, see Sect. 3.5

References

1. Andrews, D.W.K.: Heteroskedasticity and autocorrelation consistent covariance matrix estimation. Econometrica **59**(3), 817–858 (1991)
2. Berndt, D.J., Clifford, J.: Using dynamic time warping to find patterns in time series. In: Fayyad, U.M., Uthurusamy, R. (eds.) KDD Workshop, Seattle, pp. 359–370. AAAI Press, Palo Alto (1994)

3. Fawcett, T.: Introduction to ROC analysis. Pattern Recogn. Lett. **27**, 861–874 (2006). https://doi.org/10.1016/j.patrec.2005.10.010
4. Fritz, H., Garcia-Escudero, L.A., Mayo-Iscar, A.: tclust: an R package for a trimming approach to cluster analysis. J. Stat. Softw. **47**(12), 1–26 (2012). http://www.jstatsoft.org/v47/i12/
5. Galeano, P., Pena, D.: Multivariate analysis in vector time series. Resenhas, 383–404 (2000)
6. Goodfellow, I., Bengio, Y., Courville, A.: Deep Learning. MIT Press, Cambridge (2016)
7. Hall, S.: This is how COVID-19 is affecting the music industry. Technical report, World Economic Forum (2020). https://www.weforum.org/agenda/2020/05/this-is-how-covid-19-is-affecting-the-music-industry/
8. Hosmer Jr. D.W., Lemeshow, S., Sturdivant, R.X.: Applied Logistic Regression, 3rd edn. John Wiley & Sons, Inc., New Jersey (2013)
9. IFPI: Global music report. Technical report, International Federation of the Phonographic Industry (2018). https://ifpi.org/news/IFPI-GLOBAL-MUSIC-REPORT-2019
10. Lin, J., Williamson, S., Borne, K., DeBarr, D.: Pattern recognition in time series. Adv. Mach. Learn. Data Mining Astron. **1**, 617–645 (2012)
11. Müller, M.: Dynamic time warping. In: Information Retrieval for Music and Motion, pp. 69–84 (2007)
12. Rousseeuw, P., Croux, C.: Alternatives to median absolute deviation. J. Am. Stat. Assoc. **88**, 1273–1283 (1993)
13. Verbesselt, J., Hyndman, R., Newnham, G., Culvenor, D.: Detecting trend and seasonal changes in satellite image time series. Remote Sens. Environ. **114**, 106–115 (2010)
14. Verbesselt, J., Hyndman, R., Zeileis, A., Culvenor, D.: Phenological change detection while accounting for abrupt and gradual trends in satellite image time series. Remote Sens. Environ. **114**, 2970–2980 (2010)
15. Yang, L.-C., Chou, S.-Y., Liu, J.-Y., Yang, Y.-H., Chen, Y.-A.: Revisiting the problem of audio-based hit song prediction using convolutional neural networks. In: 2017 IEEE International Conference on Acoustics, Speech and Signal Processing (ICASSP), pp. 621–625. IEEE, Piscataway (2017)
16. Yohai, V.J.: High breakdown-point and high efficiency robust estimates for regression. Ann. Stat. **15**(2), 642–656 (1987)
17. Zeileis, A.: Econometric computing with HC and HAC covariance matrix estimators. J. Stat. Softw. **11**(10), 1–17 (2004). https://doi.org/10.18637/jss.v011.i10
18. Zeileis, A.: Object-oriented computation of sandwich estimators. J. Stat. Softw. **16**(9), 1–16 (2006). https://doi.org/10.18637/jss.v016.i09

Fine-Tuned Parallel Piecewise Sequential Confidence Interval and Point Estimation Strategies for the Mean of a Normal Population: Big Data Context

Nitis Mukhopadhyay and Chen Zhang

Abstract In this paper, we provide some new perspectives on sequential experimental designs for statistical inference in the context of *big data*. A fine-tuned parallel piecewise sequential procedure is developed for estimating the mean of a normal population having an unknown variance. With the help of such fine-tuning, asymptotic unbiasedness of the terminal sample size can be achieved along with the added operational efficiency as a result of utilizing the *parallel processing* or *distributed computing*. Theory and methodology will go hand-in-hand followed by illustrations from large-scale data analyses based on simulated data as well as real data from a health study.

Keywords Big data · Fine-tuning · Fixed-width confidence interval (FWCI) · Heart study · Minimum risk point estimation (MRPE) · Normal population · Parallel piecewise sampling · Purely sequential sampling · Real data illustration · Second-order asymptotic efficiency · Simulations · Stopping rule

1 Introduction

There is no denying that we now live in the era of big data. In 2011, the McKinsey Global Institute (MGI) published a report highlighting the transformational potential of big data, and it has been extensively cited since. As MGI [17] noted, "the ability to store, aggregate, and combine data and then use the results to perform deep analyses has become ever more accessible as trends such as Moore's Law in computing, its equivalent in digital storage, and cloud computing continue to lower costs and other technology barriers."

MGI [17] estimated that enterprises globally stored more than 7 exabytes of new data on disk drives in 2010 and that at the same time consumers stored more than 6

N. Mukhopadhyay (✉) · C. Zhang
Department of Statistics, University of Connecticut, Storrs, CT, USA
e-mail: nitis.mukhopadhyay@uconn.edu; chen.zhang@uconn.edu

A. Steland, K.-L. Tsui (eds.), *Artificial Intelligence, Big Data and Data Science in Statistics*, https://doi.org/10.1007/978-3-031-07155-3_3

exabytes of new data on devices such as PCs and notebooks—one exabyte of data is the equivalent of more than 4000 times the information stored in the U.S. Library of Congress which is regarded as the largest library in the world.

MGI and McKinsey Analytics [12] not only confirmed the trend but further affirmed that this progress was accelerating as a result of the convergence of several technology trends: "The volume of data continues to double every three years as information pours in from digital platforms, wireless sensors, and billions of mobile phones. Data storage capacity has increased, while its cost has plummeted. Data scientists now have unprecedented computing power at their disposal, and they are devising ever more sophisticated algorithms."

Mukhopadhyay and Zhang [29, 30] explored the *asymptotic distribution* of stopping times, that is the terminal sample sizes, from sequential sampling strategies in a wide variety of challenging estimation problems. They presented a general framework for obtaining such asymptotic distributions along with practical methodologies which also gave the rates of convergences and the sharpness of approximations validated via extensive sets of *exploratory data analysis* (EDA).

Mukhopadhyay and Zhang [29, 30] indeed provided an informative foundation as well as useful practical guidance regarding when and how one may gain better understanding of terminal sample sizes that may be *extremely large*. The present paper, on the other hand, explores the stopping time more so from the perspective of *experimental designs* for statistical inference in the context of *big data*.

For potentially very large terminal sample sizes, we are interested in exploring more efficient experimental designs for collecting data. By applying and modifying the idea of parallel processing to customary purely sequential sampling strategies with appropriate fine-tuning of each arm, our proposed methodology will achieve sizable operational efficiency in practice. Additionally, this will also overcome the (asymptotic) bias of the stopping variable as an estimate of the optimal fixed-sample size. We explore this idea in the contexts of both (i) *fixed-width confidence interval* (FWCI) estimation and (ii) *minimum risk point estimation* (MRPE) for the mean of a normal population having an unknown variance.

1.1 A Brief Literature Review

The sampling designs of multi-stage methodologies have seen a significant growth in recent decades. Stein [38, 39] gave the foundation of two-stage sampling strategy for collecting data and solved the FWCI estimation problem for the unknown mean in a normal population with preassigned coverage probability when the variance remains unknown. This led to an exact solution to this fundamental problem in statistical inference for which there is no fixed-sample size solution [6]. See also [23, Chapter 13].

Anscombe [2, 3] proposed purely sequential sampling strategies for the same problem in 1953 while in his 1952 seminal paper he developed his large-sample theories of sequential estimation by formulating the break-through *random central limit theorem* (random CLT). Ray [31] and Chow and Robbins [5] modified the stopping boundary and broadened Anscombe's [2, 3] ideas. Starr [35] developed important asymptotics. Mukhopadhyay [19] designed a modified two-stage estimation strategy whose asymptotic (first-order) efficiency property reduced the margin of oversampling associated with Stein's [38, 39] two-stage methodology. Inspired by Mukhopadhyay's [19] paper Ghosh and Mukhopadhyay [9] came up with the notion of asymptotic second-order efficiency.

In a parallel path Robbins [32] introduced the fundamental formulation of the MRPE problem for the mean of a normal population with a purely sequential stopping rule. This was later extended by Starr [36] and Starr and Woodroofe [37]. We briefly mention that [14, 15, 34, 42, 43] developed the machinery of the *non-linear renewal theory* for checking desirable second-order approximations.

Since we will focus on normal mean estimation problems, one may get an impression that the field of sequential estimation is very narrowly defined. In fact, the literature on sequential estimation problems intersects very heavily and successfully in the contexts of numerous other distributions and classes of problems, especially in the contexts of exponential models and their derivatives, as well as a wide range of nonparametric problems.

A quick glance at the following more or less chronologically listed references and taking into account the citations therein may be convincing: [1, 4, 7, 10, 11, 13, 16, 18, 20–22, 25, 27, 28, 33, 34, 41, 43]. Steland and Chang [40] have recently developed high-confidence nonparametric fixed-width uncertainty intervals and applications to projected high-dimensional data and common mean estimation.

For added operational efficiency Mukhopadhyay and Sen [26] developed *parallel processing* and introduced the notions of parallel piecewise sequential methodologies to handle both FWCI and MRPE problems. Mukhopadhyay and Datta [24] pursued an idea of *fine-tuning* a purely sequential procedure, in the context of fixed-size confidence region problems, to allow improved coverage accuracy in a variety of sequential estimation problems. This present paper marries parallel piecewise sequential methodologies with a new approach to fine-tuning, which will lead to asymptotically unbiased estimators of required optimal fixed-sample sizes.

1.2 An Outline of the Paper

In Sect. 2, we begin with brief overviews of both FWCI and MRPE problems forming the basis of the two fundamental and classical sequential inference strategies. Ghosh and Mukhopadhyay [8] reviewed some of the fundamental inference problems known at the time. Under such spirits, we propose to investigate (i) subsequent sampling strategies and inference methodologies and (ii) a number of appealing asymptotic properties of the associated stopping times in this paper.

Section 3 introduces our proposed *fine-tuned* parallel piecewise sequential strategy for estimating the mean. With appropriate fine-tuning, the asymptotic unbiasedness of the stopping variable, as an estimate of the optimal fixed-sample size, can be achieved along with the added operational efficiency as a result of the parallelization.

In Sect. 4, we present summaries from simulation studies to illustrate the usefulness of our proposed fine-tuned parallel piecewise sequential procedure by comparing (i) the purely sequential strategy, (ii) the parallel piecewise sequential strategy, and (iii) the fine-tuned parallel piecewise sequential strategy, in the contexts of both FWCI and MRPE problems.

In Sect. 5, we implement our proposed fine-tuned methodology by illustrating it with detailed analyses of real data from the Framingham Heart Study. These add a distinct flavor of the practical usefulness of our newly developed parallel piecewise sequential estimation strategies. We wrap up with brief concluding thoughts (Sect. 6).

2 An Overview of FWCI and MRPE Problems

We begin with brief overviews in Sects. 2.1 and 2.2 which, respectively, form the very core of fundamental and classical sequential inference strategies for the FWCI and MRPE problems.

2.1 A Purely Sequential FWCI Strategy

Suppose that $X_1, \cdots, X_n, \cdots, n \geq 2$, are *independent and identically distributed* (i.i.d.) random variables from a $N(\mu, \sigma^2)$ population where the mean μ and the variance σ^2 are both unknown with $-\infty < \mu < \infty$ and $0 < \sigma < \infty$. Define the customary unbiased estimators for μ and σ^2 as

$$\overline{X}_n = n^{-1}\Sigma_{i=1}^{n} X_i \text{ and } S_n^2 = (n-1)^{-1}\Sigma_{i=1}^{n}(X_i - \overline{X}_n)^2, \tag{1}$$

standing for the sample mean and the sample variance, respectively, obtained from $X_1, \cdots, X_n$. We denote $\boldsymbol{\theta} = (\mu, \sigma)$.

Given two preassigned numbers $d > 0$ and $0 < \alpha < 1$, we wish to construct a confidence interval I_n for μ such that the length of I_n is $2d$ and $P_{\boldsymbol{\theta}}\{\mu \in I_n\} \geq 1 - \alpha$ for all $\boldsymbol{\theta}$. We begin by formulating the FWCI as

$$I_n = [\overline{X}_n \pm d],$$

for the unknown mean μ.

The associated coverage probability is expressed as

$$P_\theta\{\mu \in I_n\} \equiv P_\theta\left\{|\overline{X}_n - \mu| \le d\right\} = P_\theta\left\{\sqrt{n}|\overline{X}_n - \mu|/\sigma \le \sqrt{n}d/\sigma\right\} \\ = 2\Phi(\sqrt{n}d/\sigma) - 1. \tag{2}$$

Here, we write: $\phi(x) = \{2\pi\}^{-1/2}\exp(-\frac{1}{2}x^2)$ and $\Phi(x) = \int_{-\infty}^{x}\phi(y)dy, -\infty < x < \infty$, to denote the probability density function and the cumulative distribution function of the standard normal distribution, respectively.

Observe that I_n already has the fixed width $2d$. We also require that the associated coverage probability be at least $1-\alpha$, our preassigned confidence coefficient. From (2), we can write

$$2\Phi(\sqrt{n}d/\sigma) - 1 \ge 1 - \alpha,$$

which gives

$$n \ge z_{\alpha/2}^2\sigma^2/d^2 \equiv C_d = C, \text{ say}. \tag{3}$$

Here, $z_{\alpha/2}$ is the $100(1-\frac{1}{2}\alpha)^{\text{th}}$ percentile of $N(0,1)$ and we interpret C_d as the optimal fixed-sample size required to construct the corresponding FWCI I_n for μ, had σ been known.

We note, however, that while the expression for C_d is known, its magnitude remains unknown since σ^2 is assumed unknown. Indeed, this problem has no fixed-sample size solution [6]. The stopping time associated with the ground-breaking purely sequential sampling strategies due to [2, 3, 5, 31] is stated as follows:

$$N \equiv N_d = \inf\{n \ge m : n \ge z_{\alpha/2}^2 S_n^2/d^2\}, d > 0, \tag{4}$$

where $m(\ge 2)$ is the pilot sample size. See also [35].

That is, beginning with pilot data $X_1, \cdots, X_m$ of size m, $m \ge 2$, we proceed by recording one additional observation at-a-time successively as needed until we stop according to the stopping rule (4). Termination occurs *with probability* (w.p.) 1. Upon termination, the FWCI for μ is given by

$$I_{N_d} = [\overline{X}_{N_d} \pm d], \tag{5}$$

based on the final accrued data $X_1, \cdots, X_{N_d}$ of size N_d.

A crucial set of properties of the purely sequential FWCI estimation strategy (N_d, I_{N_d}) from (4)–(5) were proved by Starr [35]. One may additionally refer to [10, Section 8.2] and [25, pp. 118–119].

2.2 A Purely Sequential MRPE Strategy

A path-breaking paper on the original formulation of a MRPE problem was due to [32]. We again assume having a $N(\mu, \sigma^2)$ population, with μ and σ^2 both unknown, from which i.i.d. observations $X_1, \cdots, X_n, \cdots$ arrive in a sequence. Let $\overline{X}_n$ and S_n^2 stand for the sample mean and the sample variance, respectively, as in (1). Recall that $\boldsymbol{\theta} = (\mu, \sigma)$.

The overall loss in estimating μ by $\overline{X}_n$ is given by:

$$L_n \equiv L_n(\mu, \overline{X}_n) = A(\overline{X}_n - \mu)^2 + cn \text{ with } A > 0 \text{ and } c > 0 \text{ both prespecified.}$$

Here, $(\overline{X}_n - \mu)^2$ represents the loss due to estimation of μ by $\overline{X}_n$, A represents the cost per unit squared error loss, and c represents the cost per unit observation. The associated fixed-sample size risk function is expressed as:

$$R_n(c) \equiv E_{\boldsymbol{\theta}}[L_n(\mu, \overline{X}_n)] = A\sigma^2 n^{-1} + cn.$$

The optimal fixed-sample size that minimizes the risk is given by

$$n_c^* \equiv n^* = (A/c)^{1/2}\sigma \text{ had } \sigma^2 \text{ been known.} \tag{6}$$

In fact, this problem also has no fixed-sample size solution [6, 25, Theorem 2.3.1]. Robbins [32] proposed the following stopping time associated with his purely sequential strategy:

$$N \equiv N_c = \inf\{n \geq m : n \geq (A/c)^{1/2} S_n\}, c > 0. \tag{7}$$

Termination occurs w.p. 1 and upon termination, we estimate:

$$\mu \text{ by } \overline{X}_{N_c}, \text{ the terminal sample mean,} \tag{8}$$

based on the final accrued data $X_1, \cdots, X_{N_c}$ of size N_c. A crucial set of properties of the purely sequential MRPE strategy $(N_c, \overline{X}_{N_c})$ from (7)–(8) were proved by Starr [36]. One may additionally refer to [10, Section 7.2] and [25, pp. 144–147].

3 Fine-Tuned Parallel Piecewise Sequential Strategies with Asymptotically Unbiased Sample Size Estimation

Implementation of a purely sequential sampling strategy can be operationally inconvenient in some situations because it only allows one to collect data one at-a-time until termination. The parallel piecewise sequential methodology, first introduced by Mukhopadhyay and Sen [26], achieves significant operational efficiency by incorporating the idea of *parallel processing*.

However, this could also result in a large (asymptotic) bias from the stopping variable relative to the optimal fixed-sample size, namely C from (3) or n^* from (6), when the number of arms (that is, the number of parallel pieces) is large. This has motivated us to introduce a suitable fine-tuning parameter, so that the asymptotic unbiasedness of the stopping variable can be achieved along with the added operational efficiency gained from the parallelization.

The FWCI and MRPE problems from Sects. 2.1 and 2.2 are revisited in Sects. 3.1 and 3.2 which include essential technical details along with our major results, Theorems 3.1 and 3.2, respectively, on approximating $E_{\theta}[N_d - C_d]$ and $E_{\theta}[N_c - n_c^*]$ up to the order $o(1)$.

3.1 Parallel Piecewise Sequential FWCI Strategies

Suppose that k investigators are collecting data from one population at the same time, but independently of each other, much in the spirit of parallel processing or distributed computing. Also, suppose that $X_{ij},\ j = 1, \cdots, n_i, ...$ are i.i.d. observations from $N(\mu, \sigma^2)$ with both parameters unknown where the subscript i corresponds to the ith investigator or the ith arm, $i = 1, \cdots, k$.

Having recorded $X_{i1}, \cdots, X_{in_i}$ from the ith arm with $n_i \geq 2$, we let $\overline{X}_{in_i}$ and $S^2_{in_i}$ be the customary sample mean and the sample variance, respectively, $i = 1, \cdots, k$. Suppose that d is half the width of the requisite confidence interval and $1 - \alpha$, $0 < \alpha < 1$, is the preassigned confidence coefficient. The optimal fixed-sample size is given by (3): $C = z^2_{\alpha/2}\sigma^2/d^2$ had σ^2 been known.

Mukhopadhyay and Sen [26] proposed their stopping time for the *parallel piecewise* sequential estimation strategy:

$$\begin{aligned} &\widetilde{N} \equiv \widetilde{N}_d = \Sigma_{i=1}^k N_{i,d} \text{ where } \widetilde{\mathbf{N}}_d = (N_{1,d}, ..., N_{k,d}) \text{ with} \\ &N_{i,d} \equiv \inf\left\{n_i \geq m : n_i \geq \tfrac{1}{k} z^2_{\alpha/2} S^2_{in_i}/d^2\right\},\ d > 0,\ i = 1, \cdots, k, \end{aligned} \tag{9}$$

where, as before, $m (\geq 2)$ is the pilot sample size on each arm.

The stopping times $N_{i,d}$ are each run independently by our k investigators. Termination occurs w.p. 1 and hence upon termination of sampling from all k arms, we obtain the combined sample mean given by:

$$\text{Estimator of } \mu : \overline{X}_{\widetilde{\mathbf{N}}_d} = \widetilde{N}_d^{-1} \Sigma_{i=1}^k N_{i,d} \overline{X}_{iN_{i,d}} \tag{10}$$

and the FWCI for μ is given by

$$I_{\widetilde{\mathbf{N}}_d} = [\overline{X}_{\widetilde{\mathbf{N}}_d} \pm d]. \tag{11}$$

Mukhopadhyay and Sen [26] exploited a number of key non-linear renewal theoretic techniques from [14, 15, 21, 42, 43] to derive the *asymptotic bias* of the stopping variable $\widetilde{N}_d$ from (9) as follows. As $d \to 0$ [26] concluded:

$$E_{\boldsymbol{\theta}}[\widetilde{N}_d - C_d] = k\eta + o(1) \text{ if } m \geq 4, \tag{12}$$

with C_d coming from (3) and

$$\eta = -\frac{1}{2} - \Sigma_{n=1}^{\infty} n^{-1} E[\max\{0, \chi_n^2 - 2n\}] \approx -1.1828. \tag{13}$$

From (12)–(13), one will clearly note that as k (= the number of independent arms or parallel pieces) becomes larger, while the data collection process becomes more efficient (less time-consuming), the asymptotic bias (undersampling) of the stopping variable $\widetilde{N}_d$ relative to the optimal fixed-sample size C_d becomes more substantial.

This motivates us to propose a modified version of the original parallel piecewise sequential strategy (9) which we refer to as the *fine-tuned* parallel piecewise sequential procedure:

$$\begin{aligned} &Q \equiv Q_d = \Sigma_{i=1}^{k} Q_{i,d} \text{ where } \mathbf{Q}_d = (Q_{1,d}, ..., Q_{k,d}) \text{ with} \\ &Q_{i,d} \equiv \inf\left\{n_i \geq m : n_i + \varepsilon \geq \tfrac{1}{k} z_{\alpha/2}^2 S_{in_i}^2 / d^2\right\}, d > 0, i = 1, \cdots, k, \end{aligned} \tag{14}$$

where $m(\geq 2)$ is the pilot sample size. We will show that the *fine-tuning parameter* $\varepsilon = \eta \approx -1.1828$. The fine-tuned parallel piecewise sequential strategy (14) overcomes the asymptotic undersampling bias of the stopping variable as stated formally in Theorem 3.1.

Theorem 3.1 *The stopping variable* $Q_d(= \Sigma_{i=1}^{k} Q_{i,d})$ *from* (14) *satisfies the asymptotic unbiasedness property, that is,*

$$E_{\boldsymbol{\theta}}[Q_d - C_d] = o(1) \textit{ as } d \to 0,$$

if $m \geq 4$, $\varepsilon \equiv \nu - 2(= \eta)$, *with* C_d, η, ν *coming from* (3), (13), *and* (19)–(20).

Proof We will prove this theorem by relying upon the non-linear renewal theory [14, 15, 42, 43]. We recall Helmert orthogonal transformation [23, pp. 197–199] and express:

$$S_{in_i}^2 = (n_i - 1)^{-1} \Sigma_{j=1}^{n_i - 1} Y_{ij}^2, \tag{15}$$

where $Y_{i1}, Y_{i2}, \cdots, Y_{in_i-1} \overset{\text{i.i.d.}}{\sim} N(0, \sigma^2), i = 1, \cdots, k$.

Then, in view of (15), observe that we can write:

$$n_i + \varepsilon \geq z_{\alpha/2}^2 S_{in_i}^2/(kd^2) \Leftrightarrow (n_i - 1)(n_i + \varepsilon) \geq (kd^2)^{-1} z_{\alpha/2}^2 \sigma^2 \Sigma_{j=1}^{n_i-1}(Y_{ij}^2/\sigma^2)$$
$$\Leftrightarrow (n_i - 1)(n_i + \varepsilon) \geq k^{-1} C_d \Sigma_{j=1}^{n_i-1} W_{ij} \Leftrightarrow \Sigma_{j=1}^{n_i-1} W_{ij} \leq k(n_i - 1)(n_i + \varepsilon)/C_d, \tag{16}$$

with $W_{ij} = Y_{ij}^2/\sigma^2$, $j = 1, 2, \cdots, n_i - 1$ which are i.i.d. χ_1^2 random variables.

Then, we can rewrite (14) and claim that $Q_{i,d} \equiv M_{i,d} + 1$ w.p. 1 where we define:

$$M_{i,d} = \inf\left\{n_i \geq m - 1 : \Sigma_{j=1}^{n_i} W_{ij} \leq kn_i^2\left[1 + (1+\varepsilon)n_i^{-1}\right]/C_d\right\}, \tag{17}$$

for $i = 1, \cdots, k$.

In order to apply the non-linear renewal theory, we match our representation from (17) with the notations from [25, pp. 446–449] to obtain:

$$\begin{aligned} h^* = k/C_d, \delta = 2, L_0 = 1 + \varepsilon, \theta = 1,\ \tau^2 = 2,\ \beta^* = 1, \\ q = \beta^{*2}\tau^2/\theta^2 = 2, m_0 = m - 1, n_0^* = (\theta/h^*)^{\beta^*} = k^{-1}C_d; \end{aligned} \tag{18}$$

with

$$\nu \equiv \beta^*(2\theta)^{-1}\{(\delta - 1)^2\theta^2 + \tau^2\} - \Sigma_{n=1}^{\infty} n^{-1} E\left[\max\left\{0, \left(\Sigma_{j=1}^{n} W_{ij}\right) - n\delta\theta\right\}\right], \tag{19}$$

that is,

$$\begin{aligned} \nu &= \tfrac{3}{2} - \Sigma_{n=1}^{\infty} n^{-1} E\left[\max\left\{0, \chi_n^2 - 2n\right\}\right] \\ &= \tfrac{3}{2} - \Sigma_{n=1}^{\infty}\{\Gamma(\tfrac{1}{2}n)\}^{-1} 2^{-n/2} n^{-1} \int_{y=2n}^{\infty} (y - 2n) e^{-y/2} y^{\frac{1}{2}n-1} dy \\ &\approx 0.8172, \end{aligned} \tag{20}$$

and also:

$$\kappa = \beta^* \nu \theta^{-1} - \beta^* L_0 - \delta\beta^{*2}\tau^2/(2\theta^2) = \nu - 3 - \varepsilon. \tag{21}$$

Additionally, we have:

$$\begin{aligned} P_\theta\{W_{11} \leq w\} &= \int_0^w (2\pi)^{-1/2} e^{-y/2} y^{-1/2} dy \leq (2\pi)^{-1/2} \int_0^w y^{-1/2} dy \\ &= (2/\pi)^{1/2} u^{1/2}, \end{aligned} \tag{22}$$

so that $r = \frac{1}{2}$. Observe that the entity "η" from (13) coincides with $\nu - 2$.

Now, given (15)–(22), by Theorem A.4.2 in [25, p. 448], we claim the following as $h^* \to 0$ $(d \to 0)$:

$$E_{\boldsymbol{\theta}}\left[Q_{i,d}\right] = E_{\boldsymbol{\theta}}\left[M_{i,d}\right] + 1 = n_0^* + \kappa + 1 + o(1) = k^{-1}C_d + (\nu - 2 - \varepsilon) + o(1),$$

for $i = 1, \dots, k$ if $m_0 > \beta^*/r$, that is, if $m \geq 4$.

Therefore, as $d \to 0$, we obtain:

$$E_{\boldsymbol{\theta}}[Q_d] = E_{\boldsymbol{\theta}}[\Sigma_{i=1}^k Q_{i,d}] = kE_{\boldsymbol{\theta}}[Q_{1,d}] = C_d + k(\nu - 2 - \varepsilon) + o(1), \tag{23}$$

if $m \geq 4$. Finally, when we fix $\varepsilon \equiv \nu - 2$, it immediately follows that $E_{\mu,\sigma}[Q_d] = C_d + o(1)$ as $d \to 0$, if $m \geq 4$. □

Remark 3.1 When $k = 1$, we observe that (14) coincides with the fine-tuned version of the original sampling strategy (4), that is, we are then back to the purely sequential sampling methodology. The choice for the fine-tuning parameter ε, however, remains exactly the same as $\nu - 2 \approx -1.1828$.

3.2 *Parallel Piecewise Sequential MRPE Strategies*

We again suppose k investigators are collecting data from one population at the same time, but independently of each other, in the spirit of parallel processing or distributed computing. Also suppose that $X_{ij}, i = 1, \cdots, k, j = 1, \cdots, n_i, \dots$ are i.i.d. observations from $N(\mu, \sigma^2)$ with both parameters unknown.

Having recorded $X_{i1}, \cdots, X_{in_i}$ from the ith arm with $n_i \geq 2$, let us recall $\overline{X}_{in_i}$ and $S^2_{in_i}$, the customary sample mean and the sample variance, respectively, $i = 1, \cdots, k$. The overall sample mean is given by $\overline{X}_n = n^{-1}\Sigma_{i=1}^k n_i \overline{X}_{in_i}$ with $n = \Sigma_{i=1}^k n_i$.

The overall loss in estimating μ by $\overline{X}_n$ is given by

$$L_n(\mu, \overline{X}_n) = A(\overline{X}_n - \mu)^2 + cn \text{ with } A > 0 \text{ and } c > 0 \text{ prespecified.}$$

The fixed-sample size risk function can be expressed as

$$R_n(c) \equiv E_{\boldsymbol{\theta}}[L_n(\mu, \overline{X}_n)] = A\sigma^2 n^{-1} + cn.$$

The optimal fixed-sample size that minimizes the risk is again given by $n_c^* = (A/c)^{1/2}\sigma$ had σ^2 been known.

On each arm, the optimal fixed-sample size is then given by $n_{i,c}^* \equiv k^{-1}(A/c)^{1/2}\sigma,\ i = 1, \cdots, k$. In the spirit of (9), Mukhopadhyay and Sen [26] proposed their stopping time for the *parallel piecewise* sequential procedure:

$$\begin{aligned}\widetilde{N} \equiv \widetilde{N}_c = \Sigma_{i=1}^k N_{i,c} \text{ where } \widetilde{\mathbf{N}}_c = (N_{1,c}, ..., N_{k,c}) \text{ with}\\ N_{i,c} \equiv \inf\left\{n_i \geq m : n_i \geq \tfrac{1}{k}(A/c)^{1/2} S_{in_i}\right\},\ c > 0,\ i = 1, \cdots, k,\end{aligned} \tag{24}$$

where $m(\geq 2)$ is the pilot sample size on each arm.

The stopping times $N_{i,c}$ are each run independently by our k investigators. Termination occurs w.p. 1 and thus upon termination of sampling from all k arms, we obtain the combined sample mean given by:

$$\overline{X}_{\widetilde{\mathbf{N}}_c} = \widetilde{N}_c^{-1} \Sigma_{i=1}^k N_{i,c} \overline{X}_{iN_{i,c}}, \tag{25}$$

as the MRPE of μ.

By exploiting a number of key results from the non-linear renewal theory from [14, 15, 42, 43] Mukhopadhyay and Sen [26] derived the *asymptotic bias* of the stopping variable $\widetilde{N}_c$ from (24) as follows: As $c \to 0$,

$$E_{\boldsymbol{\theta}}[\widetilde{N}_c - n_c^*] = k\eta^* + o(1), \text{ if } m \geq 3, \tag{26}$$

where

$$\eta^* = -\frac{1}{2}\Sigma_{n=1}^{\infty} n^{-1} E[\max\{0, \chi_n^2 - 3n\}] \approx -0.1166. \tag{27}$$

Once again, from (26), one will clearly note that as k (= the number of independent arms or parallel pieces) becomes larger, while the data collection process becomes more efficient (less time-consuming), the asymptotic bias (undersampling) of the stopping variable $\widetilde{N}_c$ relative to the optimal fixed-sample size n_c^* becomes more substantial.

This motivates us to propose a modified version of the original parallel piecewise sequential strategy (24) which we refer to as the *fine-tuned* parallel piecewise sequential procedure:

$$\begin{aligned}T \equiv T_c = \Sigma_{i=1}^k T_{i,c} \text{ where } \mathbf{T}_c = (T_{1,c}, ..., T_{k,c}) \text{ with}\\ T_{i,c} \equiv \inf\left\{n_i \geq m : n_i + \varepsilon \geq \tfrac{1}{k}(A/c)^{1/2} S_{in_i}\right\}, c > 0, i = 1, \cdots, k,\end{aligned} \tag{28}$$

where $m(\geq 2)$ is the pilot sample size. We will show that the *fine-tuning parameter* $\varepsilon \equiv \eta^* \approx -0.1166$. The fine-tuned parallel piecewise sequential strategy (28) overcomes the asymptotic undersampling bias of the stopping variable as stated formally in Theorem 3.2.

Theorem 3.2 *The stopping variable* $T_c(=\Sigma_{i=1}^k T_{i,c})$ *from* (28) *satisfies the asymptotic unbiasedness property, that is,*

$$E_{\boldsymbol{\theta}}[T_c - n_c^*] = o(1) \text{ as } c \to 0,$$

if $m \geq 3$, $\varepsilon = \frac{1}{2}\nu - \frac{3}{4}(=\eta^*)$ *with* n_c^*, η^*, ν *coming from* (6), (27) *and* (30).

Proof In the spirit of (17) and our proof of Theorem 3.1, we rewrite $T_{i,c} = M_{i,c}+1$ w.p. 1 as follows:

$$M_{i,c} \equiv \inf\left\{n_i \geq m-1 : \Sigma_{j=1}^{n_i} W_{ij} \leq k^2 n_i^3 \left[1 + 2(1+\varepsilon)n_i^{-1} + (1+\varepsilon)^2 n_i^{-2}\right]/n_c^{*2}\right\}, \tag{29}$$

where $W_{i1}, W_{i2}, \cdots, W_{in_i} \overset{\text{i.i.d.}}{\sim} \chi_1^2$, for $i = 1, \cdots, k$.

In order to apply the non-linear renewal theory, we again match this representation from (29) with the notations from [25, pp. 446–449] to obtain:

$$\begin{aligned} h^* &= k^2/n_c^{*2}, \delta = 3, L_0 = 2(1+\varepsilon), \theta = 1, \ \tau^2 = 2, \ \beta^* = \tfrac{1}{2}, \\ q &= \beta^{*2}\tau^2/\theta^2 = 2, m_0 = m-1, r = \tfrac{1}{2}, n_0^* = (\theta/h^*)^{\beta^*} = k^{-1}n_c^*; \end{aligned} \tag{30}$$

with

$$\nu = \frac{3}{2} - \Sigma_{n=1}^{\infty} n^{-1} E\left[\max\left\{0, \chi_n^2 - 3n\right\}\right] \approx 1.2669, \tag{31}$$

and

$$\kappa = \beta^*\nu\theta^{-1} - \beta^* L_0 - \delta\beta^{*2}\tau^2/(2\theta^2) = \frac{1}{2}\nu - \frac{7}{4} - \varepsilon. \tag{32}$$

Now, given (29)–(32), by Theorem A.4.2 in [25, p. 448], we claim the following as $h^* \to 0$ $(c \to 0)$:

$$E_{\boldsymbol{\theta}}\left[T_{i,c}\right] = E_{\boldsymbol{\theta}}\left[M_{i,c}\right] + 1 = n_0^* + \kappa + 1 + o(1) = k^{-1}n_c^* + \left(\frac{1}{2}\nu - \frac{3}{4} - \varepsilon\right) + o(1),$$

for $i = 1, ..., k$ if $m_0 > \beta^*/r$, that is, if $m \geq 3$.

Therefore, as $c \to 0$, we obtain:

$$E_{\boldsymbol{\theta}}[T_c] = E_{\boldsymbol{\theta}}[\Sigma_{i=1}^k T_{i,c}] = kE_{\boldsymbol{\theta}}[T_{1,c}] = n_c^* + k\left(\frac{1}{2}\nu - \frac{3}{4} - \varepsilon\right) + o(1), \tag{33}$$

if $m \geq 3$. Finally, when we fix $\varepsilon \equiv \frac{1}{2}\nu - \frac{3}{4}$, it immediately follows that $E_{\mu,\sigma}[T_c] = n_c^* + o(1)$ as $c \to 0$, if $m \geq 3$. Note that the entity η^* from (27) coincides with $\frac{1}{2}\nu - \frac{3}{4}$. □

It becomes clear that there is a stark difference between the conclusions stated in (12) or (26) and those in Theorem 3.1 or Theorem 3.2. The criticism that was labeled against (12) or (26) was the fact that $E_{\theta}[\widetilde{N}_d - C_d]$ or $E_{\theta}[\widetilde{N}_c - n_c^*]$ was visibly going further and further down as a negative number up to $o(1)$ if we made k larger under the original parallel piecewise sequential estimation strategies (9) or (24). But, the new and fine-tuned parallel piecewise sequential estimation strategy (14) with $\varepsilon \approx -1.1828$ or strategy (28) with $\varepsilon \approx -0.1166$ provides asymptotic unbiasedness of Q_d or T_c as an estimator of C_d or n_c^* up to $o(1)$ regardless of the number of parallel arms or investigators, namely k. We will substantiate these features shortly via simulations.

3.3 *Other Selected Second-Order Results*

In the case of both FWCI and MRPE problems, thus far we have devoted attention to the design of sampling strategies allowing us to come up with fine-tuned total terminal sample size Q_d from (14) or T_c from (28) in such a way that they become asymptotically unbiased for C_d or n_c^*, respectively, up to $o(1)$. But, then, what else happens to the original inference problems from Sects. 2.1 and 2.2 which led to the respective fixed-sample size C_d in (3) or n_c^* in (6)? In order to address these questions, we go back to the original formulations of the FWCI and MRPE problems briefly.

3.3.1 FWCI Problem: Asymptotic Second-Order Expansion of the Coverage Probability

Once sampling terminates via (14) with $\varepsilon \approx -1.1828$ built in, in the spirit of (11), we would propose the FWCI, namely

$$I_{\mathbf{Q}_d} = [\overline{X}_{\mathbf{Q}_d} \pm d],$$

for μ based upon the terminal total sample size Q_d after combining fully accrued data from all k arms.

Along the lines of [26, Theorem 4.1], we can claim the following second-order expansion of the associated coverage probability as $d \to 0$:

$$P_{\theta}\left\{\mu \in I_{\mathbf{Q}_d}\right\} \equiv E_{\theta}\left[2\Phi\left(Q_d^{1/2} d/\sigma\right) - 1\right] = (1-\alpha) + b^* C_d^{-1} + o(d^2) \text{ when } m \geq 7, \tag{34}$$

where b^* is a known real number involving k, $z_{\alpha/2}$, $\phi(z_{\alpha/2})$, and κ from (21).

Mukhopadhyay and Datta [24] introduced a different kind of fine-tuning parameter ε in a stopping time analogous to (14) and defined:

$$N^{'} \equiv N_{d}^{'} = \inf\{n \geq m : n + \varepsilon \geq z_{\alpha/2}^{2} S_{n}^{2}/d^{2}\}, d > 0. \tag{35}$$

They determined "ε" explicitly in order to conclude:

$$P_{\theta}\left\{\mu \in I_{N_{d}^{'}}\right\} = (1-\alpha) + o(d^{2}) \text{ when } m \geq 7, \tag{36}$$

instead of (34) but this $N_{d}^{'}$ from (35) remained asymptotically biased for estimating C_{d}. It should be clear how our present fine-tuning approach is fundamentally different from the notion adopted in [24].

3.3.2 MRPE Problem: Asymptotic Second-Order Expansion of the Regret

Once sampling terminates via (28) with $\varepsilon \approx -0.1166$ built in, in the spirit of (25), we would propose the terminal MRPE $\overline{X}_{\mathbf{T}_c}$ for μ based upon total sample size T_c after combining fully accrued data from all k arms.

Along the lines of [26, Theorem 3.2], we can claim the following asymptotic second-order expansion of the associated *regret* function, a metric put forward by Robbins [32], as $c \to 0$:

$$E_{\theta}\left[L_{\mathbf{T}_c}\left(\mu, \overline{X}_{\mathbf{T}_c}\right)\right] - R_{n_c^*}(c) \equiv cE_{\theta}\left[(T_c - n_c^*)^2/T_c\right] = \frac{1}{2}c + o(c) \text{ when } m \geq 4. \tag{37}$$

The parallel piecewise sequential strategy (24), originally introduced by Mukhopadhyay and Sen [26], had this exact same asymptotic second-order expansion of the associated regret function as $c \to 0$; however $\widetilde{N}_c$ was an asymptotically biased estimator of n_c^*. A major difference is that our fine-tuned version (28) with $\varepsilon \approx -0.1166$ provides (37) along with the fact that T_c is also an asymptotically unbiased estimator of n_c^*.

4 Simulation Data Analysis

In this section, we will summarize the performances of the FWCI strategies (4), (9), and (14) followed by those of the MRPE strategies (7), (24), and (28), all obtained from averaging over $B(= 10^6)$ independent replications under each fixed configuration. In other words, every row in every table that follows will show selected averages of performances obtained from the estimation strategies under consideration with one million replications.

One million replications have led to reliable and stable estimations of all requisite entities across the board. The simulation programs were implemented in MATLAB R2017a installed on a LENOVO Ideapad Y700 laptop with 6th Generation Intel Core i7-6700HQ CPU (2.60GHz), 16GB RAM, and Windows 10 Education system.

4.1 FWCI Strategies

Tables 2, 3, 4, 5, 6, 7, 8, 9, 10, 11, 12, and 13 compare the purely sequential procedure (4), the parallel piecewise sequential procedure (9), and the fine-tuned parallel piecewise sequential procedure (14), implemented for the FWCI estimation problem. The population used in the simulation was $N(5, 4)$ and we considered $\alpha = 0.05, 0.10$ where $1 - \alpha$ was the preassigned confidence coefficient, the initial sample size $m = 5, 10$, and the optimal fixed-sample size

$$C_d = 30, 50, 100, 200, 500, 1000, 2000, 5000, 10000,$$

with the number of replications $B = 10^6$. Two scenarios, $k = 2, 5$, were included for parallel piecewise sequential procedures (with or without fine-tuning). Having fixed σ, α, and C_d, we obtained $d = \sigma z_{\alpha/2} C_d^{-1/2}$.

Table 1 explains a generic set of notations used in our tables pretending that "N" is the stopping variable that is being implemented. Tables 2, 3, 4, 5, 6, 7, 8, 9, 10, 11, 12, and 13 will use appropriate and obvious improvisations as needed.

Table 1 The set of notations used in Tables 2, 3, 4, 5, 6, 7, 8, 9, 10, 11, 12, and 13: $B = 10^6$ replications

n_i	: terminal sample size in the ith run;
$\bar{n} = B^{-1}\Sigma_{i=1}^{B} n_i$	: average sample size, should compare with $E_\theta[N]$;
$s_{\bar{n}} = \left\{(B^2 - B)^{-1}\Sigma_{i=1}^{B}(n_i - \bar{n})^2\right\}^{1/2}$	: estimated *standard error* (s.e.) of $\bar{n}$;
$\bar{x}_{n_i}$	: terminal sample mean in the ith run;
$\bar{x} = B^{-1}\Sigma_{i=1}^{B}\bar{x}_{n_i}$	: average sample mean, should compare with μ;
$s_{\bar{x}} = \left\{(B^2 - B)^{-1}\Sigma_{i=1}^{B}(\bar{x}_{n_i} - \bar{x})^2\right\}^{1/2}$	: estimated s.e. of $\bar{x}$;
p_i	: 1 (or 0) if I_{n_i} covers (or does not cover) μ in the ith run;
$\bar{p} = B^{-1}\Sigma_{i=1}^{B} p_i$	: estimated coverage probability, should compare with $1 - \alpha$;
$s_{\bar{p}} = \left\{B^{-1}\bar{p}(1 - \bar{p})\right\}^{1/2}$	: estimated s.e. of $\bar{p}$;
Time	: completion time (in secs) for B replications in a row.

Table 2 Purely sequential strategy (4) on 95% FWCI for the mean of $N(5, 4)$ with no fine-tuning: $\varepsilon = 0,\ k = 1,\ B = 10^6$ replications

m	C	d	$\overline{n}$	$s_{\overline{n}}$	$\overline{n}/C$	$\overline{n} - C$	$\overline{p}$	$s_{\overline{p}}$	$\overline{x}$	$s_{\overline{x}}$	Time in sec
5	30	0.716	26.93	0.010	0.90	−3.07	0.905	0.000	5.00	0.00	2
	50	0.554	47.11	0.013	0.94	−2.89	0.924	0.000	5.00	0.00	4
	100	0.392	98.02	0.017	0.98	−1.98	0.941	0.000	5.00	0.00	12
	200	0.277	198.46	0.022	0.99	−1.54	0.946	0.000	5.00	0.00	37
	500	0.175	498.64	0.033	1.00	−1.36	0.949	0.000	5.00	0.00	186
	1000	0.124	998.74	0.046	1.00	−1.26	0.949	0.000	5.00	0.00	708
	2000	0.088	1998.81	0.064	1.00	−1.19	0.950	0.000	5.00	0.00	2770
	5000	0.055	4998.83	0.100	1.00	−1.17	0.950	0.000	5.00	0.00	16,511
	10,000	0.039	9998.82	0.142	1.00	−1.18	0.950	0.000	5.00	0.00	62,004
10	30	0.716	28.02	0.009	0.93	−1.98	0.925	0.000	5.00	0.00	2
	50	0.554	48.00	0.012	0.96	−2.00	0.934	0.000	5.00	0.00	4
	100	0.392	98.52	0.015	0.99	−1.48	0.944	0.000	5.00	0.00	11
	200	0.277	198.73	0.020	0.99	−1.27	0.948	0.000	5.00	0.00	35
	500	0.175	498.75	0.032	1.00	−1.25	0.949	0.000	5.00	0.00	182
	1000	0.124	998.84	0.045	1.00	−1.16	0.949	0.000	5.00	0.00	672
	2000	0.088	1998.90	0.063	1.00	−1.10	0.949	0.000	5.00	0.00	2622
	5000	0.055	4998.79	0.100	1.00	−1.21	0.950	0.000	5.00	0.00	15,757
	10,000	0.039	9998.83	0.141	1.00	−1.17	0.950	0.000	5.00	0.00	61,928

Table 3 Purely sequential strategy (4) on 90% FWCI for the mean of $N(5, 4)$ with no fine-tuning: $\varepsilon = 0,\ k = 1,\ B = 10^6$ replications

m	C	d	$\overline{n}$	$s_{\overline{n}}$	$\overline{n}/C$	$\overline{n} - C$	$\overline{p}$	$s_{\overline{p}}$	$\overline{x}$	$s_{\overline{x}}$	Time in sec
5	30	0.601	26.93	0.010	0.90	−3.07	0.850	0.000	5.00	0.00	2
	50	0.465	47.11	0.013	0.94	−2.89	0.871	0.000	5.00	0.00	4
	100	0.329	97.98	0.017	0.98	−2.02	0.890	0.000	5.00	0.00	11
	200	0.233	198.44	0.022	0.99	−1.56	0.896	0.000	5.00	0.00	35
	500	0.147	498.69	0.033	1.00	−1.31	0.898	0.000	5.00	0.00	177
	1000	0.104	998.76	0.045	1.00	−1.24	0.899	0.000	5.00	0.00	667
	2000	0.074	1998.77	0.064	1.00	−1.23	0.900	0.000	5.00	0.00	2523
	5000	0.047	4998.83	0.100	1.00	−1.17	0.900	0.000	5.00	0.00	15, 469
	10, 000	0.033	9998.56	0.141	1.00	−1.44	0.899	0.000	5.00	0.00	62, 430
10	30	0.601	28.03	0.009	0.93	−1.97	0.870	0.000	5.00	0.00	2
	50	0.465	48.01	0.012	0.96	−1.99	0.882	0.000	5.00	0.00	4
	100	0.329	98.54	0.015	0.99	−1.46	0.894	0.000	5.00	0.00	11
	200	0.233	198.70	0.020	0.99	−1.30	0.897	0.000	5.00	0.00	35
	500	0.147	498.72	0.032	1.00	−1.28	0.899	0.000	5.00	0.00	177
	1000	0.104	998.77	0.045	1.00	−1.23	0.899	0.000	5.00	0.00	668
	2000	0.074	1998.80	0.063	1.00	−1.20	0.900	0.000	5.00	0.00	2580
	5000	0.047	4998.75	0.100	1.00	−1.25	0.899	0.000	5.00	0.00	15, 686
	10, 000	0.033	9998.78	0.142	1.00	−1.22	0.900	0.000	5.00	0.00	62, 594

Table 4 Fine-tuned purely sequential strategy (14) on 95% FWCI for the mean of $N(5, 4)$: $\varepsilon = -1.1828,\ k = 1,\ B = 10^6$ replications

m	C	d	$\overline{q}$	$s_{\overline{q}}$	$\overline{q}/C$	$\overline{q} - C$	$\overline{p}$	$s_{\overline{p}}$	$\overline{x}$	$s_{\overline{x}}$	Time in sec
5	30	0.716	28.64	0.010	0.95	−1.36	0.920	0.000	5.00	0.00	2
	50	0.554	48.81	0.012	0.98	−1.19	0.933	0.000	5.00	0.00	4
	100	0.392	99.47	0.016	0.99	−0.53	0.944	0.000	5.00	0.00	12
	200	0.277	199.75	0.021	1.00	−0.25	0.948	0.000	5.00	0.00	35
	500	0.175	499.93	0.032	1.00	−0.07	0.950	0.000	5.00	0.00	182
	1000	0.124	999.93	0.045	1.00	−0.07	0.950	0.000	5.00	0.00	668
	2000	0.088	1999.98	0.064	1.00	−0.02	0.950	0.000	5.00	0.00	2554
	5000	0.055	5000.18	0.100	1.00	0.18	0.950	0.000	5.00	0.00	15, 569
	10, 000	0.039	10,000.20	0.142	1.00	0.20	0.950	0.000	5.00	0.00	62, 610
10	30	0.716	29.42	0.009	0.98	−0.58	0.933	0.000	5.00	0.00	2
	50	0.554	49.41	0.011	0.99	−0.59	0.939	0.000	5.00	0.00	4
	100	0.392	99.80	0.015	1.00	−0.20	0.946	0.000	5.00	0.00	11
	200	0.277	199.94	0.020	1.00	−0.06	0.949	0.000	5.00	0.00	35
	500	0.175	499.96	0.032	1.00	−0.04	0.950	0.000	5.00	0.00	179
	1000	0.124	999.91	0.045	1.00	−0.09	0.949	0.000	5.00	0.00	671
	2000	0.088	1999.96	0.063	1.00	−0.04	0.950	0.000	5.00	0.00	2579
	5000	0.055	5000.02	0.100	1.00	0.02	0.949	0.000	5.00	0.00	15, 676
	10, 000	0.039	9999.99	0.142	1.00	−0.01	0.950	0.000	5.00	0.00	61, 862

Table 5 Fine-tuned purely sequential strategy (14) on 90% FWCI for the mean of $N(5, 4)$: $\varepsilon = -1.1828,\ k = 1,\ B = 10^6$ replications

m	C	d	$\overline{q}$	$s_{\overline{q}}$	$\overline{q}/C$	$\overline{q} - C$	$\overline{p}$	$s_{\overline{p}}$	$\overline{x}$	$s_{\overline{x}}$	Time in sec
5	30	0.601	28.62	0.010	0.95	−1.38	0.867	0.000	5.00	0.00	2
	50	0.465	48.84	0.012	0.98	−1.16	0.882	0.000	5.00	0.00	4
	100	0.329	99.44	0.016	0.99	−0.56	0.893	0.000	5.00	0.00	12
	200	0.233	199.76	0.021	1.00	−0.24	0.898	0.000	5.00	0.00	36
	500	0.147	499.93	0.032	1.00	−0.07	0.899	0.000	5.00	0.00	181
	1000	0.104	999.89	0.045	1.00	−0.11	0.900	0.000	5.00	0.00	671
	2000	0.074	1999.92	0.064	1.00	−0.08	0.900	0.000	5.00	0.00	2582
	5000	0.047	5000.19	0.100	1.00	0.19	0.900	0.000	5.00	0.00	15, 765
	10, 000	0.033	9999.98	0.141	1.00	−0.02	0.900	0.000	5.00	0.00	61, 533
10	30	0.601	29.43	0.009	0.98	−0.57	0.881	0.000	5.00	0.00	2
	50	0.465	49.41	0.011	0.99	−0.59	0.888	0.000	5.00	0.00	4
	100	0.329	99.79	0.015	1.00	−0.21	0.896	0.000	5.00	0.00	12
	200	0.233	199.93	0.020	1.00	−0.07	0.898	0.000	5.00	0.00	36
	500	0.147	499.93	0.032	1.00	−0.07	0.899	0.000	5.00	0.00	180
	1000	0.104	1000.00	0.045	1.00	0.00	0.900	0.000	5.00	0.00	669
	2000	0.074	1999.94	0.063	1.00	−0.06	0.899	0.000	5.00	0.00	2556
	5000	0.047	5000.04	0.100	1.00	0.04	0.900	0.000	5.00	0.00	15, 621
	10, 000	0.033	10,000.08	0.141	1.00	0.08	0.900	0.000	5.00	0.00	62, 272

Table 6 Parallel piecewise sequential strategy (9) on 95% FWCI for the mean of $N(5, 4)$ with no fine-tuning: $\varepsilon = 0$, $k = 2$, $B = 10^6$ replications

m	C	d	$\overline{\overline{n}}$	$s_{\overline{\overline{n}}}$	$\overline{\overline{n}}/C$	$\overline{\overline{n}} - C$	$\overline{p}$	$s_{\overline{p}}$	$\overline{x}$	$s_{\overline{x}}$	Time in sec
5	30	0.716	26.11	0.008	0.87	−3.89	0.916	0.000	5.00	0.00	1
	50	0.554	44.14	0.013	0.88	−5.86	0.920	0.000	5.00	0.00	2
	100	0.392	94.24	0.018	0.94	−5.76	0.936	0.000	5.00	0.00	4
	200	0.277	195.99	0.023	0.98	−4.01	0.945	0.000	5.00	0.00	12
	500	0.175	496.99	0.034	0.99	−3.01	0.949	0.000	5.00	0.00	52
	1000	0.124	997.33	0.046	1.00	−2.67	0.949	0.000	5.00	0.00	179
	2000	0.088	1997.58	0.064	1.00	−2.42	0.950	0.000	5.00	0.00	673
	5000	0.055	4997.61	0.101	1.00	−2.39	0.950	0.000	5.00	0.00	3955
	10, 000	0.039	9997.64	0.142	1.00	−2.36	0.950	0.000	5.00	0.00	15, 840
10	30	0.716	29.45	0.006	0.98	−0.55	0.942	0.000	5.00	0.00	1
	50	0.554	46.54	0.011	0.93	−3.46	0.933	0.000	5.00	0.00	2
	100	0.392	96.01	0.016	0.96	−3.99	0.941	0.000	5.00	0.00	4
	200	0.277	197.03	0.021	0.99	−2.97	0.946	0.000	5.00	0.00	11
	500	0.175	497.42	0.032	0.99	−2.58	0.949	0.000	5.00	0.00	52
	1000	0.124	997.59	0.045	1.00	−2.41	0.949	0.000	5.00	0.00	183
	2000	0.088	1997.59	0.063	1.00	−2.41	0.949	0.000	5.00	0.00	684
	5000	0.055	4997.63	0.100	1.00	−2.37	0.950	0.000	5.00	0.00	3969
	10, 000	0.039	9997.43	0.141	1.00	−2.57	0.950	0.000	5.00	0.00	15, 725

Table 7 Parallel piecewise sequential strategy (9) on 90% FWCI for the mean of $N(5, 4)$ with no fine-tuning: $\varepsilon = 0$, $k = 2$, $B = 10^6$ replications

m	C	d	$\overline{\overline{n}}$	$s_{\overline{\overline{n}}}$	$\overline{\overline{n}}/C$	$\overline{\overline{n}} - C$	$\overline{p}$	$s_{\overline{p}}$	$\overline{x}$	$s_{\overline{x}}$	Time in sec
5	30	0.601	26.11	0.008	0.87	−3.89	0.858	0.000	5.00	0.00	1
	50	0.465	44.14	0.013	0.88	−5.86	0.862	0.000	5.00	0.00	2
	100	0.329	94.21	0.018	0.94	−5.79	0.882	0.000	5.00	0.00	4
	200	0.233	196.02	0.023	0.98	−3.98	0.895	0.000	5.00	0.00	12
	500	0.147	497.01	0.034	0.99	−2.99	0.898	0.000	5.00	0.00	52
	1000	0.104	997.35	0.046	1.00	−2.65	0.899	0.000	5.00	0.00	187
	2000	0.074	1997.50	0.064	1.00	−2.50	0.900	0.000	5.00	0.00	667
	5000	0.047	4997.68	0.101	1.00	−2.32	0.900	0.000	5.00	0.00	3972
	10, 000	0.033	9997.68	0.142	1.00	−2.32	0.900	0.000	5.00	0.00	15, 674
10	30	0.601	29.45	0.006	0.98	−0.55	0.890	0.000	5.00	0.00	1
	50	0.465	46.54	0.011	0.93	−3.46	0.879	0.000	5.00	0.00	2
	100	0.329	95.99	0.016	0.96	−4.01	0.888	0.000	5.00	0.00	5
	200	0.233	197.03	0.021	0.99	−2.97	0.895	0.000	5.00	0.00	14
	500	0.147	497.53	0.032	1.00	−2.47	0.899	0.000	5.00	0.00	55
	1000	0.104	997.55	0.045	1.00	−2.45	0.899	0.000	5.00	0.00	182
	2000	0.074	1997.61	0.063	1.00	−2.39	0.899	0.000	5.00	0.00	668
	5000	0.047	4997.73	0.100	1.00	−2.27	0.900	0.000	5.00	0.00	3931
	10, 000	0.033	9997.66	0.141	1.00	−2.34	0.899	0.000	5.00	0.00	15, 520

Table 8 Parallel piecewise sequential strategy (9) on 95% FWCI for the mean of $N(5,4)$ with no fine-tuning: $\varepsilon = 0$, $k = 5$, $B = 10^6$ replications

m	C	d	$\overline{\overline{n}}$	$s_{\overline{\overline{n}}}$	$\overline{\overline{n}}/C$	$\overline{\overline{n}} - C$	$\overline{p}$	$s_{\overline{p}}$	$\overline{x}$	$s_{\overline{x}}$	Time in sec
5	30	0.716	32.81	0.005	1.09	2.81	0.957	0.000	5.00	0.00	0
	50	0.554	45.75	0.009	0.91	−4.25	0.934	0.000	5.00	0.00	1
	100	0.392	87.05	0.017	0.87	−12.95	0.927	0.000	5.00	0.00	1
	200	0.277	184.58	0.026	0.92	−15.42	0.937	0.000	5.00	0.00	3
	500	0.175	489.94	0.037	0.98	−10.06	0.947	0.000	5.00	0.00	12
	1000	0.124	992.21	0.048	0.99	−7.79	0.949	0.000	5.00	0.00	35
	2000	0.088	1993.16	0.066	1.00	−6.84	0.950	0.000	5.00	0.00	121
	5000	0.055	4993.69	0.102	1.00	−6.31	0.950	0.000	5.00	0.00	677
	10, 000	0.039	9993.91	0.143	1.00	−6.09	0.950	0.000	5.00	0.00	2592
10	30	0.716	50.85	0.001	1.69	20.85	0.989	0.000	5.00	0.00	1
	50	0.554	57.61	0.005	1.15	7.61	0.964	0.000	5.00	0.00	1
	100	0.392	93.89	0.014	0.94	−6.11	0.940	0.000	5.00	0.00	1
	200	0.277	189.50	0.023	0.95	−10.50	0.941	0.000	5.00	0.00	3
	500	0.175	492.55	0.034	0.99	−7.45	0.947	0.000	5.00	0.00	12
	1000	0.124	993.57	0.046	0.99	−6.43	0.949	0.000	5.00	0.00	37
	2000	0.088	1993.89	0.064	1.00	−6.11	0.949	0.000	5.00	0.00	125
	5000	0.055	4994.08	0.100	1.00	−5.92	0.950	0.000	5.00	0.00	694
	10, 000	0.039	9993.94	0.142	1.00	−6.06	0.950	0.000	5.00	0.00	2577

Table 9 Parallel piecewise sequential strategy (9) on 90% FWCI for the mean of $N(5,4)$ with no fine-tuning: $\varepsilon = 0$, $k = 5$, $B = 10^6$ replications

m	C	d	$\overline{\overline{n}}$	$s_{\overline{\overline{n}}}$	$\overline{\overline{n}}/C$	$\overline{\overline{n}} - C$	$\overline{p}$	$s_{\overline{p}}$	$\overline{x}$	$s_{\overline{x}}$	Time in sec
5	30	0.601	32.81	0.005	1.09	2.81	0.911	0.000	5.00	0.00	0
	50	0.465	45.75	0.009	0.91	−4.25	0.879	0.000	5.00	0.00	1
	100	0.329	87.05	0.017	0.87	−12.95	0.869	0.000	5.00	0.00	1
	200	0.233	184.65	0.026	0.92	−15.35	0.883	0.000	5.00	0.00	3
	500	0.147	489.91	0.037	0.98	−10.09	0.895	0.000	5.00	0.00	12
	1000	0.104	992.23	0.049	0.99	−7.77	0.898	0.000	5.00	0.00	35
	2000	0.074	1993.09	0.066	1.00	−6.91	0.899	0.000	5.00	0.00	121
	5000	0.047	4993.79	0.101	1.00	−6.21	0.900	0.000	5.00	0.00	665
	10, 000	0.033	9994.14	0.143	1.00	−5.86	0.900	0.000	5.00	0.00	2565
10	30	0.601	50.85	0.001	1.69	20.85	0.967	0.000	5.00	0.00	1
	50	0.465	57.62	0.005	1.15	7.62	0.921	0.000	5.00	0.00	1
	100	0.329	93.88	0.014	0.94	−6.12	0.886	0.000	5.00	0.00	1
	200	0.233	189.52	0.023	0.95	−10.48	0.888	0.000	5.00	0.00	3
	500	0.147	492.61	0.033	0.99	−7.39	0.897	0.000	5.00	0.00	12
	1000	0.104	993.48	0.046	0.99	−6.52	0.899	0.000	5.00	0.00	36
	2000	0.074	1993.75	0.064	1.00	−6.25	0.899	0.000	5.00	0.00	123
	5000	0.047	4994.06	0.100	1.00	−5.94	0.899	0.000	5.00	0.00	665
	10, 000	0.033	9994.01	0.142	1.00	−5.99	0.900	0.000	5.00	0.00	2576

Table 10 Fine-tuned parallel piecewise sequential strategy (14) on 95% FWCI for the mean of $N(5,4)$: $\varepsilon = -1.1828$, $k = 2$, $B = 10^6$ replications

m	C	d	$\overline{q}$	$s_{\overline{q}}$	$\overline{q}/C$	$\overline{q}-C$	$\overline{p}$	$s_{\overline{p}}$	$\overline{x}$	$s_{\overline{x}}$	Time in sec
5	30	0.716	28.82	0.008	0.96	−1.18	0.932	0.000	5.00	0.00	1
	50	0.554	47.46	0.012	0.95	−2.54	0.932	0.000	5.00	0.00	2
	100	0.392	97.64	0.017	0.98	−2.36	0.942	0.000	5.00	0.00	4
	200	0.277	198.92	0.022	0.99	−1.08	0.948	0.000	5.00	0.00	12
	500	0.175	499.60	0.033	1.00	−0.40	0.949	0.000	5.00	0.00	52
	1000	0.124	999.83	0.046	1.00	−0.17	0.950	0.000	5.00	0.00	187
	2000	0.088	1999.84	0.064	1.00	−0.16	0.950	0.000	5.00	0.00	667
	5000	0.055	5000.10	0.101	1.00	0.10	0.950	0.000	5.00	0.00	3974
	10, 000	0.039	9999.91	0.142	1.00	−0.09	0.950	0.000	5.00	0.00	15, 681
10	30	0.716	31.36	0.007	1.05	1.36	0.949	0.000	5.00	0.00	1
	50	0.554	49.21	0.011	0.98	−0.79	0.940	0.000	5.00	0.00	2
	100	0.392	98.82	0.016	0.99	−1.18	0.945	0.000	5.00	0.00	4
	200	0.277	199.58	0.021	1.00	−0.42	0.948	0.000	5.00	0.00	11
	500	0.175	499.95	0.032	1.00	−0.05	0.949	0.000	5.00	0.00	52
	1000	0.124	999.95	0.045	1.00	−0.05	0.950	0.000	5.00	0.00	181
	2000	0.088	2000.08	0.063	1.00	0.08	0.950	0.000	5.00	0.00	670
	5000	0.055	4999.83	0.100	1.00	−0.17	0.950	0.000	5.00	0.00	3967
	10, 000	0.039	10,000.09	0.142	1.00	0.09	0.950	0.000	5.00	0.00	15, 765

Table 11 Fine-tuned parallel piecewise sequential strategy (14) on 90% FWCI for the mean of $N(5,4)$: $\varepsilon = -1.1828$, $k = 2$, $B = 10^6$ replications

m	C	d	$\overline{q}$	$s_{\overline{q}}$	$\overline{q}/C$	$\overline{q}-C$	$\overline{p}$	$s_{\overline{p}}$	$\overline{x}$	$s_{\overline{x}}$	Time in sec
5	30	0.601	28.82	0.008	0.96	−1.18	0.878	0.000	5.00	0.00	1
	50	0.465	47.45	0.012	0.95	−2.55	0.879	0.000	5.00	0.00	2
	100	0.329	97.66	0.017	0.98	−2.34	0.890	0.000	5.00	0.00	4
	200	0.233	198.90	0.022	0.99	−1.10	0.897	0.000	5.00	0.00	12
	500	0.147	499.59	0.033	1.00	−0.41	0.900	0.000	5.00	0.00	54
	1000	0.104	999.83	0.046	1.00	−0.17	0.900	0.000	5.00	0.00	181
	2000	0.074	2000.00	0.064	1.00	0.00	0.899	0.000	5.00	0.00	667
	5000	0.047	4999.84	0.100	1.00	−0.16	0.900	0.000	5.00	0.00	4012
	10, 000	0.033	10,000.21	0.142	1.00	0.21	0.900	0.000	5.00	0.00	15, 708
10	30	0.601	31.34	0.007	1.04	1.34	0.900	0.000	5.00	0.00	1
	50	0.465	49.20	0.011	0.98	−0.80	0.890	0.000	5.00	0.00	2
	100	0.329	98.84	0.016	0.99	−1.16	0.894	0.000	5.00	0.00	5
	200	0.233	199.59	0.021	1.00	−0.41	0.898	0.000	5.00	0.00	13
	500	0.147	499.93	0.032	1.00	−0.07	0.899	0.000	5.00	0.00	54
	1000	0.104	999.84	0.045	1.00	−0.16	0.900	0.000	5.00	0.00	184
	2000	0.074	2000.04	0.063	1.00	0.04	0.900	0.000	5.00	0.00	666
	5000	0.047	5000.05	0.100	1.00	0.05	0.900	0.000	5.00	0.00	3992
	10,000	0.033	9999.97	0.142	1.00	−0.03	0.900	0.000	5.00	0.00	15, 666

Table 12 Fine-tuned parallel piecewise sequential strategy (14) on 95% FWCI for the mean of $N(5, 4)$: $\varepsilon = -1.1828$, $k = 5$, $B = 10^6$ replications

m	C	d	$\overline{q}$	$s_{\overline{q}}$	$\overline{q}/C$	$\overline{q} - C$	$\overline{p}$	$s_{\overline{p}}$	$\overline{x}$	$s_{\overline{x}}$	Time in sec
5	30	0.716	36.17	0.005	1.21	6.17	0.966	0.000	5.00	0.00	0
	50	0.554	51.01	0.009	1.02	1.01	0.948	0.000	5.00	0.00	1
	100	0.392	94.79	0.017	0.95	−5.21	0.939	0.000	5.00	0.00	1
	200	0.277	193.41	0.025	0.97	−6.59	0.943	0.000	5.00	0.00	3
	500	0.175	497.36	0.035	0.99	−2.64	0.949	0.000	5.00	0.00	12
	1000	0.124	998.76	0.047	1.00	−1.24	0.950	0.000	5.00	0.00	37
	2000	0.088	1999.46	0.065	1.00	−0.54	0.950	0.000	5.00	0.00	124
	5000	0.055	4999.75	0.101	1.00	−0.25	0.950	0.000	5.00	0.00	674
	10,000	0.039	9999.69	0.142	1.00	−0.31	0.950	0.000	5.00	0.00	2561
10	30	0.716	51.53	0.002	1.72	21.53	0.990	0.000	5.00	0.00	1
	50	0.554	60.38	0.006	1.21	10.38	0.968	0.000	5.00	0.00	1
	100	0.392	99.85	0.014	1.00	−0.15	0.947	0.000	5.00	0.00	1
	200	0.277	196.72	0.023	0.98	−3.28	0.946	0.000	5.00	0.00	3
	500	0.175	499.00	0.033	1.00	−1.00	0.949	0.000	5.00	0.00	12
	1000	0.124	999.65	0.045	1.00	−0.35	0.949	0.000	5.00	0.00	36
	2000	0.088	1999.76	0.064	1.00	−0.24	0.950	0.000	5.00	0.00	125
	5000	0.055	4999.91	0.100	1.00	−0.09	0.950	0.000	5.00	0.00	667
	10,000	0.039	9999.87	0.142	1.00	−0.13	0.950	0.000	5.00	0.00	2560

Table 13 Fine-tuned parallel piecewise sequential strategy (14) on 90% FWCI for the mean of $N(5, 4)$: $\varepsilon = -1.1828$, $k = 5$, $B = 10^6$ replications

m	C	d	$\overline{q}$	$s_{\overline{q}}$	$\overline{q}/C$	$\overline{q} - C$	$\overline{p}$	$s_{\overline{p}}$	$\overline{x}$	$s_{\overline{x}}$	Time in sec
5	30	0.601	36.18	0.005	1.21	6.18	0.926	0.000	5.00	0.00	0
	50	0.465	50.99	0.009	1.02	0.99	0.898	0.000	5.00	0.00	1
	100	0.329	94.81	0.017	0.95	−5.19	0.886	0.000	5.00	0.00	1
	200	0.233	193.33	0.025	0.97	−6.67	0.891	0.000	5.00	0.00	3
	500	0.147	497.31	0.035	0.99	−2.69	0.899	0.000	5.00	0.00	12
	1000	0.104	998.76	0.047	1.00	−1.24	0.899	0.000	5.00	0.00	36
	2000	0.074	1999.40	0.065	1.00	−0.60	0.899	0.000	5.00	0.00	122
	5000	0.047	4999.88	0.101	1.00	−0.12	0.900	0.000	5.00	0.00	676
	10, 000	0.033	9999.88	0.142	1.00	−0.12	0.900	0.000	5.00	0.00	2607
10	30	0.601	51.53	0.002	1.72	21.53	0.968	0.000	5.00	0.00	1
	50	0.465	60.38	0.006	1.21	10.38	0.927	0.000	5.00	0.00	1
	100	0.329	99.85	0.014	1.00	−0.15	0.896	0.000	5.00	0.00	1
	200	0.233	196.67	0.023	0.98	−3.33	0.895	0.000	5.00	0.00	3
	500	0.147	498.93	0.033	1.00	−1.07	0.899	0.000	5.00	0.00	12
	1000	0.104	999.60	0.045	1.00	−0.40	0.900	0.000	5.00	0.00	36
	2000	0.074	1999.76	0.064	1.00	−0.24	0.900	0.000	5.00	0.00	125
	5000	0.047	4999.98	0.100	1.00	−0.02	0.900	0.000	5.00	0.00	676
	10,000	0.033	10,000.14	0.142	1.00	0.14	0.900	0.000	5.00	0.00	2577

4.1.1 Descriptions of Tables 2, 3, 4, 5, 6, 7, 8, 9, 10, 11, 12, and 13

Tables 2, 3, 4, 5, 6, 7, 8, 9, 10, 11, 12, and 13 implemented the original purely sequential estimation strategy from (4)–(5) with $\alpha = 0.05, 0.10$, respectively. $E_\theta[N]$ was estimated by $\overline{n}$ (column 4) along with its estimated s.e. $s_{\overline{n}}$ (column 5). The observed values of both $\overline{n}/C$ (column 6) and $\overline{n} - C$ (column 7) were computed to illustrate the *asymptotic first-* and *second-order efficiency* properties [25, Definition 6.2.2] of the stopping variables, respectively.

We also present the observed coverage probability $\overline{p}$ (column 8) with its estimated s.e. $s_{\overline{p}}$ (column 9) as well as the average $\overline{x}$ (column 10), the average of all terminal sample means along with its estimated s.e. $s_{\overline{x}}$ (column 11). The last column 12 in each table shows the time (in seconds) taken to complete the $B = 10^6$ replications under each configuration, that is, for the construction of each row in the tables. The entries in the remaining Tables 4, 5, 6, 7, 8, 9, 10, 11, 12, and 13 would be interpreted in the same way.

Tables 4 and 5 correspond to the fine-tuned parallel piecewise sequential strategy (14) with $k = 1$, $\varepsilon = -1.1828$ and $\alpha = 0.05, 0.10$, respectively. In other words, Tables 4 and 5 equivalently correspond to the fine-tuned purely sequential strategy (4) with $\varepsilon = -1.1828$.

Tables 6 and 7 correspond to the parallel piecewise sequential (non-fine-tuned, that is $\varepsilon = 0$) strategy (9) with $k = 2$ and $\alpha = 0.05, 0.10$, respectively. Tables 8 and 9 correspond to the parallel piecewise sequential (non-fine-tuned, that is $\varepsilon = 0$) strategy (9) with $k = 5$ and $\alpha = 0.05, 0.10$, respectively.

Tables 10 and 11 correspond to the fine-tuned parallel piecewise sequential strategy (14) with $k = 2$, $\varepsilon = -1.1828$ and $\alpha = 0.05, 0.10$, respectively. Tables 12 and 13 correspond to the fine-tuned parallel piecewise sequential strategy (14) with $k = 5$, $\varepsilon = -1.1828$ and $\alpha = 0.05, 0.10$, respectively.

4.1.2 An Overview of Comments on Simulated Performances: Tables 2, 3, 4, 5, 6, 7, 8, 9, 10, 11, 12, and 13

We note that for a parallel piecewise sequential estimation strategy (with or without fine-tuning) the data collection process is completed as soon as the last operator or arm is done sampling. The terminal estimation of the population mean can be carried out immediately after by pooling data from all arms together. Indeed, the observed runtime of our simulation program is consistent with this intuition.

The time-savings are substantial using the parallel piecewise sequential strategy and increase as k increases. For $k = 2$, we find that the parallel piecewise sequential procedure (with or without fine-tuning) for $C = 10000$ takes approximately the same time (around 1.6×10^4 s) as the purely sequential procedure for $C = 5000$, that the former for $C = 2000$ takes approximately the same time (around 670 s) as the latter for $C = 1000$, and so on. The time-savings are even more remarkable when k increases to 5— the parallel piecewise sequential procedure for $C = 10000$ now takes approximately the same time (around 2.6×10^3 s) as the purely sequential procedure for $C = 2000$, and the former for $C = 5000$ takes approximately the same time (around 670 s) as the latter for $C = 1000$, and so on.

As shown in Tables 2 and 3, 6 and 7, and 8 and 9, respectively, and consistent with (12), we find that the estimate $\overline{n} - C_d$ of $E_{\boldsymbol{\theta}}[\widetilde{N}_d - C_d]$ from the purely sequential procedure (4) converges to approximately $\eta \approx -1.1828$ as $d \to 0$, that this estimate from the parallel piecewise sequential procedure (9) with $k = 2$ arms converges to approximately $k\eta \equiv 2\eta \approx -2.37$ as $d \to 0$, and that this estimate from the parallel piecewise sequential procedure (9) with $k = 5$ arms converges to approximately $k\eta \equiv 5\eta \approx -5.91$ as $d \to 0$. The undersampling due to the asymptotic bias of the stopping variable $\widetilde{N}_d$ from (9) relative to the optimal fixed-sample size C_d becomes increasingly substantial as k increases.

On the other hand, according to Tables 4 and 5, 10 and 11, and 12 and 13, the estimate $\overline{q} - C_d$ of $E_{\boldsymbol{\theta}}[Q_d - C_d]$ from the *fine-tuned* parallel piecewise sequential procedure (14) with either $k = 1$ (the *fine-tuned* purely sequential strategy), $k = 2$ or $k = 5$ converges to 0 as $d \to 0$. This is consistent with Theorem 3.1 that the stopping variable Q_d from (14) is an asymptotically unbiased estimator of the optimal fixed-sample size C_d. We note that the point estimate $\overline{q} - C_d$ of $E_{\boldsymbol{\theta}}[Q_d - C_d]$ is accurate enough, as $s_{\overline{q}}$ appears to be reasonably small throughout.

Additionally, in all Tables 2, 3, 4, 5, 6, 7, 8, 9, 10, 11, 12, and 13, using *generic notations* from Table 1, we observe that the estimate $\overline{n}/C_d$ of $E_{\boldsymbol{\theta}}[N_d/C_d]$ converges to 1 as $d \to 0$. This is expected as the asymptotic second-order efficiency implies the asymptotic first-order efficiency. We also briefly comment that as $d \to 0$, the observed coverage probability $\overline{p}$ converges to the preassigned confidence coefficient $1 - \alpha$, whether $\alpha = 0.05$ or 0.10, which is expected according to the *asymptotic consistency* property [25, Equation (6.2.30)], and that the average $\overline{x}$ of all terminal sample means from $B = 10^6$ replications accurately estimates the population mean $\mu (= 5)$. Overall, we observe similar and consistent behaviors for different values of the confidence coefficient $1 - \alpha (= 0.90, 0.95)$ and for different values of the pilot sample size $m (= 5, 10)$.

4.2 MRPE Strategies

While we continue to use notations similar to those used in Sect. 4.1, some additional notations are laid out in Table 14 for brevity. We present Tables 15, 16, 17, 18, 19, and 20 to compare the purely sequential procedure (7), the parallel piecewise sequential procedure (24), and the fine-tuned parallel piecewise sequential procedure (28), all obtained from averaging over $B (= 10^6)$ independent replications under each fixed configuration.

The population used in the simulation was $N(5, 4)$ and we fixed $A = 10000$, $m = 5, 10$, and the optimal fixed-sample size

$$n_c^* = 30, 50, 100, 200, 500, 1000, 2000, 5000, 10000.$$

Table 14 Additional set of notations used in Tables 15, 16, 17, 18, 19, and 20: $B = 10^6$ replications

$r_i = \frac{A\sigma^2}{n_i} + cn_i$	: estimated risk in the ith run;
$\overline{r} = B^{-1}\Sigma_{i=1}^{B} r_i$	: average estimated risk, should compare with R_{n^*};
$s_{\overline{r}} = \{(B^2 - B)^{-1}\Sigma_{i=1}^{B}(r_i - \overline{r})^2\}^{1/2}$	: estimated s.e. of $\overline{r}$.

Table 15 Purely sequential strategy (7) on MRPE for the mean of $N(5, 4)$ with no fine-tuning: $\varepsilon = 0,\ k = 1,\ A = 10{,}000,\ B = 10^6$ replications

m	n_c^*	c	$\overline{n}$	$s_{\overline{n}}$	$\overline{n}/n_c^*$	$\overline{n} - n_c^*$	$\overline{r}/R_{n_c^*}$	$\overline{r} - R_{n_c^*}$	$s_{\overline{r}}$	Time in sec
5	30	44.444	29.69	0.004	0.99	−0.31	1.02	44.217	0.256	2
	50	16.000	49.81	0.005	1.00	−0.19	1.01	11.147	0.102	4
	100	4.000	99.84	0.007	1.00	−0.16	1.00	2.230	0.024	12
	200	1.000	199.86	0.010	1.00	−0.14	1.00	0.517	0.001	35
	500	0.160	499.87	0.016	1.00	−0.13	1.00	0.081	0.000	179
	1000	0.040	999.88	0.022	1.00	−0.12	1.00	0.020	0.000	660
	2000	0.010	1999.84	0.032	1.00	−0.16	1.00	0.005	0.000	2532
	5000	0.002	4999.86	0.050	1.00	−0.14	1.00	0.001	0.000	15,697
	10,000	0.000	9999.87	0.071	1.00	−0.13	1.00	0.000	0.000	61,817
10	30	44.444	29.74	0.004	0.99	−0.26	1.01	32.616	0.085	2
	50	16.000	49.82	0.005	1.00	−0.18	1.01	9.528	0.020	4
	100	4.000	99.85	0.007	1.00	−0.15	1.00	2.151	0.003	12
	200	1.000	199.86	0.010	1.00	−0.14	1.00	0.518	0.001	35
	500	0.160	499.88	0.016	1.00	−0.12	1.00	0.081	0.000	179
	1000	0.040	999.90	0.022	1.00	−0.10	1.00	0.020	0.000	665
	2000	0.010	1999.91	0.032	1.00	−0.09	1.00	0.005	0.000	2557
	5000	0.002	4999.87	0.050	1.00	−0.13	1.00	0.001	0.000	15,481
	10,000	0.000	9999.87	0.071	1.00	−0.13	1.00	0.000	0.000	61,570

Two scenarios, $k = 2, 5$, were included for parallel piecewise sequential procedures (with or without fine-tuning). Having fixed σ, A, and n_c^*, we obtained $c = A\sigma^2 n_c^{*-2}$.

4.2.1 Descriptions of Tables 15, 16, 17, 18, 19, and 20

With our generic notations, $E_{\theta}[N]$ was estimated by the average terminal sample size $\overline{n}$ with its s.e. $s_{\overline{n}}$. The observed values of both $\overline{n}/n_c^*$ and $\overline{n} - n_c^*$ were computed to illustrate the asymptotic first- and second-order efficiencies of the stopping variables, respectively.

We also present the estimated *risk efficiency* $\overline{r}/R_{n_c^*}$ and *regret* $\overline{r} - R_{n_c^*}$ with the s.e. $s_{\overline{r}}$ of the average estimated risk $\overline{r}$ provided as well. One may refer to [25, Section 6.4.1] for related details. The last column in each table is the time (in seconds) taken to complete the $B = 10^6$ replications in each case (each row in the table).

Table 16 Fine-tuned purely sequential strategy (28) on MRPE for the mean of $N(5, 4)$: $\varepsilon = -0.1166$, $k = 1$, $A = 10{,}000$, $B = 10^6$ replications

m	n_c^*	c	$\bar{t}$	$s_{\bar{t}}$	$\bar{t}/n_c^*$	$\bar{t} - n_c^*$	$\bar{r}/R_{n_c^*}$	$\bar{r} - R_{n_c^*}$	$s_{\bar{r}}$	Time in sec
5	30	44.444	29.82	0.004	0.99	−0.18	1.02	42.471	0.246	2
	50	16.000	49.92	0.005	1.00	−0.08	1.01	10.801	0.095	4
	100	4.000	99.97	0.007	1.00	−0.03	1.00	2.191	0.020	12
	200	1.000	200.00	0.010	1.00	−0.00	1.00	0.516	0.001	36
	500	0.160	500.00	0.016	1.00	−0.00	1.00	0.081	0.000	180
	1000	0.040	1000.06	0.022	1.00	0.06	1.00	0.020	0.000	666
	2000	0.010	1999.99	0.032	1.00	−0.01	1.00	0.005	0.000	2551
	5000	0.002	4999.98	0.050	1.00	−0.02	1.00	0.001	0.000	15,676
	10,000	0.000	9999.97	0.071	1.00	−0.03	1.00	0.000	0.000	62,752
10	30	44.444	29.86	0.004	1.00	−0.14	1.01	31.757	0.081	2
	50	16.000	49.94	0.005	1.00	−0.06	1.01	9.410	0.020	4
	100	4.000	99.96	0.007	1.00	−0.04	1.00	2.139	0.003	11
	200	1.000	199.99	0.010	1.00	−0.01	1.00	0.516	0.001	35
	500	0.160	499.98	0.016	1.00	−0.02	1.00	0.081	0.000	178
	1000	0.040	1000.01	0.022	1.00	0.01	1.00	0.020	0.000	665
	2000	0.010	2000.02	0.032	1.00	0.02	1.00	0.005	0.000	2569
	5000	0.002	5000.10	0.050	1.00	0.10	1.00	0.001	0.000	15,587
	10,000	0.000	10,000.05	0.071	1.00	0.05	1.00	0.000	0.000	61,893

Table 17 Parallel piecewise sequential strategy (24) on MRPE for the mean of $N(5, 4)$ with no fine-tuning: $\varepsilon = 0$, $k = 2$, $A = 10{,}000$, $B = 10^6$ replications

m	n_c^*	c	$\bar{\bar{n}}$	$s_{\bar{\bar{n}}}$	$\bar{\bar{n}}/n_c^*$	$\bar{\bar{n}} - n_c^{**}$	$\bar{r}/R_{n_c^*}$	$\bar{r} - R_{n_c^*}$	$s_{\bar{r}}$	Time in sec
5	30	44.444	28.92	0.005	0.96	−1.08	1.02	48.882	0.109	1
	50	16.000	49.23	0.006	0.98	−0.77	1.01	13.725	0.037	2
	100	4.000	99.61	0.007	1.00	−0.39	1.00	2.370	0.005	4
	200	1.000	199.71	0.010	1.00	−0.29	1.00	0.532	0.001	12
	500	0.160	499.76	0.016	1.00	−0.24	1.00	0.082	0.000	54
	1000	0.040	999.79	0.022	1.00	−0.21	1.00	0.020	0.000	180
	2000	0.010	1999.73	0.032	1.00	−0.27	1.00	0.005	0.000	666
	5000	0.002	4999.77	0.050	1.00	−0.23	1.00	0.001	0.000	3987
	10,000	0.000	9999.79	0.071	1.00	−0.21	1.00	0.000	0.000	15,640
10	30	44.444	29.68	0.004	0.99	−0.32	1.01	25.750	0.037	1
	50	16.000	49.41	0.006	0.99	−0.59	1.01	11.394	0.022	2
	100	4.000	99.62	0.007	1.00	−0.38	1.00	2.288	0.004	5
	200	1.000	199.70	0.010	1.00	−0.30	1.00	0.530	0.001	12
	500	0.160	499.75	0.016	1.00	−0.25	1.00	0.082	0.000	52
	1000	0.040	999.79	0.022	1.00	−0.21	1.00	0.020	0.000	182
	2000	0.010	1999.78	0.032	1.00	−0.22	1.00	0.005	0.000	682
	5000	0.002	4999.77	0.050	1.00	−0.23	1.00	0.001	0.000	4001
	10,000	0.000	9999.76	0.071	1.00	−0.24	1.00	0.000	0.000	15,778

Table 18 Fine-tuned parallel piecewise sequential strategy (28) on MRPE for the mean of $N(5, 4)$: $\varepsilon = -0.1166$, $k = 2$, $A = 10{,}000$, $B = 10^6$ replications

m	n_c^*	c	$\bar{t}$	$s_{\bar{t}}$	$\bar{t}/n_c^*$	$\bar{t} - n_c^*$	$\bar{r}/R_{n_c^*}$	$\bar{r} - R_{n_c^*}$	$s_{\bar{r}}$	Time in sec
5	30	44.444	29.20	0.005	0.97	−0.80	1.02	45.913	0.103	1
	50	16.000	49.50	0.006	0.99	−0.50	1.01	13.166	0.035	2
	100	4.000	99.85	0.007	1.00	−0.15	1.00	2.329	0.005	4
	200	1.000	199.94	0.010	1.00	−0.06	1.00	0.529	0.001	12
	500	0.160	499.97	0.016	1.00	−0.03	1.00	0.082	0.000	54
	1000	0.040	999.98	0.022	1.00	−0.02	1.00	0.020	0.000	184
	2000	0.010	1999.99	0.032	1.00	−0.01	1.00	0.005	0.000	661
	5000	0.002	5000.02	0.050	1.00	0.02	1.00	0.001	0.000	3962
	10, 000	0.000	9999.97	0.071	1.00	−0.03	1.00	0.000	0.000	15,650
10	30	44.444	29.91	0.004	1.00	−0.09	1.01	25.111	0.035	1
	50	16.000	49.66	0.006	0.99	−0.34	1.01	11.027	0.021	2
	100	4.000	99.87	0.007	1.00	−0.13	1.00	2.264	0.004	4
	200	1.000	199.95	0.010	1.00	−0.05	1.00	0.528	0.001	12
	500	0.160	499.97	0.016	1.00	−0.03	1.00	0.082	0.000	52
	1000	0.040	1000.00	0.022	1.00	−0.00	1.00	0.020	0.000	185
	2000	0.010	1999.98	0.032	1.00	−0.02	1.00	0.005	0.000	670
	5000	0.002	5000.06	0.050	1.00	0.06	1.00	0.001	0.000	3971
	10,000	0.000	9999.98	0.071	1.00	−0.02	1.00	0.000	0.000	15,720

Table 19 Parallel piecewise sequential strategy (24) on MRPE for the mean of $N(5, 4)$ with no fine-tuning: $\varepsilon = 0$, $k = 5$, $A = 10{,}000$, $B = 10^6$ replications

m	n_c^*	c	$\bar{\bar{n}}$	$s_{\bar{\bar{n}}}$	$\bar{\bar{n}}/n_c^*$	$\bar{\bar{n}} - n_c^*$	$\bar{r}/R_{n_c^*}$	$\bar{r} - R_{n_c^*}$	$s_{\bar{r}}$	Time in sec
5	30	44.444	31.39	0.003	1.05	1.39	1.01	14.528	0.020	0
	50	16.000	47.99	0.006	0.96	−2.01	1.01	14.009	0.023	1
	100	4.000	97.65	0.009	0.98	−2.35	1.00	3.683	0.007	1
	200	1.000	198.84	0.011	0.99	−1.16	1.00	0.615	0.001	3
	500	0.160	499.26	0.016	1.00	−0.74	1.00	0.085	0.000	12
	1000	0.040	999.35	0.023	1.00	−0.65	1.00	0.021	0.000	38
	2000	0.010	1999.38	0.032	1.00	−0.62	1.00	0.005	0.000	121
	5000	0.002	4999.39	0.050	1.00	−0.61	1.00	0.001	0.000	665
	10,000	0.000	9999.29	0.071	1.00	−0.71	1.00	0.000	0.000	2572
10	30	44.444	50.02	0.000	1.67	20.02	1.13	356.013	0.004	1
	50	16.000	54.60	0.003	1.09	4.60	1.01	8.213	0.009	1
	100	4.000	98.45	0.008	0.98	−1.55	1.00	2.816	0.005	1
	200	1.000	198.94	0.011	0.99	−1.06	1.00	0.591	0.001	3
	500	0.160	499.27	0.016	1.00	−0.73	1.00	0.084	0.000	12
	1000	0.040	999.34	0.023	1.00	−0.66	1.00	0.021	0.000	37
	2000	0.010	1999.37	0.032	1.00	−0.63	1.00	0.005	0.000	122
	5000	0.002	4999.38	0.050	1.00	−0.62	1.00	0.001	0.000	673
	10,000	0.000	9999.36	0.071	1.00	−0.64	1.00	0.000	0.000	2574

Table 20 Fine-tuned parallel piecewise sequential strategy (28) on MRPE for the mean of $N(5, 4)$: $\varepsilon = -0.1166$, $k = 5$, $A = 10{,}000$, $B = 10^6$ replications

m	n_c^*	c	$\bar{t}$	$s_{\bar{t}}$	$\bar{t}/n_c^*$	$\bar{t} - n_c^*$	$\bar{r}/R_{n_c^*}$	$\bar{r} - R_{n_c^*}$	$s_{\bar{r}}$	Time in sec
5	30	44.444	31.79	0.003	1.06	1.79	1.01	16.373	0.022	0
	50	16.000	48.62	0.006	0.97	−1.38	1.01	12.842	0.021	1
	100	4.000	98.34	0.009	0.98	−1.66	1.00	3.444	0.006	1
	200	1.000	199.44	0.011	1.00	−0.56	1.00	0.602	0.001	3
	500	0.160	499.84	0.016	1.00	−0.16	1.00	0.084	0.000	12
	1000	0.040	999.96	0.023	1.00	−0.04	1.00	0.020	0.000	37
	2000	0.010	1999.98	0.032	1.00	−0.02	1.00	0.005	0.000	123
	5000	0.002	4999.92	0.050	1.00	−0.08	1.00	0.001	0.000	688
	10,000	0.000	10,000.06	0.071	1.00	0.06	1.00	0.000	0.000	2575
10	30	44.444	50.02	0.000	1.67	20.02	1.13	356.117	0.004	1
	50	16.000	54.88	0.003	1.10	4.88	1.01	9.056	0.009	1
	100	4.000	99.09	0.008	0.99	−0.91	1.00	2.692	0.004	1
	200	1.000	199.54	0.011	1.00	−0.46	1.00	0.579	0.001	3
	500	0.160	499.87	0.016	1.00	−0.13	1.00	0.084	0.000	12
	1000	0.040	999.95	0.023	1.00	−0.05	1.00	0.020	0.000	36
	2000	0.010	1999.98	0.032	1.00	−0.02	1.00	0.005	0.000	121
	5000	0.002	4999.93	0.050	1.00	−0.07	1.00	0.001	0.000	681
	10,000	0.000	9999.96	0.071	1.00	−0.04	1.00	0.000	0.000	2573

Table 15 corresponds to the original purely sequential strategy (7). Table 16 corresponds to the fine-tuned parallel piecewise sequential strategy (28) with $k = 1$, $\varepsilon = -0.1166$ which is equivalent to the fine-tuned version of (7).

Table 17 corresponds to the parallel piecewise sequential (non-fine-tuned, that is $\varepsilon = 0$) strategy (24) with $k = 2$. Table 18 corresponds to the fine-tuned parallel piecewise sequential strategy (28) with $k = 2$, $\varepsilon = -0.1166$.

Table 19 corresponds to the parallel piecewise sequential (non-fine-tuned, that is $\varepsilon = 0$) strategy (24) with $k = 5$. Table 20 corresponds to the fine-tuned parallel piecewise sequential strategy (28) with $k = 5$, $\varepsilon = -0.1166$.

4.2.2 An Overview of Comments on Simulated Performances: Tables 15, 16, 17, 18, 19, and 20

We again note that for a parallel piecewise sequential procedure (with or without fine-tuning) the data collection process is completed as soon as the last operator is done sampling and the final estimation of the population mean can be provided immediately by pooling data from all arms together. Indeed, the observed runtime of our simulation program is consistent with this intuition. The time-savings are substantial using the parallel piecewise sequential strategy and increase as k increases.

For $k = 2$, we find that the parallel piecewise sequential procedure (with or without fine-tuning) for $n_c^* = 10000$ takes approximately the same time (around 1.6×10^4 s) as the purely sequential procedure for $n_c^* = 5000$, that the former for $n_c^* = 2000$ takes approximately the same time (around 670 s) as the latter for $n_c^* = 1000$, and so on. The time-savings are even more remarkable when k increases to 5—the parallel piecewise sequential procedure for $n_c^* = 10000$ now takes approximately the same time (around 2.6×10^3 s) as the purely sequential procedure for $n_c^* = 2000$, and the former for $n_c^* = 5000$ takes approximately the same time (around 670 s) as the latter for $n_c^* = 1000$, and so on.

As shown in Tables 15, 17, and 19, respectively, and consistent with (26), we find that the estimate $\overline{n} - n_c^*$ of $E_{\boldsymbol{\theta}}[\widetilde{N}_c - n_c^*]$ from the purely sequential procedure (7) converges to approximately $\eta^* \approx -0.12$ as $c \to 0$, that this estimate from the parallel piecewise sequential procedure (24) with $k = 2$ arms converges to approximately $2\eta^* \approx -0.23$ as $c \to 0$, and that this estimate from the parallel piecewise sequential procedure (24) with $k = 5$ arms converges to approximately $5\eta^* \approx -0.58$ as $c \to 0$. The undersampling due to the asymptotic bias of the stopping variable $\widetilde{N}_c$ from (24) relative to the optimal fixed-sample size n_c^* becomes increasingly noticeable as k increases and can be substantial for large k.

On the other hand, according to Tables 16, 18, and 20, the estimate $\overline{t} - n_c^*$ of $E_{\boldsymbol{\theta}}[T_c - n_c^*]$ from the fine-tuned parallel piecewise sequential procedure (28) with either $k = 1$ (the fine-tuned purely sequential procedure), $k = 2$ or $k = 5$ converges to 0 as $c \to 0$. This is consistent with Theorem 3.2 that the stopping variable T_c from (28) is an asymptotically unbiased estimator of the optimal fixed-sample size n_c^*. We note that the point estimate $\overline{t} - n_c^*$ of $E_{\boldsymbol{\theta}}[T_c - n_c^*]$ is accurate enough, as $s_{\overline{t}}$ appears to be reasonably small throughout.

Additionally, in all Tables 15, 16, 17, 18, 19, and 20, using generic notations, we observe that the estimate $\overline{n}/n_c^*$ of $E_{\boldsymbol{\theta}}[N/n_c^*]$ converges to 1 as $c \to 0$. This is expected as the asymptotic second-order efficiency implies the asymptotic first-order efficiency. We also briefly comment that as $c \to 0$, the estimated risk efficiency $\overline{r}/R_{n_c^*}$ and regret $\overline{r} - R_{n_c^*}$, respectively, converge to 1 and $\frac{1}{2}c$, as expected, due to the asymptotic *first-* and *second-order risk efficiencies* [25, Equations (6.4.8) and (6.4.14)]. Overall, we observe similar and consistent behaviors across different values of the pilot sample size $m(= 5, 10)$.

5 A Real Data Illustration: The Framingham Heart Study

As an illustration, we consider the Framingham Heart Study, which is a long-term, ongoing cardiovascular cohort study on residents of the city of Framingham, Massachusetts. Much of the now-common knowledge concerning heart disease, such as the effects of diet, exercise, and customary medications such as aspirin, is based on this longitudinal study. One may refer to Wikipedia:

https://en.wikipedia.org/wiki/Framingham_Heart_Study

for the history and many other details of this study.

Admittedly, this real data set is not very "big" although it is reasonably large. In this section, we simply demonstrate the efficiency and broad applicability of our proposed methodologies with an interesting real data illustration. One should quickly realize that our proposed methodologies are readily applicable to much bigger data sets with even more substantial time-savings as practicalities may demand.

We illustrate the fine-tuned parallel piecewise sequential procedures (14) and (28) using a data set from the Framingham Heart Study developed by the NIH National Heart, Lung, and Blood Institute. One may request the data set from

https://biolincc.nhlbi.nih.gov/teaching/.

This data set contains three clinic examinations and 20-year follow-up data on a large subset of the original Framingham cohort participants and it is made publicly available especially for graduate education.

High blood cholesterol is one of the key risk factors for heart disease. It can build up in arteries and restrict or even block blood flow. As a result, the heart may not get as much oxygen-rich blood as it needs, increasing the risk of a heart attack. The variable that we are focused on here is the (*natural*) *logarithm* of the serum total cholesterol (mg/dL) for males during the third examination period. For the purpose of this illustration, we treated the data set of size 1312 (excluding missing values) as our population, which had a mean of 5.403 and a standard deviation of 0.181. The five-number summary is given as follows:

$$\min = 4.868,\ Q_1 = 5.288,\ \text{median} = 5.403,\ Q_3 = 5.529,\ \text{and}\ \max = 6.023.$$

Figure 1 is a histogram of the population superimposed with the $N(5.403, (0.181)^2)$ density curve. The population did not appear to contradict a normal distribution as confirmed via the Shapiro-Wilk (p-value $= 0.733$) and Anderson–Darling (p-value $= 0.547$) normality tests.

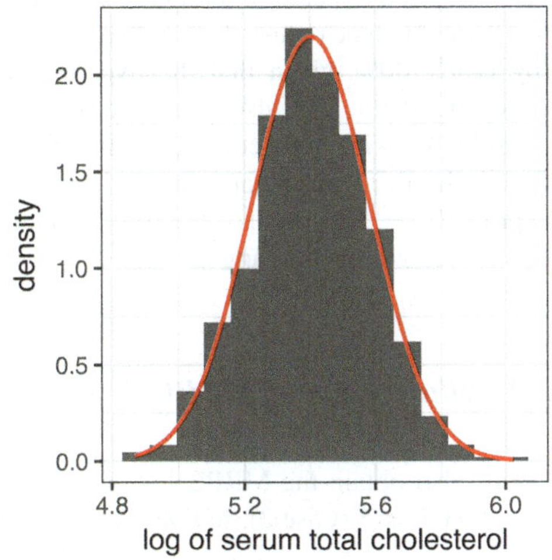

Fig. 1 Histogram of (natural) log serum total cholesterol for males during the third examination period, superimposed with $N(5.403, (0.181)^2)$ density curve

5.1 *Illustration of FWCI*

A 99% FWCI for the mean with $d = 0.020$, $m = 10$, and $k = 3$ turned out to be [5.386, 5.426] and indeed, the true population mean of 5.403 was contained in this interval. Let us also interpret this interval on the original scale (mg/dL). That is, with 99% confidence the *median* serum total cholesterol for males during the third examination period lay between $\exp(5.386) \approx 218$ mg/dL and $\exp(5.426) \approx 227$ mg/dL—the true median serum total cholesterol of 222 mg/dL was indeed within this range.

To give the reader some idea of serum total cholesterol (mg/dL) for males during the third examination period (that is, the original scale, without log-transformation), we may provide its five-number summary as follows:

$$\min = 130,\ Q_1 = 198,\ \text{median} = 222,\ Q_3 = 252,\ \text{and}\ \max = 413.$$

In this case, three researchers collected data independently (and simultaneously) with $n_1 = 222$, $n_2 = 158$, $n_3 = 165$, $n = n_1 + n_2 + n_3 = 545$, $C_d = 544.23$, and $n - C_d \approx 1$. The final sample size was only one observation more than the optimal fixed-sample size and the length of time required for data collection was reduced by nearly two-thirds (compared to a purely sequential strategy) given that the three researchers were about equally efficient in collecting data.

We obtained another 99% FWCI for the mean with $d = 0.015$, $m = 10$, and $k = 5$, in which case the true population mean was estimated to lie between 5.390 and 5.420. This interval also contained the true population mean of 5.403. Once again, we note that while the interval estimation became more precise with the reduced margin of error (from 0.020 to 0.015), the required final sample size became much larger—in this case, five researchers collected data independently (and simultaneously) with $n_1 = 233$, $n_2 = 204$, $n_3 = 176$, $n_4 = 171$, $n_5 = 182$, $n = n_1 + n_2 + n_3 + n_4 + n_5 = 966$, $C_d = 967.52$, and $n - C_d \approx -1.5$.

However, compared to a purely sequential strategy, the parallel processing with five "arms" helped reduce the length of time required for data collection significantly (by nearly 80%, given that the five researchers were about equally efficient in collecting data). We also interpret the interval on the original scale (mg/dL) as follows—with 99% confidence the *median* serum total cholesterol for males during the third examination period lay between $\exp(5.390) \approx 219$ mg/dL and $\exp(5.420) \approx 226$ mg/dL—the true median serum total cholesterol of 222 mg/dL was indeed within this range.

5.2 *Illustration of MRPE*

Let us also obtain the MRPE for the mean with $c = 0.05, k = 3$ (Case 1) and $c = 0.03, k = 5$ (Case 2). In both cases, we fixed $A = 10^6$ and $m = 10$. In Case 1, the point estimate turned out to be 5.408 (which was very close to the true population mean 5.403) with $n_1 = 289, n_2 = 257, n_3 = 259, n = n_1 + n_2 + n_3 = 805, n_c^* = 810.07$, and $n - n_c^* \approx -5$. Back to the original scale (mg/dL), the point estimate of $\exp(5.408) = 223$ mg/dL for the *median* serum total cholesterol for males during the third examination period was very close to the true population median serum total cholesterol of 222 mg/dL. The length of time required for data collection was reduced by nearly two-thirds (compared to a purely sequential strategy) given that the three researchers were about equally efficient in collecting data.

In Case 2, we ended up with a larger sample, as expected, as the cost per unit observation reduced from 0.05 to 0.03. Indeed, we observed $n_1 = 230, n_2 = 214, n_3 = 200, n_4 = 196, n_5 = 202, n = n_1 + n_2 + n_3 + n_4 + n_5 = 1042, n_c^* = 1045.79, n - n_c^* \approx -4$. In this case, the point estimate for the mean was 5.405. On the original scale (mg/dL), the point estimate $\exp(5.405) = 222.5$ mg/dL for the *median* serum total cholesterol for males during the third examination period came very close to the true median serum total cholesterol of 222 mg/dL. Once again, we mention that compared to a purely sequential strategy, the five parallel "pieces" helped reduce the length of time required for data collection significantly (by nearly 80%, given that the five researchers were about equally efficient in collecting data).

Remark 5.1 A reader may genuinely get confused since we have indeed mentioned both true *median* and *mean* serum total cholesterol. The following clarification may help on the median vs. mean issue in the Framingham Heart Study data: If a population ($Y = \ln(X)$) is normally distributed on the log scale of Y which implies that mean(Y) = median(Y), then the population has a positively skewed distribution on the original scale of X where median(X) < mean(X) and so the median and the mean of X are not equal. Since the ln (and exp) transformation is a monotonic transformation which preserves the order of the data, we have exp(mean(Y)) = exp(median(Y)) = median(exp(Y)) = median(X). That is, if a confidence interval for the mean(Y), which is essentially also the median(Y), is constructed on the log scale of Y, then by taking the "exp" of the lower and upper limits of the interval we will end up with a confidence interval for the median(X), rather than the mean(X), on the original scale of X.

6 Concluding Thoughts

One referee remarked: The authors have presented the paper in the context of big data. It is unclear what the role of stopping times is in the context of big data. Generally, big data is synonymous with data sets, which have no cap on their size.

We truly appreciated the genuine matter of inquiry raised by the referee. We have provided our perspective with respect to the relationship between stopping times and big data (and data sets in general), in the context of (sequential) experimental designs for statistical inference. We begin by agreeing that when we talk about big data, we generally mean a data set whose size may be extremely large.

From the experimental designs (for statistical inference) perspective, either in the conventional (non-sequential) setting or in the sequential situation (addressed in our paper here), however, an appropriate sample size will need to be determined first in order for one to collect an appropriate amount of data through the experiment. This will then lead to subsequent statistical inference (e.g., estimation or hypothesis testing) eventually carried out to achieve the predetermined level of precision (for estimation problems) or level of significance (for hypothesis testing problems). Such relevance is aptly validated by the recent contributions of [40] among other sources.

In the context of sequential experimental designs particularly addressed in our paper, a stopping time (or stopping rule) is a way of determining the most appropriate sample size to achieve the set goal in a statistical inference problem, more specifically, to arrive at a preassigned level of precision (i.e., half-width) in an FWCI problem or to achieve a minimized risk in an MRPE problem with predetermined "cost" per data point (unit observation). Generally speaking, if the stopping time (final sample size) is less than the theoretical optimal fixed-sample size, we would be unable to achieve the level of precision required for the FWCI problem or the minimum risk for the MRPE problem. On the other hand, if the stopping time is greater than the optimal fixed-sample size, we would be wasting unnecessary resources (e.g., time and manpower in the traditional sense, data storage and data processing time and cost in the era of big data and cloud computing) by oversampling to achieve the set goal in a statistical inference problem that could have otherwise been addressed with less effort, resources, or cost. Therefore, it is critical for one to be able to determine the most appropriate final sample size (stopping time) that is "just right" by developing a carefully designed stopping rule, on which our paper is focused, in order to most efficiently and effectively perform our hoped statistical inference with the data collected.

Acknowledgments We remain indebted to Professor Ansgar Steland and the referees for critically evaluating this invited contribution. Their feedback has improved an original version of our work. We take this opportunity to thank them.

References

1. Aerts, M., Geertsema, J.C.: Bounded length confidence intervals in nonparametric regression. Seq. Anal. **9**, 171–192 (1990)
2. Anscombe, F.J.: Large-sample theory of sequential estimation. Proc. Camb. Philol. Soc. **48**, 600–607 (1952)
3. Anscombe, F.J.: Sequential estimation. J. R. Stat. Soc. Ser. B **15**, 1–29 (1953)
4. Aoshima, M., Mukhopadhyay, N., Kobayashi, Y.: Two-stage procedures for estimating the difference of means when the sampling cost is different. Seq. Anal. **30**, 160–171 (2011)
5. Chow, Y.S., Robbins, H.: On the asymptotic theory of fixed width confidence intervals for the mean. Ann. Math. Stat. **36**, 457–462 (1965)
6. Dantzig, G.B.: On the non-existence of tests of student's hypothesis having power functions independent of σ. Ann. Math. Stat. **11**, 186–192 (1940)

7. Geertsema, J.C.: Sequential confidence intervals based on rank tests. Ann. Math. Stat. **41**, 1016–1026 (1970)
8. Ghosh, M., Mukhopadhyay, N.: On two fundamental problems of sequential estimation. Sankhyā Ser. B **38**, 203–218 (1976)
9. Ghosh, M., Mukhopadhyay, N.: Consistency and asymptotic efficiency of two-stage and sequential procedures. Sankhyā Ser. A **43**, 220–227 (1981)
10. Ghosh, M., Mukhopadhyay, N., Sen, P.K.: Sequential Estimation. Wiley, New York (1997)
11. Ghosh, B.K., Sen, P.K.: Handbook of Sequential Analysis, edited volume. Dekker, New York (1991)
12. Henke, N., Bughin, J., Chui, M., Manyika, J., Saleh, T., Wiseman, B., Sethupathy, G.: The age of analytics: competing in a data-driven world. McKinsey Global Institute report (2016). https://www.mckinsey.com/business-functions/mckinsey-analytics/our-insights/the-age-of-analytics-competing-in-a-data-driven-world
13. Jurečkovā, J., Sen, P.K.: Robust Statistical Procedures. Wiley, New York (1996)
14. Lai, T.L., Siegmund, D.: A nonlinear renewal theory with applications to sequential analysis I. Ann. Stat. **5**, 946–954 (1977)
15. Lai, T.L., Siegmund, D.: A nonlinear renewal theory with applications to sequential analysis II. Ann. Stat. **7**, 60–76 (1979)
16. Lombard, F., Swanepoel, J.W.H.: On finite and infinite confidence sequences. South African Stat. J. **12**, 1–24 (1978)
17. Manyika, J., Chui, M., Brown, B., Bughin, J., Dobbs, R., Roxburgh, C., Byers, A.H.: Big data: the next frontier for innovation, competition, and productivity. McKinsey Global Institute report (2011). https://www.mckinsey.com/business-functions/digital-mckinsey/our-insights/big-data-the-next-frontier-for-innovation
18. Mukhopadhyay, N.: Sequential estimation of location parameters in exponential distributions. Calcutta Statist. Assoc. Bull. **23**, 85–93 (1974)
19. Mukhopadhyay, N.: A consistent and asymptotically efficient two-stage procedure to construct fixed-width confidence intervals for the mean. Metrika **27**, 281–284 (1980)
20. Mukhopadhyay, N.: A study of the asymptotic regret while estimating the location of an exponential distribution. Calcutta Statist. Assoc. Bull. **31**, 207–213 (1982)
21. Mukhopadhyay, N.: Sequential estimation problems for negative exponential populations. Commun. Stat. Theory Methods Ser. A **17**, 2471–2506 (1988)
22. Mukhopadhyay, N.: Two-stage and multi-stage estimation. In: Balakrishnan, N., Basu, A.P. (eds.) The Exponential Distribution: Theory, Methods and Application, Chapter 26, pp. 429–452. Gordon and Breach, Amsterdam (1995)
23. Mukhopadhyay, N.: Probability and Statistical Inference. Dekker, New York (2000)
24. Mukhopadhyay, N., Datta, S.: On fine-tuning a purely sequential procedure and the associated second-order properties. Sankhyā Ser. A **57**, 100–117 (1995)
25. Mukhopadhyay, N., de Silva, B.M.: Sequential Methods and Their Applications. Chapman & Hall/CRC, Boca Raton (2009)
26. Mukhopadhyay, N., Sen, P.K.: Replicated piecewise stopping numbers and sequential analysis. Seq. Anal. **12**, 179–197 (1993)
27. Mukhopadhyay, N., Solanky, T.K.S.: Multistage Selection and Ranking Procedures. Dekker, New York (1994)
28. Mukhopadhyay, N., Vik, G.: Asymptotic results for stopping times based on U-statistics. Seq. Anal. **4**, 83–110 (1985)
29. Mukhopadhyay, N., Zhang, C.: EDA on the asymptotic normality of the standardized sequential stopping times, part-I: parametric models. Seq. Anal. **37**, 342–374 (2018)
30. Mukhopadhyay, N., Zhang, C.: EDA on the asymptotic normality of the standardized sequential stopping times, part-II: distribution-free models. Seq. Anal. **39**, 367–398 (2020)
31. Ray, W.D.: Sequential confidence intervals for the mean of a normal population with unknown variance. J. R. Stat. Soc. Ser. B **19**, 133–143 (1957)

32. Robbins, H.: Sequential estimation of the mean of a normal population. In: Cramér volume, H., Grenander, U. (eds.) Probability and Statistics, pp. 235–245. Almquist and Wiksell, Uppsala (1959)
33. Sen, P.K.: Sequential Nonparametrics. Wiley, New York (1981)
34. Siegmund, D.: Sequential Analysis: Tests and Confidence Intervals. Springer, New York (1985)
35. Starr, N.: The performance of a sequential procedure for fixed-width interval estimate. Ann. Math. Stat. **36**, 36–50 (1966)
36. Starr, N.: On the asymptotic efficiency of a sequential procedure for estimating the mean. Ann. Math. Stat. **37**, 1173–1185 (1966)
37. Starr, N., Woodroofe, M.: Remarks on sequential point estimation. Proc. Natl. Acad. Sci. USA **63**, 285–288 (1969)
38. Stein, C.: A Two sample test for a linear hypothesis whose power is independent of the variance. Ann. Math. Stat. **16**, 243–258 (1945)
39. Stein, C.: Some problems in sequential estimation (abstract). Econometrica **17**, 77–78 (1949)
40. Steland, A., Chang, Y.-T.: High-confidence nonparametric fixed-width uncertainty intervals and applications to projected high-dimensional data and common mean estimation. Seq. Anal. **40**, 97–124 (2021)
41. Swanepoel, J.W.H., van Wyk, J.W.J.: Fixed width confidence intervals for the location parameter of an exponential distribution. Commun. Stat. Theory Methods **11**, 1279–1289 (1982)
42. Woodroofe, M.: Second order approximations for sequential point and interval estimation. Ann. Stat. **5**, 984–995 (1977)
43. Woodroofe, M.: Nonlinear Renewal Theory in Sequential Analysis, CBMS #39. SIAM, Philadelphia (1982)

Statistical Learning for Change Point and Anomaly Detection in Graphs

Anna Malinovskaya, Philipp Otto, and Torben Peters

Abstract Complex systems which can be represented in the form of static and dynamic graphs arise in different fields, e.g., communication, engineering and industry. One of the interesting problems in analysing dynamic network structures is monitoring changes in their development. Statistical learning, which encompasses both methods based on artificial intelligence and traditional statistics, can be used to progress in this research area. However, the majority of approaches apply only one or the other framework. In this chapter, we discuss the possibility of bringing together both disciplines in order to create enhanced network monitoring procedures focussing on the example of combining statistical process control and deep learning algorithms. Together with the presentation of change point and anomaly detection in network data, we propose to monitor the response time of ambulance service, applying jointly the control chart for quantile function values and a graph convolutional network.

Keywords Network monitoring · Statistical process control · Control charts · Neural networks · Machine learning on graphs · Graph convolutional networks

1 Introduction

Network representation is fascinating. It conveys complexity by introducing a relational structure between objects and enables the incorporation of various information. The field was founded in 1735 when Leonhard Euler solved the Königsberg bridge problem, and since then, network science has developed into a significant area of study. The broad interest of the statistical community in

A. Malinovskaya · P. Otto (✉)
Leibniz University Hannover, Institute of Cartography and Geoinformatics, Hannover, Germany
e-mail: anna.malinovskaya@ikg.uni-hannover.de; philipp.otto@ikg.uni-hannover.de

T. Peters (✉)
ETH Zürich, Institute of Geodesy and Photogrammetry, Zürich, Switzerland
e-mail: torben.peters@geod.baug.ethz.ch

A. Steland, K.-L. Tsui (eds.), *Artificial Intelligence, Big Data and Data Science in Statistics*, https://doi.org/10.1007/978-3-031-07155-3_4

graph-based data analysis arose in the last century with the development of Erdös-Rényi-Gilbert model [1, 2] introducing a probabilistic view on the problem. Another significant factor in the growing popularity of networks is the availability of extensive data sources. The present era of Big Data provides a unique opportunity to gain remarkable insight into molecular, social, economic and many other structures (cf. [3–5]). Consequently, it shifts perspective to new analytical methods, which are not solely developed in the traditional statistical framework but also involve recent inventions in machine learning. Although the integration of artificial intelligence in network data analysis has achieved impressive results, the research in this area is in its early stages. Many existing machine learning algorithms cannot be applied directly to graphs because of their specific structural properties. To be precise, a standard vector representation of graphs does not exist compared to such data types as images or audio, which can be defined on regular lattices. This and other aspects related to the methods' generalisation and their evaluation challenge the development of machine learning approaches for network data but also offer the possibility to identify and unify the benefits of artificial intelligence and the traditional statistical framework for analysing graph-structured data efficiently.

Considering dynamic networks, those which develop over time, and one of the main interests in their study—the detection of anomalous behaviour, many powerful techniques based on statistical inference exist to perform network monitoring. However, the identification of the subsequent timestamp when the network system started deviating from its target state is not a solution to the issue as the inspection and possible improvement of the system are the following necessary steps to undertake. If we decide to perform monitoring entirely by a machine learning algorithm together with conducting the inspection step, we might encounter certain obstacles. Possible problems include the detection speed, data amount, interpretability and reliability, which remain poorly understood. Also, it will probably overcomplicate the monitoring process, especially when the network is stable and does not experience any sudden changes. However, a joint approach could combine benefits from both classical and modern learning foundations and, as a result, improve network surveillance and statistical learning in general. The expression "statistical learning" is a broader term for defining the approaches to learn from data, including algorithms based on artificial intelligence as well as methods that are purely statistical (cf. [6]).

An example of a well-known monitoring tool from the classical statistical framework is the control chart. It belongs to Statistical Process Control (SPC), being an effective instrument for detecting process deviation from the in-control or target state. Its universality and efficient technique to monitor the process online, meaning in real time, cannot be easily outperformed by novel approaches. However, as soon as a chart detects a change, often the investigation of a possible reason happens manually. If we consider the surveillance of graph-structured data that can be considerably voluminous and challenging to process in its raw format, the task to identify the cause of the signal and, if applicable, to resolve the issue may become extremely time-consuming. To improve the actions in the post-monitoring phase, we propose an enhanced application of the control charts which, in case of a signal, is followed by a graph machine learning algorithm that can operate on graphs to classify the cases, i.e., identify the reasons which led to the out-of-control state.

It is worth mentioning that the idea to extend the functionality of the control chart by combining it with machine learning algorithms is already an existing approach. Several publications propose different possibilities to bring together the two areas and generally evaluate the usage of machine learning in SPC (cf. [7–13]). However, the authors are not aware of the introduction to the topic from the point of network data monitoring, which currently obtains rapidly growing attention.

In this chapter, we propose a monitoring method that consolidates the control chart for quantile function values and a graph convolutional network. In Sect. 2 we introduce the mathematical definition of a graph, followed by the description of the change point and anomaly detection problems with the respective literature overview and the presentation of the control chart in Sect. 3. Section 4 focuses on the advancements of the graph learning representation, including the description of neural networks in general and graph convolutional networks. The simulation study of monitoring compliance with the response time prescribed to ambulance services is described in Sect. 5, bringing together SPC, graph theory and deep learning. We conclude with a discussion of possibilities to expand the joint applications of machine learning with the classical statistical tools and present several directions for future research.

2 What Is a Graph?

A graph (interchangeably called "network") $G = (V, E)$ is defined by nodes (also known as "vertices") $v_i \in V$, where $i = 1, \ldots, |V|$ with $|V|$ representing the total number of nodes, and edges $e_{i,j} \in E$ with $e_{i,j}$ being an edge (also called "link" or "tie") between vertices v_i and v_j, $j \neq i$. Usually, the network is defined by a binary or weighted adjacency matrix $\boldsymbol{A} \in \mathbb{R}^{|V|\times|V|}$. Two vertices are adjacent if they are connected by an edge. If we consider a binary adjacency matrix, then $A_{ij} = 1$, otherwise, $A_{ij} = 0$. In case of an undirected network, $\boldsymbol{A}$ is symmetric.

To illustrate these definitions, we design a small social network consisting of four colleagues (Fig. 1, left side). Consequently, if we consider the graph representation to display the interactions within this group, we would obtain a network positioned in the centre of Fig. 1. In this case, we interpret the edges as connections between two colleagues if they work on the same project. The same network can be expressed in form of an adjacency matrix (Fig. 1, right side). Additionally, we can assign nodal or edge attributes which are described by $\boldsymbol{X}^V$ and $\boldsymbol{L}^E$ so that $\boldsymbol{x}_i$ defines attributes of node v_i and $\boldsymbol{l}_{i,j}$ contains attributes of edge $e_{i,j}$. If the graph is weighted, we can incorporate the weights as one of the edge attributes into $\boldsymbol{L}^E$ or the representation of $\boldsymbol{A}$ directly. For example, if we decide to create a weighted network of the case presented in Fig. 1, the straightforward extension of our graph could be to include the number of projects the colleagues work on jointly as the edge weight.

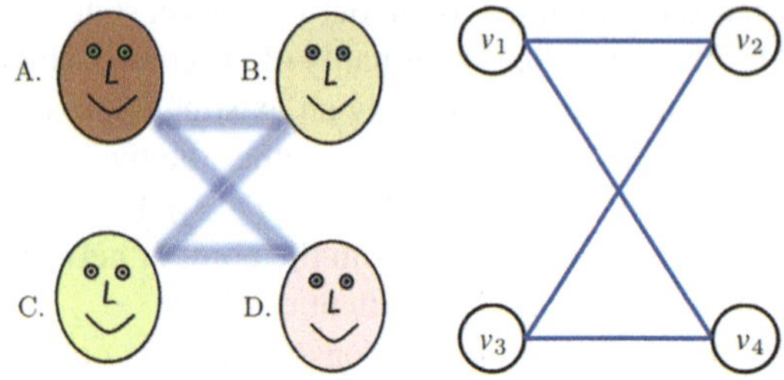

	v_1	v_2	v_3	v_4
v_1	0	1	0	1
v_2	1	0	1	0
v_3	0	1	0	1
v_4	1	0	1	0

Fig. 1 An example of a social network that consists of four colleagues (left side). In the centre, the graph represents the dynamics between the colleagues, where an edge defines working in the same project. Here, the vertex set consists of $V = \{v_1, v_2, v_3, v_4\}$, meaning v_1 is the colleague A up to v_4 being the colleague D. Consequently, the edge set is defined as $E = \{e_{1,2}, e_{1,4}, e_{2,3}, e_{3,4}\}$. On the right side is the representation of the social network as a binary adjacency matrix

Since graphs are powerful abstractions, there are numerous applications of them, including semantic, transportation, document citation, protein-protein interactions networks and many others (cf. [14, 15]). Consequently, the focus of the statistical analysis of networks differs from concentrating on the descriptive properties of the graph up to implementing inferential modelling and beyond.

To distinguish between a network as graph-structured data and the neural network approach, we use a full name for the latter, e.g., "graph convolutional network" or "neural network".

3 Change Point and Anomaly Detection in Network Data

In this section, we demonstrate the application of network modelling and other statistical approaches which enable us to analyse the graph-structured data over time. To be precise, we discuss the change point and anomaly detection in network data.

3.1 *What Is a Change Point?*

Network monitoring is a form of online surveillance procedure to detect a change point when the network system starts deviating from a so-called in-control state, i.e., the state when no unaccountable variation of the process is present. In other words, consider $\boldsymbol{s}_t = (s_1(G_t), \ldots, s_p(G_t))'$ as a collection of p network statistics which are derived from G at timepoint t. Following, let F_0 be the in-control or target distribution and F_τ the out-of-control distribution. We call τ a changepoint for a stochastic process s_t, if

$$s_t \sim \begin{cases} F_0 & \text{if } t < \tau \\ F_\tau & \text{if } t \geq \tau \end{cases}.$$

3.2 Methods for Network Monitoring

The approaches to monitor network data can be mainly subdivided into hypothesis testing methods, Bayesian methods and scan methods. The first category is dominated by the application of different forms of control charts, which represent the leading SPC method. This graphical technique records over time the behaviour of a control (or test) statistic, which is derived from one or more relevant process characteristics. Figure 2 illustrates a typical control chart for monitoring process mean, which includes a central line (CL) that defines the process average and two horizontal red lines being the upper control limit (UCL) and the lower control limit (LCL) [16]. When the process is in control, the control statistic is plotted close to the CL within the area defined by the control limits. As soon as unusual variability occurs, the observations start appearing on or outside the control limits. This practice is known as "signalling", meaning that the control chart "signals" an out-of-control state, informing us about the need for an investigation of the process.

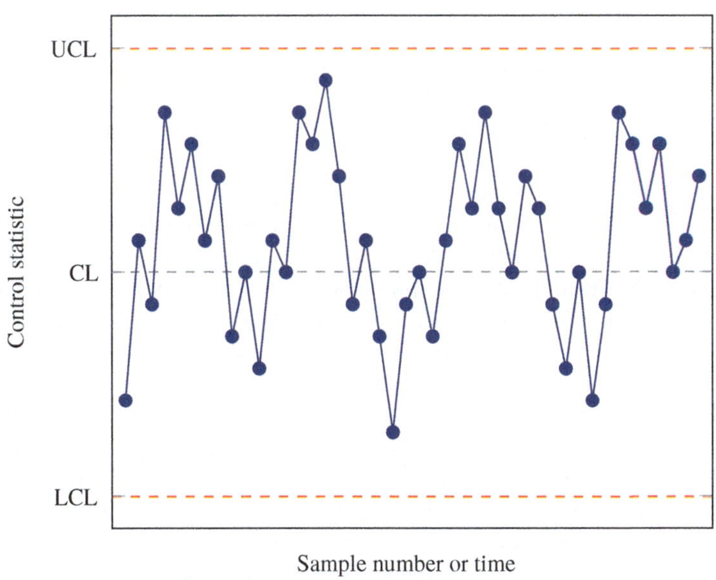

Fig. 2 A typical control chart [16]

There is a strong connection between control charts and hypothesis testing, as it repeatedly tests at different points of time t the null hypothesis $H_{0,t}$ against the alternative $H_{1,t}$. If we are interested in monitoring the deviation of the network statistics s_t from its expected value s_0, then we specify

$$H_{0,t} : \mathbb{E}(s_t) = s_0 \qquad \text{against} \qquad H_{1,t} : \mathbb{E}(s_t) \neq s_0 \, .$$

A hypothesis $H_{0,t}$ is rejected if the control statistic is equal to or exceeds the value of the control limit.

Usually, the application of control charts is divided into Phase I and Phase II, which have two distinct objectives. The data collected in Phase I serve as a baseline for estimation of the parameters such as expected value and calculation of control chart limits. In other words, we calibrate the control chart based on the observations that were collected under the assumption that the process is in its target state. In Phase II, we start monitoring the system, which is assumed to stay in control and examine the functionality of the control chart in respect to detected anomalies, i.e., out-of-control states, and to false alarms—when no abnormality is presented but the chart signals a change.

There are several ways to classify control charts. In terms of the number of variables, there are univariate ($c = 1$) and multivariate ($c > 1$) control charts. It is worth noting that for the second type, the LCL = 0 because the control charts are typically based on distances from the in-control state. Thus, only the UCL has to be computed. In the simulation study described in Sect. 5, we apply a multivariate control chart for quantile function values. To be precise, we consider the control chart introduced by Grimshaw and Alt [17]. Other examples of control charts that involve the application of quantile function and recommendations about the chart's calibration can be found in [18–21]. To estimate quantile function values of a random variable at the time point t, we need to obtain a sample of respective observations. In other words, the choice of a control chart for quantile function values indicates that outcomes from a specified period of the random variable are aggregated at each timestamp. In our case, we are interested in the response time of ambulance service, meaning we derive sample quantile function values of interest exactly from this quantity. Although we employ only one process characteristic, the control chart is a multivariate chart due to the specification of two quantile values (i.e., $c = 2$) for the calculation of the test statistic in this application.

For both univariate and multivariate control charts there exist various examples of applications in network surveillance. For instance, McCulloh and Carley [22] monitor the topology statistics of military networks applying the Cumulative Sum (CUSUM) chart. The Shewhart and Exponentially Weighted Moving Average (EWMA) charts were used by Wilson et al. [23] in combination with the dynamic Degree-Corrected Stochastic Block Model (DCSBM) to generate the networks and then perform surveillance over the Maximum Likelihood (ML) estimates. The application of EWMA and CUSUM to degree measures for detecting outbreaks on a weighted undirected network was introduced in [24]. Sparks and Wilson [25] detect communication outbreaks by designing an adaptive EWMA control chart. Malinovskaya and Otto [26] apply multivariate EWMA (MEWMA) and CUSUM (MCUSUM) together with the Temporal Exponential Random Graph Model (TERGM) and detect the beginning of the national lock-down period due to the COVID-19 pandemic by monitoring daily flights in the United States. Farahani et al. [27] evaluate the combination of the former charts used with the Poisson regression model for monitoring social networks. An overview of further control chart-based studies can be found in [28].

Regarding other hypothesis testing methods, Azarnoush et al. [29] propose monitoring network attributes instead of measures derived from its connectivity structure applying a logistic regression model and a likelihood-ratio test. Another method that shares similar assumptions as Azarnoush et al. [29] and incorporates vertex attributes is described in Miller et al. [30].

The Bayesian framework presented by Heard et al. [31] applies a two-stage approach using control chart limits based on a Bayesian predictive distribution. This technique concentrates on identifying anomalous behaviour between pairs of nodes (stage 1) which later are monitored as a sub-network (stage 2). Technically, for each pair of vertices, a communication trend is developed with the increments of the process following a Bayesian probability model. The node pair is considered anomalous if the p-value, which is derived throughout time, falls below the defined threshold (e.g., 0.05). Scan-based monitoring is known in the engineering literature as "moving window analysis" and is based on the concept of scanning a particular region of data by calculating a standardised metric for each window. This idea is applied in Priebe et al. [32] for detecting anomalies in the directed graphs (digraphs). Woodall et al. [33] provide a broader discussion of these methods.

3.3 *How Can We Specify Anomaly Detection in Terms of Network Monitoring?*

It is often the case that a monitoring statistic is aggregated from several observations which were collected within a specific time frame. That means, if the change point is detected, it is possible that only a few samples were anomalous and the rest not. Thus, we would need to perform anomaly detection as a postprocessing step to resolve a possible issue in the network system.

It is worth mentioning that there is no unique definition of the problem "anomaly detection". Akoglu et al. [34] use "change point detection" as a synonym for the anomaly detection problem for dynamic graphs, emphasising the existing difference between methods for static and temporal graphs in the survey. On the contrary, Ranshous et al. [35], who also provide an extensive methodological overview, introduce the change point detection as a subcategory of anomaly detection problem. The reason for the considerably different points of view is that the meaningful definition can only be established after a context and particular application are specified, otherwise, the interpretation is ambiguous. Here, under "anomaly" we understand an abnormal activity being a sudden and a significant change in the interaction patterns of a network [36]. Consequently, "anomaly detection" defines the task to find the networks which significantly differ from the majority of the reference networks and, if applicable, to ascertain the type of anomalous behaviour.

In contrast to the introduced definition of an anomalous observation in form of a whole graph, we can also define the anomaly detection problem in terms of edges or vertices. In other words, the aim is to find a subset of nodes/edges such that every element in this subset presents an uncommon evolution compared to other nodes/edges in a network. Another possible task is to identify anomalous subgraphs. Recent advancements in the area of machine learning on graphs led to impressive results in solving the specified problems by applying the graph convolutional network (GCN) framework (cf. [37–39]). However, to extract necessary information and provide substantial results, the neural network needs the graph data to be constructed as a set of low-dimensional learned continuous vectors (called "embeddings") without neglecting the relational structure and corresponding attributes. This task can be fulfilled by the graph representation learning techniques, which are briefly discussed together with the GCN in the following section.

4 Graph Representation Learning

Undeniably, the hand-engineered graph statistics s_t are useful in analysing the graph-structured data in terms of interpretation and computational costs. However, the manual selection of which features should be incorporated into the metrics, and further determination of statistics can be a time-consuming process. Moreover, this approach is restrictive because neither the selection of features nor metrics can be adapted through a learning process, which crucially constrains the effectiveness of machine learning-based algorithms. An alternative that encodes the network structure compactly and without losing any relevant information is Graph Representation Learning (GRL). In contrast to conventional methods, where we see the selection and design of graph statistics as a preprocessing step, GRL techniques regard the problem to learn embeddings as a machine learning task. To be more precise, the goal is to learn and optimise a mapping that embeds vertices, edges or entire (sub)graphs as points in a low-dimensional vector space $\mathbb{R}^d$ such that geometric relationships in this latent space reflect the structure of the initial graph [40]. Subsequently, the learned representation can be used as input for the main machine learning task, for example, classification. Hamilton [41] comprehensively reviews traditional as well as modern learning approaches over graph-structured data.

If we consider node embedding, the main purpose is to find a projection

$$f_\Theta : v_i \to z_i \in \mathbb{R}^d,$$

where $d \ll |V|$ and $z_i = \{z_1, z_2, \ldots, z_d\}$ represents the embedded vector that captures the graph position of node v_i and the structure of its local graph neighbourhood, and f_Θ is a mapping function parametrised by Θ. Depending on the embedding method, the incorporation of edge and nodal attributes into the latent representation z_i of the node v_i is also possible. Further encoding techniques which do not only focus on node representation together with the discussion of recent challenges in GRL can be found in [40, 42–44].

4.1 *Shallow Embedding Methods*

Shallow embedding approaches define the "encoder" mapping function f as an embedding-lookup. In this case, the set of trainable parameters is optimised directly, meaning that $\Theta = \mathbf{Z}$, with $\mathbf{Z} \in \mathbb{R}^{d \times |V|}$ being a matrix, where each column defines node embeddings z_i for each vertex v_i. The best-known techniques are either based on matrix factorisation (e.g., Laplacian eigenmaps) or random-walk statistics (e.g., DeepWalk and node2vec) [40]. However, shallow embedding approaches have some considerable limitations. The first issue is that there is a unique embedding for each node in the graph, meaning that no parameters are shared across vertices resulting in the absence of generalisation. Another problem is the ability to generate embeddings only for nodes that were present during the learning process. That means the graph structure should remain unchanged for the method to work correctly, which is highly unrealistic in many applications. To overcome these limitations, an alternative framework was proposed, which is explained in the next section.

4.2 *Graph Convolutional Networks*

To understand what graph convolutional networks are, it is important to define what "deep learning" and "neural networks" mean. "Deep learning" is a group of machine learning algorithms that can learn gradually a large number of parameters in an architecture composed of multiple non-linear transformations. An important example of these is neural networks, which we discuss subsequently.

4.2.1 Feedforward Fully Connected Neural Networks and Convolutional Neural Networks

Artificial neural networks represent a wide class of advanced computational models whose (often multi-layer) structure was initially inspired by the biological brain, although the current understanding and development of deep learning algorithms go beyond this neuroscientific perspective [45].

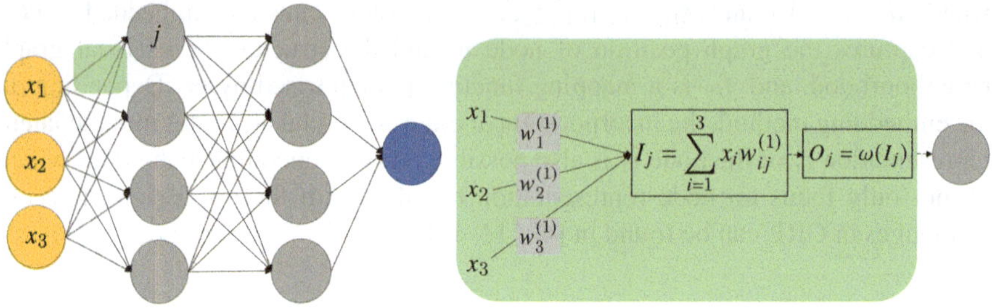

Fig. 3 A feedforward neural network consisting of four fully connected layers (left side) and the computation of a value for the neuron j in the first hidden layer (right side). For simplicity, the bias term is not included

In Fig. 3 we have an example of a simple feedforward neural network consisting of four fully connected layers. Each layer is composed of separate processing elements that are known as "neurons". Here, each neuron in one layer is connected to every neuron in the subsequent layer. Regarding Fig. 3, the input layer (yellow) has three neurons, the two consecutive hidden layers (grey) have four neurons, and the output layer (blue) consists of one neuron. The goal of a neural network is to process the incoming data that are entered as the input layer up to the output layer, where a corresponding result known as the target (e.g., a class label) is returned.

The right side of Fig. 3 illustrates calculations for obtaining a value of a particular neuron in a hidden layer. The input I_j of the neuron j corresponds to the weighted sum (applying the weights $w_{ij}^{(1)}$, where the (1) subscript refers to the layer being calculated, i defines the input neuron and j describes the hidden layer neuron) of values from neurons in the previous layer. Next, a non-linear function $\omega(\cdot)$ also known as "activation function" is used, with the final output of the neuron j being $O_j = \omega(I_j)$. To minimise the error between the desired and computed outputs in the final layer of a neural network, the parameters (in case of Fig. 3, the weights displayed on the arrows connecting respective neurons) are estimated during the training phase. A general overview of methods, including the principles of constructing and training a neural network model, can be found in [46].

In terms of graphs, the adaptation of such algorithms would enable us to generate a generalised representation of nodes that would depend on both the structure of the graph and additional feature information. The pioneer framework is known as "Graph Neural Networks" (GNNs) [47] which establishes the idea of including the neighbourhood information of a node into its latent representation, applying neural message passing form. The earliest GNN variations were limited in covering edge features and also were restricted in the choice of trainable parameters. Consequently, many advanced models arose, one of the examples being Graph Convolutional Networks (GCNs) [48]. To facilitate the understanding of the GCN, we explain first how a conventional Convolutional Neural Network (CNN) works, using the most popular data format they are applied on, namely images.

Consider a binary classification of the letters into vowels and consonants. Figure 4 provides a schematic illustration of a CNN for this machine learning task. First of all, our input is an image displaying the letter "A". Following, to start extracting necessary features for learning whether this letter is a vowel or a consonant, we apply a convolutional layer that is represented by four filters (also known as "feature detectors" or "kernels") with different weights of size 5 × 5. These filters are applied sequentially to the image, sliding over all pixels and performing convolution of the filter weights and respective neighbouring pixels. After the introduction of nonlinearity by using a suitable activation function (this happens after each layer in the neural network), our interim output consists of four feature maps combined into a single three dimensional representation. As you might notice, the width and height of our initial image do not coincide with the size of the new output. The reason is the standard application of a pooling operation after the convolutional layer that reduces the size of the produced feature maps, accelerating the data processing and affirming some of the detected features. Next, we have another convolutional layer, this time with five filters of size 3 × 3 × 4. The convolution is performed in the same way as in the previous layer, taking into consideration that the kernel size must be adjusted to the size of the previously produced interim output. As result, we obtain five feature maps connected into a single tensor. Subsequently, the convolutional operations are followed by the usage of an activation function and pooling operation (these steps are not represented in Fig. 4 due to the space limitation).

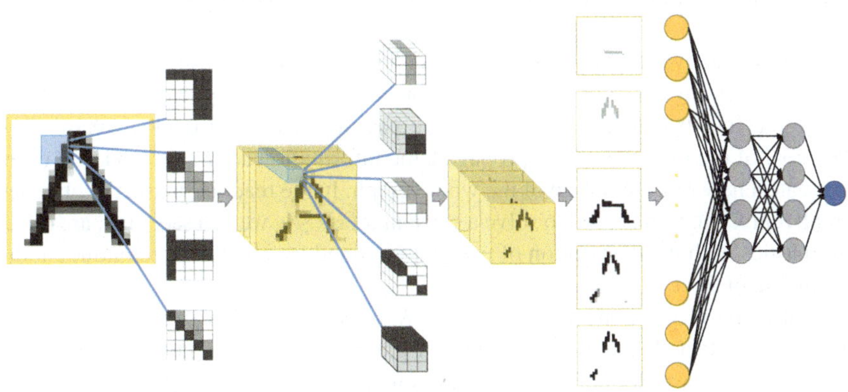

Fig. 4 An example of a convolutional neural network solving a binary classification task. The last four layers coincide with the neural network architecture presented in Fig. 3. The input to this part of the network is a long array, which is obtained by flattening the tensor consisting of final feature maps

Now feature extraction is completed, and we come to the classification step. To proceed, we need the final feature maps to have a one-dimensional form due to the subsequent application of fully connected layers. To obtain the desired dimension, we apply the operation called "flatten", reshaping the obtained tensor to an array (the yellow neurons in Fig. 4). The motivation of applying exactly this structure for determining the class label is the global consideration of the extracted features: the information flow between the fully connected neurons enables the mixing of signals. In contrast, the convolutional layers are particularly useful for data preprocessing due to their focus on the regions of adjacent pixels. Reaching the output layer of the CNN in Fig. 4, as we have a binary classification, only one neuron (coloured blue) is provided, which would return a class label, meaning the letter "A" is either a vowel or a consonant.

Coming back to the network structure, we wish to exchange the image with a graph input. Consequently, instead of pixels, we consider nodes and their local neighbourhoods. However, a graph is an example of non-Euclidean data [49], therefore, the CNN cannot be directly applied to network data. Thus, the field "geometric deep learning" has emerged, whose aim is to develop deep learning models for irregular data structures [50]. Examples of such models are the GNNs and the GCNs.

Both GNNs and GCNs belong to a more general category which is Message Passing Neural Networks (MPNNs) [51]. For readers who are interested in other neural network architectures which operate on graphs, we suggest Wu et al. [52] as a reference.

4.2.2 Application Phases of Graph Convolutional Networks

The GCN framework generalises the concept of CNN that is especially popular in image processing, as shown in the previous section. The application of GCNs is structured into two main phases: message passing phase and readout phase [53]. The goal of the first phase is to propagate the information across the nodes in order to create a new representation of the whole graph. In the readout phase, the obtained graph representation is used to solve a particular task. As we can see, there are direct similarities with the description of two steps defined for any GRL techniques at the beginning of Sect. 4.

In the first phase, consider $k = 0, \ldots, K$ to be the number of message passing iterations. In fact, K equals the number of graph convolutional layers in the neural network. Next, we define a set $N(i)$ to contain the neighbouring nodes of v_i. In contrast to an image input as presented in Fig. 4 where each pixel has a constant number of neighbouring pixels (see the blue areas), the size of the sets $N(i)$ may vary for each node as shown in Fig. 5. Similar to the feature map as a latent representation of an image, in the GCN we have a collection of feature vectors $\boldsymbol{H}^{(k)} \in \mathbb{R}^{d \times |V|}$, where $\boldsymbol{h}_i^{(k)}$ defines a d-dimensional hidden embedding of node v_i. At each iteration k, $\boldsymbol{h}_i^{(k)}$ incorporates the aggregated information (called the

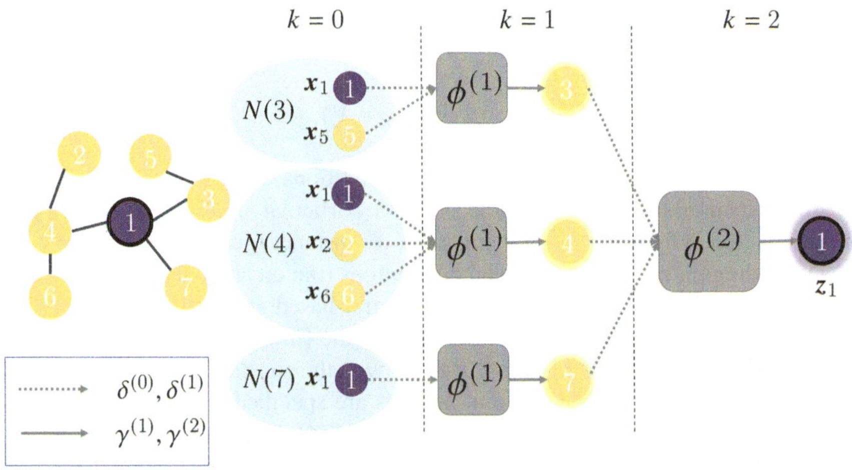

Fig. 5 An example of the procedure to generate the embedding z_1 of the vertex v_1 during the message passing phase. The attributes of node v_i are specified by $\boldsymbol{x}_i$. The sets $N(3)$, $N(4)$ and $N(7)$ define the neighbourhood of the nodes v_3, v_4 and v_7, respectively. The $\delta^{(k)}$ functions correspond to the message aggregation functions, the $\phi^{(k)}$ functions are the message creation functions and the $\gamma^{(k)}$ defines the update functions

"message") from $N(i)$. As it can be seen in Fig. 5, initially at $k = 0$ $\boldsymbol{h}_i^{(0)} = \boldsymbol{x}_i$ that represents the input features of the node v_i. For example, considering node v_1 in Fig. 5, we iterate the aggregation and update process of the node embedding for $k = 2$, so that the final learned representation is $z_1 = \boldsymbol{h}_1^{(2)}$, which includes the information about the 2-hop neighbourhood.

To summarise, we have a step that creates a message for a vertex v_i based on the knowledge about $N(i)$

$$\boldsymbol{m}_i^k = \delta^k\big(\phi^k(\boldsymbol{h}_i^{k-1}, \boldsymbol{h}_j^{k-1}, \boldsymbol{l}_{i,j}) : j \in N(i)\big),$$

where δ^k defines a differentiable, permutation invariant function of the kth convolutional layer, e.g., sum or average, and $\phi^{(k)}$ is a differentiable function that creates messages between the vertex i and the nodes in $N(i)$, incorporating the edge features $\boldsymbol{l}_{i,j}$. For instance, this could be a multi-layer perceptron or a complex filter function defined by Gaussian mixture models. In Fig. 5, these functions are represented by grey boxes. Similar to a filter in the CNN displayed in Fig. 4, the weights of the function $\phi^{(k)}$ stay unchanged for the whole input in the layer k. The message passing is followed by the update step

$$\boldsymbol{h}_i^k = \gamma^k(\boldsymbol{h}_i^{k-1}, \boldsymbol{m}_i^k),$$

where γ specifies another differentiable function–the activation function such as rectified linear unit (ReLU).

In the readout phase, we aggregate node features from the final iteration to obtain the entire graph representation $\boldsymbol{h}_G$

$$\boldsymbol{h}_G = \zeta(\boldsymbol{h}_i^K, v_i \in V),$$

where function ζ should satisfy the same conditions as δ, e.g., being invariant to graph isomorphism. An example could a particular pooling operation. This representation in then used for the final task, for instance, graph classification.

Depending on the types of graph convolutions (the creation and propagation of messages), we can categorise GCN into spectral-based and spatial-based models. The GCNs defined in the spectral domain (e.g., the Chebyshev Spectral GCN) are based on the graph Fourier transform, starting with the construction of the frequency filtering, whereas the spatial domain methods are specified directly on the graph, operating on groups of spatially close neighbours [54]. In the following section, we apply a spatial-based GCN with the Gaussian mixture model convolutional operator.

5 Simulation Study

To illustrate a possible combination of classical statistical tools together with machine learning algorithms for creating an enhanced network monitoring procedure, we consider an important and complex problem: the compliance of ambulance stations with the maximum allowed response time.

5.1 *Motivation*

There are examples of networks where failure could lead to irreparable harm. These networks can be defined as "system-relevant", where particular nodes represent ambulance or fire and rescue stations. To guarantee proper functionality, these services are obliged to satisfy a strict policy regarding the time limit for arriving at the accident epicentre. For instance, in emergency medical cases, the ambulance must reach the patient without exceeding the legally prescribed response time. Its fundamental part is appointed to the travelling time and in some places is not allowed to exceed 12 min in 95% of cases. The monitoring of the compliance with this rule can be performed by using the control chart for quantile function [17]. One of the potential choices is to create a test statistic based on the 0.95 and 0.97 quantiles so that the monitoring procedure corresponds to an early warning system and possible deviation towards the maximum limit is detected quicker. However, in case of a signal, it remains unclear what led to its occurrence unless we inspect the network state. One of the supportive methods in this task would be a GCN which can classify the network states in predefined categories, providing the first insight into a possible issue. This motivation guides the following application demonstrated on the simulated graph-structured data.

5.2 Network Definition

Consider a simplified road network shown in Fig. 6 whose topology is based on an existing city map. It can be graphically represented by $|V| = 18$ vertices and $|E| = 25$ edges. To differentiate between an ambulance node and a patient region, we introduce the vertex attribute "Role". We assume that the ambulance station can serve only one patient at once and that two fixed vertices in total define the ambulance stations. Also, which patient needs help from an ambulance is decided randomly. Thus, we include another vertex attribute "Involvement in an accident" that describes whether an ambulance station provides help (in this case, the value is set to 1) or is free (the value equals 0). Considering the patient nodes, as soon as the patient is involved in an accident, the value is set to 1. It remains 0 if no help is needed or obtained from the ambulance service. Both attributes are contained in $\boldsymbol{X}^V \in \mathbb{R}^2$ and can be found in Table 1.

Regarding the edges, we model two characteristics $\boldsymbol{L}^E \in \mathbb{R}^2$ that reflect distinct types of roads. continuous attribute defines travelling time in minutes (min) L_1^E (we select 1, 2, 3 and 5 min to be the expected values for passing respective roads), which is generated by applying lognormal distribution with different μ and σ. The selected values of μ and σ parameters displayed in Table 1 reflect the target state of the network. The second attribute L_2^E defines the level of construction works on the roads. Here the in-control state is dominated by the attribute with values 0 or 1, which means no or minor roadworks are observed.

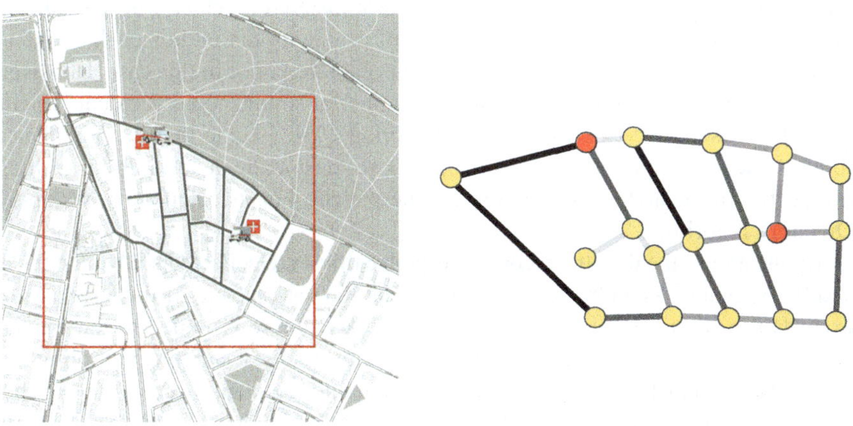

Fig. 6 An example of a small road network which is derived from an existing city map on Stamen Maps with two arbitrary placed ambulance stations (left) and its undirected graph representation (right). Different colours of edges replicate the travelling time along the road, where a darker colour means longer travelling time. The red nodes define ambulance stations

Table 1 Edge and nodal attributes

Type	Attribute	Value
Edge	Travelling time (in minutes)	$\mathbb{E}(1) \sim \mathscr{F}(0.1, 0.05^2)$
		$\mathbb{E}(2) \sim \mathscr{F}(0.7, 0.05^2)$
		$\mathbb{E}(3) \sim \mathscr{F}(1.1, 0.05^2)$
		$\mathbb{E}(5) \sim \mathscr{F}(1.6, 0.05^2)$
		with $\mathscr{F}(\cdot) = \text{Lognormal}(\mu, \sigma^2)$
	Level of road blocking due to construction work	Free: 0
		Low: 1
		Middle: 2
		High: 3
Node	Role	0: Patient
		1: Ambulance
	Involvement in an accident	0: No involvement
		1: Help is provided, obtained or needed

5.3 Generation of the Response Time Data

For monitoring the travelling time from the ambulance station to the patients, we make some assumptions considering the simulation of the response time data. Daily, as soon as there is a patient call with the need for help, the travelling time of the ambulance which is closer to a patient is registered together with the current network situation. Sometimes, the number of accidents can be higher than one at the same time, so that the network situation is captured once in this case. However, if the network is in control, the maximum number of simultaneous accidents equals two, otherwise, there exists a personnel shortage. We assume that the ambulance follows the most efficient route, being the shortest path in terms of travelling time between the ambulance station and the patient. For its calculation, we apply Dijkstra's algorithm.

By the end of the day, the recorded response times are collected so that the 95 % and 97 % quantiles can be derived for defining the test statistic. If the test statistic exceeds the control limit, the collected network data are provided to the trained GCN that classifies the scenes into four different groups: a stable condition of the road network (label 0), an unstable condition due to the manpower shortage (label 1), an unstable condition due to the construction works (label 2) and an unstable condition due to the traffic jams (label 3). It is important to include the label 0 graphs which dominate in the definition of the in-control state for the identification of possible false alarms. To proceed with the application of the control chart itself, we first will explain how different label groups were designed.

5.4 Road Conditions

To discuss the four condition classes, we should concentrate on the problem from the angle of what the neural network should learn in terms of the main differences between the specified scenarios. Considering the reliable state, the neural network needs to distinguish between the problems on the roads which affect the travelling time so that it potentially leads to the out-of-control case and which not, as the patient was still reached on time. It means, despite the obstacles on the roads that can be modelled by increased values of edge attributes, each time the patient was reached by ambulance in or under 12 min (simply because the route to the patient was not affected considerably), the graph obtains the label "0".

The class with the label "1" defines the problem of manpower shortage, meaning the reason for longer travelling times is due to the imbalance in the capacity of the ambulance service and the number of patients who needs help. In this case, we generate a higher number of patients, i.e., more nodes with "Role = 0" are involved into an accident ($X_2^V = 1$). As soon as both ambulance stations provide help and some further patients are not treated yet, the travelling time to these nodes is calculated as $\mathbb{E}(L_1^E) \cdot 2 \cdot 2$, where $\mathbb{E}(L_1^E)$ defines the expected value of edge feature "Travelling time" multiplied with the number of roads to pass on average and the need to travel first to the ambulance station and then to the patient. In this case, the network is out-of-control due to the considerably increased travelling time to the third and further patients.

Creating the unreliable situation on roads due to construction works (label 2), a particular group of roads which come from one or several distributions considering the travelling time is selected and higher values of the second edge attribute L_2^E are assigned to these roads. Thus, the μ and σ parameters are changed so that the higher attribute value corresponds to a longer travelling time along the road.

For modelling a traffic jam (label 3), L_2^E is set low (0 or 1), and some values describing the travelling time of specific roads are generated from a different distribution that implies their increase. The examples from the groups with labels 2 and 3 do not necessarily lead to an out-of-control state if the patients are still reached under the critical time prescription.

After defining the network composition and specifying possible in- and out-of-control scenarios, we can collect the quantile observations for the calibration of the control chart and the graph representations of different classes in order to train the neural network.

5.5 Calibration of the Control Chart for Quantile Function Values

To calculate daily 95% and 97% quantile values, we randomly simulate between 10 and 100 accidents which are repeatedly assigned to different patients. Next, the shortest paths between an available ambulance station and the selected patient are found and saved. Using the control chart presented in [17], the test statistic is defined as follows

$$a_t = (\hat{\boldsymbol{Q}}_t - \boldsymbol{Q}_0)'\boldsymbol{\Sigma}_0^{-1}(\hat{\boldsymbol{Q}}_t - \boldsymbol{Q}_0),$$

where $\hat{\boldsymbol{Q}}_t = \left(\hat{Q}_{0.95,t}, \hat{Q}_{0.97,t}\right)'$ and the length of $\hat{\boldsymbol{Q}}_t$ is denoted by c.

In Phase I, the expected value $\boldsymbol{Q}_0$ is estimated by the mean $\bar{\boldsymbol{Q}}$ and $\boldsymbol{\Sigma}_0$ by the sample covariance matrix $\boldsymbol{S}$ with 2500 in-control samples. For sufficiently large number of samples, a_t follows the χ^2 distribution with c degrees of freedom, if the sample at time point t corresponds to the specified in-control state. Hence, the control limit can be defined by $\chi^2_\alpha(c)$, selecting α with respect to the in-control average run length (ARL) using $\alpha = 1/ARL$. Here, we choose ARL = 1000, therefore, $\chi^2_{0.001}(2) = 13.816$.

5.6 *Construction and Training of the Graph Convolutional Network*

In this simulation study, we are interested in the classification of collected graphs that belong to a change point. The goal is to assign a given graph to one of the predefined categories by learning the feature representation from provided training data which contain class labels. Consequently, we have to define the GCN architecture so that it can solve the specified task. Also, our graph convolutional operator should be capable to integrate the node, as well as edge attributes into the message passing process because they encompass valuable information about the network's condition.

Figure 7 presents the architecture of the applied GCN. The first three graph convolutional layers, each encoding the input in a feature vector of size 18×10, perform three propagation steps and effectively convolve the 3rd-order neighbourhood of every node. We chose the Gaussian mixture model convolutional operator described in Monti et al. [55] which is implemented in the programming framework provided by Fey and Lenssen [56]. Each convolutional layer is followed by the ReLU activation function. Afterwards, the dropout operation is applied, which randomly sets the processed input units to 0 with a specified frequency ξ (in our case, $\xi = 25\%$) during the training time, preventing the model from overfitting, i.e., learning from the training dataset without its generalisation. Before the following convolution begins, we normalise the inputs across the features; this technique is known as "layer normalisation" (cf. [57]).

After the message passing phase, a readout layer that is defined by a global mean pooling operation transforms the latent vertex representations to a graph representation as a fixed-size vector. Here, the interim output is averaged across each hidden node dimension so that the graph-level output size is 1×10. Next, we attach

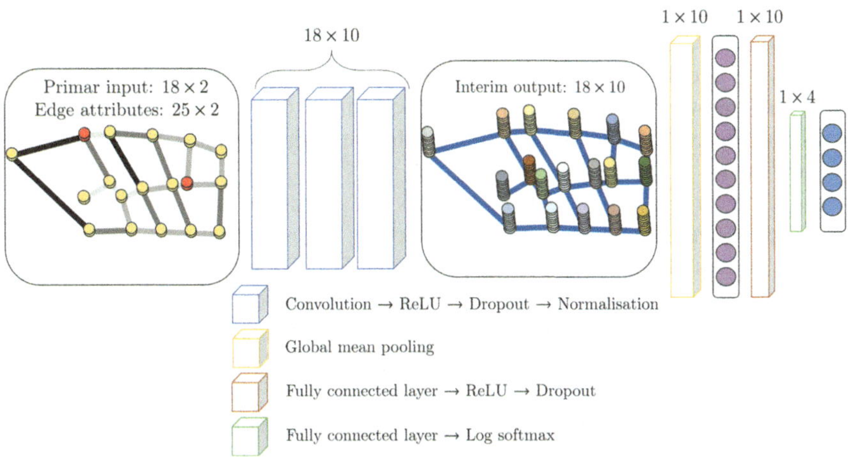

Fig. 7 The schematic architecture of the applied GCN. Each block represents a single layer where the first stage (blue blocks) contains graph convolutional layers with layer normalisations for learning the feature representation and the second stage (yellow, orange and green blocks) consists of dense layers for classification

two fully connected layers to increase the ability to learn a complex function and solve the classification task. Consequently, the second layer predicts the final class probability distribution of size 1×4 followed by the softmax activation function. We cannot apply the ReLU activation function as it provides continuous output in range $[0; \infty]$. In the final stage, we need the output to be in the finite range $[0; 1]$ for interpreting its results as probabilities, with the highest value corresponding to the predicted class.

After defining the architecture of the GCN, we can start with training or fitting the neural network. This procedure involves the usage of a training dataset to update the model parameters (weights and biases) so that we obtain a reliable mapping between input (graph) and output (class label). For the training dataset, we generate 2500 graphs. It is important to avoid class imbalance during the training process, therefore, each label is represented by the same number of examples. Another vital part of the training process is the loss function. It calculates the difference between the computed output from the input data (this process is known as "forward pass") and the value provided as ground truth. Here, we choose the negative log-likelihood loss, which is appropriate for a multiclass classification problem. It defines the objective function that we minimise by updating the model parameters.

The results that are provided by the loss function are applied in the optimisation step of our parameters, which are based on gradient computation (known as "backpropagation" or "backward pass"). The negative log-likelihood is minimised using the Adaptive Moment Estimation (Adam) function [58] with a learning rate of 10^{-3}.

The execution of the backward and forward pass together defines one iteration. During one iteration, we usually pass a subset of the dataset known as "mini-batch". In case we decide to pass all data at once, it is called "batch". Here, we train the neural network using the mini-batch with size 16, i.e., in every iteration, 16 graphs are processed together. As soon as the entire dataset was passed, one epoch is completed.

As a performance metric that supports the selection of the best model, we compute the weighted F-score after each epoch. Figure 8 (left side) illustrates the training and validation history of the applied GCN. To test how well the network generalises to unseen data, we apply the holdout validation method. The validation set, which contains more complex samples, i.e., new examples which belong to the classes but are not included in the training dataset, was designed with a size of 800 graphs.

In order not to overtrain the network, we use early stopping after the 100th epoch was reached with respect to the F-score improvement of the validation dataset, which terminates the training process if the value has not increased within ten epochs. We find the optimal model to be at epoch 101 with 93.4% and 87.5% the weighted F-score of the training and the validation dataset, respectively. However, to see whether the model operates correctly, we need to test it on a new dataset coming from the monitoring procedure.

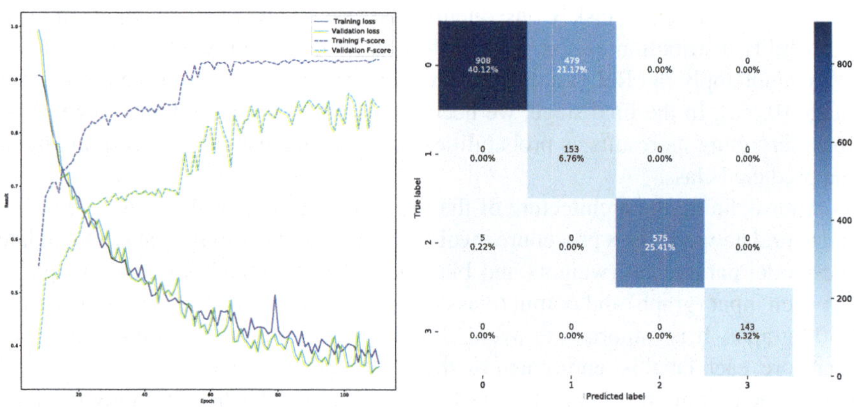

Fig. 8 The training progress (left) shown on the training (blue curves) and validation sets (green curves). The confusion matrix (right) presents the performance of the trained GCN in Phase II. The numbers on the diagonal define the proportions of correctly classified examples (compared to the size of the complete test dataset), and the off-diagonal entries correspond to the proportions of the misclassified graphs

5.7 *Phase II Analysis*

Here, we combine the implementation of Phase II with testing the trained GCN. We define the length of the monitoring period to be 100 days, where the network in the first 30 and the last 10 days is considered to be in control. The out-of-control period is designed in the remaining days, where the process is exposed to the personnel shortage (10 days), excessive construction works (30 days) and increase of traffic jams (20 days). After simulating the cases and calculating the quantiles, we obtained the control chart presented in Fig. 9. In terms of potential false signals, there are several points that are close to the control limit. The possible reason may be the high variance in the in-control data. We can also notice that not all the test statistics show the out-of-control state in the period when the network was exposed to an increased number of traffic jams. However, they do not define missing signals; as it was mentioned in Sect. 5.4, if the ambulance services were still able to reach the patients within the allowed time, then no out-of-control state is given.

Normally, we would apply the neural network only in the out-of-control state for gaining insight into the cause. Nevertheless, the primary aim here is to evaluate the performance of the trained GCN to classify provided graph observations in general. Hence, we create a test dataset using the data from Phase II displayed in Fig. 9, i.e., from the 100 generated days that include both in-control and out-of-control periods

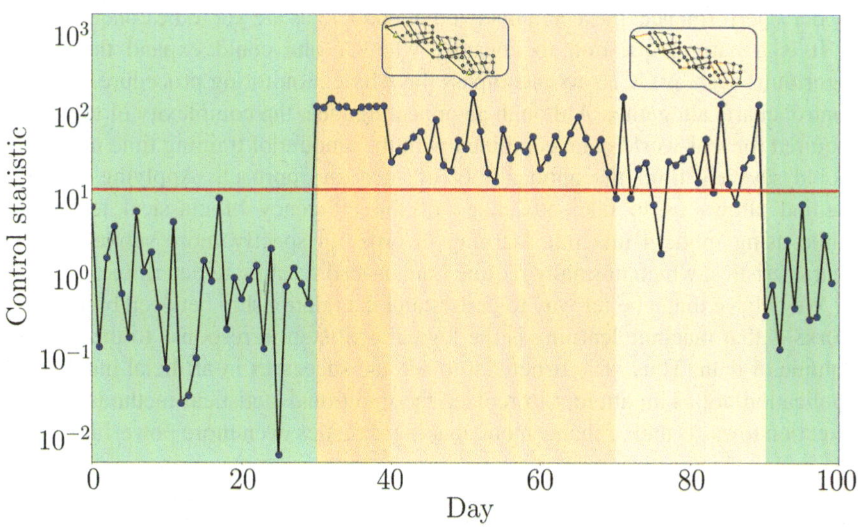

Fig. 9 The control chart for quantile function values on a logarithmic scale. The horizontal red line corresponds to the control limit. The green areas are designed by the label 0 cases, the dark orange by the label 1, followed by the label 2 and 3. The incorporation of the graphs with different properties such as construction works (triangular symbols) or traffic jams (orange coloured edges) defines the availability of additional information to understand the reason of the detected change point

and examples from each of the four classes. As we can see in Fig. 8 (right side), the GCN can almost flawlessly identify the classes 1, 2 and 3. However, class 0 seems to be not well learned, possibly due to lacking clarity in its representation. Overall, the model achieves the weighted F-score of 82.9% being an encouraging result.

6 Conclusion and Discussion

In many applications, treating the underlying data as a graph can achieve greater efficiency. However, data representation in the form of graphs is still novel for both machine learning and statistics. Hence, it is particularly important to use synergy effects between two different statistical learning frameworks to develop efficient and modern analytical approaches. In this chapter, we uncover the possibility to bring together statistical process control and deep learning algorithms to monitor graph-structured data.

Learnable models which operate on graphs are only a stepping stone on the path toward a significant expansion in understanding the environment. Besides the topic of how to unify both frameworks, from the statistical perspective, there are many other open questions in this area. How to represent the graph data and convolve the information in a unified form, how to identify a corresponding approach to a specific problem, how to decide on the proportions of both fields and how to measure the model's performance: these and many other challenges are yet to be conquered.

It is a natural question to consider whether one could expand the use of algorithms such as GCNs to encompass the whole monitoring procedure, omitting control charts altogether. Although an appealing idea, the complexity of the model required for real-world data, combined with the amount of training time necessary, would severely limit the applicability of such an approach. Applying a hybrid method allows us to take advantage of the efficiency of classical techniques while using modern machine learning in order to specify more subtle network characteristics, which normally require human-lead scrutiny to determine.

We believe that a better way to understand the relationship between both frameworks is that machine learning is the logical next step in response to the growing volume of data. Thus, it is beneficial to see the successes in artificial intelligence application not as an attempt to replace the traditional statistical methods but as a direction towards their enhancement, making statistics even more powerful.

Acknowledgments This research is supported by the German Research Foundation within the project 412992257. The authors thank Artem Leichter and Thomas Cope for valuable insight and fruitful discussions.

References

1. Erdös, P., Rényi, A.: On random graphs, I. Publ. Math. **6**, 290–297 (1959)
2. Gilbert, E.N.: Random graphs. Ann. Math. Stat. **30**, 1141–1144 (1959)
3. Grennan, K.S., Chen, C., Gershon, E.S., Liu, C.: Molecular network analysis enhances understanding of the biology of mental disorders. Bioessays Wiley Online Library **36**(6), 606–616 (2014)
4. O'malley, A.J., Marsden, P.V.: The analysis of social networks. Health Serv. Outcome Res. Methodol. **8**(4), 222–269 (2008)
5. Schweitzer, F., Fagiolo, G., Sornette, D., Vega-Redondo, F., Vespignani, A. and White, D. R.: Economic networks: the new challenges. Sci. Am. Assoc. Adv. Sci. **325**(5939), 422–425 (2009)
6. Hastie T., Tibshirani R., Friedman J.: The Elements of Statistical Learning: Data Mining, Inference, and Prediction. Springer Science & Business Media, Berlin (2009)
7. Kang, B.S., Park, S.C.: Integrated machine learning approaches for complementing statistical process control procedures. Decis. Support. Syst. **29**(1), 59–72 (2000)
8. Psarakis, S.: The use of neural networks in statistical process control charts. Qual. Reliab. Eng. Int. **27**(5), 641–650 (2011)
9. Fountoulaki, A., Karacapilidis, N., Manatakis, M.: Augmenting statistical quality control with machine learning techniques: an overview. Int. J. Bus. Syst. Res. **5**(6), 610–626 (2011)
10. Demircioglu Diren, D., Boran, S., Cil, I.: Integration of machine learning techniques and control charts for multivariate processes. Sci. Iran. **27**(6), 3233–3241 (2020)
11. Khoza, S.C., Grobler, J.: Comparing machine learning and statistical process control for predicting manufacturing performance. In: EPIA Conference on Artificial Intelligence, pp. 108–119. Springer, Berlin (2019)
12. Zan, T., Liu, Z., Su, Z., Wang, M., Gao, X., Chen, D.: Statistical process control with intelligence based on the deep learning model. Appl. Sci. **10**(1), 308 (2020)
13. Apsemidis, A., Psarakis, S.: Support vector machines: a review and applications in statistical process monitoring. In: Data Analysis and Applications 3: Computational, Classification, Financial, Statistical and Stochastic Methods, vol. 5, pp. 123–144 (2020)
14. Chen, L., Yu, T., Liu, M.: A semantic graph model. In: OTM Confederated International Conferences "On the Move to Meaningful Internet Systems", pp. 378–396. Springer, Berlin (2015)
15. Brevier, G., Rizzi, R., Vialette, S.: Pattern matching in protein-protein interaction graphs. In: International Symposium on Fundamentals of Computation Theory. Springer, Berlin (2007)
16. Montgomery, D.C.: (2009) Introduction to Statistical Quality Control. John Wiley & Sons, Inc., Hoboken
17. Grimshaw, S.D., Alt, F.B.: Control charts for quantile function values. J. Qual. Technol. **29**(1), 1–7 (1997)
18. Kanji, G.K., Arif, O.H.: Median rankit control chart by the quantile approach. J. Appl. Stat. **27**(6), 757–770 (2000)
19. Ning, X., Wu, C.: Improved design of quantile-based control charts. J. Chin. Inst. Indust. Eng. **28**(7), 504–511 (2011)
20. Park, K., Jung, D., Kim, J.M.: Control charts based on randomized quantile residuals. Appl. Stoch. Model. Bus. Ind. **36**(4), 716–729 (2020)
21. Hwang, W.Y.: Quantile-based control charts for poisson and gamma distributed data. J. Korean Stat. Society **50**(4), 1129–1146 (2021)
22. McCulloh, I., Carley, K.M.: Detecting change in longitudinal social networks. Technical report. Military Academy West Point NY Network Science Center (NSC) (2011)
23. Wilson J.D., Stevens, N.T., Woodall W.H.: Modeling and detecting change in temporal networks via the degree corrected stochastic block model. Qual. Reliab. Eng. Int. **35**(5), 1363–1378 (2019)
24. Hosseini, S.S., Noorossana, R.: Performance evaluation of EWMA and CUSUM control charts to detect anomalies in social networks using average and standard deviation of degree measures. Qual. Reliab. Eng. Int. **34**(4), 477–500 (2018)

25. Sparks R., Wilson J.D.: Monitoring communication outbreaks among an unknown team of actors in dynamic networks. J. Qual. Technol. **51**(4), 353–374 (2019)
26. Malinovskaya, A., Otto, P.: Online network monitoring. Stat. Methods Appl. **30**(5), 1337–1364 (2021)
27. Farahani E.M., Baradaran Kazemzadeh R., Noorossana R., Rahimian G.: A statistical approach to social network monitoring. Commun. Stat. Theory Methods **46**(22), 11,272–11,288 (2017)
28. Noorossana R., Hosseini, S.S., Heydarzade, A.: An overview of dynamic anomaly detection in social networks via control charts. Qual. Reliab. Eng. Int. **34**(4), 641–648 (2018)
29. Azarnoush, B., Paynabar, K., Bekki, J., Runger, G.: Monitoring temporal homogeneity in attributed network streams. J. Qual. Technol. **48**(1), 28–43 (2016)
30. Miller, B.A., Arcolano, N., Bliss, N.T.: Efficient anomaly detection in dynamic, attributed graphs: emerging phenomena and big data. In: 2013 IEEE International Conference on Intelligence and Security Informatics, pp. 179–184. IEEE, Piscataway (2013)
31. Heard, N.A., Weston, D.J., Platanioti, K., Hand, D.J.: Bayesian anomaly detection methods for social networks. Ann. Appl. Stat. **4**(2), 645–662 (2010)
32. Priebe, C.E., Conroy, J.M., Marchette, D.J., Park, Y.: Scan statistics on enron graphs. Comput. Math. Organ. Theory **11**(3), 229–247 (2005)
33. Woodall, W.H., Zhao, M.J., Paynabar, K., Sparks, R., Wilson, J.D.: An overview and perspective on social network monitoring. IISE Trans. **49**(3), 354–365 (2017)
34. Akoglu, L., Tong, H., Koutra, D.: Graph based anomaly detection and description: a survey. Data Min. Knowl. Disc. **29**(3), 626–688 (2015)
35. Ranshous, S., Shen, S., Koutra, D., Harenberg, S., Faloutsos, C., Samatova, N.F.: Anomaly detection in dynamic networks: a survey. Wiley Interdiscip. Rev. Comput. Stat. **7**(3), 223–247 (2015)
36. Zhao, M.J., Driscoll, A.R., Sengupta, S., Stevens, N.T., Fricker Jr, R.D., Woodall, W.H.: The effect of temporal aggregation level in social network monitoring. PLOS One **13**(12) (2018)
37. Wang, X., Du, Y., Cui, P., Yang, Y.: OCGNN: one-class classification with graph neural networks. arXiv preprint: 2002.09594 (2020)
38. Zheng, L., Li, Z., Li, J., Li, Z., Gao, J.: AddGraph: anomaly detection in dynamic graph using attention-based temporal GCN. IJCAI 4419–4425 (2019)
39. Kumagai, A., Iwata, T., Fujiwara, Y.: Semi-supervised anomaly detection on attributed graphs. In: 2021 International Joint Conference on Neural Networks (IJCNN), pp. 1–8. IEEE, Piscataway (2021)
40. Hamilton, W.L., Ying, R., Leskovec, J.: Representation learning on graphs: methods and applications. arXiv preprint: 1709.05584 (2017)
41. Hamilton, W.L.: Graph representation learning. Synth. Lect. Artif. Intell. Mach. Learn. **14**(3), 1–159 (2020)
42. Cai, H., Zheng, V.W., Chang, K.C.C.: A comprehensive survey of graph embedding: problems, techniques, and applications. IEEE Trans. Knowl. Data Eng. **30**(9), 1616–1637 (2018)
43. Chen, F., Wang, Y.C., Wang, B., Kuo, C.C.J.: Graph representation learning: a Survey. APSIPA Trans. Signal Inform. Process. **9** (2020)
44. Gogoglou, A., Bruss, C.B., Nguyen, B., Sarshogh, R., Hines, K.E.: Quantifying challenges in the application of graph representation learning. In: 2020 19th IEEE International Conference on Machine Learning and Applications (ICMLA), pp. 1519–1526. IEEE, Piscataway (2020)
45. Goodfellow, I., Bengio, Y., Courville, A.: Deep Learning. MIT Press, Cambridge (2016)
46. Calin, O.: Deep Learning Architectures. Springer International Publishing, New York City (2020)
47. Scarselli, F., Gori, M., Tsoi, A.C., Hagenbuchner, M., Monfardini, G.: The graph neural network model. IEEE Trans. Neural Netw. **20**(1), 61–80 (2008)
48. Kipf, T.N., Welling, M.: Semi-supervised classification with graph convolutional networks. In: International Conference on Learning Representations, ICLR (2017)
49. Asif, N.A., Sarker, Y., Chakrabortty, R.K., Ryan, M.J., Ahamed, M.H., Saha, D.K., Badal, F.R., Das, S.K., Ali, M.F., Moyeen, S.I., Islam, M.R.: Graph neural network: a comprehensive review on Non-euclidean space. IEEE Access **9**, 60588–60606 (2021)

50. Bronstein, M.M., Bruna, J., LeCun, Y., Szlam, A., Vandergheynst, P.: Geometric deep learning: going beyond Euclidean data. IEEE Signal Process. Mag. **34**(4), 18-42 (2017)
51. Gilmer, J., Schoenholz, S.S., Riley, P.F., Vinyals, O., Dahl, G.E.: Neural message passing for quantum chemistry. In: International Conference on Machine Learning (2017)
52. Wu, Z., Pan, S., Chen, F., Long, G., Zhang, C., Philip, S.Y.: A comprehensive survey on graph neural networks. IEEE Trans. Neural Netw. Learn. Syst. **32**(1), 4–24 (2020)
53. Zhou, J., Cui, G., Hu, S., Zhang, Z., Yang, C., Liu, Z., Wang, L., Li, C., Sun, M.: Graph neural networks: a review of methods and applications. AI Open **1**, 57–81 (2020)
54. Zhang, S., Tong, H., Xu, J., Maciejewski, R.: Graph convolutional networks: a comprehensive review. Comput. Soc. Netw. **6**(1), 11 (2019)
55. Monti, F., Boscaini, D., Masci, J., Rodola, E., Svoboda, J., Bronstein, M.M.: Geometric deep learning on graphs and manifolds using mixture model CNNs. In: Proceedings of the IEEE Conference on Computer Vision and Pattern Recognition, pp. 5115–5124 (2017)
56. Fey, M., Lenssen, J.E.: Fast graph representation learning with PyTorch Geometric. arXiv preprint: 1903.02428 (2019)
57. Ba, J.L., Kiros, J.R., Hinton, G.E.: Layer normalization. arXiv preprint: 1607.06450 (2016)
58. Kingma, D.P., Ba, J.L.: Adam: a method for stochastic optimization. In: Proceedings of the 3rd International Conference on Learning Representations (ICLR) (2014)

[illegible]

On the Robustness of Kernel-Based Pairwise Learning

Patrick Gensler and Andreas Christmann

Abstract It is shown that many results on the statistical robustness of kernel-based pairwise learning can be derived under basically no assumptions on the input and output spaces. In particular, neither moment conditions on the conditional distribution of Y given $X = x$ nor the boundedness of the output space is needed. We obtain results on the existence and boundedness of the influence function and show qualitative robustness of the kernel-based estimator. The present paper generalizes results by Christmann and Zhou [11] by allowing the prediction function to take two arguments and can thus be applied in a variety of situations such as ranking, similarity learning and distance metric learning.

Keywords Kernel methods · Machine learning · Support vector machines · Robust statistics

1 Introduction

Stute [45, 46] showed the (universal) consistency of conditional U-statistics under weak conditions. Based on these results [14] reused U-statistics in the field of statistical learning theory and more precisely for the ranking problem. The present paper is connected with these papers in the field of statistical machine learning from a general point of a view. That is, for general pairwise loss functions and general kernels.

As mentioned above, an example of pairwise learning is ranking based problems that can be simplified to a situation where profiles of entities are given and have to be compared against each other to find the order of these entities in a particular case. For instance, an employer is interested in two candidates and wants to select the "better" one for the company considering their applications and the experience

P. Gensler (✉) · A. Christmann
Department of Mathematics, University of Bayreuth, Bayreuth, Germany
e-mail: patrick.gensler@uni-bayreuth.de; andreas.christmann@uni-bayreuth.de

A. Steland, K.-L. Tsui (eds.), *Artificial Intelligence, Big Data and Data Science in Statistics*, https://doi.org/10.1007/978-3-031-07155-3_5

of the employer from former employees. The same problem in another setting can be found in a lot of fields such as insurance companies, banks, product marketing, etc.

In the field of statistical machine learning theory, one approach is kernel-based methods such as support vector machines, see, e.g., [47, 48] for classical textbooks on this subject. The field of kernel-based learning methods has been widely researched, we refer to, for instance [15, 16, 41, 44]. Kernel-based pairwise learning methods were studied by e.g. [11]. They showed the statistical robustness of pairwise learning methods in the sense of bounded influence functions and qualitative robustness, as introduced by Hampel [28], Hampel et al. [29] and generalized by Cuevas [17]. The difference compared to classic support vector machines is that the loss function does not only have three arguments $(x, y, f(x))$, where x is from the input space $\mathcal{X}$, y from the output space $\mathcal{Y}$ and the prediction $f(x)$, but rather six x, y, x', y' as pairwise components –what explains the modified name pairwise loss functions– and the predictions $f(x)$ and $f(x')$.

In this article, we analyze several statistical robustness properties of kernel-based pairwise learning methods based on pairwise loss functions $L(x, y, x', y', f(x, x'))$ that take five arguments with the real valued prediction function taking two arguments $f(x, x')$. This is an additional generalization compared to [11] as the difference $\tilde{f}(x) - \tilde{f}(x') = f(x, x')$ investigated in the mentioned paper is a special case for a *"bivariate" prediction function* with $\tilde{f}$ being a suitable function.

Other branches where the generalized theory discussed in this paper can be applied are *similarity learning* and *distance metric learning*, see Example 2.4 and for references see, e.g., [38], in the classification setting [24] and for results on clustering [50]. The before mentioned ranking problem analyzed by Clémençon et al. [14] yields the motivation to learn functions $f(x, x')$ that induce ranking rules $r(x, x')$, hence deciding for x or x' depending on the sign of the value of f. The context between similarity learning and ranking has been studied by, for instance [9, 34].

Rejchel [39, p. 1375] investigated the ranking problem in a similar way, but considered a uniformly bounded function class with $f(x, x') = -f(x', x)$ or parametrized ranking rules $f(x, x') = \theta^T(x - x')$ in the regression setting with $\mathcal{X} = \mathbb{R}^d, \mathcal{Y} = \mathbb{R}$ and parameter $\theta \in \mathbb{R}^d$ ([40, p. 6]). In the context of online learning, we refer to [25, 51] and for metric learning to [2, 7] and the references cited therein.

Using *shifted loss functions*, see Definition 2.16, to tackle the robustness problem for support vector machines in the case of heavy-tailed distributions was already done by Christmann et al. [12] using an idea going back at least to Huber [31] and is applied here in the context of regularized pairwise learning in order to be able to compute prediction functions without any moment assumption on the output variable Y. To be more precise, shifted loss functions are a technical tool to avoid moment conditions for the conditional distribution of Y given $X = x$ without changing the estimator, if the estimator exists based on the unshifted loss function.

The paper is organized as follows. Section 2 introduces the necessary mathematical prerequisites. Readers familiar with kernel-based pairwise learning can skip this section. Section 3 presents the main results: a general representer theorem and the robustness of the kernel-based regularized pairwise learning method. Section 4 gives a discussion and an outlook for further research topics in this field. Additional theorems and lemmas, as well as proofs for our results are listed in the Appendix A.1 or Appendix A.2, respectively.

2 Mathematical Prerequisites

In this section we collect some definitions and results which are useful to study regularized pairwise learning. If not mentioned otherwise, we equip topological spaces $(\mathcal{Z}, \tau_{\mathcal{Z}})$ with their Borel σ-algebras $\mathcal{B}(\mathcal{Z})$. For brevity we denote the set of all Borel probability measures on a topological space $(\mathcal{Z}, \tau_{\mathcal{Z}})$ as $\mathcal{M}_1(\mathcal{Z})$ instead of $\mathcal{M}_1(\mathcal{Z}, \mathcal{B}(\mathcal{Z}))$. We denote the set of all continuous bounded function $f : \mathcal{Z} \to \mathbb{R}$ by $C_b(\mathcal{Z})$.

As mentioned in the introduction, let the input space $\mathcal{X}$, and the output space $\mathcal{Y}$, be topological spaces. Let (X, Y) and (X_i, Y_i), for $i \in \mathbb{N}$, be independent and identically distributed pairs of random quantities with values in $\mathcal{X} \times \mathcal{Y}$ and unknown distribution $\mathrm{P} \in \mathcal{M}_1(\mathcal{X} \times \mathcal{Y})$. We denote the realizations of (X_i, Y_i) by (x_i, y_i) for $i \in \mathbb{N}$. A data set is given by $D_n = \big((x_1, y_1), \ldots, (x_n, y_n)\big) \in (\mathcal{X} \times \mathcal{Y})^n$. Please note that we denote a data set as an n-tuple and not as a set because we allow that not all data points are different.

We define the set of measurable functions $f : \mathcal{Z} \to \mathbb{R}$ by $\mathcal{L}_0(\mathcal{Z})$. The set of all measurable functions $f : \mathcal{X} \times \mathcal{Y} \to \mathbb{R}$ satisfying $\int |f| \,\mathrm{dP} < \infty$, and therefore P-integrable, is defined by $\mathcal{L}_1(\mathcal{X} \times \mathcal{Y}, \mathrm{P})$ (or short $\mathcal{L}_1(\mathrm{P})$, if the domain is obvious from the context) with $\mathrm{P} \in \mathcal{M}_1(\mathcal{X} \times \mathcal{Y})$. For all $f \in \mathcal{L}_0(\mathcal{X} \times \mathcal{Y})$ that are almost surely bounded, given a probability measure $\mathrm{P} \in \mathcal{M}_1(\mathcal{X} \times \mathcal{Y})$, we write $\mathcal{L}_\infty(\mathcal{X} \times \mathcal{Y}, \mathrm{P})$ (or short $\mathcal{L}_\infty(\mathrm{P})$).

Definition 2.1 Let $(\mathcal{X}, \mathcal{A})$ be a measurable space and $\mathcal{Y} \subset \mathbb{R}$ be a closed subset. Then a measurable function $L : (\mathcal{X} \times \mathcal{Y})^2 \times \mathbb{R} \to [0, \infty)$ is called a **pairwise loss function** or **pairwise loss** in short.

Example 2.2 An example of a pairwise loss function is given by Clémençon et al. [14, p. 863] utilizing an auxiliary measurable function $\phi : \mathbb{R} \to [0, \infty)$ satisfying the following two conditions: $\phi(0) = 1$ and $\phi(x) \geq 1$ for all $x \in [0, \infty)$. It is then possible to define the loss function

$$L(x, y, x', y', f(x, x')) := \phi\big(-\operatorname{sign}\big(y - y'\big)\, f(x, x')\big),$$

with $\operatorname{sign}(0) := 0$ and ϕ chosen as, for instance, the exponential function $\exp(x)$, the function $\log_2(1 + \exp(x))$ or the hinge loss function with $\max\{0, 1 + x\}$. Replacing the sign function by a differentiable surrogate function leads to the following "smoothed" pairwise loss function

$$L_\sigma(x, y, x', y', f(x, x')) := \phi\big(-\tanh\big(\sigma^{-1}(y - y')\big) f(x, x')\big),$$

with an arbitrary small $\sigma > 0$.

Example 2.3 Another example for a pairwise loss function is the **least squares ranking loss** used by Chen et al. [10, p. 55]

$$L(x, y, x', y', f(x, x')) := (y - y' - f(x, x'))^2,$$

with $f(x, x') := \tilde{f}(x) - \tilde{f}(x')$ for a univariate prediction function $\tilde{f} : \mathcal{X} \to \mathbb{R}$.

Example 2.4 The advantage of learning functions with two arguments can be seen in the similarity learning and distance metric learning, see, for instance [38].

In similarity learning functions $f : \mathcal{X}^2 \to \mathbb{R}$ of the form $f(x, x') = x^T M x'$ with M a matrix have to be learned, whereas in most methods in distance based learning the functions have the form $f(x, x') = (x - x')^T N (x - x')$ with N a positive semidefinite matrix.

Both functions cannot be expressed as a difference like in the example above which emphasizes the further generalization in the present paper.

We now define several quantities we will need to introduce our kernel-based pairwise learning.

Definition 2.5 Let $L : (\mathcal{X} \times \mathcal{Y})^2 \times \mathbb{R} \to [0, \infty)$ be a pairwise loss function, $\mathrm{P} \in \mathcal{M}_1(\mathcal{X} \times \mathcal{Y})$, and $\mathrm{P}^2 = \mathrm{P} \otimes \mathrm{P}$ denoting the product measure of P.

(a) Then, for a measurable function $f : \mathcal{X}^2 \to \mathbb{R}$, the L-**risk** is defined by

$$\begin{aligned}\mathcal{R}_{L,\mathrm{P}}(f) &:= \mathbb{E}_{\mathrm{P}^2}\left[L(X, Y, X', Y', f(X, X'))\right] \\ &= \int_{(\mathcal{X}\times\mathcal{Y})^2} L(x, y, x', y', f(x, x'))\mathrm{d}\mathrm{P}^2(x, y, x', y').\end{aligned}$$

(b) Given a data set $D_n = \big((x_1, y_1), \ldots, (x_n, y_n)\big) \in (\mathcal{X} \times \mathcal{Y})^n$ and the corresponding empirical measure $\mathrm{D}_n = \frac{1}{n}\sum_{i=1}^n \delta_{(x_i, y_i)}$, the **empirical** L-**risk** is defined by

$$\begin{aligned}\mathcal{R}_{L,\mathrm{D}_n}(f) &= \mathbb{E}_{\mathrm{D}_n^2}\left[L(X, Y, X', Y', f(X, X'))\right] \\ &= \frac{1}{n^2}\sum_{i=1}^n\sum_{j=1}^n L(x_i, y_i, x_j, y_j, f(x_i, x_j)).\end{aligned}$$

(c) The minimal L-risk

$$\mathcal{R}^*_{L,\mathrm{P}} := \inf_{f \in \mathcal{L}_0(\mathcal{X}^2)} \mathcal{R}_{L,\mathrm{P}}(f)$$

is called the **Bayes risk** and a measurable minimizer $f_{L,\mathrm{P}} : \mathcal{X}^2 \to \mathbb{R}$ is called a **Bayes decision function**, if it exists.

Please note that D_n denotes a data set and D_n denotes the corresponding empirical measure. The random empirical measure $\mathbb{D}_n$ is given by $\mathbb{D}_n := \frac{1}{n}\sum_{i=1}^n \delta_{(X_i,Y_i)}$.

In later sections of this paper, we will need finite L-risks which can be achieved by using bounded and measurable kernels. Furthermore, shifted loss functions, which will be defined in Definition 2.16, help us to avoid *any* moment conditions on Y and therefore allow us to define and investigate our pairwise machine learning methods for *all* probability measures P.

Remark 2.6 If $(\mathcal{X}, \tau)$ is a Polish space (with topology τ) and $\mathcal{Y} \subset \mathbb{R}$ is closed, then $\mathcal{X} \times \mathcal{Y}$ is a Polish space and so is $(\mathcal{X} \times \mathcal{Y})^2$ as a countable product of Polish spaces, see, e.g., [32, p. 13]. Hence we can split up P into the conditional probability of Y given X and the marginal distribution P_X, i.e.,

$$\begin{aligned}\mathcal{R}_{L,\mathrm{P}}(f) &= \int_{(\mathcal{X}\times\mathcal{Y})^2} L(x, y, x', y', f(x, x'))\mathrm{dP}^2(x, y, x', y')\\ &= \int_{\mathcal{X}}\int_{\mathcal{Y}}\int_{\mathcal{X}}\int_{\mathcal{Y}} L(x, y, x', y', f(x, x'))\mathrm{P}(\mathrm{d}y|x)\mathrm{P}_X(\mathrm{d}x)\mathrm{P}(\mathrm{d}y'|x')\mathrm{P}_X(\mathrm{d}x'),\end{aligned}$$

see [20, Section 10.2].

The following definition and remarks provide a very short introduction to reproducing kernel Hilbert spaces, see [3] for a classical textbook on this subject, in order to motivate the regularized pairwise learning method described shortly thereafter.

Definition 2.7 Let $\mathcal{X} \neq \emptyset$ and $\mathcal{H}$ be an $\mathbb{R}$-Hilbert space over $\mathcal{X}^2$ containing functions mapping from $\mathcal{X}^2$ to $\mathbb{R}$.

(a) A function $k : \mathcal{X}^2 \times \mathcal{X}^2 \to \mathbb{R}$ is called a **kernel** on $\mathcal{X}^2$ if there exists an $\mathbb{R}$-Hilbert space $\mathcal{H}$ and a map $\Phi : \mathcal{X}^2 \to \mathcal{H}$ such that for all $(x, x'), (\tilde{x}, \tilde{x}') \in \mathcal{X}^2$ we have

$$k\big((x, x'), (\tilde{x}, \tilde{x}')\big) = \big\langle \Phi(\tilde{x}, \tilde{x}'), \Phi(x, x')\big\rangle_{\mathcal{H}} \overset{\text{in } \mathbb{R}}{=} \big\langle \Phi(x, x'), \Phi(\tilde{x}, \tilde{x}')\big\rangle_{\mathcal{H}}.$$

(b) A kernel $k : \mathcal{X}^2 \times \mathcal{X}^2 \to \mathbb{R}$ is called **bounded** if

$$\|k\|_\infty := \sup_{(x,x')\in\mathcal{X}^2} \sqrt{k((x, x'), (x, x'))} < \infty.$$

(c) A function $k : \mathcal{X}^2 \times \mathcal{X}^2 \to \mathbb{R}$ is called a **reproducing kernel** of $\mathcal{H}$ if we have $k\big((\cdot,\cdot),(x,x')\big) \in \mathcal{H}$ for all $(x,x') \in \mathcal{X}^2$ and the reproducing property

$$f(x,x') = \big\langle f, k\big((\cdot,\cdot),(x,x')\big)\big\rangle_{\mathcal{H}}$$

holds for all $f \in \mathcal{H}$ and all $(x,x') \in \mathcal{X}^2$.

(d) The space $\mathcal{H}$ is called a **reproducing kernel Hilbert space (RKHS)** over $\mathcal{X}^2$ if for all $(x,x') \in \mathcal{X}^2$ the Dirac functional $\delta_{(x,x')} : \mathcal{H} \to \mathbb{R}$ defined by

$$\delta_{(x,x')}(f) := f(x,x'), \qquad f \in \mathcal{H},$$

is continuous.

(e) The **canonical feature map** of the RKHS $\mathcal{H}$ with kernel $k : \mathcal{X}^2 \times \mathcal{X}^2 \to \mathbb{R}$ is defined as follows

$$\Phi : \mathcal{X}^2 \to \mathcal{H},\ \Phi(x,x') := k\big((\cdot,\cdot),(x,x')\big).$$

An RKHS $\mathcal{H}$ is a Hilbert space and is therefore endowed with an inner product $\langle\cdot,\cdot\rangle_{\mathcal{H}} : \mathcal{H} \times \mathcal{H} \to \mathbb{R}$. The associated norm is denoted by $\|\varphi\|_{\mathcal{H}} := \sqrt{\langle\varphi,\varphi\rangle_{\mathcal{H}}}$ for all $\varphi \in \mathcal{H}$.

It is well-known that, if $k : \mathcal{X}^2 \times \mathcal{X}^2 \to \mathbb{R}$ is a bounded and measurable kernel with RKHS $\mathcal{H}$, $\Phi : \mathcal{X}^2 \to \mathcal{H}$ the canonical feature map and $f \in \mathcal{H}$ a function, then for all $(x,x') \in \mathcal{X}^2$

$$\big\|\Phi(x,x')\big\|_{\mathcal{H}} \leq \|k\|_\infty \tag{1}$$

$$\big\|\Phi(x,x')\big\|_\infty \leq \|k\|_\infty^2 \tag{2}$$

$$\|f\|_\infty \leq \|f\|_{\mathcal{H}} \, \|k\|_\infty , \tag{3}$$

where we used $\|k\|_\infty$ defined in Definition 2.7 (b).

It can be shown that there is a one-to-one correspondence between reproducing kernel Hilbert spaces and kernel functions, see, e.g., [44, Theorems 4.20 and 4.21].

Computing the infimum of the risk over the set of all measurable functions for empirical distributions D_n instead of P (as defined in 2.5) is in general not doable and might lead to overfitting. In order to reduce the danger of overfitting, one approach is to introduce a regularizing term to penalize such estimated predictor functions. Another modification that can be made is to restrict the set that the risk is minimized over from all measurable functions to a reproducing kernel Hilbert space (RKHS) $\mathcal{H}$ of a measurable kernel $k : \mathcal{X}^2 \times \mathcal{X}^2 \to \mathbb{R}$ in order to simplify the computation. If a universal kernel, such as the Gaussian RBF or the Laplacian kernel, is chosen, then every continuous prediction function can be arbitrarily approximated due to the denseness of the corresponding RKHS in the space of continuous functions (see, e.g., [44, p. 152ff.]). Both ways are used in the setting of support vector machines and regularized pairwise learning.

The remarks above lead to the introduction of a regularized version of the risk.

Definition 2.8 Let $L : (\mathcal{X} \times \mathcal{Y})^2 \times \mathbb{R} \to [0, \infty)$ be a pairwise loss function and $\mathrm{P} \in \mathcal{M}_1(\mathcal{X} \times \mathcal{Y})$. Then, for $f \in \mathcal{H}$ and $\lambda > 0$, the **regularized** L**-risk** is defined by

$$\mathcal{R}^{\mathrm{reg}}_{L,\mathrm{P},\lambda}(f) := \mathcal{R}_{L,\mathrm{P}}(f) + \lambda \, \| f \|^2_{\mathcal{H}}.$$

The corresponding minimizer, if it exists, is then abbreviated with $f_{L,\mathrm{P},\lambda} : \mathcal{X}^2 \to \mathbb{R}$,

$$f_{L,\mathrm{P},\lambda} = \arg\inf_{f \in \mathcal{H}} \mathcal{R}^{\mathrm{reg}}_{L,\mathrm{P},\lambda}(f).$$

The following definitions, theorems, and lemmas are taken from [12] and [11].

Lemma 2.9 *Let L be a pairwise loss function and $\mathcal{F} \subset \mathcal{L}_0(\mathcal{X}^2)$ be a subset that is equipped with a complete and separable metric d and its corresponding Borel σ-algebra. Assume that the metric d dominates pointwise convergence, i.e.,*

$$\lim_{n\to\infty} d(f_n, f) = 0 \Longrightarrow \lim_{n\to\infty} f_n(x, x') = f(x, x') \; \forall (x, x') \in \mathcal{X}^2, \; \forall f, f_n \in \mathcal{F}.$$

Then the evaluation map $\mathcal{F} \times \mathcal{X}^2 \to \mathbb{R}$ defined by $(f, (x, x')) \mapsto f(x, x')$ is measurable and consequently the map $(x, y, x', y', f) \mapsto L(x, y, x', y', f(x, x'))$ defined on $(\mathcal{X} \times \mathcal{Y})^2 \times \mathcal{F}$ is also measurable. Finally, given $\mathrm{P} \in \mathcal{M}_1(\mathcal{X} \times \mathcal{Y})$, the risk functional $\mathcal{R}_{L,P} : \mathcal{F} \to [0, \infty)$ is measurable.

Definition 2.10 A pairwise loss function L is called

(i) **(strictly) convex**, **continuous**, or **differentiable**, if $L(x, y, x', y', \cdot) : \mathbb{R} \to [0, \infty)$ is (strictly) convex, continuous, or differentiable for all $(x, y, x', y') \in (\mathcal{X} \times \mathcal{Y})^2$, respectively. We denote the partial Fréchet derivative with respect to the fifth argument by $D_5 L$, if it exists.

(ii) **locally Lipschitz continuous**, if, for all $b \geq 0$, there exists a constant $c_b \geq 0$ such that, for all $t, t' \in [-b, b]$, we have

$$\sup_{\substack{x,x' \in \mathcal{X} \\ y,y' \in \mathcal{Y}}} \left| L(x, y, x', y', t) - L(x, y, x', y', t') \right| \leq c_b \left| t - t' \right|.$$

Moreover, for $b \geq 0$, the smallest such constant c_b is denoted by $|L|_{b,1}$.

(iii) **Lipschitz continuous**, if there exists a constant $|L|_1 \in [0, \infty)$ such that, for all $t, t' \in \mathbb{R}$,

$$\sup_{\substack{x,x' \in \mathcal{X} \\ y,y' \in \mathcal{Y}}} \left| L(x, y, x', y', t) - L(x, y, x', y', t') \right| \leq |L|_1 \left| t - t' \right|.$$

Example 2.11 The loss function from Example 2.2 is Lipschitz continuous and convex, if the auxiliary function ϕ is as well. Let $t_1, t_2 \in \mathbb{R}$ and $|\phi|_1$ the Lipschitz constant of ϕ, then we have

$$\begin{aligned} \left| L(x, y, x', y', t_1) - L(x, y, x', y', t_2) \right| &= \left| \phi\left(-\operatorname{sign}\left(y - y'\right) t_1\right) - \phi\left(-\operatorname{sign}\left(y - y'\right) t_2\right) \right| \\ &\leq |\phi|_1 \left| -\operatorname{sign}\left(y - y'\right) \right| |t_1 - t_2| \\ &\leq |\phi|_1 \, |t_1 - t_2| \,. \end{aligned}$$

Choosing, for instance, $\phi(x) = \log_2(1 + e^x)$, then the Lipschitz continuity follows from the boundedness of its derivative $\phi'(x) = \frac{e^x}{\log(2)(1+e^x)}$. For ϕ the exponential function we can only obtain local Lipschitz continuity.

For convex ϕ it follows with $\upsilon \in [0, 1]$

$$\begin{aligned} & L(x, y, x', y', \upsilon t_1 + (1 - \upsilon) t_2) \\ &= \phi\left(-\operatorname{sign}\left(y - y'\right)\left(\upsilon t_1 + (1 - \upsilon) t_2\right)\right) \\ &\leq \upsilon \phi\left(-\operatorname{sign}\left(y - y'\right) t_1\right) - (1 - \upsilon) \phi\left(-\operatorname{sign}\left(y - y'\right) t_2\right) \\ &= \upsilon L(x, y, x', y', \upsilon t_1) + (1 - \upsilon) L(x, y, x', y', (1 - \upsilon) t_2). \end{aligned}$$

Both properties are also satisfied by the smoothed version in Example 2.2. Differentiability with respect to the fifth argument is guaranteed by both examples and only depends on the differentiability of the auxiliary function.

Lemma 2.12 *Let L be a (strictly) convex loss function and* $\mathrm{P} \in \mathcal{M}_1(\mathcal{X} \times \mathcal{Y})$. *Then* $\mathcal{R} : \mathcal{L}_0(\mathcal{X}^2) \to [0, \infty]$ *is (strictly) convex.*

Lemma 2.13 *Let* $\mathrm{P} \in \mathcal{M}_1(\mathcal{X} \times \mathcal{Y})$ *and L be a locally Lipschitz continuous pairwise loss function. Then for all* $B \geq 0$ *and all* $f, g \in \mathcal{L}_\infty(\mathrm{P}_X^2)$ *with* $\|f\|_\infty \leq B$ *and* $\|g\|_\infty \leq B$, *we have*

$$\left| \mathcal{R}_{L,\mathrm{P}}(f) - \mathcal{R}_{L,\mathrm{P}}(g) \right| \leq |L|_{B,1} \, \|f - g\|_{L_1(\mathrm{P}_X^2)} \,.$$

Furthermore, the risk functional $\mathcal{R}_{L,\mathrm{P}} : \mathcal{L}_\infty(\mathrm{P}_X^2) \to [0, \infty)$ *is well-defined and continuous.*

For $f, g \in \mathcal{H}$ and a Lipschitz continuous pairwise loss function L, Lemma 2.13 and inequality (3) yield

$$\left|\mathcal{R}_{L,\mathrm{P}}(f) - \mathcal{R}_{L,\mathrm{P}}(g)\right| \le |L|_1 \, \|f-g\|_{\mathcal{L}_1(\mathrm{P}_X^2)} \le |L|_1 \, \|f-g\|_\infty \le |L|_1 \, \|k\|_\infty \, \|f-g\|_{\mathcal{H}} .$$

Definition 2.14 A pairwise loss function $L : (\mathcal{X} \times \mathcal{Y})^2 \times \mathbb{R} \to [0, \infty)$ is called a **pairwise Nemitski loss function** if a measurable function $b : (\mathcal{X} \times \mathcal{Y})^2 \to [0, \infty)$ and a monotonically increasing function $h : [0, \infty) \to [0, \infty)$ exist, such that

$$L(x, y, x', y', t) \le b(x, y, x', y') + h(|t|), \qquad (x, y, x', y', t) \in (\mathcal{X} \times \mathcal{Y})^2 \times \mathbb{R}.$$

Lemma 2.15 *Let* $\mathrm{P} \in \mathcal{M}_1(\mathcal{X} \times \mathcal{Y})$ *and* L *be a differentiable pairwise loss function such that* $|D_5 L|$ *is a* P*-integrable Nemitski loss function. Then the risk function* $\mathcal{R}_{L,\mathrm{P}} : \mathcal{L}_\infty(\mathrm{P}_X^2) \to [0, \infty)$ *is Fréchet differentiable and its derivative at* $f \in \mathcal{L}_\infty(\mathrm{P}_X^2)$ *is the bounded linear operator* $\mathcal{R}'_{L,\mathrm{P}} : \mathcal{L}_\infty(\mathrm{P}_X^2) \to \mathbb{R}$ *with*

$$\mathcal{R}'_{L,\mathrm{P}}(f)g = \int_{(\mathcal{X}\times\mathcal{Y})^2} D_5 L(x, y, x', y', f(x, x'))g(x, x')\mathrm{d}\mathrm{P}^2(x, y, x', y').$$

If in Lemma 2.15 the derivative of the pairwise loss function with respect to the fifth argument is continuous and uniformly bounded for all $x, x' \in \mathcal{X}$ and all $y, y' \in \mathcal{Y}$ by a constant $c_L \in [0, \infty)$, then the upper assertion follows immediately from the lemma, because

$$\left|D_5 L(x, y, x', y', t)\right| \le c_L, \quad \forall (x, y, x', y', t) \in (\mathcal{X} \times \mathcal{Y})^2 \times \mathbb{R}$$

and thus the condition that $|D_5 L|$ is a P^2-integrable pairwise Nemitski loss function follows, because we can set $b(x, y, x', y') \equiv c_L$ and $h(|t|) \equiv 0$.

One problem with the L-risk is that it can easily happen that $\mathcal{R}_{L,\mathrm{P}}(f) = \infty$ even for bounded functions f. One example is the least squares ranking loss (see Example 2.3) with $f \equiv 0$. Then, $\mathcal{R}_{L,\mathrm{P}}(0) = \mathbb{E}_{\mathrm{P}^2}\left[(Y - Y')^2\right]$ and we will need some moment conditions on P to achieve $\mathcal{R}_{L,\mathrm{P}}(0) < \infty$.

To overcome the need for moment assumptions, a technical tool is to shift the loss function which goes back at least to Huber [31] in the context of M-estimation.

Definition 2.16 Let L be a pairwise loss function.

(a) The corresponding **shifted pairwise loss function** $L^* : (\mathcal{X} \times \mathcal{Y})^2 \times \mathbb{R} \to \mathbb{R}$ is defined by

$$L^*(x, y, x', y', t) := L(x, y, x', y', t) - L(x, y, x', y', 0).$$

(b) The L^***-risk** is defined by

$$\mathcal{R}_{L^*,\mathrm{P}}(f) := \mathbb{E}_{\mathrm{P}^2}\left[L^*(X, Y, X', Y', f(X, X'))\right].$$

(c) For $\lambda > 0$, the **regularized** L^***-risk** is defined by

$$\mathcal{R}^{\mathrm{reg}}_{L^*,\mathrm{P},\lambda}(f) := \mathcal{R}_{L^*,\mathrm{P}}(f) + \lambda \, \|f\|^2_{\mathcal{H}}.$$

Remark 2.17 By using the shifted loss function of a Lipschitz continuous pairwise loss it is possible under weak assumptions on L and k to avoid *any* moment assumptions on the conditional distribution Y given $X = x$ when computing the L^*-risk, as can be seen from the following argument:

$$\begin{aligned}
\left|\mathcal{R}_{L^*,\mathrm{P}}(f)\right| &\le \mathbb{E}_{\mathrm{P}^2}\left[\left|L(X, Y, X', Y', f(X, X')) - L(X, Y, X', Y', 0)\right|\right] \\
&\le \mathbb{E}_{\mathrm{P}^2}\left[|L|_1 \left|f(X, X') - 0\right|\right] = |L|_1 \, \mathbb{E}_{\mathrm{P}^2}\left[\left|f(X, X')\right|\right] \\
&\le |L|_1 \, \|f\|_{\mathcal{L}_\infty(\mathcal{X}^2, \mathrm{P}^2_X)} < \infty,
\end{aligned}$$

with P_X denoting the marginal distribution of X. The last inequality can be guaranteed by, for instance, choosing a measurable and bounded kernel $k : \mathcal{X}^2 \times \mathcal{X}^2 \to \mathbb{R}$, and $f \in \mathcal{H}$ with $\mathcal{H}$ the corresponding reproducing kernel Hilbert space of k, see [44, Lemma 4.23].

Example 2.3 (continued): The shifted least squares ranking loss is given by

$$\begin{aligned}
L^*\big(x, y, x', y', f(x, x')\big) &= L\big(x, y, x', y', f(x, x')\big) - L\big(x, y, x', y', 0\big) \\
&= \big(y - y' - f(x, x')\big)^2 - (y - y')^2 \\
&= \big[f(x, x')\big]^2 - 2(y - y')f(x, x').
\end{aligned}$$

From this, it is easily seen that some moment assumption on the conditional distribution of Y given $X = x$ is in general necessary to obtain $\left|\mathcal{R}_{L^*,P}(f)\right| < \infty$. To be more precise, $\left|\mathcal{R}_{L^*,P}(f)\right| < \infty$ is satisfied in this case, if $\mathbb{E}_{\mathrm{P}}\left[|Y|\right] < \infty$, and if $\sup_{(x,x') \in \mathcal{X}^2} \left|f(x, x')\right| < \infty$. The latter can easily be obtained by considering the reproducing kernel Hilbert space of a bounded and measurable kernel, as mentioned in the remark above. Hence, the shifted least squares ranking loss is not covered by our main results as it is not a Lipschitz continuous pairwise loss function and therefore the above mentioned moment assumption will in general be unavoidable.

We will now prove several lemmas to be able to show the uniqueness and existence of (regularized) Risk-minimizing functions in the pairwise learning setting.

Lemma 2.18 *Let L be a pairwise loss function. Then the following statements concerning the corresponding shifted loss function L^* are valid.*

(i) L^* is (strictly) convex, if L is (strictly) convex.
(ii) L^* is Lipschitz continuous, if L is Lipschitz continuous. Furthermore, both Lipschitz constants are equal, i.e., $|L|_1 = |L^*|_1$.

The absolute value of a shifted Lipschitz continuous pairwise loss function $|L^*|$ is a pairwise Nemitski loss function, because it follows

$$\left|L^*(x, y, x', y', t)\right| = \left|L(x, y, x', y', t) - L(x, y, x', y', 0)\right| \leq |L|_1\,|t|$$

and thus $|L^*|$ fulfills the property of a Nemitski loss function with $b(x, y, x', y') \equiv 0$ and $h(|t|) = |L|_1\,|t|$. If $f \in \mathcal{L}_1(\mathrm{P}_X^2)$, then $|L^*|$ is a P^2-integrable pairwise Nemitski loss function with $t \equiv f(x, x')$.

Lemma 2.19 *The following assertions are valid for shifted pairwise loss functions L^*. Let $\lambda \in (0, \infty)$.*

(i)

$$\inf_{t \in \mathbb{R}} L^*(x, y, x', y', t) \leq 0. \tag{4}$$

(ii) If L is a Lipschitz continuous loss function, then, for all $f \in \mathcal{H}$,

$$\left|\mathcal{R}_{L^*,\mathrm{P}}(f)\right| \leq |L|_1\,\mathbb{E}_{\mathrm{P}^2}\left[\left|f(X, X')\right|\right], \tag{5}$$

$$\left|\mathcal{R}^{\mathrm{reg}}_{L^*,\mathrm{P},\lambda}(f)\right| \leq |L|_1\,\mathbb{E}_{\mathrm{P}^2}\left[\left|f(X, X')\right|\right] + \lambda\,\|f\|^2_{\mathcal{H}}, \tag{6}$$

(iii) $\inf_{f \in \mathcal{H}} \mathcal{R}^{\mathrm{reg}}_{L^*,\mathrm{P},\lambda}(f) \leq 0$ and $\inf_{f \in \mathcal{H}} \mathcal{R}_{L^*,\mathrm{P}}(f) \leq 0$.
(iv) Let L be a Lipschitz continuous loss function and assume that $f_{L^*,\mathrm{P},\lambda}$ exists. Then we have

$$\lambda\,\left\|f_{L^*,\mathrm{P},\lambda}\right\|^2_{\mathcal{H}} \leq -\mathcal{R}_{L^*,\mathrm{P}}(f_{L^*,\mathrm{P},\lambda}) \leq \mathcal{R}_{L,\mathrm{P}}(0), \tag{7}$$

$$0 \leq -\mathcal{R}^{\mathrm{reg}}_{L^*,\mathrm{P},\lambda}(f) \leq \mathcal{R}_{L,\mathrm{P}}(0), \tag{8}$$

$$\lambda\,\left\|f_{L^*,\mathrm{P},\lambda}\right\|^2_{\mathcal{H}} \leq \min\left\{\,|L|_1\,\mathbb{E}_{\mathrm{P}^2}\left[\left|f_{L^*,\mathrm{P},\lambda}(X, X')\right|\right], \mathcal{R}_{L,\mathrm{P}}(0)\right\}. \tag{9}$$

If the kernel $k : \mathcal{X}^2 \times \mathcal{X}^2 \to \mathbb{R}$ is additionally bounded, then

$$\left\|f_{L^*,\mathrm{P},\lambda}\right\|_{\infty} \leq \lambda^{-1}\,|L|_1\,\|k\|^2_{\infty} < \infty, \tag{10}$$

$$\left|\mathcal{R}_{L^*,\mathrm{P}}(f_{L^*,\mathrm{P},\lambda})\right| \leq \lambda^{-1}\,|L|_1^2\,\|k\|^2_{\infty} < \infty. \tag{11}$$

(v) *If the partial Fréchet derivatives with respect to the fifth argument of L and L^* exist for* $(x, y, x', y') \in (\mathcal{X} \times \mathcal{Y})^2$, *then*

$$D_5 L^*(x, y, x', y', t) = D_5 L(x, y, x', y', t), \qquad \forall t \in \mathbb{R}. \tag{12}$$

Lemma 2.20 *Let L be a Lipschitz continuous pairwise loss function and* $f \in \mathcal{L}_1(\mathrm{P}_X^2)$. *Then* $\mathcal{R}_{L^*,\mathrm{P}}(f) \notin \{-\infty, \infty\}$. *Moreover, we have* $\mathcal{R}^{\mathrm{reg}}_{L^*,\mathrm{P},\lambda}(f) > -\infty$ *for all* $f \in \mathcal{L}_1(\mathrm{P}_X^2) \cap \mathcal{H}$.

Definition 2.21 Let $\lambda \in (0, \infty)$. A **regularized pairwise learning operator** is a function S defined by

$$S : \mathcal{M}_1(\mathcal{X} \times \mathcal{Y}) \to \mathcal{H}, \quad \mathrm{P} \mapsto S(\mathrm{P}) := f_{L^*,\mathrm{P},\lambda} := \arg\inf_{f \in \mathcal{H}} \mathcal{R}_{L^*,\mathrm{P}}(f) + \lambda \, \|f\|^2_{\mathcal{H}}.$$

A function $f_{L^*,\mathrm{P},\lambda}$ is called **RPL estimator** or **minimizing prediction function** if it satisfies the condition above.

In the following assertions the classic problem of existence and uniqueness of such minimizers is taken care of.

Theorem 2.22 (Uniqueness of Minimizer) *Let L be a convex pairwise loss function. Assume that*

(i) $\mathcal{R}_{L^*,\mathrm{P}}(f) < \infty$ *for some* $f \in \mathcal{H}$ *and* $\mathcal{R}_{L^*,\mathrm{P}}(f) > -\infty$ *for all* $f \in \mathcal{H}$

or

(ii) *L is Lipschitz continuous and* $f \in \mathcal{L}_1(\mathrm{P}_X^2)$ *for all* $f \in \mathcal{H}$.

Then, for all $\lambda > 0$, *there exists at most one solution* $f_{L^*,\mathrm{P},\lambda}$.

Theorem 2.23 (Existence of Minimizer) *Let L be a Lipschitz continuous, convex pairwise loss function and* $\mathcal{H}$ *be the RKHS of a bounded measurable kernel k on* $\mathcal{X}^2$. *Then, for all* $\lambda > 0$, *there exists a minimizing prediction function* $f_{L^*,\mathrm{P},\lambda}$.

The theorem above shows the existence of a Bayes decision function for regularized pairwise learning methods, if the loss function is convex. The Minimum Error Entropy (MEE) loss function, see, e.g., [37] and [11, p. 5f.] and the references cited therein, is a leading example for a non-convex loss function and therefore it is relevant to prove the existence of a minimizer in such a case as well.

Theorem 2.24 *If L is a Lipschitz continuous pairwise loss function,* $\mathrm{P} \in \mathcal{M}_1(\mathcal{X} \times \mathcal{Y})$, $\mathcal{R}_{L,\mathrm{P}}(f_0) < \infty$ *for some* $f_0 \in \mathcal{H}$, *and* $\mathcal{H}$ *the RKHS of a bounded and measurable kernel k on* $\mathcal{X}^2$, *then a minimizer* $f_{L,\mathrm{P},\lambda} \in \mathcal{H}$ *exists for any* $\lambda > 0$.

For a general representer theorem, a couple of notational remarks and the introduction of the subdifferential are necessary.

Let E be a Banach space, E' its dual space, and $x \in E, x' \in E'$. A common notation is the so-called **dual pairing** $\langle x', x \rangle_{E',E} := x'(x)$.

Let $f : E \to \mathbb{R} \cup \{\infty\}$ be a convex function and $w \in E$ with $f(w) < \infty$. Then the **subdifferential** of f at w is defined by

$$\begin{aligned} \partial f(w) :&= \{w' \in E' : \langle w', v - w \rangle_{E',E} \leq f(v) - f(w) \quad \forall v \in E\} \\ &= \{w' \in E' : w'(v - w) \leq f(v) - f(w) \quad \forall v \in E\}. \end{aligned}$$

3 Main Results

In this section we will give our main results: a general representer theorem, bounds for bias, and (qualitative) robustness for the regularized pairwise learning method. The proofs are given in the Appendix A.2. We use techniques from [12] and [11]. However, here we treat the more general case of prediction functions $f : \mathcal{X} \times \mathcal{X} \to \mathbb{R}$ instead of the well-investigated case $f : \mathcal{X} \to \mathbb{R}$ by the authors mentioned above. Our main goal is to show that no moment assumptions on the conditional distribution of Y given $X = x$ are needed under weak and *provable* (i.e., data-independent) assumptions on L and k.

Robustness is an important statistical property as small deviations in the probability measures (or the underlying data sets), due to noise or because these measures have been obtained by approximation, should only have little influence on the results, i.e., the minimizing prediction function in this case. The discipline of statistical robustness has a long tradition and goes back at least to Huber [30].

J. W. Tukey, one of the pioneers in robust statistics, motivated the issue and mentioned in 1960 (see [29, p. 21]):

> A tacit hope in ignoring deviations from ideal models was that they would not matter; that statistical procedures which were optimal under the strict model would still be approximately optimal under the approximate model. Unfortunately, it turned out that this hope was often drastically wrong; even mild deviations often have much larger effects than were anticipated by most statisticians.

If not mentioned otherwise, we will assume for the rest of this section that Assumption 3.1 is valid. Please note that we do *not* make any assumptions on the unknown probability measure P.

Assumption 3.1

(i) Let $\mathcal{Y} \subset \mathbb{R}$ be a closed subset and $\mathcal{X}$ a complete separable metric space. Let (X, Y) and (X_i, Y_i), $i \in \mathbb{N}$, be independent and identically distributed pairs of random quantities with values in $\mathcal{X} \times \mathcal{Y}$. Let $\mathrm{P} \in \mathcal{M}_1(\mathcal{X} \times \mathcal{Y})$ be the joint distribution of (X_i, Y_i).

(ii) Let $k : \mathcal{X}^2 \times \mathcal{X}^2 \to \mathbb{R}$ be a continuous and bounded kernel and $\Phi : \mathcal{X}^2 \to \mathcal{H} : \Phi(x, x') := k\big((\cdot, \cdot), (x, x')\big)$ with $(x, x') \in \mathcal{X}^2$ being the canonical feature map.

(iii) Let L be a Lipschitz continuous, differentiable pairwise loss function for which the first and second partial derivatives with respect to the fifth argument are continuous and uniformly bounded such that:

- $\sup_{\substack{x,x'\in\mathcal{X}\\ y,y'\in\mathcal{Y}}} \left|D_5 L(x,y,x',y',\cdot)\right| \le c_{L,1} \in (0,\infty)$
- $\sup_{\substack{x,x'\in\mathcal{X}\\ y,y'\in\mathcal{Y}}} \left|D_5 D_5 L(x,y,x',y',\cdot)\right| \le c_{L,2} \in (0,\infty)$.

(iv) Let L be a convex pairwise loss function.

We mention that Assumption 3.1(i) and (ii) imply that the RKHS $\mathcal{H}$ is separable, see, e.g., [44, Lemma 4.33].

The first result in this section is the following representer theorem which can eventually be proven with the same methods as in the case of support vector machines.

Theorem 3.2 (Representer Theorem) *Let L^* be the corresponding shifted pairwise loss function of L. Then, for all $\lambda > 0$, there exists an $h_{\mathrm{P}} \in \mathcal{L}_\infty((\mathcal{X}\times\mathcal{Y})^2, \mathrm{P}^2)$ such that*

(i) $h_{\mathrm{P}}(x,y,x',y') \in \partial L^*(x,y,x',y',f_{L^*,\mathrm{P},\lambda}(x,x')), \quad \forall (x,y,x',y') \in (\mathcal{X}\times\mathcal{Y})^2$,
(ii) $f_{L^*,\mathrm{P},\lambda} = -(2\lambda)^{-1}\mathbb{E}_{\mathrm{P}^2}[h_{\mathrm{P}}\Phi]$,
(iii) $\|h_{\mathrm{P}}\|_\infty \le |L^*|_1$,
(iv) $\left\|f_{L^*,\mathrm{P},\lambda} - f_{L^*,\mathrm{Q},\lambda}\right\|_{\mathcal{H}} \le \lambda^{-1}\left\|\mathbb{E}_{\mathrm{P}^2}[h_{\mathrm{P}}\Phi] - \mathbb{E}_{\mathrm{Q}^2}[h_{\mathrm{P}}\Phi]\right\|_{\mathcal{H}}, \quad \forall \mathrm{Q} \in \mathcal{M}_1(\mathcal{X}\times\mathcal{Y})$.

The following result gives an upper bound for the $\mathcal{H}$-norm of the difference between minimizers of the probability measure P and a contaminated probability measure P_ε, which is a mixture of P and another probability measure Q.

Theorem 3.3 (Bounds for Bias) *For all $\lambda > 0$, all $\varepsilon \in (0,1)$, and all probability measures* $\mathrm{P},\mathrm{Q} \in \mathcal{M}_1(\mathcal{X}\times\mathcal{Y})$, *we have, for all* $\mathrm{P}_\varepsilon = (1-\varepsilon)\mathrm{P}+\varepsilon\mathrm{Q} \in \mathcal{M}_1(\mathcal{X}\times\mathcal{Y})$,

$$\left\|f_{L^*,\mathrm{P},\lambda} - f_{L^*,\mathrm{P}_\varepsilon,\lambda}\right\|_{\mathcal{H}} \le c_{\mathrm{P},\mathrm{Q}}\varepsilon,$$

where $c_{\mathrm{P},\mathrm{Q}} = \frac{8}{\lambda}\|k\|_\infty |L|_1$.

Hence, Theorem 3.3 shows that $\left\|f_{L^*,\mathrm{P},\lambda} - f_{L^*,\mathrm{P}_\varepsilon,\lambda}\right\|_{\mathcal{H}}$ increases *at most* linearly for increasing radii $\varepsilon \in (0,1)$.

Theorem 3.4 *Let $\lambda \in (0,\infty)$. For all probability measures* $\mathrm{P},\mathrm{Q} \in \mathcal{M}_1(\mathcal{X}\times\mathcal{Y})$, *the regularized pairwise learning operator (**RPL** operator).*

$$S : \mathcal{M}_1(\mathcal{X}\times\mathcal{Y}) \to \mathcal{H}, \qquad S(P) := f_{L^*,\mathrm{P},\lambda},$$

has a bounded Gâteaux derivative $S'_G(\mathrm{P})$ *at* P *and*

$$S'_G(\mathrm{P})(\mathrm{Q}) = -M(\mathrm{P})^{-1}T(\mathrm{Q};\mathrm{P}).$$

To shorten the notation, we write

$$L'_{f_{L^*,\mathrm{P},\lambda}}(X,Y,X',Y') := D_5L(X,Y,X',Y',f_{L^*,\mathrm{P},\lambda}(X,X')).$$

Then,

$$\begin{aligned}
&T(\mathrm{Q};\mathrm{P})\\
&= -2\mathbb{E}_{\mathrm{P}^2}\left[L'_{f_{L^*,\mathrm{P},\lambda}}(X,Y,X',Y')\Phi(X,X')\right] + \mathbb{E}_{\mathrm{P}\otimes\mathrm{Q}}\left[L'_{f_{L^*,\mathrm{P},\lambda}}(X,Y,X',Y')\Phi(X,X')\right]\\
&\quad + \mathbb{E}_{\mathrm{Q}\otimes\mathrm{P}}\left[L'_{f_{L^*,\mathrm{P},\lambda}}(X,Y,X',Y')\Phi(X,X')\right]
\end{aligned}$$

equals the gradient of the regularized risk and

$$M(P) = 2\lambda\,\mathrm{id}_{\mathcal{H}} + \mathbb{E}_{\mathrm{P}^2}\left[D_5L'_{f_{L^*,\mathrm{P},\lambda}}(X,Y,X',Y')\langle\Phi(X,X'),\cdot\rangle_{\mathcal{H}}\Phi(X,X')\right].$$

We obtain an important special case of Theorem 3.4 for Q being a Dirac measure, which leads to the influence function, which is one of the most important notions in robust statistics. For a more thorough introduction of the influence function, we refer to [27–29].

Definition 3.5 The **influence function** $\mathrm{IF}: \mathcal{X}\times\mathcal{Y}\to\mathcal{H}$ of $S: \mathcal{M}_1(\mathcal{X}\times\mathcal{Y})\to\mathcal{H}$ at a point (x_0,y_0) for a distribution $\mathrm{P}\in\mathcal{M}_1(\mathcal{X}\times\mathcal{Y})$ is given by

$$\mathrm{IF}\big((x_0,y_0);S,\mathrm{P}\big) = \lim_{\varepsilon\downarrow 0}\frac{S\big((1-\varepsilon)\mathrm{P}+\varepsilon\delta_{(x_0,y_0)}\big)-S(\mathrm{P})}{\varepsilon}$$

in those $(x_0,y_0)\in\mathcal{X}\times\mathcal{Y}$ where the limit exists.

Corollary 3.6 (Bounded Influence Function) *For all* $\mathrm{P}\in\mathcal{M}_1(\mathcal{X}\times\mathcal{Y})$*, for all* $(x,y,x',y')\in(\mathcal{X}\times\mathcal{Y})^2$*, and for all* $\lambda\in(0,\infty)$*, the influence function of* $S: \mathcal{M}_1(\mathcal{X}\times\mathcal{Y})\to\mathcal{H}$ *with* $S(\mathrm{P}) := f_{L^*,\mathrm{P},\lambda}$ *is bounded. It holds*

$$\mathrm{IF}((x_0,y_0);S,\mathrm{P}) = -M(\mathrm{P})^{-1}T(\delta_{(x_0,y_0)};\mathrm{P}),$$

where $\delta_{(x_0,y_0)}$ *denotes the Dirac distribution in the point* $(x_0,y_0)\in\mathcal{X}\times\mathcal{Y}$*, and* $L'_{f_{L^*,\mathrm{P},\lambda}}$*,* $T(\delta_{(x_0,y_0)};\mathrm{P})$ *as well as* $M(\mathrm{P})$ *are given by Theorem 3.4. Here*

$T(\delta_{(x_0,y_0)}; \mathrm{P})$ *simplifies to*

$$\begin{aligned} &T(\delta_{(x_0,y_0)}; \mathrm{P}) \\ &= -2\mathbb{E}_{\mathrm{P}^2}\left[L'_{f_{L^*,\mathrm{P},\lambda}}(X,Y,X',Y')\Phi(X,X')\right] \\ &\quad + \mathbb{E}_{\mathrm{P}}\left[L'_{f_{L^*,\mathrm{P},\lambda}}(X,Y,x_0,y_0)\Phi(X,x_0)\right] + \mathbb{E}_{\mathrm{P}}\left[L'_{f_{L^*,\mathrm{P},\lambda}}(x_0,y_0,X',Y')\Phi(x_0,X')\right]. \end{aligned}$$

The definition of qualitative robustness was given by Hampel [28, p. 1890] and generalized by Cuevas [17, Def. 1, p. 278]. We refer to [18] for the qualitative robustness of empirical bootstrap approximations.

Definition 3.7 A sequence of estimators $(S_n)_{n\in\mathbb{N}}$ is called **qualitatively robust** at a probability measure P if and only if

$$\forall \varepsilon > 0\, \exists \delta > 0\, \forall \mathrm{Q} \in \mathcal{M}_1(\mathcal{X} \times \mathcal{Y}) : \quad \left[d_*(\mathrm{Q},\mathrm{P}) < \delta \Longrightarrow d_*(\mathcal{L}_{\mathrm{Q}}(S_n), \mathcal{L}_{\mathrm{P}}(S_n)) < \varepsilon\ \forall n \in \mathbb{N}\right],$$

with $\mathcal{L}_P(S_n)$ and $\mathcal{L}_Q(S_n)$ denoting the image measures $S_n \circ \mathrm{P}^n$ and $S_n \circ \mathrm{Q}^n$, respectively, and d_* being either the bounded Lipschitz metric or the Prohorov metric.

Please note that originally the Prohorov metric was used by Hampel [28]. Due to the equivalence of the Prohorov metric d_{Pro} and the bounded Lipschitz metric d_{BL} for separable metric spaces, see e.g. [20, Thm. 11.3.3, Cor. 11.6.5], Assumption 3.1 allows us to use the bounded Lipschitz metric which is easier to use in our situation, see also [21].

We set $\mathbb{D}_n := \frac{1}{n}\sum_{i=1}^{n} \delta_{(X_i,Y_i)}$ the random empirical probability measure, and denote the distribution of the $\mathcal{H}$-valued RPL estimator $f_{L^*,\mathbb{D}_n,\lambda}$ by $\mathcal{L}_n(S; \mathrm{P})$ for $n \in \mathbb{N}$, if all (X_i, Y_i) are i.i.d. from P. Similarly, we denote the distribution of the bootstrap approximated $\mathcal{H}$-valued RPL estimator $f_{L^*,\mathbb{D},\lambda}$, when all pairs $(X_i^{(b)}, Y_i^{(b)}) \sim \mathbb{D}_n$ are independent and identically distributed by $\mathcal{L}_n(S; \mathbb{D}_n)$ for $n \in \mathbb{N}$ and $1 \le b \le B \in \mathbb{N}$ for some fixed $B \in \mathbb{N}$.

Theorem 3.8 *For all Borel probability measures* $\mathrm{P} \in \mathcal{M}_1(\mathcal{X} \times \mathcal{Y})$ *and all* $\lambda \in (0, \infty)$, *we have:*

(i) The RPL operator $S : \mathcal{M}_1(\mathcal{X} \times \mathcal{Y}) \to \mathcal{H}$, *where* $S(\mathrm{P}) = f_{L^*,\mathrm{P},\lambda}$, *is continuous with respect to the weak topology on* $\mathcal{M}_1(\mathcal{X} \times \mathcal{Y})$ *and the norm topology on* $\mathcal{H}$.
(ii) The operator $S : \mathcal{M}_1(\mathcal{X} \times \mathcal{Y}) \to C_b(\mathcal{X}^2)$, *where* $S(\mathrm{P}) = f_{L^*,\mathrm{P},\lambda}$, *is continuous with respect to the weak topology on* $\mathcal{M}_1(\mathcal{X} \times \mathcal{Y})$ *and the norm topology on* $C_b(\mathcal{X}^2)$.

Corollary 3.9 *For any data set* $D_n \in (\mathcal{X}\times\mathcal{Y})^n$ *denote the corresponding empirical measure by* $\mathrm{D}_n := \frac{1}{n}\sum_{i=1}^{n} \delta_{(x_i,y_i)}$. *Then, for every* $\lambda \in (0, \infty)$ *and every* $n \in \mathbb{N}$, *the* *mapping*

$$S_n : \left((\mathcal{X} \times \mathcal{Y})^n, d_{(\mathcal{X}\times\mathcal{Y})^n}\right) \to (\mathcal{H}, d_{\mathcal{H}}), \quad S_n(D_n) = f_{L^*,\mathrm{D}_n,\lambda},$$

is continuous.

Theorem 3.10 (Qualitative Robustness) *For all* $\lambda \in (0, \infty), n \in \mathbb{N}$ *and* $\mathbb{D}_n := \frac{1}{n}\sum_{i=1}^{n} \delta_{(X_i,Y_i)}$, *we have:*

(i) The sequence of $\mathcal{H}$*-valued RPL estimators* $(S_n)_{n\in\mathbb{N}}$, *where* $S_n := f_{L^*,\mathbb{D}_n,\lambda}$, *is qualitatively robust for all Borel probability measures* $\mathrm{P} \in \mathcal{M}_1(\mathcal{X} \times \mathcal{Y})$.
(ii) If the metric space $\mathcal{X} \times \mathcal{Y}$ *is additionally compact, then the sequence* $\mathscr{L}_n(S; \mathbb{D}_n), n \in \mathbb{N}$, *of empirical bootstrap approximations of* $\mathscr{L}_n(S; \mathrm{P})$ *is qualitatively robust for all Borel probability measures* $\mathrm{P} \in \mathcal{M}_1(\mathcal{X} \times \mathcal{Y})$.

We mention that it is not possible to replace λ in Theorem 3.10 by a null sequence $(\lambda_n)_{n\in\mathbb{N}}$ in general, as there is a goal conflict between qualitative robustness and universal consistency, see [26, Counterexample 5.2].

4 Discussion

We showed that kernel-based pairwise learning methods have good statistical robustness properties without making moment assumptions on the conditional distribution of Y given $X = x$ or boundedness assumptions on the input or output spaces. This is valid for convex Lipschitz continuous shifted loss functions and kernels which are continuous and bounded. This shows that such kernel-based pairwise learning methods can be applied even for heavy-tailed distributions such as Student's t-distributions or the Cauchy distribution. Distributions with heavy tails often occur in insurance projects. The results can be applied in a variety of fields such as ranking, metric, and online learning, we refer to, for instance [2, 39, 40, 51]. The techniques we used are tied to those of solving nonparametric regression or classification problems with support vector machines.

Our work extends the results of [11] to the use of prediction functions $f : \mathcal{X} \times \mathcal{X} \to \mathbb{R}$ with two arguments instead of restricting ourselves on the special case $f(x, x') = \tilde{f}(x) - \tilde{f}(x')$ with $\tilde{f} : \mathcal{X} \to \mathbb{R}$ being a well-investigated univariate prediction function.

As the present paper is on statistical robustness properties, an investigation of learning rates is beyond the scope of this paper. This also applies to the case of multivariate ranking which was already mentioned by Clémençon et al. [14, Rem. 3, p. 847] as an important problem for future research.

Another problem for which the theory described above could be applied to is localized learning in the same manner as for support vector machines to make such kernel methods applicable for the so-called Big Data situation. Optimal learning rates for localized support vector machines have been studied by Meister and Steinwart [35]. The learning rates for localized classification under margin conditions have recently been improved by Blaschzyk [5]. Dumpert and Christmann [22] have shown consistency and robustness results, and Köhler and Christmann [33] have shown generalized stability results for the case of localized support vector machines without moment assumptions.

Appendix

The appendix consists of one section providing definitions, theorems, and lemmas which are needed for the proofs of the assertions in this paper in the second section of the appendix.

A.1 Important Definitions, Theorems, and Lemmas

The following results, see, e.g., [44, Lemmas 4.23, 4.24 and A.5.9], are well-known and are only given here, because we are going to use them later, and to improve the readability of this manuscript.

Lemma A.1.1 *Let $\mathcal{X}$ be a set and k be a kernel on $\mathcal{X}^2$ with RKHS $\mathcal{H}$. Then k is bounded if and only if every $f \in \mathcal{H}$ is bounded. Moreover, in this case the inclusion* $\mathrm{id} : \mathcal{H} \to \ell_\infty(\mathcal{X}^2)$ *is continuous and we have* $\left\| \mathrm{id} : \mathcal{H} \to \ell_\infty(\mathcal{X}^2) \right\| = \|k\|_\infty$.

Lemma A.1.2 *Let $\mathcal{X}$ be a measurable space and k be a kernel on $\mathcal{X}^2$ with RKHS $\mathcal{H}$. Then all $f \in \mathcal{H}$ are measurable if and only if $k\big((\cdot,\cdot),(x,x')\big) : \mathcal{X}^2 \to \mathbb{R}$ is measurable for all $(x,x') \in \mathcal{X}^2$.*

Lemma A.1.3 *Let $\mathcal{H}$ be a Hilbert space with inner product $\langle \cdot,\cdot \rangle : \mathcal{H} \times \mathcal{H} \to \mathbb{R}$. Then for all $f, g \in \mathcal{H}$, we have*

$$4\,\langle f, g\rangle = \|f+g\|_{\mathcal{H}}^2 - \|f-g\|_{\mathcal{H}}^2,$$

$$\|f+g\|_{\mathcal{H}}^2 + \|f-g\|_{\mathcal{H}}^2 = 2\,\|f\|_{\mathcal{H}}^2 + 2\,\|g\|_{\mathcal{H}}^2.$$

Definition A.1.4 We define the **local modulus of continuity** for the second order derivative of a loss function L with respect to the last argument as

$$\omega(h)_r := \sup\Big\{ \Big| D_5 D_5 L(x, y, x', y', f(x, x')) - D_5 D_5 L(x, y, x', y', \tilde{f}(x, x')) \Big| :$$

$$(x, y, x', y') \in (\mathcal{X} \times \mathcal{Y})^2,\ f(x, x'),\ \tilde{f}(x, x') \in [-r, r],\ \Big| f(x, x') - \tilde{f}(x, x') \Big| \le h \Big\}.$$

The next lemma which is a consequence of [23, Prop II.4.6] is necessary for the existence and uniqueness of a risk-minimizing prediction function.

Lemma A.1.5 *Let E be a Banach space and $f : E \to \mathbb{R} \cup \{\infty\}$ be a convex function. If f is continuous and $\lim_{\|x\|_E \to \infty} f(x) = \infty$, then f has a minimizer. Moreover if f is strictly convex, then f has a unique minimizer in E.*

The following proposition is a slightly modified version of Proposition 23 from [12, p. 318] and can be proven with the same techniques.

Proposition A.1.6 *Let $\hat{L} : (\mathcal{X} \times \mathcal{Y})^2 \times \mathbb{R} \to \mathbb{R}$ be a measurable function which is both convex and Lipschitz continuous with respect to its fifth argument, P be a distribution on $\mathcal{X} \times \mathcal{Y}$ and $p \in [1, \infty)$. Assume that $R : \mathcal{L}_p(\mathrm{P}^2) \to \mathbb{R} \cup \{-\infty, \infty\}$ defined by*

$$R(g) := \int_{(\mathcal{X}\times\mathcal{Y})^2} \hat{L}(x, y, x', y', g(x, y, x', y'))\mathrm{dP}^2(x, y, x', y')$$

exists for all $g \in \mathcal{L}_p(\mathrm{P}^2)$ and define p' by $\frac{1}{p} + \frac{1}{p'} = 1$. If $|R(g)| < \infty$ for at least one $g \in \mathcal{L}_p(\mathrm{P}^2)$, then, for all $g \in \mathcal{L}_p(\mathrm{P}^2)$, we have

$$\partial R(g) = \{h \in \mathcal{L}_{p'}(\mathrm{P}^2) : h(x, y, x', y') \in \partial \hat{L}(x, y, x', y', g(x, y, x', y'))$$
$$\text{for } \mathrm{P}^2\text{-almost all } (x, y, x', y')\},$$

where $\partial \hat{L}(x, y, x', y', t)$ denotes the subdifferential of $\hat{L}(x, y, x', y', \cdot)$ at the point t.

The next statements provide all necessities to work with subdifferentials, see [36, Prop 1.11]

Proposition A.1.7 *Let $f : E \to \mathbb{R} \cup \{\infty\}$ be a convex function and $w \in E$ such that $f(w) < \infty$. If f is continuous at w, then the subdifferential $\partial f(w)$ is a non-empty, convex, and weak*-compact subset of E'. In addition, if $c \ge 0$ and $\delta > 0$ are constants satisfying $|f(v) - f(w)| \le c \|v - w\|_E$, $v \in w + \delta B_E$, then we have $\|w'\|_E \le c$ for all $w' \in \partial f(w)$.*

Lemma A.1.8 *Let $f, g : E \to \mathbb{R} \cup \{\infty\}$ be convex functions, $\lambda \geq 0$ and $A : F \to E$ be a bounded linear operator. We then have:*

(i) *For all $w \in E$ with $f(x) < \infty$, we have $\partial(\lambda f)(w) = \lambda \partial f(w)$.*
(ii) *If there exists a $w_0 \in E$ at which f is continuous, then, for all $w \in E$ satisfying both $f(w) < \infty$ and $g(w) < \infty$, we have $\partial(f + g)(w) = \partial f(w) + \partial g(w)$.*
(iii) *If there exists a $v_0 \in F$ such that f is finite and continuous at Av_0, then, for all $v \in F$ satisfying $f(Av) < \infty$, we have $\partial(f \circ A)(v) = A'\partial f(Av)$, where $A' : E' \to F'$ denotes the adjoint operator of A.*
(iv) *The function f has a global minimum at $w \in E$ if and only if $0 \in \partial f(w)$.*
(v) *If f is finite and continuous at all $w \in E$, then ∂f is a monotone operator, i.e., for all $v, w \in E$ and $v' \in \partial f(v)$, $w' \in \partial f(w)$, we have $\langle v' - w', v - w \rangle \geq 0$.*

The following theorem has been taken from [1, Thm. 2.6] and will be used in the proof of Theorem 3.4.

Theorem A.1.9 *Let E_1, E_2, and F be Banach spaces, $U_1 \subset E_1$ and $U_2 \subset E_2$ be open subsets and $G : U_1 \times U_2 \to F, (x_1, x_2) \mapsto G(x_1, x_2)$ be a continuous map. Then G is continuously differentiable, if and only if G is partially Fréchet differentiable and the partial derivatives $\frac{\partial G}{\partial x_1}$ and $\frac{\partial G}{\partial x_2}$ are continuous. In this case, the derivative of G at $(x_1, x_2) \in U_1 \times U_2$ is given by*

$$G'(x_1, x_2)(y_1, y_2) = \frac{\partial G}{\partial x_1}(x_1, x_2)y_1 + \frac{\partial G}{\partial x_2}(x_1, x_2)y_2, \quad (y_1, y_2) \in E_1 \times E_2.$$

The next theorem is a version of the classical implicit function theorem and can be found in Corollary 4.2 of [1].

Theorem A.1.10 *Let E, F be Banach spaces and $G : E \times F \to F$ be a continuously differentiable map. Suppose that we have $(x_0, y_0) \in E \times F$ such that $G(x_0, y_0) = 0$ and $\frac{\partial G}{\partial y}(x_0, y_0)$ is invertible. Then there exists a $\delta > 0$ and a continuously differentiable map $f : x_0 + \delta B_E \to y_0 + \delta B_F$ such that for all $x \in x_0 + \delta B_E$, $y \in y_0 + \delta B_F$ we have $G(x, y) = 0$ if and only if $y = f(x)$. Moreover, the derivative of f is given by*

$$f'(x) = -\left(\frac{\partial G}{\partial y}(x, f(x))\right)^{-1} \frac{\partial G}{\partial x}(x, f(x)).$$

In order to show the qualitative robustness for the RPL estimator, the next theorem by Cuevas [17, Thm. 2], which has been adapted to our notation, is useful.

Theorem A.1.11 *Let $(S_n)_{n \in \mathbb{N}}$ be a sequence of measurable estimators such that there exists an operator $S : \mathcal{M}_1(\mathcal{X} \times \mathcal{Y}) \to \mathcal{W}$ with $\mathcal{W}$ being a complete separable metric space verifying $S_n(D_n) = S(\mathrm{D}_n)$ for all possible sets $((x_1, y_1), \ldots, (x_n, y_n)) = D_n \in (\mathcal{X} \times \mathcal{Y})^n$ and $\mathrm{D}_n = n^{-1} \sum_{i=1}^{n} \delta_{(x_i, y_i)}$. If S is continuous on $\mathcal{M}_1(\mathcal{X} \times \mathcal{Y})$, then the sequence $(S_n)_{n \in \mathbb{N}}$ is qualitatively robust for all $\mathrm{P} \in \mathcal{M}_1(\mathcal{X} \times \mathcal{Y})$.*

In order to show qualitative robustness for empirical bootstrap approximations, the following result by Christmann et al. [13, Cor. 16.1, p. 271] will be used.

Theorem A.1.12 *Let $(\Omega, \mathcal{A}, \mu)$ be a probability space, $(\mathcal{Z}, d_{\mathcal{Z}})$ be a compact metric space. Let $S : (\mathcal{M}_1(\mathcal{Z}), d_{BL}) \to (\mathcal{W}, d_W)$ be a statistical operator with $(\mathcal{W}, d_W)$ being a complete, separable metric space. Let $Z_n : (\Omega, \mathcal{A}, \mu) \to (\mathcal{Z}, \mathcal{B}(\mathcal{Z})), n \in \mathbb{N}$, be independent and identically distributed random quantities and denote the image measure by $\mathrm{P} := Z_n \circ \mu$. Let $S_n : (\mathcal{Z}^n, d_{\mathcal{Z}^n}) \to (\mathcal{W}, d_{\mathcal{W}})$ be a statistic defined by $S_n(\mathcal{Z}_1, \ldots, \mathcal{Z}_n) = S(\mathbb{D}_n)$ with $\mathbb{D}_n = \frac{1}{n}\sum_{i=1}^{n} \delta_{\mathcal{Z}_i}$ being the corresponding (random) empirical measure. Then, if S is a continuous operator, the sequence $\mathcal{L}_n(S; \mathbb{D}_n), n \in \mathbb{N}$, of empirical bootstrap approximations of $\mathcal{L}_n(S; \mathrm{P})$ is qualitatively robust for all $\mathrm{P} \in \mathcal{M}_1(\mathcal{Z})$.*

A.2 Proofs

We are now ready to give our proofs.

***Proof** (of **Lemma 2.9**)* Define $e_{(x,x')} : \mathcal{F} \to \mathbb{R}, f \mapsto f(x,x')$, the evaluation map at $(x,x') \in \mathcal{X}^2$. Let $(f_n)_{n\in\mathbb{N}} \subset \mathcal{F}$ be a convergent sequence, such that $d(f_n, f) \to 0$ for some $f \in \mathcal{F}$. Since d dominates the pointwise convergence, it follows that $f_n(x,x') \to f(x,x')$. This yields the continuity of $e_{(x,x')}$, as $e_{(x,x')}(f_n) = f_n(x,x') \to f(x,x') = e_{(x,x')}(f)$.

Furthermore, the assumption $\mathcal{F} \subset \mathcal{L}_0(\mathcal{X}^2)$ implies that, for any $f \in \mathcal{F}$ the real valued map $(x,x') \mapsto f(x,x')$ defined on $\mathcal{X}^2$ is measurable. After applying Lemma III.14 due to [8, p. 70], we then obtain the first assertion. The second assertion now follows from the measurability statement in Tonelli-Fubini's theorem, see [20, p. 148]. □

***Proof** (of **Lemma 2.12**)* Let $\beta \in [0,1]$ and $f, g \in \mathcal{L}_0(\mathcal{X}^2)$, we have

$$\begin{aligned}
&\mathcal{R}_{L,\mathrm{P}}(\beta f + (1-\beta)g)\\
&= \int_{(\mathcal{X}\times\mathcal{Y})^2} L(x,y,x',y',\beta f(x,x') + (1-\beta)g(x,x'))\mathrm{dP}^2(x,y,x',y')\\
&\le \int_{(\mathcal{X}\times\mathcal{Y})^2} \beta L(x,y,x',y',f(x,x')) + (1-\beta)L(x,y,x',y',g(x,x'))\mathrm{dP}^2(x,y,x',y')\\
&= \beta\mathcal{R}_{L,\mathrm{P}}(f) + (1-\beta)\mathcal{R}_{L,\mathrm{P}}(g).
\end{aligned}$$

In the strictly convex case, the inequality turns into a sharp one. □

Proof (of Lemma 2.13) Firstly, we show the inequality

$$\begin{aligned}
&\left|\mathcal{R}_{L,\mathrm{P}}(f) - \mathcal{R}_{L,\mathrm{P}}(g)\right| \\
&= \left|\int L(x, y, x', y', f(x, x')) - L(x, y, x', y', g(x, x'))\mathrm{dP}^2(x, y, x', y')\right| \\
&\le \int \left|L(x, y, x', y', f(x, x')) - L(x, y, x', y', g(x, x'))\right| \mathrm{dP}^2(x, y, x', y') \\
&\le \int |L|_{B,1} \left|f(x, x') - g(x, x')\right| \mathrm{dP}^2(x, y, x', y') \\
&= |L|_{B,1} \, \|f - g\|_{\mathcal{L}_1(\mathrm{P}_X^2)} \, .
\end{aligned}$$

Using the inequality above, the continuity of the risk functional follows immediately. The risk functional is well-defined as L is measurable and only takes values in $[0, \infty)$ as $f, g \in \mathcal{L}_\infty(\mathrm{P}_X^2)$. □

Proof (of Lemma 2.15) Define $L_{z,z'}(t) := L(x, y, x', y', t)$ with $z = (x, y)$ and $z' = (x', y')$ since we consider L as a function of its last argument and the other four arguments are held fixed. Now let $f \in \mathcal{L}_\infty(\mathrm{P}_X^2)$ and $(f_n)_{n\in\mathbb{N}} \subset \mathcal{L}_\infty(\mathrm{P}_X^2)$ be a sequence with $f_n \neq 0, n \geq 1$ and $\|f_n\|_\infty \to 0$ for $n \to \infty$. Without loss of generality, we assume that $\|f_n\|_\infty \leq 1$ for all $n \geq 1$. For $n \geq 1$, we define

$$G_n(z, z') := \frac{L_{z,z'}(f(x, x') + f_n(x, x')) - L_{z,z'}(f(x, x'))}{f_n(x, x')} - D_5 L_{z,z'}(f(x, x')),$$

if $f_n(x, x') \neq 0$, $G_n(z, z') = 0$ else. It now follows that, for all $n \in \mathbb{N}$,

$$\begin{aligned}
&\left|\frac{\mathcal{R}_{L,\mathrm{P}}(f + f_n) - \mathcal{R}_{L,\mathrm{P}}(f) - \mathcal{R}'_{L,\mathrm{P}}(f) f_n}{\|f_n\|_\infty}\right| \\
&\le \int \left|\frac{L_{z,z'}(f(x, x') + f_n(x, x')) - L_{z,z'}(f(x, x')) - f_n(x, x') D_5 L_{z,z'}(f(x, x'))}{\|f_n\|_\infty}\right| \mathrm{dP}^2(z, z') \\
&\le \int \left|G_n(z, z')\right| \mathrm{dP}^2(z, z').
\end{aligned}$$

Since L is differentiable with respect to its fifth argument, $G_n \to 0$ for $n \to \infty$ by definition of f_n. Moreover, the mean value theorem yields for $f_n(x, x') \neq 0$ that there exists a function $g_n : (\mathcal{X} \times \mathcal{Y})^2 \to \mathbb{R}$ with $\left|g_n(z, z')\right| \in [0, f_n(x, x')] \subset [0, 1]$ and

$$\frac{L_{z,z'}(f(x, x')) + f_n(x, x')) - L_{z,z'}(f(x, x'))}{f_n(x, x')} = D_5 L_{z,z'}\big(f(x, x') + g_n(z, z')\big).$$

Since $|D_5 L|$ is a P-integrable Nemitski loss, it follows, for all $(x, y, x', y') \in (\mathcal{X} \times \mathcal{Y})^2$, that

$$\left|D_5 L(x, y, x', y', t)\right| \leq b(x, y, x', y') + h(|t|),$$

with $b \in \mathcal{L}_1(\mathrm{P}^2)$ and $h : [0, \infty) \to [0, \infty)$ an increasing function. Combining these two considerations, we get

$$\begin{aligned}
&\left|\frac{L_{z,z'}(f(x, x')) + f_n(x, x')) - L_{z,z'}(f(x, x'))}{f_n(x, x')}\right| \\
&\leq b(x, y, x', y') + h\left(\left|f(x, x') + g_n(x, y, x', y')\right|\right) \\
&\leq b(x, y, x', y') + h\left(\|f\|_\infty + 1\right),
\end{aligned}$$

for all $n \geq 1$ with $f_n(x, x') \in (0, 1]$. It follows for all $(x, y, x', y') \in (\mathcal{X} \times \mathcal{Y})^2$ that

$$0 \leq G_n(x, y, x', y') \leq 2b(x, y, x', y') + 2h\left(\|f\|_\infty + 1\right).$$

The assertion now follows from Lebesgue's theorem of dominated convergence.

□

Proof (of Lemma 2.18) Follows immediately from the definition of a convex or Lipschitz continuous pairwise loss function, respectively. □

Proof (of Lemma 2.19)

(i) We immediately obtain, for all $x, x' \in \mathcal{X}$, $y, y' \in \mathcal{Y}$,

$$\inf_{t \in \mathbb{R}} L^*(x, y, x', y', t) \leq L^*(x, y, x', y', 0) = L(x, y, x', y', 0) - L(x, y, x', y', 0) = 0.$$

(ii) For all $f \in \mathcal{H}$, we have

$$\begin{aligned}
\left|\mathcal{R}_{L^*,\mathrm{P}}(f)\right| &= \left|\mathbb{E}_{\mathrm{P}^2}\left[L^*(X, Y, X', Y', f(X, X'))\right]\right| \\
&\leq \mathbb{E}_{\mathrm{P}^2}\left[\left|L(X, Y, X', Y', f(X, X')) - L(X, Y, X', Y', 0)\right|\right] \\
&\leq |L|_1 \, \mathbb{E}_{\mathrm{P}^2}\left[\left|f(X, X')\right|\right],
\end{aligned}$$

which proves (5). The inequality (6) follows from Definition 2.8 and the calculations given above.

(iii) As $0 \in \mathcal{H}$, we obtain

$$\inf_{f \in \mathcal{H}} \mathcal{R}^{\mathrm{reg}}_{L^*,\mathrm{P},\lambda}(f) \leq \mathcal{R}^{\mathrm{reg}}_{L^*,\mathrm{P},\lambda}(0) = 0 = \mathcal{R}_{L^*,\mathrm{P}}(0).$$

Hence $\inf_{f \in \mathcal{H}} \mathcal{R}_{L^*,\mathrm{P}}(f) \leq 0$.

(iv) Due to (iii) $\mathcal{R}^{\text{reg}}_{L^*,\mathrm{P},\lambda}(f_{L^*,\mathrm{P},\lambda}) \leq 0$. As L is a non-negative function, we obtain

$$\begin{aligned}\lambda \left\| f_{L^*,\mathrm{P},\lambda} \right\|_{\mathcal{H}}^2 &\leq -\mathcal{R}_{L^*,\mathrm{P}}(f_{L^*,\mathrm{P},\lambda}) \\ &= \mathbb{E}_{\mathrm{P}^2}\left[L(X,Y,X',Y',0) - L(X,Y,X',Y',f_{L^*,\mathrm{P},\lambda}(X,X'))\right] \\ &\leq \mathbb{E}_{\mathrm{P}^2}\left[L(X,Y,X',Y',0)\right] = \mathcal{R}_{L,\mathrm{P}}(0)\end{aligned}$$

and thus (7) follows. To prove (8), we consider

$$\begin{aligned}0 &\leq -\mathcal{R}^{\text{reg}}_{L^*,\mathrm{P},\lambda}(f_{L^*,\mathrm{P},\lambda}) \\ &= \mathbb{E}_{\mathrm{P}^2}\left[L(X,Y,X',Y',0) - L(X,Y,X',Y',f_{L,\mathrm{P},\lambda}(X,X'))\right] - \lambda \left\| f_{L^*,\mathrm{P},\lambda} \right\|_{\mathcal{H}}^2 \\ &\overset{L \geq 0}{\leq} \mathbb{E}_{\mathrm{P}^2}\left[L(X,Y,X',Y',0)\right] = \mathcal{R}_{L,\mathrm{P}}(0).\end{aligned}$$

Furthermore, we obtain

$$-\left|L\right|_1 \mathbb{E}_{\mathrm{P}^2}\left[\left|f_{L^*,\mathrm{P},\lambda}(X,X')\right|\right] + \lambda \left\| f_{L^*,\mathrm{P},\lambda} \right\|_{\mathcal{H}}^2 \leq \mathcal{R}^{\text{reg}}_{L^*,\mathrm{P},\lambda}(f_{L^*,\mathrm{P},\lambda}) \leq \mathcal{R}^{\text{reg}}_{L^*,\mathrm{P},\lambda}(0) = 0,$$

which yields (9). Using (9) and the reproducing property, we get for $f_{L^*,\mathrm{P},\lambda} \neq 0$ that

$$\begin{aligned}\left\| f_{L^*,\mathrm{P},\lambda} \right\|_\infty \leq \|k\|_\infty \left\| f_{L^*,\mathrm{P},\lambda} \right\|_{\mathcal{H}} &\leq \|k\|_\infty \sqrt{\lambda^{-1} \left|L\right|_1 \mathbb{E}_{\mathrm{P}^2}\left[\left|f_{L^*,\mathrm{P},\lambda}(X,X')\right|\right]} \\ &\leq \|k\|_\infty \sqrt{\lambda^{-1} \left|L\right|_1 \left\| f_{L^*,\mathrm{P},\lambda} \right\|_\infty},\end{aligned}$$

which is finite as k is a bounded kernel. Hence $\left\| f_{L^*,\mathrm{P},\lambda} \right\|_\infty \leq \|k\|_\infty^2 \lambda^{-1} \left|L\right|_1$. The case $f_{L^*,\mathrm{P},\lambda} = 0$ is trivial. If $f_{L^*,\mathrm{P},\lambda} \neq 0$, the inequality (11) now follows, immediately, as

$$\begin{aligned}\mathcal{R}_{L^*,\mathrm{P}}(f_{L^*,\mathrm{P},\lambda}) &= \mathbb{E}_{\mathrm{P}^2}\left[L(X,Y,X',Y',f_{L^*,\mathrm{P},\lambda}(X,X')) - L(X,Y,X',Y',0)\right] \\ &\leq \mathbb{E}_{\mathrm{P}^2}\left[\left|L\right|_1 \left|f_{L^*,\mathrm{P},\lambda}(X,X') - 0\right|\right] \\ &= \left|L\right|_1 \mathbb{E}_{\mathrm{P}^2}\left[\left|f(X,X')\right|\right] \\ &\leq \left|L\right|_1 \left\| f_{L^*,\mathrm{P},\lambda} \right\|_\infty \\ &\leq \lambda^{-1} \left|L\right|_1^2 \|k\|_\infty^2 .\end{aligned}$$

(v) We have, for all $(x, y, x, y', t) \in (\mathcal{X} \times \mathcal{Y})^2 \times \mathbb{R}$,

$$\begin{aligned}
&D_5 L^*(x, y, x', y', t) \\
&= \lim_{\substack{h\to 0\\ h\neq 0}} \frac{L^*(x, y, x', y', t+h) - L^*(x, y, x', y', t)}{h} \\
&= \lim_{\substack{h\to 0\\ h\neq 0}} \frac{L(x, y, x', y', t+h) - L(x, y, x', y', 0) - L(x, y, x', y', t) + L(x, y, x', y', 0)}{h} \\
&= \lim_{\substack{h\to 0\\ h\neq 0}} \frac{L(x, y, x', y', t+h) - L(x, y, x', y', t)}{h} \\
&= D_5 L(x, y, x', y', t).
\end{aligned}$$

□

***Proof** (of Lemma 2.20)* Using (5) from Lemma 2.19, we have

$$\left|\mathcal{R}_{L^*,\mathrm{P}}(f)\right| \leq |L|_1 \, \mathbb{E}_{\mathrm{P}^2}\left[\left|f(X, X')\right|\right] < \infty,$$

for any $f \in \mathcal{L}_1(\mathrm{P}^2_X)$. Inequality (6) yields that $\mathcal{R}^{\mathrm{reg}}_{L^*,\mathrm{P},\lambda}(f) > -\infty$. □

***Proof** (of Theorem 2.22)* Let us assume that the mapping $f \mapsto \lambda \|f\|^2_{\mathcal{H}} + \mathcal{R}_{L^*,\mathrm{P}}(f)$ has two minimizers $f_1, f_2 \in \mathcal{H}$ with $f_1 \neq f_2$.

(i) By the parallelogram identity, we then find

$$\left\|\frac{1}{2}(f_1 + f_2)\right\|^2_{\mathcal{H}} < \frac{1}{2}\left(\|f_1\|^2_{\mathcal{H}} + \|f_2\|^2_{\mathcal{H}}\right).$$

As L is a convex pairwise loss function, L^* is a convex shifted pairwise loss function due to Lemma 2.18, and Lemma 2.12 yields that $\mathcal{R}_{L^*,\mathrm{P}}$ is also a convex function. The convexity of the map $f \mapsto \mathcal{R}_{L^*,\mathrm{P}}(f)$ and

$$\lambda \|f_1\|^2_{\mathcal{H}} + \mathcal{R}_{L^*,\mathrm{P}}(f_1) = \lambda \|f_2\|^2_{\mathcal{H}} + \mathcal{R}_{L^*,\mathrm{P}}(f_2)$$

yield for $f^* := \frac{1}{2}(f_1 + f_2)$ that

$$\begin{aligned}
\lambda \|f^*\|^2_{\mathcal{H}} + \mathcal{R}_{L^*,\mathrm{P}}(f^*) &< \frac{\lambda}{2}\left(\|f_1\|^2_{\mathcal{H}} + \|f_2\|^2_{\mathcal{H}}\right) + \frac{1}{2}\mathcal{R}_{L^*,\mathrm{P}}(f_1) + \frac{1}{2}\mathcal{R}_{L^*,\mathrm{P}}(f_2) \\
&< \lambda \|f_1\|^2_{\mathcal{H}} + \mathcal{R}_{L^*,\mathrm{P}}(f_1),
\end{aligned}$$

i.e., f_1 is not a minimizer of $f \mapsto \lambda \|f\|^2_{\mathcal{H}} + \mathcal{R}_{L^*,\mathrm{P}}(f)$. Consequently, the assumption that there are two minimizers is false by contradiction.

(ii) This condition implies $\left|\mathcal{R}_{L^*,\mathrm{P}}(f)\right| < \infty$ due to Lemma 2.20 and the assertion follows from (i).

□

***Proof** (of Theorem 2.23)* Since the kernel k is measurable, its RKHS $\mathcal{H}$ consists of measurable functions. Moreover, k is bounded and thus $\mathrm{id} : \mathcal{H} \to \mathcal{L}_\infty(\mathrm{P}_X^2)$ is continuous. Additionally, L is non-negative and hence $-\infty < L^*(x, y, x', y', t) < \infty$ for all $(x, y, x', x', t) \in (\mathcal{X} \times \mathcal{Y})^2 \times \mathbb{R}$. Thus L^* is continuous by the convexity of L^* with respect to the fifth argument. Therefore, Lemma 2.13 yields that $\mathcal{R}_{L^*,\mathrm{P}} : \mathcal{L}_\infty(\mathrm{P}_X^2) \to \mathbb{R}$ is continuous and hence $\mathcal{R}_{L^*,\mathrm{P}} : \mathcal{H} \to \mathbb{R}$ is continuous, because $\mathcal{H} \subset \mathcal{L}_\infty(\mathrm{P}_X^2)$. Furthermore, Lemma 2.12 provides the convexity of this mapping. It follows that $f \mapsto \lambda \|f\|_{\mathcal{H}}^2 + \mathcal{R}_{L^*,\mathrm{P}}(f)$ is convex, because $f \mapsto \lambda \|f\|_{\mathcal{H}}^2$ is convex. Lemma A.1.5 shows that if $\mathcal{R}^{\mathrm{reg}}_{L^*,\mathrm{P},\lambda}(\cdot)$ is convex and continuous and additionally $\mathcal{R}^{\mathrm{reg}}_{L^*,\mathrm{P},\lambda}(f) \to \infty$ for $\|f\|_{\mathcal{H}} \to \infty$, then $\mathcal{R}^{\mathrm{reg}}_{L^*,\mathrm{P},\lambda}(\cdot)$ has a minimizer. Therefore it is only left to show that this limit is infinite. We have

$$\begin{aligned}
\mathcal{R}^{\mathrm{reg}}_{L^*,\mathrm{P},\lambda}(f) &\overset{(6)}{\geq} -|L|_1 \, \mathbb{E}_{\mathrm{P}^2}\left[\left|f(X, X')\right|\right] + \lambda \|f\|_{\mathcal{H}}^2 \\
&\geq -|L|_1 \, \|f\|_\infty + \lambda \|f\|_{\mathcal{H}}^2 \\
&\overset{(3)}{\geq} -|L|_1 \, \|k\|_\infty \, \|f\|_{\mathcal{H}} + \lambda \|f\|_{\mathcal{H}}^2 \to \infty,
\end{aligned}$$

for $\|f\|_{\mathcal{H}} \to \infty$, as $|L|_1 \, \|k\|_\infty \in [0, \infty)$ and $\lambda > 0$. □

We need the following auxiliary lemma in order to prove the existence of minimizers in the non-convex case.

Lemma A.2.1 *Let $r \in (0, \infty)$. If $f_0 \in \mathcal{H}$ and if the sequence $(f_\ell)_{\ell\in\mathbb{N}} \subset B_r(f_0) := \{f \in \mathcal{H} : \|f - f_0\|_{\mathcal{H}} \leq r\}$, then there exists a subsequence $(f_{\eta(\ell)})_{\ell\in\mathbb{N}} \subset (f_\ell)_{\ell\in\mathbb{N}} \subset B_r(f_0)$ with $\eta : \mathbb{N} \to \mathbb{N}$ increasing and $f^* \in B_r(f_0)$ such that*

$$\left\|f^*\right\|_{\mathcal{H}} \leq \liminf_{\ell\to\infty} \left\|f_{\eta(\ell)}\right\|_{\mathcal{H}}$$

and

$$\lim_{\ell\to\infty} f_{\eta(\ell)}(x, x') = f^*(x, x'), \quad \forall (x, x') \in \mathcal{X}^2.$$

Proof The closed ball $B_r(f_0) \subset \mathcal{H}$ is weakly compact and hence there exists a subsequence $(f_{\eta(\ell)})_{\ell\in\mathbb{N}} \subset B_r(f_0)$ weakly converging to some $f^* \in B_r(f_0)$, i.e.,

$$\lim_{\ell\to\infty} \left\langle f_{\eta(\ell)}, f\right\rangle_{\mathcal{H}} = \left\langle f^*, f\right\rangle_{\mathcal{H}}, \quad \forall f \in \mathcal{H}.$$

Let $f = f^*$, then by using the Cauchy–Schwarz inequality, it follows

$$\|f^*\|_{\mathcal{H}}^2 = \langle f^*, f^* \rangle_{\mathcal{H}} = \lim_{\ell\to\infty} \langle f_{\eta(\ell)}, f^* \rangle_{\mathcal{H}} \le \liminf_{\ell\to\infty} \|f_{\eta(\ell)}\|_{\mathcal{H}} \|f^*\|_{\mathcal{H}},$$

which implies $\|f^*\|_{\mathcal{H}} \le \liminf_{\ell\to\infty} \|f_{\eta(\ell)}\|_{\mathcal{H}}$. Let $f = \Phi(x, x'), (x, x') \in \mathcal{X}^2$, then the reproducing property yields the remaining assertion

$$\lim_{\ell\to\infty} f_{\eta(\ell)}(x, x') = \lim_{\ell\to\infty} \langle f_{\eta(\ell)}, \Phi(x, x') \rangle_{\mathcal{H}} = \langle f^*, \Phi(x, x') \rangle_{\mathcal{H}} = f^*(x, x').$$

□

Proof (of Theorem 2.24) For every $\ell \in \mathbb{N}$, set $f_\ell \in \mathcal{H}$ such that

$$\mathcal{R}^{\mathrm{reg}}_{L,\mathrm{P},\lambda}(f_\ell) = \mathcal{R}_{L,\mathrm{P}}(f_\ell) + \lambda \|f_\ell\|_{\mathcal{H}}^2 \le \inf_{f\in\mathcal{H}} \mathcal{R}_{L,\mathrm{P}}(f) + \lambda \|f\|_{\mathcal{H}}^2 + \frac{1}{\ell}. \tag{13}$$

Taking $f = f_0$, we conclude that

$$\lambda \|f_\ell\|_{\mathcal{H}}^2 \le \mathcal{R}_{L,\mathrm{P}}(f_0) + \lambda \|f_0\|_{\mathcal{H}}^2 + 1$$

and thus $f_\ell \in B_r(0) = \{f \in \mathcal{H} : \|f\|_{\mathcal{H}} \le r\}$ with $r := \lambda^{-1}\big(\mathcal{R}_{L,\mathrm{P}}(f_0) + \lambda \|f_0\|_{\mathcal{H}}^2 + 1\big)$. Application of Lemma A.2.1 yields that there exists a subsequence $(f_{\eta(\ell)})_{\ell\in\mathbb{N}} \subset B_r(0)$ and some $f^* \in B_r(0)$ such that $\|f^*\|_{\mathcal{H}} \le \liminf_{\ell\to\infty} \|f_{\eta(\ell)}\|_{\mathcal{H}}$ and $f_{\eta(\ell)}(x, x') \to f^*(x, x')$ for all $(x, x') \in \mathcal{X}^2$. By the Lipschitz continuity of L, it follows

$$\begin{aligned}
&\left|L(x, y, x', y', f_{\eta(\ell)}(x, x')) - L(x, y, x', y', f_0(x, x'))\right| \\
&\le |L|_1 \left|f_{\eta(\ell)}(x, x') - f_0(x, x')\right| \\
&\le |L|_1 \cdot \left(\left|f_{\eta(\ell)}(x, x')\right| + \left|f_0(x, x')\right|\right) \\
&\le |L|_1 \left(\left\|f_{\eta(\ell)}\right\|_\infty + \|f_0\|_\infty\right) \\
&\le |L|_1 \left(\left\|f_{\eta(\ell)}\right\|_{\mathcal{H}} + \|f_0\|_{\mathcal{H}}\right) \cdot \|k\|_\infty \\
&\le 2r\,|L|_1\,\|k\|_\infty .
\end{aligned}$$

Therefore

$$\begin{aligned}
L(x, y, x', y', f_{\eta(\ell)}(x, x')) &\le L(x, y, x', y', f_0(x, x')) + 2r\,|L|_1\,\|k\|_\infty < \infty \\
L(x, y, x', y', f_{\eta(\ell)}(x, x')) &\ge -L(x, y, x', y', f_0(x, x')) - 2r\,|L|_1\,\|k\|_\infty > -\infty
\end{aligned}$$

with the upper and lower bound being P^2-integrable. Since $f_{\eta(\ell)} \to f^*$ pointwise for every $(x, x') \in \mathcal{X}^2$, we have by the continuity of L that, for all $(x, x') \in \mathcal{X}^2$ and all $(y, y') \in \mathcal{Y}^2$,

$$\lim_{\ell\to\infty} L(x, y, x', y', f_{\eta(\ell)}(x, x')) = L(x, y, x', y', f^*(x, x')).$$

Lebesgue's theorem of dominated convergence yields $\lim_{\ell\to\infty} \mathcal{R}_{L,\mathrm{P}}(f_{\eta(\ell)}) = \mathcal{R}_{L,\mathrm{P}}(f^*)$. Taking the limit inferior on both sides of inequality (13) gives the result

$$\mathcal{R}^{\mathrm{reg}}_{L,\mathrm{P},\lambda}(f^*) \le \inf_{f\in\mathcal{H}} \mathcal{R}_{L,\mathrm{P}}(f) + \lambda \, \|f\|^2_{\mathcal{H}},$$

which means that f^* is a minimizer for the regularized risk. □

Let us denote the partial derivative of L with respect to the fifth argument by $D_5 L$.

Lemma A.2.2 *Let Assumption 3.1 (i)–(iii) be satisfied. Let $f_{L^*,\mathrm{P},\lambda} \in \mathcal{H}$ be any fixed minimizer of $\inf_{f\in\mathcal{H}} \left(\mathcal{R}_{L^*,\mathrm{P}}(f) + \lambda \, \|f\|^2_{\mathcal{H}}\right)$. Then we have, for any $g \in \mathcal{H}$,*

$$\mathbb{E}_{\mathrm{P}^2}\left[D_5 L(X, Y, X', Y', f_{L^*,\mathrm{P},\lambda}(X, X'))g(X, X')\right] + 2\lambda \left\langle f_{L^*,\mathrm{P},\lambda}, g\right\rangle_{\mathcal{H}} = 0.$$

Proof To shorten the notation, we write $f_{\mathrm{P}} := f_{L^*,\mathrm{P},\lambda}$, as we consider L^* and λ to be fixed in this proof. Let $g \in \mathcal{H}$. We define

$$\tilde{G} : [-1, 1] \to \mathbb{R}, \qquad \tilde{G}(t) = \mathcal{R}_{L^*,\mathrm{P}}(f_{\mathrm{P}} + tg) + \lambda \, \|f_{\mathrm{P}} + tg\|^2_{\mathcal{H}}.$$

$\tilde{G}$ is continuous as it is a composition of continuous functions. Recall that the derivatives of L and L^* with respect to the fifth argument are identical, because L and L^* only differ by the term $L(x, y, x', y', 0)$. For $t \neq 0$, we obtain by using the Lipschitz continuity of L that

$$
\begin{aligned}
&\frac{\tilde{G}(t) - \tilde{G}(0)}{t} \\
&= \frac{1}{t}\left(\mathcal{R}_{L^*,\mathrm{P}}(f_{\mathrm{P}} + tg) + \lambda \, \|f_{\mathrm{P}} + tg\|^2_{\mathcal{H}} - \mathcal{R}_{L^*,\mathrm{P}}(f_{\mathrm{P}}) - \lambda \, \|f_{\mathrm{P}}\|^2_{\mathcal{H}}\right) \\
&= \frac{1}{t}\int_{(\mathcal{X}\times\mathcal{Y})^2} L\big(x, y, x', y', f_{\mathrm{P}}(x, x') + tg(x, x')\big) - L\big(x, y, x', y', f_{\mathrm{P}}(x, x')\big) \mathrm{dP}^2(x, y, x', y') \\
&\quad + \frac{1}{t}\lambda \, \langle f_{\mathrm{P}} + tg, f_{\mathrm{P}} + tg\rangle_{\mathcal{H}} - \frac{1}{t}\lambda \, \langle f_{\mathrm{P}}, f_{\mathrm{P}}\rangle_{\mathcal{H}} \qquad (14) \\
&= \frac{1}{t}\int_{(\mathcal{X}\times\mathcal{Y})^2} L\big(x, y, x', y', f_{\mathrm{P}}(x, x') + tg(x, x')\big) - L\big(x, y, x', y', f_{\mathrm{P}}(x, x')\big) \mathrm{dP}^2(x, y, x', y') \\
&\quad + 2\lambda \, \langle f_{\mathrm{P}}, g\rangle_{\mathcal{H}} + t \, \|g\|^2_{\mathcal{H}} \\
&\le \frac{1}{|t|}\int_{(\mathcal{X}\times\mathcal{Y})^2} |L|_1 \left|tg(x, x')\right| \mathrm{dP}^2(x, y, x', y') + 2\lambda \, \langle f_{\mathrm{P}}, g\rangle_{\mathcal{H}} + t \, \|g\|^2_{\mathcal{H}}
\end{aligned}
$$

$$= |L|_1 \int_{(\mathcal{X}\times\mathcal{Y})^2} |g(x,x')|\, \mathrm{dP}^2(x,y,x',y') + 2\lambda \langle f_{\mathrm{P}}, g\rangle_{\mathcal{H}} + t\, \|g\|_{\mathcal{H}}^2$$
$$\leq |L|_1 \|g\|_\infty + 2\lambda \langle f_{\mathrm{P}}, g\rangle_{\mathcal{H}} + t\, \|g\|_{\mathcal{H}}^2$$
$$< \infty.$$

Furthermore, we have for all $(x, y, x', y') \in (\mathcal{X} \times \mathcal{Y})^2$,

$$\lim_{t\to 0} \frac{1}{t}\Big(L\big(x,y,x',y', f_{\mathrm{P}}(x,x') + tg(x,x')\big) - L\big(x,y,x',y', f_{\mathrm{P}}(x,x')\big)\Big)$$
$$= D_5L(x,y,x',y', f_{\mathrm{P}}(x,x'))g(x,x').$$

Therefore, (14) and an application of Lebesgue's theorem of dominated convergence yield

$$\lim_{t\to 0} \frac{\tilde{G}(t) - \tilde{G}(0)}{t}$$
$$= \int_{(\mathcal{X}\times\mathcal{Y})^2} D_5L(x,y,x',y', f_{\mathrm{P}}(x,x'))g(x,x')\mathrm{dP}^2(x,y,x',y') + 2\lambda \langle f_{\mathrm{P}}, g\rangle_{\mathcal{H}}.$$

We know from Lemma 2.19(iii) that

$$\tilde{G}(0) = \mathcal{R}^{\mathrm{reg}}_{L^*,\mathrm{P},\lambda}(f_{\mathrm{P}}) = \inf_{f\in\mathcal{H}} \mathcal{R}^{\mathrm{reg}}_{L^*,\mathrm{P},\lambda}(f) \leq 0$$

and therefore $\tilde{G}(t) \geq \tilde{G}(0)$ which yields $\tilde{G}(t) - \tilde{G}(0) \geq 0$. This inequality also holds for the function $-g$. Hence the desired identity follows. □

Theorem A.2.3 *Let Assumption 3.1 be satisfied and let* $\mathrm{P}, \mathrm{Q} \in \mathcal{M}_1(\mathcal{X} \times \mathcal{Y})$. *The function* $G : \mathbb{R} \times \mathcal{H} \to \mathcal{H}$ *defined by*

$$G(\varepsilon, f) := 2\lambda f + \mathbb{E}_{\mathrm{P}^2_\varepsilon}\left[D_5L(X,Y,X',Y', f(X,X'))\Phi(X,X')\right]$$

with $\mathrm{P}_\varepsilon = (1-\varepsilon)\mathrm{P} + \varepsilon\mathrm{Q}$ *is continuously differentiable and the partial derivative* $\frac{\partial G}{\partial f}(0, f)$ *is invertible for all* $f \in \mathcal{H}$.

Proof We use Theorem A.1.9 and will show that $\frac{\partial G}{\partial \varepsilon}$ and $\frac{\partial G}{\partial f}$ are continuous. To shorten the notation in the proof, set $L'_f(X,Y,X',Y') := D_5L(X,Y,X',Y', f(X,X'))$ and $L''_f(X,Y,X',Y') := D_5L'_f(X,Y,X',Y')$. Note that for $\varepsilon \in \mathbb{R}$ and $f \in \mathcal{H}$,

$$\begin{aligned}
&\frac{\partial G}{\partial \varepsilon}(\varepsilon, f) \\
&= -2(1-\varepsilon)\mathbb{E}_{\mathrm{P}^2}\left[L'_f(X,Y,X',Y')\Phi(X,X')\right] \\
&+ (1-2\varepsilon)\mathbb{E}_{\mathrm{P}\otimes\mathrm{Q}}\left[L'_f(X,Y,X',Y')\Phi(X,X')\right] \\
&+ (1-2\varepsilon)\mathbb{E}_{\mathrm{Q}\otimes\mathrm{P}}\left[L'_f(X,Y,X',Y')\Phi(X,X')\right] \\
&+ 2\varepsilon\mathbb{E}_{\mathrm{Q}^2}\left[L'_f(X,Y,X',Y')\Phi(X,X')\right].
\end{aligned}$$

For $\varepsilon, \tilde{\varepsilon} \in \mathbb{R}$ and $f, \tilde{f} \in \mathcal{H}$, we have

$$\begin{aligned}
\frac{\partial G}{\partial \varepsilon}(\varepsilon, f) - \frac{\partial G}{\partial \varepsilon}(\tilde{\varepsilon}, \tilde{f}) &= \left(\frac{\partial G}{\partial \varepsilon}(\varepsilon, f) - \frac{\partial G}{\partial \varepsilon}(\varepsilon, \tilde{f})\right) + \left(\frac{\partial G}{\partial \varepsilon}(\varepsilon, \tilde{f}) - \frac{\partial G}{\partial \varepsilon}(\tilde{\varepsilon}, \tilde{f})\right) \\
&=: \partial G_1 + \partial G_2. \qquad (15)
\end{aligned}$$

Here ∂G_1 equals

$$\begin{aligned}
&-2(1-\varepsilon)\mathbb{E}_{\mathrm{P}^2}\left[(L'_f - L'_{\tilde{f}})(X,Y,X',Y')\Phi(X,X')\right] + \\
&\qquad (1-2\varepsilon)\mathbb{E}_{\mathrm{P}\otimes\mathrm{Q}}\left[(L'_f - L'_{\tilde{f}})(X,Y,X',Y')\Phi(X,X')\right] \\
&+ (1-2\varepsilon)\mathbb{E}_{\mathrm{Q}\otimes\mathrm{P}}\left[(L'_f - L'_{\tilde{f}})(X,Y,X',Y')\Phi(X,X')\right] + \\
&\qquad 2\varepsilon\mathbb{E}_{\mathrm{Q}^2}\left[(L'_f - L'_{\tilde{f}})(X,Y,X',Y')\Phi(X,X')\right].
\end{aligned}$$

Set $Z := \{(x,y,x',y') \in (\mathcal{X}\times\mathcal{Y})^2 : f(x,x') \neq \tilde{f}(x,x')\}$. We compute the expectation with respect to probability measures $\mathrm{P}_1, \mathrm{P}_2 \in \mathcal{M}_1(\mathcal{X}\times\mathcal{Y})$ first, in order to simplify the term given above. An application of the mean value theorem (MVT) and the boundedness of the second derivative of the pairwise loss function L yield, for all $\mathrm{P}_1, \mathrm{P}_2 \in \mathcal{M}_1(\mathcal{X}\times\mathcal{Y})$, that

$$\begin{aligned}
&\left\|\mathbb{E}_{\mathrm{P}_1\otimes\mathrm{P}_2}[(L'_f - L'_{\tilde{f}})(X,Y,X',Y')\Phi(X,X')]\right\|_{\mathcal{H}} \\
&\le \mathbb{E}_{\mathrm{P}_1\otimes\mathrm{P}_2}\left[\left\|(L'_f - L'_{\tilde{f}})(X,Y,X',Y')\Phi(X,X')\right\|_{\mathcal{H}}\right] \\
&= \int_{(\mathcal{X}\times\mathcal{Y})^2} \left|(L'_f - L'_{\tilde{f}})(x,y,x',y')\right| \|\Phi(x,x')\|_{\mathcal{H}} \mathrm{d}(\mathrm{P}_1\otimes\mathrm{P}_2)(x,y,x',y')
\end{aligned}$$

$$= \int_Z \left| f(x,x') - \tilde{f}(x,x') \right| \left| \frac{(L'_f - L'_{\tilde{f}})(x,y,x',y')}{f(x,x') - \tilde{f}(x,x')} \right| \left\| \Phi(x,x') \right\|_{\mathcal{H}} \mathrm{d}(\mathrm{P}_1 \otimes \mathrm{P}_2)(x,y,x',y')$$

$$\stackrel{\text{MVT}}{\leq} \int_Z \left| f(x,x') - \tilde{f}(x,x') \right| c_{L,2} \left\| \Phi(x,x') \right\|_{\mathcal{H}} \mathrm{d}(\mathrm{P}_1 \otimes \mathrm{P}_2)(x,y,x',y')$$

$$\leq \int_Z c_{L,2} \left\| f - \tilde{f} \right\|_{\infty} \|k\|_{\infty} \mathrm{d}(\mathrm{P}_1 \otimes \mathrm{P}_2)(x,y,x',y')$$

$$\stackrel{(3)}{\leq} c_{L,2} \left\| f - \tilde{f} \right\|_{\mathcal{H}} \|k\|_{\infty}^2 < \infty.$$

As this upper bound is valid for all $\mathrm{P}_1, \mathrm{P}_2 \in \mathcal{M}_1(\mathcal{X} \times \mathcal{Y})$, the desired result for ∂G_1 from (15) follows. We have

$$\begin{aligned}
&\|\partial G_1\|_{\mathcal{H}} \\
&\leq (2\,|1-\varepsilon| + 2\,|1-2\varepsilon| + 2\,|\varepsilon|) \cdot \left\| \mathbb{E}_{\mathrm{P}_1 \otimes \mathrm{P}_2}\left[(L'_f - L'_{\tilde{f}})(X,Y,X',Y')\Phi(X,X')\right] \right\|_{\mathcal{H}} \\
&\leq (2(1+|\varepsilon|) + 2(1+2\,|\varepsilon|) + 2\,|\varepsilon|) \cdot \left(c_{L,2} \|k\|_{\infty}^2 \, ||f - \tilde{f}||_{\mathcal{H}}\right) \\
&= (4 + 8\,|\varepsilon|) \cdot \left(c_{L,2} \|k\|_{\infty}^2 \, ||f - \tilde{f}||_{\mathcal{H}}\right).
\end{aligned}$$

Moreover, the term ∂G_2 in (15) equals

$$\begin{aligned}
&2(\varepsilon - \tilde{\varepsilon})\mathbb{E}_{\mathrm{P}^2}\left[L'_{\tilde{f}}(X,Y,X',Y')\Phi(X,X')\right] + 2(\tilde{\varepsilon} - \varepsilon)\mathbb{E}_{\mathrm{P}\otimes\mathrm{Q}}\left[L'_{\tilde{f}}(X,Y,X',Y')\Phi(X,X')\right] \\
&+ 2(\tilde{\varepsilon} - \varepsilon)\mathbb{E}_{\mathrm{Q}\otimes\mathrm{P}}\left[L'_{\tilde{f}}(X,Y,X',Y')\Phi(X,X')\right] + 2(\varepsilon - \tilde{\varepsilon})\mathbb{E}_{\mathrm{Q}^2}\left[L'_{\tilde{f}}(X,Y,X',Y')\Phi(X,X')\right].
\end{aligned}$$

Hence, we have by the boundedness of the first derivative using the same approach as above

$$\begin{aligned}
\|\partial G_2\|_{\mathcal{H}} &\leq 8\,|\varepsilon - \tilde{\varepsilon}|\,\mathbb{E}_{\mathrm{P}_1 \otimes \mathrm{P}_2}\left[\left\| L'_{\tilde{f}}(X,Y,X',Y')\Phi(X,X') \right\|_{\mathcal{H}}\right] \\
&\leq 8\,|\varepsilon - \tilde{\varepsilon}|\,\mathbb{E}_{\mathrm{P}_1 \otimes \mathrm{P}_2}\left[\left| L'_{\tilde{f}}(X,Y,X',Y') \right| \left\| \Phi(X,X') \right\|_{\mathcal{H}}\right] \\
&\stackrel{(1)}{\leq} 8\,|\varepsilon - \tilde{\varepsilon}|\,c_{L,1}\,\|k\|_{\infty}\,.
\end{aligned}$$

Thus we obtain from (15) that

$$\left\| \frac{\partial G}{\partial \varepsilon}(\varepsilon, f) - \frac{\partial G}{\partial \varepsilon}(\tilde{\varepsilon}, \tilde{f}) \right\|_{\mathcal{H}} \leq (4 + 8\,|\varepsilon|) \cdot \left(c_{L,2} \|k\|_{\infty}^2 \, ||f - \tilde{f}||_{\mathcal{H}}\right) + 8 c_{L,1}\,|\varepsilon - \tilde{\varepsilon}|\,\|k\|_{\infty}\,.$$

From this, we obtain for $\varepsilon \to \tilde{\varepsilon}$ and $f \to \tilde{f}$ the continuity of the partial derivative $\frac{\partial G}{\partial \varepsilon}$.

The partial derivative $\frac{\partial G}{\partial f}$ can be expressed as

$$\frac{\partial G}{\partial f}(\varepsilon, f) = 2\lambda \operatorname{id}_{\mathcal{H}} + \mathbb{E}_{\mathrm{P}^2_\varepsilon}\left[D_5 L'_f(X, Y, X', Y')\langle\Phi(X, X'), \cdot\rangle_{\mathcal{H}}\Phi(X, X')\right],$$

for $\varepsilon \in \mathbb{R}$ and $f \in \mathcal{H}$.To prove its continuity, we first observe, for any $\tilde{f} \in \mathcal{H}$,

$$\begin{aligned}
&\frac{\partial G}{\partial f}(\varepsilon, f) - \frac{\partial G}{\partial f}(\varepsilon, \tilde{f}) \\
&= \mathbb{E}_{\mathrm{P}^2_\varepsilon}\big[D_5 L'_f(X, Y, X', Y')\langle\Phi(X, X'), \cdot\rangle_{\mathcal{H}}\Phi(X, X') \\
&\quad - D_5 L'_{\tilde{f}}(X, Y, X', Y')\langle\Phi(X, X'), \cdot\rangle_{\mathcal{H}}\Phi(X, X')\big] \\
&= \mathbb{E}_{\mathrm{P}^2_\varepsilon}\left[D_5(L'_f(X, Y, X', Y') - L'_{\tilde{f}}(X, Y, X', Y'))\langle\Phi(X, X'), \cdot\rangle_{\mathcal{H}}\Phi(X, X')\right].
\end{aligned}$$

By the definition of the local modulus of continuity for the second order derivatives of L, see Definition A.1.4, and the Cauchy–Schwarz inequality, we obtain, for all $(x, x') \in \mathcal{X}^2$, that

$$\begin{aligned}
\left\|\langle\Phi(x, x'), \cdot\rangle\,\Phi(x, x')\right\|_{\mathcal{L}(\mathcal{H},\mathcal{H})} &= \sup_{\substack{h\in\mathcal{H}\\ \|h\|_{\mathcal{H}}\leq 1}} \left\|\langle\Phi(x, x'), h\rangle_{\mathcal{H}}\Phi(x, x')\right\|_{\mathcal{H}} \\
&\leq \sup_{\substack{h\in\mathcal{H}\\ \|h\|_{\mathcal{H}}\leq 1}} \left\|\Phi(x, x')\right\|_{\mathcal{H}}\|h\|_{\mathcal{H}}\left\|\Phi(x, x')\right\|_{\mathcal{H}} \\
&= \left\|\Phi(x, x')\right\|^2_{\mathcal{H}} \overset{(1)}{\leq} \|k\|^2_\infty .
\end{aligned}$$

Hence, for $f, \tilde{f} \in \{g \in \mathcal{H} : \|g\|_{\mathcal{H}} \leq r\}$, we obtain the upper bound

$$\begin{aligned}
&\left\|\frac{\partial G}{\partial f}(\varepsilon, f) - \frac{\partial G}{\partial f}(\varepsilon, \tilde{f})\right\|_{\mathcal{L}(\mathcal{H},\mathcal{H})} \\
&= \left\|\mathbb{E}_{\mathrm{P}^2_\varepsilon}\left[D_5(L'_f(X, Y, X', Y') - L'_{\tilde{f}}(X, Y, X', Y'))\langle\Phi(X, X'), \cdot\rangle_{\mathcal{H}}\Phi(X, X')\right]\right\|_{\mathcal{L}(\mathcal{H},\mathcal{H})} \\
&\leq \mathbb{E}_{\mathrm{P}^2_\varepsilon}\left[\left|D_5(L'_f(X, Y, X', Y') - L'_{\tilde{f}}(X, Y, X', Y'))\right|\left\|\langle\Phi(X, X'), \cdot\rangle_{\mathcal{H}}\Phi(X, X')\right\|_{\mathcal{L}(\mathcal{H},\mathcal{H})}\right] \\
&\leq \|k\|^2_\infty\,\mathbb{E}_{\mathrm{P}^2_\varepsilon}\left[\left|D_5(L'_f(X, Y, X', Y') - L'_{\tilde{f}}(X, Y, X', Y'))\right|\right] \\
&\leq \|k\|^2_\infty\,\omega\left(\|k\|_\infty\left\|f - \tilde{f}\right\|_{\mathcal{H}}\right)_{r\|k\|_\infty} .
\end{aligned}$$

The second difference of partial derivatives we need to consider is the following, in which the integrands are the same, but the probability measures differ. We denote by $T_{\tilde{f}}(x, y, x', y')$ the term $D_5 L'_{\tilde{f}}(x, y, x', y')\langle\Phi(x, x'), \cdot\rangle_{\mathcal{H}}\Phi(x, x')$. We then obtain by elementary calculations,

$$
\begin{aligned}
&\frac{\partial G}{\partial f}(\varepsilon, \tilde{f}) - \frac{\partial G}{\partial f}(\tilde{\varepsilon}, \tilde{f}) \\
&= \mathbb{E}_{\mathrm{P}^2_{\varepsilon}}\left[D_5 L'_{\tilde{f}}(X, Y, X', Y') \langle \Phi(X, X'), \cdot \rangle_{\mathcal{H}} \Phi(X, X')\right] \\
&\quad - \mathbb{E}_{\mathrm{P}^2_{\tilde{\varepsilon}}}\left[D_5 L'_{\tilde{f}}(X, Y, X', Y') \langle \Phi(X, X'), \cdot \rangle_{\mathcal{H}} \Phi(X, X')\right] \\
&= (1-\varepsilon)^2 \mathbb{E}_{\mathrm{P}^2}\left[T_{\tilde{f}}(X, Y, X', Y')\right] + (1-\varepsilon)\varepsilon \mathbb{E}_{\mathrm{P}\otimes\mathrm{Q}}\left[T_{\tilde{f}}(X, Y, X', Y')\right] \\
&\quad + \varepsilon(1-\varepsilon)\mathbb{E}_{\mathrm{Q}\otimes\mathrm{P}}\left[T_{\tilde{f}}(X, Y, X', Y')\right] + \varepsilon^2 \mathbb{E}_{\mathrm{Q}^2}\left[T_{\tilde{f}}(X, Y, X', Y')\right] \\
&\quad - (1-\tilde{\varepsilon})^2 \mathbb{E}_{\mathrm{P}^2}\left[T_{\tilde{f}}(X, Y, X', Y')\right] - (1-\tilde{\varepsilon})\tilde{\varepsilon} \mathbb{E}_{\mathrm{P}\otimes\mathrm{Q}}\left[T_{\tilde{f}}(X, Y, X', Y')\right] \\
&\quad - \tilde{\varepsilon}(1-\tilde{\varepsilon})\mathbb{E}_{\mathrm{Q}\otimes\mathrm{P}}\left[T_{\tilde{f}}(X, Y, X', Y')\right] - \tilde{\varepsilon}^2 \mathbb{E}_{\mathrm{Q}^2}\left[T_{\tilde{f}}(X, Y, X', Y')\right] \\
&= (\tilde{\varepsilon}-\varepsilon)(2-\tilde{\varepsilon}-\varepsilon)\mathbb{E}_{\mathrm{P}^2}\left[T_{\tilde{f}}(X, Y, X', Y')\right] + (\varepsilon-\tilde{\varepsilon})(1-\varepsilon-\tilde{\varepsilon})\mathbb{E}_{\mathrm{P}\otimes\mathrm{Q}}\left[T_{\tilde{f}}(X, Y, X', Y')\right] \\
&\quad + (\varepsilon-\tilde{\varepsilon})(1-\varepsilon-\tilde{\varepsilon})\mathbb{E}_{\mathrm{Q}\times\mathrm{P}}\left[T_{\tilde{f}}(X, Y, X', Y')\right] + (\varepsilon-\tilde{\varepsilon})(\varepsilon+\tilde{\varepsilon})\mathbb{E}_{\mathrm{Q}^2}\left[T_{\tilde{f}}(X, Y, X', Y')\right].
\end{aligned}
$$

Due to the boundedness of the second derivative and the inequality

$$
\left\| \langle \Phi(x, x'), \cdot \rangle_{\mathcal{H}} \Phi(x, x') \right\|_{\mathcal{L}(\mathcal{H},\mathcal{H})} \le \|k\|_\infty^2 \quad \forall (x, x') \in \mathcal{X}^2,
$$

it follows that

$$
\begin{aligned}
\left\| \frac{\partial G}{\partial f}(\varepsilon, \tilde{f}) - \frac{\partial G}{\partial f}(\tilde{\varepsilon}, \tilde{f}) \right\|_{\mathcal{L}(\mathcal{H},\mathcal{H})} &\le c_{L,2} \|k\|_\infty^2 |\varepsilon - \tilde{\varepsilon}| \,(4 + 4|\varepsilon| + 4|\tilde{\varepsilon}|) \\
&= 4 c_{L,2} \|k\|_\infty^2 |\varepsilon - \tilde{\varepsilon}| \,(1 + |\varepsilon| + |\tilde{\varepsilon}|).
\end{aligned}
$$

Hence

$$
\begin{aligned}
&\left\| \frac{\partial G}{\partial f}(\varepsilon, f) - \frac{\partial G}{\partial f}(\tilde{\varepsilon}, \tilde{f}) \right\|_{\mathcal{L}(\mathcal{H},\mathcal{H})} \\
&= \left\| \frac{\partial G}{\partial f}(\varepsilon, f) - \frac{\partial G}{\partial f}(\varepsilon, \tilde{f}) + \frac{\partial G}{\partial f}(\varepsilon, \tilde{f}) - \frac{\partial G}{\partial f}(\tilde{\varepsilon}, \tilde{f}) \right\|_{\mathcal{L}(\mathcal{H},\mathcal{H})} \\
&\le \left\| \frac{\partial G}{\partial f}(\varepsilon, \tilde{f}) - \frac{\partial G}{\partial f}(\tilde{\varepsilon}, \tilde{f}) \right\|_{\mathcal{L}(\mathcal{H},\mathcal{H})} + \left\| \frac{\partial G}{\partial f}(\varepsilon, f) - \frac{\partial G}{\partial f}(\varepsilon, \tilde{f}) \right\|_{\mathcal{L}(\mathcal{H},\mathcal{H})} \\
&\le \|k\|_\infty^2 \, \omega\Big(\|k\|_\infty \, \|f - \tilde{f}\|_{\mathcal{H}} \Big)_{r\|k\|_\infty} + 4 c_{L,2} \|k\|_\infty^2 |\varepsilon - \tilde{\varepsilon}| \,(1 + |\varepsilon| + |\tilde{\varepsilon}|),
\end{aligned}
$$

which yields the continuity of the partial derivative $\frac{\partial G}{\partial f}$ and thus the continuous differentiability of G. Let $f \in \mathcal{H}$ and consider the linear operator $\frac{\partial G}{\partial f}(0, f)$. We obtain

$$\frac{\partial G}{\partial f}(0, f) = 2\lambda \,\mathrm{id}_{\mathcal{H}} + \mathbb{E}_{\mathrm{P}^2}\left[D_5 L'_f(X, Y, X', Y')\langle \Phi(X, X'), \cdot\rangle_{\mathcal{H}} \Phi(X, X')\right].$$

Hence, for all $g, \tilde{g} \in \mathcal{H}$,

$$\begin{aligned}
&\left\langle \frac{\partial G}{\partial f}(0, f)(g), \tilde{g}\right\rangle_{\mathcal{H}} \\
&= 2\lambda \langle g, \tilde{g}\rangle_{\mathcal{H}} + \mathbb{E}_{\mathrm{P}^2}\left[D_5 L'_f(X, Y, X', Y')\langle \Phi(X, X'), g\rangle_{\mathcal{H}} \langle \Phi(X, X'), \tilde{g}\rangle_{\mathcal{H}}\right] \\
&= 2\lambda \langle g, \tilde{g}\rangle_{\mathcal{H}} + \mathbb{E}_{\mathrm{P}^2}\left[D_5 L'_f(X, Y, X', Y') g(X, X')\tilde{g}(X, X')\right].
\end{aligned}$$

Therefore, the linear operator $\left\langle \frac{\partial G}{\partial f}(0, f)(g), \tilde{g}\right\rangle_{\mathcal{H}}$ is symmetric. Hence its spectrum lies in the closed interval $[a, b]$ where

$$a := \inf_{\|g\|_{\mathcal{H}}=1} \left\langle \frac{\partial G}{\partial f}(0, f)(g), g\right\rangle_{\mathcal{H}}, \qquad b := \sup_{\|g\|_{\mathcal{H}}=1} \left\langle \frac{\partial G}{\partial f}(0, f)(g), g\right\rangle_{\mathcal{H}}.$$

Due to Assumption 3.1(iv), L is a convex pairwise loss function. This implies that the second derivative of L with respect to the fifth argument is non-negative. Hence, we obtain, for all $g \in \mathcal{H}$, by the convexity of L

$$\begin{aligned}
&\left\langle \frac{\partial G}{\partial f}(0, f)(g), g\right\rangle_{\mathcal{H}} \\
&= 2\lambda \langle g, g\rangle_{\mathcal{H}} + \mathbb{E}_{\mathrm{P}^2}\left[D_5 L'_f(X, Y, X', Y')\langle \Phi(X, X'), g\rangle_{\mathcal{H}} \langle \Phi(X, X'), g\rangle_{\mathcal{H}}\right] \\
&= 2\lambda \|g\|^2_{\mathcal{H}} + \mathbb{E}_{\mathrm{P}^2}\Big[\underbrace{D_5 L'_f(X, Y, X', Y') g^2(X, X')}_{\geq 0}\Big] \\
&\geq 2\lambda \|g\|^2_{\mathcal{H}}.
\end{aligned}$$

Thus it also applies for normalized functions, hence $a \geq 2\lambda > 0$. This shows that the operator $\frac{\partial G}{\partial f}(0, f)$ is invertible. □

Proof (of Theorem 3.2) The existence and uniqueness of $f_{L^*,\mathrm{P},\lambda}$ follow from Theorem 2.22 and Theorem 2.23. As k is bounded, Lemma 2.19(iv) is applicable and inequalities (10) and (11) are valid. Furthermore, due to Lemma 2.18(ii) L^* is a Lipschitz continuous pairwise loss function, because L is given as such. Define the risk functional $R : \mathcal{L}_1(\mathrm{P}^2) \to \mathbb{R}$ by

$$R(g) := \int_{(\mathcal{X}\times\mathcal{Y})^2} L^*(x, y, x', y', g(x, y, x', y'))\mathrm{dP}^2(x, y, x', y').$$

The operator R is well-defined, because due to the Lipschitz continuity of L^* with respect to its fifth argument, we obtain

$$|R(g)| \leq |L|_1 \int_{(\mathcal{X}\times\mathcal{Y})^2} \left|g(x, y, x', y')\right| \mathrm{dP}^2(x, y, x', y') < \infty$$

as $g \in \mathcal{L}_1(\mathrm{P}^2)$ by definition of R. The continuity of R can be shown as follows. Fix $\delta > 0$ and let $f_1, f_2 \in \mathcal{L}_1(\mathrm{P}^2)$ with $\|f_1 - f_2\|_{\mathcal{L}_1(\mathrm{P}^2)} < \delta$. The Lipschitz continuity of L^* yields

$$\begin{aligned} &|R(f_1) - R(f_2)| \\ &\leq \int_{(\mathcal{X}\times\mathcal{Y})^2} \left|L^*(x, y, x', y', f_1(x, x')) - L^*(x, y, x', y', f_2(x, x'))\right| \mathrm{dP}^2(x, y, x', y') \\ &\leq |L|_1 \int_{(\mathcal{X}\times\mathcal{Y})^2} \left|f_1(x, y, x', y') - f_2(x, y, x', y')\right| \mathrm{dP}^2(x, y, x', y') \\ &< \delta\, |L|_1 , \end{aligned}$$

and so the continuity of R. We can now apply Proposition A.1.6 with $p = 1$, because $R(f)$ exists and is well-defined for all $g \in \mathcal{L}_1(\mathrm{P}^2)$. The subdifferential of R can thus be computed by

$$\begin{aligned} \partial R(g) = \{h \in \mathcal{L}_\infty(\mathrm{P}^2) : h(x, y, x', y') \in \partial L^*(x, y, x', y', g(x, y, x', y')) \\ \text{for } \mathrm{P}^2\text{-almost all } (x, y, x', y')\}. \end{aligned}$$

Now, we infer from [44, p. 172 and Lemma 4.23] that the inclusion map $I : \mathcal{H} \to \mathcal{L}_1(\mathrm{P}^2)$ defined by

$$(If)(x, y, x', y') := f(x, x')$$

is a bounded linear operator. Furthermore, $S : \mathcal{H} \to \mathbb{R}$, $S(g) := \langle f, g\rangle_{\mathcal{H}}$ is a bounded linear operator and it follows that $S(\mathbb{E}_{\mathrm{P}^2}[g]) = \mathbb{E}_{\mathrm{P}^2}[S(g)]$ for bounded linear operators and Bochner integrals, see, e.g., [19, Thm. 3.10.16]. Recall that $p = 1$. Hence $\frac{1}{p} + \frac{1}{p'} = 1$ yields that $p' = \infty$. Moreover, for all $h \in \mathcal{L}_\infty(\mathrm{P}^2)$ and all $f \in \mathcal{H}$, the reproducing property yields with $\Phi : \mathcal{X}^2 \to \mathcal{H} : \Phi(x, x') :=$

$k\big((\cdot,\cdot),(x,x')\big)$ the canonical feature map:

$$
\begin{aligned}
\langle h, If\rangle_{\mathcal{L}_\infty(\mathrm{P}^2),\mathcal{L}_1(\mathrm{P}^2)} &= \mathbb{E}_{\mathrm{P}^2}\left[hIf\right] \\
&= \int_{(\mathcal{X}\times\mathcal{Y})^2} h(x,y,x',y')(If)(x,y,x',y')\mathrm{dP}^2(x,y,x',y') \\
&= \int_{(\mathcal{X}\times\mathcal{Y})^2} h(x,y,x',y')f(x,x')\mathrm{dP}^2(x,y,x',y') \\
&= \int_{(\mathcal{X}\times\mathcal{Y})^2} h(x,y,x',y')\big\langle f,\Phi(x,x')\big\rangle_{\mathcal{H}}\mathrm{dP}^2(x,y,x',y') \\
&= \mathbb{E}_{\mathrm{P}^2}\left[h\left\langle f,\Phi\right\rangle_{\mathcal{H}}\right] = \big\langle f, \mathbb{E}_{\mathrm{P}^2}\left[h\Phi\right]\big\rangle_{\mathcal{H}} = \big\langle \iota\mathbb{E}_{\mathrm{P}^2}\left[h\Phi\right], f\big\rangle_{\mathcal{H}',\mathcal{H}},
\end{aligned}
$$

with $\iota : \mathcal{H} \to \mathcal{H}'$ the Fréchet–Riesz isomorphism, see, e.g., [49, Thm. V.3.6]. Thus the adjoint operator I' of I is given by

$$I'h = \iota\mathbb{E}_{\mathrm{P}^2}\left[h\Phi\right], \quad h \in \mathcal{L}_\infty(\mathrm{P}^2).$$

Moreover, the L^*-risk functional $\mathcal{R}_{L^*,\mathrm{P}} : \mathcal{H} \to \mathbb{R}$ satisfies

$$\mathcal{R}_{L^*,\mathrm{P}} = R \circ I$$

and hence the chain rule for subdifferentials, Lemma A.1.8(iii), see also [19, Thm. 5.3.33], yields

$$\partial\mathcal{R}_{L^*,\mathrm{P}}(f) = \partial(R\circ I)(f) = I'\partial R(If), \quad \forall f \in \mathcal{H}.$$

Applying the formula for $\partial R(f)$ thus yields, for all $f \in \mathcal{H}$,

$$
\begin{aligned}
&\partial\mathcal{R}_{L^*,\mathrm{P}}(f) \\
&= \{\iota\mathbb{E}_{\mathrm{P}^2}\left[h\Phi\right] : h \in \mathcal{L}_\infty(\mathrm{P}^2) \text{ with } h(x,y,x',y') \in \partial L^*(x,y,x',y',f(x,x'))\ \mathrm{P}^2\text{-a.s.}\}.
\end{aligned}
$$

In addition, $f \mapsto \|f\|_{\mathcal{H}}^2$ is Fréchet-differentiable and its derivative at f is $2\iota f$ for all $f \in \mathcal{H}$. By picking suitable representations of $h \in \mathcal{L}_\infty(\mathrm{P}^2)$, Lemma A.1.8 thus gives, for all $f \in \mathcal{H}$,

$$
\begin{aligned}
\partial\mathcal{R}^{\mathrm{reg}}_{L^*,\mathrm{P},\lambda}(f) = 2\lambda\iota f + \{\iota\mathbb{E}_{\mathrm{P}^2}\left[h\Phi\right] : h \in \mathcal{L}_\infty(\mathrm{P}^2)& \\
\text{with } h(x,y,x',y') \in \partial L^*(x,y,x',y',f(x,x'))& \\
\forall (x,y,x',y') \in (\mathcal{X}\times\mathcal{Y})^2\}&
\end{aligned}
$$

for all $f \in \mathcal{H}$. Now recall that $\mathcal{R}^{\mathrm{reg}}_{L^*,\mathrm{P},\lambda}(\cdot)$ has a minimum at $f_{L^*,\mathrm{P},\lambda} \in \mathcal{H}$ and therefore we have $0 \in \partial\mathcal{R}^{\mathrm{reg}}_{L^*,\mathrm{P},\lambda}(f_{L^*,\mathrm{P},\lambda})$ by Lemma A.1.8(iv). This together with the injectivity of ι yields the assertions (i) and (ii). Of course, h will depend on P in general, thus we often write h_{P} instead of h. Let us now show that (iii) is valid. As k is a bounded kernel, we have by inequality (10)

$$\left\| f_{L^*,\mathrm{P},\lambda} \right\|_\infty \le \lambda^{-1} |L|_1 \, \|k\|_\infty^2 =: B_\lambda < \infty .$$

Now part (i) and Proposition A.1.7 with $\delta := 1$ yield, for all $(x, y, x', y') \in (\mathcal{X} \times \mathcal{Y})^2$,

$$\left| h_{\mathrm{P}}(x, y, x', y') \right| \le \sup_{(x,y,x',y') \in (\mathcal{X}\times\mathcal{Y})^2} \left| \partial L^*(x, y, x', y', f_{L^*,\mathrm{P},\lambda}(x, x')) \right| \le |L|_1 .$$

Hence $h_{\mathrm{P}} \in \mathcal{L}_\infty(\mathrm{P}_X^2)$ and the assertion (iii) follows.

To prove (iv), we use (i) and the definition of the subdifferential to obtain, for all $(x, y, x', y') \in (\mathcal{X} \times \mathcal{Y})^2$,

$$\begin{aligned} & h_{\mathrm{P}}(x, y, x', y')(f_{L^*,\mathrm{Q},\lambda}(x, x') - f_{L^*,\mathrm{P},\lambda}(x, x')) \\ & \le L^*(x, y, x', y', f_{L^*,\mathrm{Q},\lambda}(x, x')) - L^*(x, y, x', y', f_{L^*,\mathrm{P},\lambda}(x, x')). \end{aligned}$$

By integrating with respect to Q, we hence obtain

$$\left\langle f_{L^*,\mathrm{Q},\lambda} - f_{L^*,\mathrm{P},\lambda}, \mathbb{E}_{\mathrm{Q}^2}\left[h_{\mathrm{P}}\Phi\right]\right\rangle_{\mathcal{H}} \le \mathcal{R}_{L^*,\mathrm{Q}}(f_{L^*,\mathrm{Q},\lambda}) - \mathcal{R}_{L^*,\mathrm{Q}}(f_{L^*,\mathrm{P},\lambda}).$$

Moreover, an easy calculation yields

$$\begin{aligned} & \left\langle f_{L^*,\mathrm{Q},\lambda} - f_{L^*,\mathrm{P},\lambda}, \mathbb{E}_{\mathrm{Q}^2}\left[h_{\mathrm{P}}\Phi\right] + 2\lambda f_{L^*,\mathrm{P},\lambda}\right\rangle_{\mathcal{H}} + \lambda \left\| f_{L^*,\mathrm{P},\lambda} - f_{L^*,\mathrm{Q},\lambda} \right\|_{\mathcal{H}}^2 \\ & \le \mathcal{R}^{\mathrm{reg}}_{L^*,\mathrm{Q},\lambda}(f_{L^*,\mathrm{Q},\lambda}) - \mathcal{R}^{\mathrm{reg}}_{L^*,\mathrm{Q},\lambda}(f_{L^*,\mathrm{P},\lambda}) \le 0, \end{aligned}$$

and consequently using the representation $f_{L^*,\mathrm{P},\lambda} = -(2\lambda)^{-1}\mathbb{E}_{\mathrm{P}^2}\left[h_{\mathrm{P}}\Phi\right]$, it follows after using the Cauchy–Schwarz inequality that

$$\begin{aligned} \lambda \left\| f_{L^*,\mathrm{P},\lambda} - f_{L^*,\mathrm{Q},\lambda} \right\|_{\mathcal{H}}^2 & \le \left\langle f_{L^*,\mathrm{P},\lambda} - f_{L^*,\mathrm{Q},\lambda}, \mathbb{E}_{\mathrm{Q}^2}\left[h_{\mathrm{P}}\Phi\right] - \mathbb{E}_{\mathrm{P}^2}\left[h_{\mathrm{P}}\Phi\right]\right\rangle_{\mathcal{H}} \\ & \le \left\| f_{L^*,\mathrm{P},\lambda} - f_{L^*,\mathrm{Q},\lambda} \right\|_{\mathcal{H}} \left\| \mathbb{E}_{\mathrm{Q}^2}\left[h_{\mathrm{P}}\Phi\right] - \mathbb{E}_{\mathrm{P}^2}\left[h_{\mathrm{P}}\Phi\right] \right\|_{\mathcal{H}} . \end{aligned}$$

This yields the last assertion. We like to mention that h_{P} depends on P but on Q. □

Proof (of Theorem 3.3) Denote with $T(x, y, x', y') := h_{\mathrm{P}}(x, y, x', y')\Phi(x, x')$. Using Fubini's theorem and the inequality

$$\left\|T(x, y, x', y')\right\|_{\mathcal{H}} = \left\|h_{\mathrm{P}}(x, y, x', y')\Phi(x, x')\right\|_{\mathcal{H}} \leq \|h_{\mathrm{P}}\|_{\infty} \left\|\Phi(x, x')\right\|_{\mathcal{H}} \leq |L|_1 \|k\|_{\infty},$$

it follows by rearranging terms

$$\begin{aligned}
&\lambda \left\| f_{L^*,\mathrm{P},\lambda} - f_{L^*,\mathrm{P}_\varepsilon,\lambda} \right\|_{\mathcal{H}} \\
&\leq \left\| \mathbb{E}_{\mathrm{P}^2}\left[h_{\mathrm{P}}\Phi\right] - \mathbb{E}_{\mathrm{P}_\varepsilon^2}\left[h_{\mathrm{P}}\Phi\right] \right\|_{\mathcal{H}} \\
&= \Bigg\| \varepsilon \int\int T(x, y, x', y')\mathrm{dP}(x, y)\mathrm{d}(\mathrm{P}-\mathrm{Q})(x', y') \\
&\qquad + \varepsilon \int\int T(x, y, x', y')\mathrm{d}(\mathrm{P}-\mathrm{Q})(x, y)\mathrm{dP}(x', y') \\
&\qquad\quad + \varepsilon^2 \int\int T(x, y, x', y')\mathrm{dP}(x, y)\mathrm{d}(\mathrm{Q}-\mathrm{P})(x', y') \\
&\qquad\qquad + \varepsilon^2 \int\int T(x, y, x', y')\mathrm{dQ}(x, y)\mathrm{d}(\mathrm{P}-\mathrm{Q})(x', y') \Bigg\|_{\mathcal{H}} \\
&\leq 4\varepsilon\, |L|_1 \|k\|_{\infty} \underbrace{\|\mathrm{P}-\mathrm{Q}\|_{TV}}_{\leq 2} \\
&\leq 8\varepsilon\, |L|_1 \|k\|_{\infty},
\end{aligned}$$

where we used that $\varepsilon \in (0, 1)$ and $\|\mathrm{P}-\mathrm{Q}\|_{TV}$ denotes the norm of total variation, see, e.g., [42, Prop 2.2, p. 543] or [43, p. 1519], i.e.,

$$\|\mathrm{P}-\mathrm{Q}\|_{TV} := \sup_{\substack{\|g\|_\infty \leq 1 \\ g:(\mathcal{X}\times\mathcal{Y})^2 \to \mathbb{R}}} \left| \int g\mathrm{dP} - \int g\mathrm{dQ} \right|.$$

Obviously $\|\mathrm{P}-\mathrm{Q}\|_{TV} \in [0, 2]$ for all $\mathrm{P}, \mathrm{Q} \in \mathcal{M}_1(\mathcal{X} \times \mathcal{Y})$. In conclusion

$$\left\| f_{L^*,\mathrm{P},\lambda} - f_{L^*,\mathrm{P}_\varepsilon,\lambda} \right\|_{\mathcal{H}} \leq \frac{8}{\lambda} |L|_1 \|k\|_{\infty} \varepsilon.$$

□

Proof (of Theorem 3.4) Fix $\mathrm{Q} \in \mathcal{M}_1(\mathcal{X} \times \mathcal{Y})$ and $\lambda \in (0, \infty)$. Denote $\mathrm{P}_\varepsilon := (1-\varepsilon)\mathrm{P} + \varepsilon\mathrm{Q}$ with $\varepsilon \in (0, 1)$. The function $G : \mathbb{R} \times \mathcal{H}$, defined by

$$G(\varepsilon, f) = 2\lambda\, \mathrm{id}_{\mathcal{H}} + \mathbb{E}_{\mathrm{P}_\varepsilon^2}\left[D_5 L(X, Y, X', Y', f(X, X'))\Phi(X, X')\right]$$

plays an important role in this proof. Since k is bounded, all functions $f \in \mathcal{H}$ in the corresponding RKHS fulfill $\|f\|_\infty < \infty$. Additionally the partial derivative $D_5 L$ is bounded by Assumption 3.1. It follows, for all $\varepsilon \in \mathbb{R}$, and all $f \in \mathcal{H}$, that

$$\begin{aligned}\|G(\varepsilon, f)\|_{\mathcal{H}} &\leq 2\lambda \|f\|_{\mathcal{H}} + \mathbb{E}_{\mathrm{P}_\varepsilon^2}\left[\left|D_5 L(X, Y, X', Y', f(X, X'))\right| \left\|\Phi(x, x')\right\|_\infty\right] \\ &\overset{(2)}{\leq} 2\lambda \|f\|_{\mathcal{H}} + c_{L,1} \|k\|_\infty^2 < \infty.\end{aligned}$$

Therefore, the map G is well-defined and bounded with respect to the $\mathcal{H}$-norm. Hence,

$$\|G(\varepsilon, f)\|_\infty \overset{(3)}{\leq} \|G(\varepsilon, f)\|_{\mathcal{H}} \|k\|_\infty \leq (2\lambda \|f\|_{\mathcal{H}} + c_{L,1} \|k\|_\infty^2) \|k\|_\infty < \infty.$$

Note that for $\varepsilon \notin [0, 1]$ the $\mathcal{H}$-valued Bochner integral is with respect to a signed measure. Hence Lemma 2.15 yields, for all $\varepsilon \in [0, 1]$, that

$$G(\varepsilon, f) = \frac{\partial(\mathcal{R}_{L^*,\mathrm{P}_\varepsilon}(\cdot) + \lambda \|\cdot\|_{\mathcal{H}})}{\partial f}(f).$$

Since L is convex, the map $f \mapsto \mathcal{R}_{L^*,\mathrm{P}_\varepsilon}(f) + \lambda \|f\|_{\mathcal{H}}^2$ is continuous and convex for all $\varepsilon \in [0, 1]$. The equation above shows that we have $G(\varepsilon, f) = 0$ if and only if $f = f_{L^*,\mathrm{P}_\varepsilon,\lambda}$ for such ε. We now want to show the existence of a differentiable function $\varepsilon \mapsto f_\varepsilon$ on a small interval $(-\delta, \delta)$ for some $\delta > 0$ that satisfies $G(\varepsilon, f_\varepsilon) = 0$ for all $\varepsilon \in (-\delta, \delta)$. According to Theorem A.1.10, we have to check that G is continuously differentiable and that $\frac{\partial G}{\partial f}(0, f_{\mathrm{P},\lambda})$ is invertible which was proven in Theorem A.2.3. Hence we can apply the implicit function theorem to see that the map $\varepsilon \mapsto f_\varepsilon$ is differentiable on a small non-empty interval $(-\delta, \delta)$. In conclusion, we obtain

$$S_G'(\mathrm{P})(\mathrm{Q}) = \frac{\partial f_\varepsilon}{\partial \varepsilon}(0) = -\left(\frac{\partial G}{\partial f}(0, f_{L^*,\mathrm{P},\lambda})\right)^{-1} \circ \frac{\partial G}{\partial \varepsilon}(0, f_{L^*,\mathrm{P},\lambda}) = -M(\mathrm{P})^{-1} T(\mathrm{Q}; \mathrm{P}),$$

which yields the assertion. □

Proof (of Corollary 3.6) The assertion follows immediately by setting Q as the Dirac measure $\delta_{(x_0, y_0)}$ in Theorem 3.4. □

Proof (of Theorem 3.8) To (i). Let $\mathrm{P} \in \mathcal{M}_1(\mathcal{X} \times \mathcal{Y})$ be fixed. As L^* and λ are fixed, we denote with

$$\begin{aligned}{L^*}'_{f_{L^*,\mathrm{P},\lambda}}(X, Y, X', Y') &\overset{2.19(\mathrm{v})}{=} L'_{f_{L^*,\mathrm{P},\lambda}}(X, Y, X', Y') \\ &:= D_5 L^*(X, Y, X', Y', f_{L^*,\mathrm{P},\lambda}(X, X')).\end{aligned}$$

Let $(\mathrm{P}_n)_{n \in \mathbb{N}} \subset \mathcal{M}_1(\mathcal{X} \times \mathcal{Y})$ be a weakly convergent sequence with $\mathrm{P}_n \rightsquigarrow \mathrm{P}$. We know that due to the separability of $\mathcal{X} \times \mathcal{Y}$, weak convergence of probability measures is equivalent to $d_{BL}(\mathrm{P}_n, \mathrm{P}) \to 0$, where d_{BL} denotes the bounded Lipschitz metric, see [20, Thm. 11.3.3]. Hence the metric space $(\mathcal{X} \times \mathcal{Y})^2$ is separable and thus guarantees

$$\mathrm{P}_n \rightsquigarrow \mathrm{P} \Longleftrightarrow \mathrm{P}_n^2 \rightsquigarrow \mathrm{P}^2 \quad (n \to \infty)$$

see [4, Thm. 3.8(ii), p. 23]. The definition of weak convergence guarantees that

$$\lim_{n\to\infty} \int g d\mathrm{P}_n^2 = \int g d\mathrm{P}^2$$

for all continuous and bounded real-valued functions $g : (\mathcal{X} \times \mathcal{Y})^2 \to \mathbb{R}$. However, we need a corresponding result for $\mathcal{H}$-valued Bochner integrals. The fourth part of the representer theorem, see Theorem 3.2, yields

$$\begin{aligned} &\|S(\mathrm{P}_n) - S(\mathrm{P})\|_{\mathcal{H}} \\ &= \left\| f_{L^*,\mathrm{P}_n,\lambda} - f_{L^*,\mathrm{P},\lambda} \right\|_{\mathcal{H}} \\ &\le \frac{1}{\lambda} \left\| \mathbb{E}_{\mathrm{P}_n^2}\left[h_{\mathrm{P}}(X,Y,X',Y')\Phi(X,X')\right] - \mathbb{E}_{\mathrm{P}^2}\left[h_{\mathrm{P}}(X,Y,X',Y')\Phi(X,X')\right] \right\|_{\mathcal{H}}. \end{aligned}$$

As k is a continuous and bounded kernel, the canonical feature map Φ is also continuous and bounded. Furthermore, as the shifted loss function L^* is twice continuously differentiable and the partial derivative is bounded, it follows that, for every fixed $\mathrm{P} \in \mathcal{M}_1(\mathcal{X} \times \mathcal{Y})$ and every fixed $\lambda \in (0, \infty)$, the function

$$\psi_P : ((\mathcal{X} \times \mathcal{Y})^2, d_{(\mathcal{X}\times\mathcal{Y})^2}) \to (\mathcal{H}, d_{\mathcal{H}}), \quad \psi_{\mathrm{P}}(x, y, x', y') := h_{\mathrm{P}}(x, y, x', y')\Phi(x, x')$$

is continuous and bounded, where $d_{\mathcal{H}}$ denotes the metric induced by the norm $\|\cdot\|_{\mathcal{H}}$. We thus obtain from [6, p. III.40], see also [26, Thm. A.1], the following convergence result for Bochner integrals

$$\mathrm{P}_n^2 \rightsquigarrow \mathrm{P}^2 \Longrightarrow \lim_{n\to\infty} \int \psi_{\mathrm{P}} d\mathrm{P}_n^2 = \int \psi_{\mathrm{P}} d\mathrm{P}^2,$$

which implies that $\mathrm{P}_n \rightsquigarrow \mathrm{P}$, which is equivalent to $d_{\mathrm{BL}}(\mathrm{P}_n, \mathrm{P}) \to 0$ due to [20, Thm. 11.3.3], leads to $\|S(\mathrm{P}_n) - S(\mathrm{P})\|_{\mathcal{H}} \to 0$ and therefore (i) is proven.

The proof for (ii) follows immediately from part (i) and the fact that the inclusion map $\mathrm{id} : \mathcal{H} \to C_b(\mathcal{X}^2)$ is continuous and bounded, see [44, Lemma 4.28]. □

Proof (of Corollary 3.9) Let $(D_{n,m})_{m\in\mathbb{N}} \subset (\mathcal{X} \times \mathcal{Y})^n$ be a sequence of tuples $\left((x_1^{(m)}, y_1^{(m)}), \ldots, (x_n^{(m)}, y_n^{(m)})\right)$ which converges to some $D_{n,0} = \left((x_1^{(0)}, y_1^{(0)}), \ldots,\right.$

$\left(x_n^{(0)}, y_n^{(0)}\right)\right) \in (\mathcal{X} \times \mathcal{Y})^n$ for $m \to \infty$. Then let $\mathrm{D}_{n,m} = \frac{1}{n}\sum_{i=1}^n \delta_{\left(x_i^{(m)}, y_i^{(m)}\right)}$ and $\mathrm{D}_{n,0} = \frac{1}{n}\sum_{i=1}^n \delta_{\left(x_i^{(0)}, y_i^{(0)}\right)}$ be the corresponding empirical measures, and $g \in C_b(\mathcal{X} \times \mathcal{Y})$ a continuous and bounded real-valued function. Hence, it follows that:

$$\begin{aligned} 0 \leq \left|\int g\mathrm{dD}_{n,m} - \int g\mathrm{dD}_{n,0}\right| &= \left|\frac{1}{n}\sum_{i=1}^n g\left(x_i^{(m)}, y_i^{(m)}\right) - \frac{1}{n}\sum_{i=1}^n g\left(x_i^{(0)}, y_i^{(0)}\right)\right| \\ &\leq \frac{1}{n}\sum_{i=1}^n \left|g\left(x_i^{(m)}, y_i^{(m)}\right) - g\left(x_i^{(0)}, y_i^{(0)}\right)\right| \\ &\xrightarrow{m\to\infty} 0, \end{aligned}$$

as g is a continuous function and $D_{n,m} \to D_{n,0}$ for $m \to \infty$ and therefore $\mathrm{D}_{n,m} \rightsquigarrow \mathrm{D}_{n,0}$. Hence, the assertion follows from Theorem 3.8 and $S(\mathrm{D}_n) = f_{L^*,\mathrm{D}_n,\lambda} = S_n(D_n)$. □

Proof (of Theorem 3.10) Fix $\lambda \in (0, \infty)$. For any $D_n := \left((x_1, y_1), \ldots, (x_n, y_n)\right) \in (\mathcal{X} \times \mathcal{Y})^n$ denote its empirical measure by $\mathrm{D}_n := \frac{1}{n}\sum_{i=1}^n \delta_{(x_i, y_i)}$. According to Corollary 3.9, the function

$$S_n : ((\mathcal{X} \times \mathcal{Y})^n, d_{(\mathcal{X}\times\mathcal{Y})^n}) \to (\mathcal{H}, d_{\mathcal{H}}), \quad S_n(D_n) = f_{L^*,\mathrm{D}_n,\lambda}$$

is continuous and therefore measurable with respect to the corresponding Borel σ-algebras for every $n \in \mathbb{N}$. Theorem 3.8 yields that

$$S : (\mathcal{M}_1(\mathcal{X} \times \mathcal{Y}), d_{\mathrm{BL}}) \to (\mathcal{H}, d_{\mathcal{H}}), \quad S(\mathrm{P}) = f_{L^*,\mathrm{P},\lambda}$$

is a continuous operator. Furthermore S_n and S satisfy by definition the condition $S_n(D_n) = S(\mathrm{D}_n)$ for all $D_n \in (\mathcal{X} \times \mathcal{Y})^n$ and all $n \in \mathbb{N}$. As $\mathcal{H}$ is a separable RKHS, which is implied by Assumption 3.1(i) and (ii), and [44, Lemma 4.33], $(\mathcal{H}, d_{\mathcal{H}})$ is a complete and separable metric space. Theorem A.1.11 yields that for the random measure $\mathbb{D}_n = \frac{1}{n}\sum_{i=1}^n \delta_{(X_i, Y_i)}$ the sequence of RPL estimators $(f_{L^*,\mathbb{D}_n,\lambda})_{n\in\mathbb{N}}$ is qualitatively robust for all $\mathrm{P} \in \mathcal{M}_1(\mathcal{X} \times \mathcal{Y})$. Hence the assertion of part (i) is shown.

Part (ii) can be proven as follows. Theorem 3.8 yields that the operator S is continuous for all $\mathrm{P} \in \mathcal{M}_1(\mathcal{X} \times \mathcal{Y})$. Hence all assumptions for Theorem A.1.12 are satisfied, because $\mathcal{Z} := \mathcal{X} \times \mathcal{Y}$ is a compact metric space by assumption and $\mathcal{W} := \mathcal{H}$ is a complete and separable metric space. This yields the assertion. □

Acknowledgments Authors Patrick Gensler and Andreas Christmann were partially supported by DFG grant CH 291/3-1.

References

1. Akerkar, R.: Nonlinear Functional Analysis. Narosa Publishing House, New Delhi (1999)
2. Bellet, A., Habrard, A.: Robustness and generalization for metric learning. Neurocomputing **151**, 259–267 (2015)
3. Berlinet, A., Thomas-Agnan, C.: Reproducing Kernel Hilbert Spaces in Probability and Statistics. Springer US, Boston (2004)
4. Billingsley, P.: Convergence of probability measures, 2nd edn. In: Wiley Series in Probability and Statistics: Probability and Statistics. John Wiley & Sons Inc., New York (1999)
5. Blaschzyk, I.: Improved Classification Rates for Localized Algorithms under Margin Conditions. Springer, Wiesbaden (2020)
6. Bourbaki, N., Berberian, S.: Integration I: Chapters 1–6. Springer, Berlin (2004)
7. Cao, Q., Guo, Z.-C., Ying, Y.: Generalization bounds for metric and similarity learning. Mach. Learn. **102**(1), 115–132 (2016)
8. Castaing, C., Valadier M.: Convex Analysis and Measurable Multifunctions. Lecture Notes in Mathematics. Springer, Berlin (1977)
9. Chechik, G., et al.: Large scale online learning of image similarity through ranking. J. Mach. Learn. Res. **11**, 1109–1135 (2010)
10. Chen, H., Pan, Z., Li, L.: Learning performance of coefficient-based regularized ranking. Neurocomputing **133**, 54–62 (2014)
11. Christmann, A., Zhou, D.X.: On the robustness of regularized pairwise learning methods based on kernels. J. Complex. **37**, 1–33 (2016)
12. Christmann, A., van Messem, A., Steinwart, I.: On consistency and robustness properties of support vector machines for heavy-tailed distributions. Stat. Inter. **2**(3), 311–327 (2009)
13. Christmann, A., Salibián-Barrera, M., van Aelst, S.: Qualitative robustness of bootstrap approximations for kernel based methods. In: Robustness and Complex Data Structures: Festschrift in Honour of Ursula Gather, pp. 263–278. Springer, New York (2013)
14. Clémençon, S., Lugosi, G., Vayatis, N.: Ranking and empirical minimization of U-statistics. Ann. Stat. **36**(2), 844–874 (2008)
15. Cristianini, N., Shawe-Taylor, J.: An Introduction to Support Vector Machines and Other Kernel-Based Learning Methods. Cambridge University Press, Cambridge (2000)
16. Cucker, F., Zhou, D.X.: Learning Theory: An Approximation Theory Viewpoint. Cambridge University Press, New York (2007)
17. Cuevas, A.: Qualitative robustness in abstract inference. J. Stat. Plan. Infer. **18**, 277–289 (1988)
18. Cuevas, A., Romo, J.: On robustness properties of bootstrap approximations. J. Stat. Plan. Infer. **37**(2), 181–191 (1993)
19. Denkowski, Z., Migórski, S., Papageorgiou, N.S.: An Introduction to Nonlinear Analysis: Theory. Kluwer Academic/Plenum Publishers, New York (2003)
20. Dudley, R.M.: Real Analysis and Probability, 2nd ed. Cambridge Studies in Advanced Mathematics. Cambridge University Press, Cambridge (2002)
21. Dudley, R.M., Giné, E., Zinn, J.: Uniform and universal Glivenko-Cantelli classes. J. Theor. Probab. **4**(3), 485–510 (1991)
22. Dumpert, F., Christmann, A.: Universal consistency and robustness of localized support vector machines. Neurocomputing **315**, 96–106 (2018)
23. Ekeland, I., Turnbull, T.: Infinite-Dimensional Optimization and Convexity. Chicago Lectures in Mathematics. University of Chicago Press, Chicago (1983)
24. Guo, Z.-C., Ying, Y.: Guaranteed classification via regularized similarity learning. Neural Comput. **26**(3), 497–522 (2014)
25. Guo, Z.-C., Ying, Y., Zhou, D.X.: Online regularized learning with pairwise loss functions. Adv. Comput. Math. **43**(1), 127–150 (2017)
26. Hable, R., Christmann, A.: Qualitative robustness of support vector machines. J. Multivar. Anal. **102**, 993–1007 (2011)

27. Hampel, F.R.: Contributions to the Theory of Robust Estimation. University of California, Berkeley (1968)
28. Hampel, F.R.: A general qualitative definition of robustness. Ann. Math. Stat. **42**(6), 1887–1896 (1971)
29. Hampel, F.R., et al.: Robust Statistics: The Approach Based on Influence Functions. Wiley Series in Probability and Statistics. Wiley, Hoboken (1986)
30. Huber P.J.: Robust estimation of a location parameter. Ann. Math. Stat. **35**(1), 73–101 (1964)
31. Huber P.J.: The behavior of maximum likelihood estimates under nonstandard conditions. In: Proceedings of the Fifth Berkeley Symposium on Mathematical Statistics and Probability, vol. 1, pp. 221–233 (1967)
32. Kechris, A.: Classical Descriptive Set Theory. Graduate Texts in Mathematics. Springer, New York (1995)
33. Köhler H., Christmann, A.: Total stability of SVMs and localized SVMs. J. Mach. Learn. Res. **23**(100), 1–41 (2022)
34. Lim, D., Lanckriet, G.: Efficient learning of mahalanobis metrics for ranking. In: Proceedings of the 31st International Conference on Machine Learning, vol. 32. Proceedings of Machine Learning Research 2, pp. 1980–1988. PMLR, Beijing (2014)
35. Meister, M., Steinwart, I.: Optimal learning rates for localized SVMs. J. Mach. Learn. Res. **17**(1), 6722–6765 (2016)
36. Phelps, R.: Convex Functions, Monotone Operators and Differentiability. Lecture Notes in Mathematics. Springer, Berlin (1993)
37. Principe, J.C.: Information Theoretic Learning: Renyi's Entropy and Kernel Perspectives. Springer, New York (2010)
38. Qian, Q., et al.: Similarity learning via adaptive regression and its application to image retrieval. arXiv: 1512.01728 (2015)
39. Rejchel, W.: On ranking and generalization bounds. J. Mach. Learn. Res. **13**, 1373–1392 (2012)
40. Rejchel, W.: Model selection consistency of U-statistics with convex loss and weighted lasso penalty. J. Nonparamet. Stat. **29**, 1–24 (2017)
41. Schölkopf, B., Smola, A.J.: Learning with Kernels: Support Vector Machines, Regularization, Optimization, and Beyond. MIT Press, Cambridge (2001)
42. Shorack, G.R.: Probability for Statisticians. Springer, New York (2000)
43. Sriperumbudur, B.K., et al.: Hilbert space embeddings and metrics on probability measures. J. Mach. Learn. Res. **11**, 1517–1561 (2010)
44. Steinwart, I., Christmann, A.: Support Vector Machines. Information Science and Statistics. Springer, New York (2008)
45. Stute, W.: Conditional U-statistics. Ann. Probab. **19**(2), 812–825 (1991)
46. Stute, W.: Universally consistent conditional U-statistics. Ann. Stat. **22**(1), 460–473 (1994)
47. Vapnik, V.N.: The Nature of Statistical Learning Theory. Springer, Berlin (1995)
48. Vapnik, V.N.: Statistical Learning Theory. Wiley-Interscience, New York (1998)
49. Werner, D.: Funktionalanalysis. Springer-Lehrbuch. Springer, Berlin (2011)
50. Xing, E.P., et al.: Distance metric learning, with application to clustering with side-information. Neural Inf. Proces. Syst. 521–528 (2002)
51. Ying, Y., Zhou, D.X.: Online pairwise learning algorithms with kernels. Neural Comput. **28**(4), 743–777 (2015)

Global Sensitivity Analysis for the Interpretation of Machine Learning Algorithms

Sonja Kuhnt and Arkadius Kalka

Abstract Global sensitivity analysis aims to quantify the importance of model input variables for a model response. We highlight the role sensitivity analysis can play in interpretable machine learning and provide a short survey on sensitivity analysis with a focus on global variance-based sensitivity measures like Sobol' indices and Shapley values. We discuss the Monte Carlo estimation of various Sobol' indices as well as their graphical presentation in the so-called FANOVA graphs. Global sensitivity analysis is applied to an analytical example, a Kriging model of a piston simulator and a neural net model of the resistance of yacht hulls.

Keywords Interpretable machine learning · Global sensitivity analysis · Sobol' indices · Shapley values · FANOVA graph · Kriging

1 Introduction

Machine learning is a set of methods that improve automatically through experience, i.e. it is based on data. Popular machine learning methods are, e.g. support vector machines (SVMs [2]), artificial neural networks (ANNs) and random forests (RFs). Machine learning algorithms are increasingly applied in science and business and have achieved impressive performances in diverse tasks, outperforming humans. However, for several machine learning algorithms it is hard to tell what the machine has actually learned from the data. For example, in the case of ANNs, what was learned is hidden in the weights and biases of the neurons involved. If a machine learning model performs well, one might simply trust the model and ignore why it made a certain decision. However, such an attitude goes against human curiosity and thirst for knowledge. This raises the issue of interpretability [24, 25]. The straightforward way to achieve interpretability in statistical learning is to use only

S. Kuhnt (✉) · A. Kalka
University of Applied Sciences and Arts, Dortmund, Germany
e-mail: sonja.kuhnt@fh-dortmund.de; arkadius.kalka@fh-dortmund.de

A. Steland, K.-L. Tsui (eds.), *Artificial Intelligence, Big Data and Data Science in Statistics*, https://doi.org/10.1007/978-3-031-07155-3_6

interpretable models. Interpretable models are, e.g. linear models, generalized linear models [23], generalized additive models [12], decision trees and rules. On the other hand, model-agnostic interpretation methods are more flexible and can be applied to any machine learning algorithm. Graphical model-agnostic methods are, e.g. Partial Dependence Plots (PDP) and Individual Conditional Expectation (ICE) plots. We suggest to provide model-agnostic methods to evaluate the influence of different regressors and their interactions by applying methods from the statistical field of global sensitivity analysis (GSA). Sensitivity analysis is the study of how the uncertainty in the output of a mathematical model or function can be divided and allocated to different sources of uncertainty in its inputs [14, 29]. Given any real-valued function on several variables –whether analytical or given by a black-box– one wants to know which input variables affect the variability of the function the most. GSA has proven to be a valuable tool in analysing expensive to evaluate computer models with a surrogate model, e.g. a Kriging model, build first. Cheng et al. [3] use support vector regression as surrogate model within GSA, whereas [37] built a new feature selection approach upon GSA. Like in [4] we suggest to achieve an understanding and interpretability of, e.g. ANNs, SVMs and RFs by combining GSA and visualization.

2 Global Sensitivity Analysis

This section reviews sensitivity analysis with a focus on global variance-based sensitivity measures, but we also discuss derivative-based global sensitivity measures briefly.

2.1 *Global Sensitivity Indices*

Consider a function $f : \Delta \subseteq \mathbb{R}^d \to \mathbb{R}$ that is square integrable w.r.t. a d-dimensional product measure μ. The functional analysis of variance (FANOVA) decomposition (also called Hoeffding-Sobol' decomposition) of $f \in L^2(\mu)$ is the unique decomposition

$$f(X) = f_0 + \sum_i f_i(X_i) + \sum_{i<j} f_{i,j}(X_i, X_j) + \cdots + f_{1,\dots,d}(X_1, \dots, X_d) \tag{1}$$

such that $E(f_I(X_I) \mid X_J) = 0$ for all $J \subset I \subseteq [d] := \{1, \dots, d\}$. In particular, we have $E(f_I(X_I)) = 0$ for all $\emptyset \neq I \subseteq [d]$, see e.g. [6]. Furthermore, this implies orthogonality of all summands in the decomposition, i.e. $E\left(f_I(X_I) f_J(X_J)\right) = 0$ for all $I \neq J \subseteq [d]$. The FANOVA decomposition can be computed recursively by

$$f_0 = E(f(X)), \text{ and } f_I(X_I) = E(f(X) \mid X_I) - \sum_{J \subset I} f_J(X_J). \tag{2}$$

Orthogonality allows for an ANOVA like decomposition of the variance of f:

$$D = Var(f(X)) = \sum_{I \subseteq [d]} Var(f_I(X_I)). \tag{3}$$

The variance of each term, $D_I = Var(f_I(X_I))$, gives a sensitivity index of its effect. The standardized indices

$$S_I = D_I / D \tag{4}$$

are known as *Sobol' indices* [31]. Especially, Sobol' indices $S_i = S_{\{i\}}$ for individual variables are referred to as first-order indices and $S_{ij} = S_{\{i,j\}}$ as second-order indices. The same holds for the unstandardized versions D_i and D_{ij}. Sobol' introduced the *closed sensitivity index* to describe the influence of a group of variables:

$$D_I^{cl} = \sum_{J \subseteq I} D_J = Var(E(f(X)|X_I)). \tag{5}$$

The *total sensitivity index* by Homma and Saltelli [13] describes the total contribution of a set of variables including all interactions of any order and is defined by all partial variances containing at least one of the variables, i.e.

$$D_I^T = \sum_{I \cap J \neq \emptyset} D_J, \quad S_I^T = \frac{D_I^T}{D}. \tag{6}$$

For $I = \{i\}$, this total sensitivity index is defined by considering all supersets. An extension [21] of the concept of *superset importance* is given by

$$D_I^{sup} = \sum_{J \supseteq I} D_J. \tag{7}$$

In particular, the unnormalized and normalized *total interaction* indices (TIIs) [9] are given by

$$D_{i,j}^{sup} = \sum_{J \supseteq \{i,j\}} D_J \quad \text{and} \quad S_{i,j}^{sup} = \sum_{J \supseteq \{i,j\}} \frac{D_J}{D}. \tag{8}$$

So, each of these indices characterizes a different aspect of the sensitivity of the model response to individual input variables or interactions between them.

2.2 Shapley Values

A similar problem to the FANOVA decomposition has been studied in game theory and economics, namely the problem of attributing the value created in a team effort to individual team members. Consider the setting where one can measure the value $val(I) \in \mathbb{R}$ created by any subset $I \subseteq [d]$ of the d-member team. In that case the so-called *Shapley values* ϕ_i are the unique choice that satisfy the following four natural criteria [30, 36].

1. (Efficiency) $\sum_i^d \phi_i = val([d])$.
2. (Symmetry) $val(I \cup i) = val(I \cup j) \quad \forall I \subseteq [d] \setminus \{i, j\}$ implies $\phi_i = \phi_j$.
3. (Dummy) $val(I \cup i) = val(I) \quad \forall I \subseteq [d]$ implies $\phi_i = 0$.
4. (Additivity) The game with value $val^{(1)} + val^{(2)}$ has Shapley values $\phi^{(1)} + \phi^{(2)}$ with $\phi^{(1)} = \phi(val^{(1)})$ and $\phi^{(2)} = \phi(val^{(2)})$.

Then the Shapley value of an individual variable is given by

$$\phi_i = \frac{1}{d} \sum_{I \subseteq [d] \setminus \{i\}} \binom{d-1}{|I|}^{-1} (val(I \cup i) - val(I)). \tag{9}$$

Shapley values are connected to the FANOVA decomposition by Owen [27]. In that context, for any subset I of input variables, their combined value $val(I)$ is the "variance explained" in the FANOVA decomposition. More precisely, the choice in [27] is $val(I) = D_I^{cl}$. Then, using the properties (1) – (4), it can be shown that the Shapley value is

$$\phi_i = \sum_{I \subseteq [d], i \in I} \frac{D_I}{|I|} \tag{10}$$

according to Theorem 1 in [27]. The Shapley value does not coincide with any first-order Sobol' index, but it is bracketed between the closed and total sensitivity index [27]:

$$D_i^{cl} \leq \phi_i \leq D_i^T. \tag{11}$$

A normalized Shapley value may be defined as $\phi_i^* = \phi_i / D$. Because these indices are comparatively easy to compute, Sobol' indices provide effectively computable bounds for the Shapley value. An exact computation of the Shapley value is computationally expensive because there are 2^d subsets of $[d]$, representing coalitions of variables. Štrumbelj and Kononenko [34] and Song et al. [33] propose effective algorithms to estimate Shapley values using Monte Carlo sampling.

2.3 Derivative-Based Global Sensitivity Measures

Based on the work of [32] derivative-based global sensitivity measures (DGSM) were introduced by Kucherneko et al. [20] as

$$\nu_i = \int \left(\frac{\partial f}{\partial x_i}(x) \right)^2 d\mu. \tag{12}$$

A normalized DGSM can be defined by $\nu_i^* = \nu_i / \sum_j^d v_j$. DGSMs are not associated with a functional decomposition, but they are connected to total sensitivity indices by the inequality $D_i^T \leq C(\mu_i)\nu_i$ if for the measure μ the Poincare inequality

$$\int g(x)^2 \mu \leq C(\mu) \int ||\nabla g(x)||^2 d\mu \tag{13}$$

holds for all centred functions $g \in L^2(\mu)$ with $\int g(x) d\mu = 0$ and $||\nabla g|| \in L^2(\mu)$. Friedman and Popescu [7] introduced crossed DSGMs, in particular, for interactions:

$$\nu_{i,j} = \int \left(\frac{\partial^2 f}{\partial x_i \partial_j}(x) \right)^2 d\mu. \tag{14}$$

Roustant et al. [28] provide an inequality to link crossed DGSMs to superset importance.

2.4 Estimation of Indices

For analytically tractable test functions, the indices above may be calculated by evaluating the integrals involved. In general, the function f is not known analytically and will be treated as black-box function. In *Monte Carlo estimation*, we take a high number of n samples $x^{(1)}, \ldots, x^{(n)}$ from the distribution μ and approximate the integral by

$$\frac{1}{n} \sum_{k=1}^{n} f(x^{(k)}) \quad \stackrel{n \to \infty}{\longrightarrow} \int f(x) d\mu = E(f(X)). \tag{15}$$

The approximation is unbiased and convergent with probability one according to the law of large numbers. For the estimation, we require a representation of the sensitivity indices that is suitable for Monte Carlo integration. A popular choice is based on the pick-and-freeze formula $D_I^{cl} = E(f(X)f(X_I, Z_{-I})) - f_0^2$ which gives the pick-and-freeze Monte Carlo estimator

$$\hat{D}_I^{cl} = \frac{1}{n}\sum_{k=1}^{n} f(x^{(k)}) f\left(x_I^{(k)}, z_{-I}^{(k)}\right) - f_0^2. \tag{16}$$

Here Z is an independent copy of the random variable X, and $-I$ denotes the complement set $[d] \setminus I$. Since, the pick-and-freeze estimator gets a large variance if f_0 is large, other formulas have been suggested that avoid the subtraction of f_0^2 [18, 31]. In particular, the total sensitivity index can be computed using the *Jansen formula* $D_I^T = 1/2E((f(X) - f(Z_I, X_{-I}))^2)$.

Computationally cheaper than Monte Carlo estimation, but also slightly biased, are frequency-based estimation methods. The first frequency-based estimation method was the so-called Fourier amplitude sensitivity test (FAST) by Cukier et al. [5]. TIIs can be easily estimated via the relationship with closed sensitivity indices using pick-and-freeze. Of particular interest is a direct method using the formula of [21]:

$$D_{i,j}^{sup} = \frac{1}{4}E((f(X_i, X_j, X_{-\{i,j\}}) - f(X_i, Z_j, X_{-\{i,j\}}) \tag{17}$$

$$-f(Z_i, X_j, X_{-\{i,j\}}) + f(Z_i, Z_j, X_{-\{i,j\}}))^2). \tag{18}$$

The corresponding Liu-Owen Monte Carlo estimator is unbiased, and it is non-negative since it is a sum of squares. This implies that if the true TII is zero, then the estimator is zero as well.

3 Visualizing Interaction Structures by FANOVA Graphs

In this section the FANOVA graph, an intuitive tool to visualize the most valuable information of the FANOVA decomposition, is introduced [8, 10, 26]. Estimation and thresholding of FANOVA graphs is discussed, and we apply GSA to a standard non-linear test function.

3.1 General Idea of FANOVA Graphs

Usually, it is infeasible to look at all $2^d - 1$ terms of the decomposition of a function with d input variables individually, and quite often only main effect Sobol' indices are considered. The primal intention of FANOVA graphs is to overcome this problem and to visualize the interaction structure contained in the FANOVA decomposition by a mathematical graph [26]. The so-called *FANOVA graph* is defined as a graph $G = (V, E)$ where each of the d input variables is identified by an element of the vertex set $V = \{1, \dots, d\}$. An edge is included in the edge set $(i, j) \in E$ iff there exists a superset $J \supseteq \{i, j\}$ such that $f_J(X_J) \neq 0$. That is, the pair of input variables (X_i, X_j) has a non-zero two-way interaction or is involved in a higher order non-zero interaction. Equivalently an edge (i, j) is not in G iff all Sobol' indices $S_J = 0$ for $J \supseteq \{i, j\}$. This is exactly captured by a non-zero TII, i.e. $S_{i,j}^{sup} \neq 0$.

A FANOVA graph can be further enhanced by displaying the thicknesses of each edges (i, j) proportional to the strength of the TII of the two involved input variables. In the same way, each vertex i can be displayed by circles with lines proportional in strength to the main effect Sobol' index S_i.

Let us consider the so-called Ishigami function which is frequently used for illustrating sensitivity analysis [16]. The function, given by

$$f(X_1, X_2, X_3) = \sin(X_1) + 7\sin^2(X_2) + 0.1X_3^4\sin(X_1), \tag{19}$$

depends on three input variables (X_1, X_2, X_3) and obviously contains a non-linear interaction between X_1 and X_3 (see Fig. 1c). For this test function Sobol' indices can be computed analytically. Assuming a uniform distribution on $[-\pi, \pi]$ for each input variable, analytical calculation of the FANOVA decomposition and Sobol' indices gives us the following values

$$D_1 = 4.346,\ D_2 = 6.125,\ D_3 = 0,\ D_{12} = 0,\ D_{13} = 3.374,\ D_{23} = 0,\ D_{123} = 0. \tag{20}$$

This leads to the following first-order Sobol' indices and normalized TIIs

$$S_1 = 0.314,\ S_2 = 0.442,\ S_3 = 0,\ S_{12}^{sup} = 0,\ S_{13}^{sup} = 0.244,\ S_{23}^{sup} = 0. \tag{21}$$

Figure 2 shows a bar plot and the FANOVA graph displaying the Sobol' indices and TIIs for the Ishigami function. Main effect stands for the normalized first-order Sobol' indices and interaction is the difference between the scaled total sensitivity

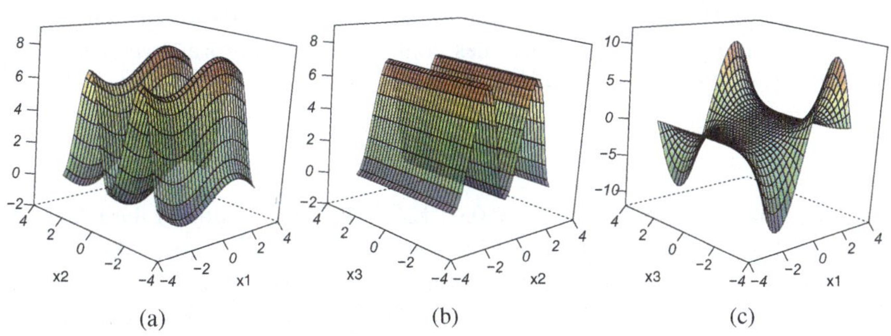

Fig. 1 3d-plots for the Ishigami function. (**a**) $f(X_1, X_2, 0)$. (**b**) $f(0, X_2, X_3)$. (**c**) $f(X_1, 0, X_3)$

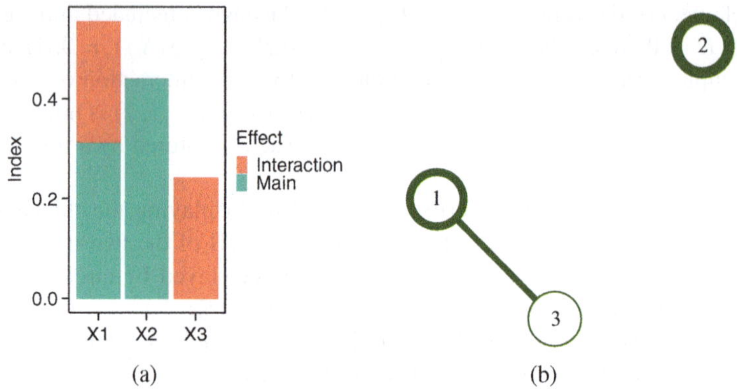

Fig. 2 Bar plot and FANOVA graph displaying Sobol' indices and TIIs for the Ishigami function. (**a**) Bar plot. (**b**) FANOVA graph

index and the Sobol' index. From the FANOVA graph it becomes immediately obvious that the input variable X_2 has the highest impact on the response, followed by X_1. Also, the interaction between X_1 and X_2 is easily detected.

In summary, FANOVA graphs visualize both first- and second-order GSA. First-order analysis in the sense of detecting the inputs X_i for which S_i is very small or even zero. Second-order in the sense of looking at pairs of input variables and detecting influential interactions and their strength, i.e. the pairs $\{i, j\}$ with $S_{i,j}^{sup} > 0$.

3.2 Estimation and Thresholding

In practice, S_i and $S_{i,j}^{sup}$ are usually not analytically available and replaced by estimates $\hat{S}_{i,j}^{sup}$ and $\hat{S}_i$. Moreover, we often even apply GSA not to the actual black-box model but a meta-model or surrogate model of it. Then, estimates are typically not exactly equal to zero even if the "true" or analytically calculated sensitivity index would be. The resulting graph becomes confusing and uninformative. Therefore, edges (i, j) may be included into the graph only if

$$\hat{S}_{i,j}^{sup} > \delta \tag{22}$$

for some small threshold δ, e.g. $\delta = 0.01$ [26]. The computation of the FANOVA graph has been implemented in the R package `fanovaGraph`, providing several estimation methods as well as a thresholding functionality [8, 10].

To exemplify, let us now assume that we cannot analyse the Ishigami function analytically. Based on a random Latin hypercube design with 100 design points, we build a Kriging model of the Ishigami function. Our Kriging model has the usual

Table 1 Sobol' indices (first order) and TIIs for the Kriging model of the Ishigami function

	$\hat{S}_i$	$\hat{\phi}_i^*$	$\hat{S}_i^T$		$\hat{S}_{i,j}^{sup}$
X_1	0.300	0.447	0.603	X_1X_2	0.008
X_2	0.391	0.406	0.420	X_1X_3	0.273
X_3	0.009	0.148	0.285	X_2X_3	0.010

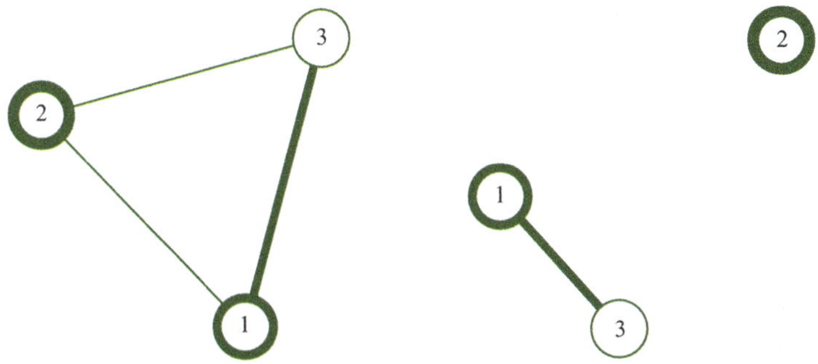

Fig. 3 Fanova graphs without and with thresholding for the Kriging model of the Ishigami function. (**a**) Fanova graph. (**b**) Fanova graph with threshold $\delta = 0.025$

Matern 5/2 covariance structure, no trend and no nugget effect. Table 1 shows the results for the estimators of the first-order Sobol' indices $\hat{S}_i$, normalized Shapley values $\hat{\phi}_i^*$ and TIIs $\hat{S}_{i,j}^{sup}$ of the Kriging model. These estimators have been computed using the R packages `fanovaGraph` [10] and `sensitivity` [15]. Remember that the inequalities

$$S_i^{cl} \leq \phi_i^* \leq S_i^T \tag{23}$$

hold and that $S_i^{cl} = S_i$, which is reflected by the order of the values in Table 1. Comparison also shows that the estimates slightly deviate from the true values given above.

Figure 3 displays on the left hand side the resulting pure FANOVA graph. This is a complete graph as all estimated TIIs are different from zero, even if only slightly. Therefore, we threshold the values by $\delta = 0.025$ and gain the graph on the right hand side, which is the same as for the analytical evaluation of the Ishigami function. The TII's and the FANOVA graph help to discover an underlying block-additive structure of the function f, i.e. we can find a decomposition into cliques of input variables such that variables in different cliques do not interact. As outlined in [26], the detected interaction structure by the FANOVA graph can be a valuable aid in constructing block-additive Kriging models. Therefore, the `fanovaGraph` package also contains methods for block-additive Kriging analysis. The block-additive decomposition provided by the FANOVA graph can also be used in a parallelized global optimization procedure [17].

4 Fields of Applications

The interpretation of a machine learning model by global sensitivity indices and FANOVA graph is in general applicable to any kind of model with a continuous response variable. We show examples of a Kriging model of a piston simulator and an ANN of resistance of sailing yachts.

4.1 Kriging Model of a Piston Simulator

As an example for the application in the field of the design and analysis of computer experiments we are using the piston simulator from the `mistat` package in R [1], first presented in [19]. A piston is moving within a cylinder. The piston's performance is measured by the time it takes to complete one cycle, in seconds. Here, we take the mean of 50 cycles as response, since the cycle time of the piston fluctuates strongly. The following factors can affect the piston's performance. The ranges, in which these factors are varied uniformly in our sensitivity analysis, are given in brackets.

m	The impact pressure determined by the piston mass (30–60) $[kg]$.
S	The piston surface area (0.005–0.020) $[m^2]$.
V_0	The initial volume of the gas inside the piston (0.002-0.010) $[m^3]$.
k	The spring coefficient (1000–5000) $[N/m^3]$.
p_0	The atmospheric pressure $(9 \cdot 10^4 - 11 \cdot 10^4)$ $[N/m^2]$.
T	The surrounding ambient temperature (290–296) $[K]$.
T_0	The filling gas temperature (340–360) $[K]$.

Based on a random Latin hypercube design with 70 design points, we build a Kriging model of the piston simulator. The Kriging model has a Matern 5/2 covariance kernel, no trend and no nugget effect. Table 2 shows the results for the Sobol' indices (first-order and total) and the Shapley values of the piston simulator. The slightly negative value for, e.g. $\hat{\phi}_6^*$ is of course an artefact of the estimation method. We observe that the piston surface $X_2 = S$ and the spring coefficient $X_4 = k$ have the largest effect on the cycle time.

Figure 4 displays the FANOVA graph for the Kriging model of the piston simulator after thresholding by $\delta = 0.005$.

In Fig. 4a both the edges as well as the vertices of the graph are presented by lines proportional to the values of the respective indices. It becomes obvious that $X_2 = S$ and $X_4 = k$ have the highest impact, followed by $X_1 = m$. However, as the values of the TIIs are noticeably smaller than the first-order Sobol' indices, it is not possible to detect which interactions are the largest. Therefore, in the FANOVA

Table 2 Sobol' indices for the Kriging model of the piston simulator

	Sobol' $\hat{S}_i$	Shapley $\hat{\phi}_i^*$	Total Sobol' $\hat{S}_i^T$
X_1	0.109	0.091	0.103
X_2	0.375	0.416	0.423
X_3	0.062	0.069	0.082
X_4	0.353	0.401	0.414
X_5	−0.000	0.014	0.003
X_6	−0.002	−0.009	0.007
X_7	0.026	0.018	0.036

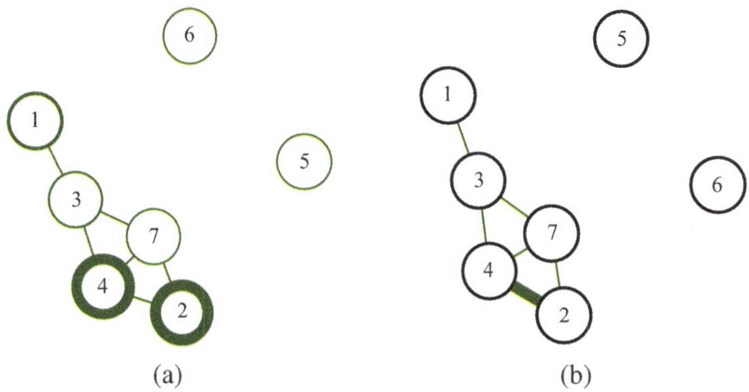

Fig. 4 FANOVA graph with threshold $\delta = 0.005$ for Kriging model of the piston simulator. (**a**) FANOVA graph. (**b**) FANOVA graph (only TIIs displayed)

graph in Fig. 4b we only vary the edges in strength proportional to the values of the TIIs. The largest TII is observed for the interaction $X_2X_4 = Sk$ with $\hat{S}_{2,4}^{sup} = 0.033$.

4.2 *Neural Net Model of Resistance of Sailing Yachts*

The residuary resistance of a ship is its total resistance minus the viscous resistance. In this section we are studying the residual resistance of sailing yachts in dependence of their hull geometry and the yacht velocity.

The Delft systematic yacht hull series data set [11] comprises $308 = 22 \cdot 14$ experiments with yacht models of scale 6.25 performed at the Delft Ship Hydromechanics Laboratory. In total, 22 different hull forms were tested with 14 different velocities. Based on the Delft series, semi-empirical models were developed [11] which are widely used in the yacht industry [22]. The Delft data set has 6 regressors and one dependent variable, all of which are dimensionless, i.e. their unit is 1 or % or ‰. Let the weight displacement Δ be the weight of water equivalent to the immersed volume of the hull. Then the dependent variable is the ratio R_r/Δ of the residuary resistance R_r to the weight displacement, given in ‰. The independent variables are as follows.

X_1 The longitudinal centre of buoyancy (LCB) is the longitudinal distance, given in % of some characteristic length, from a point of reference (often midships) to the centre of the displaced volume of water.

X_2 The prismatic coefficient $C_p = \nabla/L_{WL}A_m$ with A_m the cross-sectional area of the underwater slice at midships. C_p displays the ratio of the immersed volume of the hull to a volume of a prism with equal length and cross-sectional area A_m.

X_3 The length-displacement ratio $L_{WL}/\nabla^{1/3}$ where the volume displacement ∇ is the volume of water displaced by the hull.

X_4 The beam-draught ratio B_{WL}/T where the draught T is the maximal distance from the water line to the bottom of the keel.

X_5 The length-beam ratio L_{WL}/B_{WL} is the ratio of length to maximal width at water line.

X_6 The Froude number $Fr = u/\sqrt{gL_{WL}}$. Here u is the flow velocity relative to the yacht, g the gravitational acceleration, and L_{WL} is the length of the hull at water line.

We train a single hidden layer ANN to learn the relationship between input and output variables in the Delft data set. Such ANNs are implemented in the `nnet` package in R [35]. For regression, we choose an ANN with a linear activation function to the output neuron. The data set is divided into training, validation and testing subsets, containing 50%, 25% and 25% of the samples, respectively. The hyperparameter to be tuned is the number n of neurons in the hidden layer. We choose the ANN with lowest validation error, i.e. highest R^2-value for the validation data. That is according to Table 3 an ANN with 8 hidden neurons, i.e. a 6-8-1 net with $6 \cdot 8 + 8 = 56$ weights and $8 + 1 = 9$ biases.

For the chosen ANN as black-box function we perform a GSA. We compute the Sobol' indices as well as the scaled TIIs using the Liu-Owen method with $n = 100,000$ Monte Carlo samples with the help of the fanovaGraph package in R. Table 4 displays these sensitivity indices with the following coding of the regressors $X_1 = LCB$, $X_2 = C_p$, $X_3 = L_{WL}/\nabla^{1/3}$, $X_4 = B_{WL}/T$, $X_5 = L_{WL}/B_{WL}$ and $X_6 = Fr$. The Sobol' indices and scaled TIIs are graphically displayed in the bar plot in Fig. 5a.

Table 3 R^2 values for trainings, validation and test data for ANNs with different number of hidden neurons

m	2	3	4	5	6	7	8	9
R^2_{train}	0.9969	0.9992	0.9991	0.9998	0.9997	0.9998	0.9998	0.9997
R^2_{valid}	0.9972	0.9970	0.9859	0.9989	0.9985	0.9969	0.9990	0.9988
R^2_{test}	0.9962	0.9943	0.9939	0.9971	0.9924	0.9954	0.9957	0.9970

Table 4 Sobol' and Shapley values for the neural net model

	Sobol' $\hat{S}_i$	Shapley $\hat{\phi}_i^*$	Total Sobol' $\hat{S}_i^T$
X_1	0.024	0.020	0.142
X_2	0.006	0.042	0.037
X_3	0.033	0.022	0.071
X_4	0.028	0.163	0.237
X_5	0.021	0.031	0.087
X_6	0.591	0.722	0.688

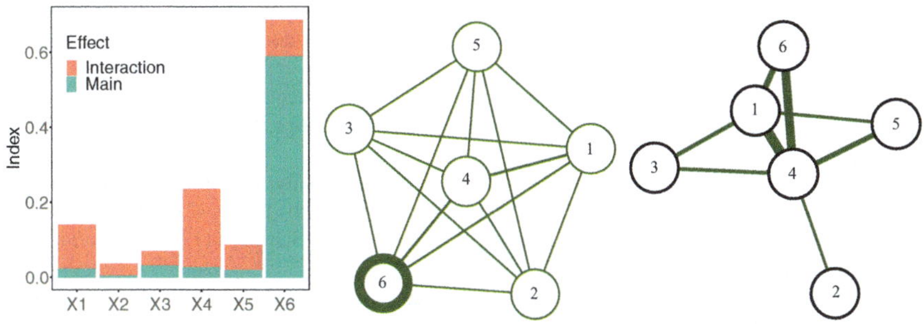

(a) (b) (c)

Fig. 5 Bar plot and FANOVA graphs without and with thresholding for the ANN model. (**a**) Bar plot of first-order and total Sobol' indices. (**b**) FANOVA graph. (**c**) FANOVA graph with threshold $\delta = 0.025$ (only TIIs displayed)

Figure 5 displays the FANOVA graph for the ANN model with and without thresholding. The Froude number, a proxy for velocity, has by far the largest impact on the residuary resistance. The largest interactions are X_1X_4, X_4X_6 and X_1X_6.

5 Summary

We have discussed the usefulness of GSA as a tool for interpretable machine learning. Global sensitivity indices based on Sobol' indices, Shapley values as well as derivative-based global sensitivity measures are revisited. FANOVA graphs allow for a very intuitive visualization of interaction structures and the strength of first-order Sobol' indices and TIIs. The approach is exemplified with a Kriging meta-model for a piston simulator and an ANN model for the resistance of yachts.

Acknowledgments The financial support of the Deutsche Forschungsgemeinschaft (SFB 823, project B1) is gratefully acknowledged.

References

1. Amberti, D.: mistat: Data Sets, Functions and Examples from the Book: "Modern Industrial Statistics" by Kenett, Zacks and Amberti (2018). https://CRAN.R-project.org/package=mistat. R package version 1.0-5
2. Boser, B.E., Guyon, I.M., Vapnik, V.N.: A training algorithm for optimal margin classifiers. In: Haussler, D. (ed.) Proceedings of the Fifth Annual Workshop on Computational Learning Theory – COLT '92, pp. 144–152. ACM Press, New York (1992). https://doi.org/10.1145/130385.130401
3. Cheng, K., Lu, Z., Zhou, Y., Shi, Y., Wei, Y.: Global sensitivity analysis using support vector regression. Appl. Math. Model. **49**(4), 587–598 (2017). https://doi.org/10.1016/j.apm.2017.05.026
4. Cortez, P., Embrechts, M.J.: Using sensitivity analysis and visualization techniques to open black box data mining models. Inf. Sci. **225**(1), 1–17 (2013). https://doi.org/10.1016/j.ins.2012.10.039
5. Cukier, R., Levine, H., Shuler, K.: Nonlinear sensitivity analysis of multiparameter model systems. J. Comput. Phys. **26**(1), 1–42 (1978). https://doi.org/10.1016/0021-9991(78)90097-9
6. Efron, B., Stein, C.: The jackknife estimate of variance. Ann. Stat. **9**(3) (1981). https://doi.org/10.1214/AOS/1176345462
7. Friedman, J.H., Popescu, B.E.: Predictive learning via rule ensembles. Ann. Appl. Stat. **2**(3), 916–954 (2008). https://doi.org/10.1214/07-AOAS148
8. Fruth, J., Roustant, O., Muehlenstaedt, T.: The fanovaGraph Package: Visualization of Interaction Structures and Construction of Block-additive Kriging Models (2013). https://hal.archives-ouvertes.fr/hal-00795229
9. Fruth, J., Roustant, O., Kuhnt, S.: Total interaction index: a variance-based sensitivity index for second-order interaction screening. J. Stat. Plann. Infer. **147**, 212–223 (2014). https://doi.org/10.1016/j.jspi.2013.11.007
10. Fruth, J., Muehlenstaedt, T., Roustant, O., Jastrow, M., Kuhnt, S.: fanovaGraph: building Kriging Models from FANOVA Graphs (2020). https://CRAN.R-project.org/package=fanovaGraph. R package version 1.5
11. Gerritsma, J., Onnink, R., Versluis, A.: Geometry, resistance and stability of the delft systematic yacht hull series. Int. Shipbuild. Prog. **28**(328), 276–297 (1981). https://doi.org/10.3233/ISP-1981-2832801
12. Hastie, T., Tibshirani, R.J.: Generalised additive models. In: Monographs on Statistics and Applied Probability, vol. 43. Chapman and Hall, London (1990)
13. Homma, T., Saltelli, A.: Importance measures in global sensitivity analysis of nonlinear models. Reliab. Eng. Syst. Saf. **52**(1), 1–17 (1996). https://doi.org/10.1016/0951-8320(96)00002-6
14. Iooss, B., Lemaître, P.: A review on global sensitivity analysis methods. In: Dellino, G., Meloni, C. (eds.) Uncertainty Management in Simulation-Optimization of Complex Systems, Operations Research/Computer Science Interfaces Series, vol. 59, pp. 101–122. Springer US, Boston (2015). https://doi.org/10.1007/978-1-4899-7547-8_5
15. Iooss, B., Veiga, S.D., Janon, A., Pujol, G., with contributions from Baptiste Broto, Boumhaout, K., Delage, T., Amri, R.E., Fruth, J., Gilquin, L., Guillaume, J., Le Gratiet, L., Lemaitre, P., Marrel, A., Meynaoui, A., Nelson, B.L., Monari, F., Oomen, R., Rakovec, O., Ramos, B., Roustant, O., Song, E., Staum, J., Sueur, R., Touati, T., Weber, F.: Sensitivity: global Sensitivity Analysis of Model Outputs (2020). https://CRAN.R-project.org/package=sensitivity. R package version 1.22.1
16. Ishigami, T., Homma, T.: An importance quantification technique in uncertainty analysis for computer models. In: 1990 Proceedings of First International Symposium on Uncertainty Modeling and Analysis, pp. 398–403. IEEE Computer Society Press, Washington (1990). https://doi.org/10.1109/ISUMA.1990.151285

17. Ivanov, M., Kuhnt, S.: A parallel optimization algorithm based on FANOVA decomposition. Qual. Reliab. Eng. Int. **30**(7), 961–974 (2014). https://doi.org/10.1002/qre.1710
18. Jansen, M.J.: Analysis of variance designs for model output. Comput. Phys. Commun. **117**(1–2), 35–43 (1999). https://doi.org/10.1016/S0010-4655(98)00154-4
19. Kenett, R., Zacks, S., Amberti, D.: Modern Industrial Statistics: With Applications in R, MINITAB and JMP, 2nd edn. Statistics in Practice. Wiley, Chichester (2014)
20. Kucherenko, S., Rodriguez-Fernandez, M., Pantelides, C., Shah, N.: Monte Carlo evaluation of derivative-based global sensitivity measures. Reliab. Eng. Syst. Saf. **94**(7), 1135–1148 (2009). https://doi.org/10.1016/j.ress.2008.05.006
21. Liu, R., Owen, A.B.: Estimating mean dimensionality of analysis of variance decompositions. J. Am. Stat. Assoc. **101**(474), 712–721 (2006). https://doi.org/10.1198/016214505000001410
22. Lopez Gonzalez, R.: Neural networks for variational problems in engineering. PhD Thesis. Technical University of Catalonia (2008)
23. McCullagh, P., Nelder, J.A.: Generalized linear models. Monographs on Statistics and Applied Probability, vol. 37, 2nd edn. Chapman and Hall, London (1989)
24. Molnar, C.: Interpretable Machine Learning. lulu.com (2020)
25. Molnar, C., Casalicchio, G., Bischl, B.: Quantifying model complexity via functional decomposition for better post-hoc interpretability. In: Cellier, P., Driessens, K. (eds.) Machine Learning and Knowledge Discovery in Databases. Communications in Computer and Information Science, vol. 1167, pp. 193–204. Springer International Publishing, Cham (2020). https://doi.org/10.1007/978-3-030-43823-4_17
26. Muehlenstaedt, T., Roustant, O., Carraro, L., Kuhnt, S.: Data-driven kriging models based on Fanova-decomposition. Stat. Comput. **22**(3), 723–738 (2012). https://doi.org/10.1007/s11222-011-9259-7
27. Owen, A.B.: Sobol' indices and Shapley value. SIAM/ASA J. Uncertain. Quantif. **2**(1), 245–251 (2014). https://doi.org/10.1137/130936233
28. Roustant, O., Fruth, J., Iooss, B., Kuhnt, S.: Crossed-derivative based sensitivity measures for interaction screening. Math. Comput. Simul. **105**, 105–118 (2014). https://doi.org/10.1016/j.matcom.2014.05.005
29. Saltelli, A., Ratto, M., Andres, T., Campolongo, F., Cariboni, J., Gatelli, D., Saisana, M., Tarantola, S.: Global Sensitivity Analysis. The Primer. John Wiley & Sons, Ltd, Chichester (2007). https://doi.org/10.1002/9780470725184
30. Shapley, L.S.: A value for n-person games. In: Kuhn, H.W., Tucker, A.W. (eds.) Contributions to the Theory of Games (AM-28), vol. II, pp. 307–318. Princeton University Press, Princeton (1953). https://doi.org/10.1515/9781400881970-018
31. Sobol, I.M.: Sensitivity analysis for non-linear mathematical models. Math. Modeling Comput. Experiment **1**(4), 407–414 (1993)
32. Sobol, I., Gershman, A.: On an alternative global sensitivity estimators. In: Proceedings of SAMO 1995, Belgirate, pp. 40–42 (1995)
33. Song, E., Nelson, B.L., Staum, J.: Shapley effects for global sensitivity analysis: theory and computation. SIAM/ASA J. Uncertain. Quantif. **4**(1), 1060–1083 (2016). https://doi.org/10.1137/15M1048070
34. Štrumbelj, E., Kononenko, I.: Explaining prediction models and individual predictions with feature contributions. Knowl. Inf. Syst. **41**(3), 647–665 (2014). https://doi.org/10.1007/s10115-013-0679-x
35. Venables, W.N., Ripley, B.D.: Modern Applied Statistics with S, 4th edn. Springer, New York (2002). ISBN 0-387-95457-0. http://www.stats.ox.ac.uk/pub/MASS4
36. Winter, E.: The Shapley value. In: Handbook of Game Theory with Economic Applications, vol. 3, pp. 2025–2054. Elsevier, Amsterdam (2002). https://doi.org/10.1016/S1574-0005(02)03016-3
37. Zhang, P.: A novel feature selection method based on global sensitivity analysis with application in machine learning-based prediction model. Appl. Soft Comput. **85**, 105859 (2019). https://doi.org/10.1016/j.asoc.2019.105859

Improving Gaussian Process Emulators with Boundary Information

Zhaohui Li and Matthias Hwai Yong Tan

Abstract Gaussian process (GP) models are widely used as emulators of time-consuming deterministic simulators, which are mostly computer codes that solve partial differential equation (PDE) models of physical systems numerically. In many cases, the functional relationship between the inputs and output of the simulator at parts of the boundary of the experiment domain or input domain can be determined using mathematical analysis, logical reasoning based on physical laws, or a cheap-to-compute low-fidelity simulator, as those subsets of the boundary correspond to simplified physical processes. However, this information is not taken into account in standard stationary GP priors used to construct GP emulators. This chapter considers the problem of constructing a GP emulator that reproduces known input–output relationships of a simulator at some boundary faces of the experiment/input domain, called boundary information/constraints. The proposed boundary modified GP (BMGP) emulator, which employs a nonstationary GP prior with specific forms for the mean and variance functions chosen so that the GP prior satisfies given boundary constraints, is shown to outperform the standard GP emulator based on a stationary GP prior and alternative emulators that satisfy given boundary constraints in two realistic examples.

Keywords Boundary information · Constrained Gaussian process emulator · Uncertainty quantification

Z. Li
School of Data Science, City University of Hong Kong, Hong Kong, China
e-mail: zhaohuili4-c@my.cityu.edu.hk

M. H. Y. Tan (✉)
School of Data Science, City University of Hong Kong, Hong Kong, China

Hong Kong Institute for Data Science (HKIDS), City University of Hong Kong, Hong Kong, China
e-mail: matthtan@cityu.edu.hk

A. Steland, K.-L. Tsui (eds.), *Artificial Intelligence, Big Data and Data Science in Statistics*, https://doi.org/10.1007/978-3-031-07155-3_7

1 Introduction

Computer models of physical systems, also called simulators, are increasingly used in practice. These simulators are often constructed from PDE models [7] that are solved using numerical methods such as the finite element method [13]. Time-consuming simulators are often approximated with cheap-to-evaluate surrogate models, also called emulators, built with data from computer experiments [11, 34] to save computation time.

GP models [33] are widely used as emulators to approximate time-consuming deterministic simulators. The standard GP modeling approach is to use a stationary GP with a parameterized correlation function as a prior for the input–output relationship represented by the simulator. The stationarity assumption corresponds to a priori indifference about the input–output relationship over the entire experiment domain and the correlation function is typically from a family that gives a GP with mean square partial derivatives up to a certain order. Popular choices of the correlation function are the product Gaussian and product Matérn correlation functions [34]. To build the GP model/emulator, experiment data are obtained by running the simulator at points given by an experiment design. Then, the prior GP is updated with the data, which gives a posterior GP (i.e., the emulator) that is used for predicting the simulator output. To ensure the emulator predicts well over the whole experiment region, the experiment design should be a space-filling design such as a Latin hypercube design (LHD) that optimizes a distance-based criterion (see [21], chapter 2 of [11] and chapter 5 of [34]) or a model-based optimal design (see chapter 6 of [34]).

In standard implementation of the GP emulator, parameters of the stationary prior GP such as its mean, variance, and correlation parameters are usually estimated by the maximum likelihood method. The data-driven method for parameter estimation and the stationarity assumption imply that the simulator is essentially treated as a black box. In practice, strong prior information about the simulator in the form of constraints is often available and there are some existing works on utilizing these constraints to improve the prediction performance of the GP emulator. For example, [14, 37, 42] propose methods to modify the GP emulator to incorporate monotonicity constraints. Tan [40] builds a GP model based on Green's function representation of a PDE solution. Alvarez et al. [2], Raissi et al. [31], and Chen et al. [6] propose GP models for physical data that incorporate information from a PDE. Other related works include [19, 20, 22, 36, 44].

In many practical problems, the value of the output y at a set of boundary points of the experiment region or input domain is known. For example, for a simulator with two inputs (x_1, x_2) and experiment region $[0, 1]^2$, one may know from physical reasoning or mathematical analysis that $y(0, x_2) = b_{01}(x_2)\ \forall x_2 \in [0, 1]$, where $b_{01}(\cdot)$ is a known function (see Sect. 2). We call this kind of information *boundary information/constraints and we call $b_{01}(\cdot)$ a boundary function*. Note that boundary information can be obtained from physical considerations. This often involves letting a nonnegative input equal to zero or infinity (which are boundary points of

$[0, \infty)$). We shall give a few examples. The first example concerns modeling of the temperature of a metal cube with five insulated faces, and one face that is raised to and fixed at temperature $\mathcal{T}$ instantly at time 0, via the heat equation (a parabolic PDE). Then, the temperature $y(x)$ at any point in the cube at a fixed time $\tau > 0$ converges to $\mathcal{T}$ as the thermal resistivity (the inverse thermal conductivity, see page 620 of [17]) x goes to zero, i.e., $\lim_{x \to 0} y(x) = \mathcal{T}$, since heat is transferred without resistance if $x = 0$. The second example concerns the bending of a plate. For this problem, it is known from physical reasoning that the maximum displacement of a plate approaches zero as the thickness of the plate becomes infinitely large, and this boundary information should be satisfied by all reasonable PDE models of plate bending such as the Mindlin and Kirchhoff plate models [43].

Boundary information can also be found from mathematical simplifications made to a PDE model (i.e., a PDE with boundary and initial conditions), which gives as its solution a physical quantity P as a function of space and time [12]. Specifically, if a PDE or a boundary condition is nonlinear, one can linearize it to make the PDE model easier to solve. This often involves setting to zero/infinity a nonnegative parameter. For instance, in a heat equation model for the temperature $P = T$ in a solid body, the thermal conductivity κ may be modeled by an increasing linear function of T [27], i.e., $\kappa = x_1 + x_2(T - \bar{T})$, where x_1 and $x_2 \geq 0$ are input parameters, and $\bar{T}$ is a reference temperature. This makes the PDE nonlinear. To linearize the PDE, we can set $\kappa = x_1$ (a constant over the space–time domain), which makes it possible to obtain the PDE solution explicitly ([17], Chapter 3). As $\kappa = x_1$ is obtained by letting $x_2 = 0$, an explicit solution found by letting $x_2 = 0$ provides boundary information. A related example is the elimination of a nonlinear radiation heat transfer term ([17], page 19) in a boundary condition for the heat equation model by setting the emissivity to zero. Finally, a PDE can be simplified by letting the length of the physical system in a spatial direction go to zero/infinity so that P is constant in that direction, as is common in structural mechanics (see [5], Chapter 19).

Commonly employed stationary GP priors do not take boundary information into account. It is anticipated that if the GP prior is constrained by the known boundary functions, it will give better prediction performance. In this chapter, we build a GP emulator that reproduces known boundary functions of a *deterministic* simulator at boundary faces of hypercube experiment regions. The proposed GP model, called boundary modified GP (BMGP) model, is numerically shown to predict better than the standard GP model and other existing GP models that satisfy boundary constraints. In many cases, one may obtain boundary information at a boundary of a domain larger than the experiment region, which is often the set of input values where the simulator can be run. A common example is where the experiment domain for an input x is $[0, 1]$ but the input domain in which the simulator can be run is $[0, \infty)$, and boundary information is available at $x = \infty$. We shall also discuss how this case can be handled.

This chapter is based on the work in [39], which proposes the BMGP model for exploiting boundary information. We employ the general BMGP model in [39] to exploit known input–output relationships at the boundary of compact hypercube experiment domains while [39] only apply the BMGP model to problems with boundary information at the boundary of the input domain $[0, \infty)^d$ that contains the experiment region as a proper subset. We give an improved model parameter estimation method and additional explanations, theoretical analysis, and examples to more clearly elucidate the ideas in [39]. Moreover, we compare the BMGP model with alternative GP models (some of them recently developed) that satisfy boundary constraints on hypercube domains. Note that use of boundary information to improve GP emulator construction is a topic of recent interest in the uncertainty quantification literature (e.g., [38, 39]). Nevertheless, the related problem of estimating a PDE solution with noisy data using a GP model that satisfies boundary and initial conditions has been studied as early as [15]. In contrast to [39], which considers boundary constraints in the simulator input space [38] considers imposing the Dirichlet boundary and initial conditions of a PDE model in building a GP emulator to predict the solution of the PDE model. Ding et al. [9] provide a theoretical analysis of the convergence rate of a GP emulator that satisfies known boundary constraints, called BdryGP emulator. However, the emulator uses a specific covariance function that gives continuous but non-differentiable sample paths. Vernon et al. [41] and Jackson and Vernon [18] propose analytical methods to update the standard GP emulator with data points placed along the boundary. Nevertheless, this can only give a GP that satisfies boundary constraints approximately when the boundary consists of an uncountable set of points. Solin and Kok [35] proposes a method for imposing a boundary constraint on a GP. However, they only consider a constant boundary function, while our proposed BMGP model works for any continuous boundary functions.

Note that our work in this chapter *does not* involve use of GP models to infer solutions of linear PDEs with Dirichlet boundary conditions based on noisy data, which is a problem solved in [15, 16, 23]. These authors use a prior GP that satisfies a linear PDE (approximately only in the first two references) and linear boundary conditions. Indeed, the boundary information referred to in this chapter does not refer to Dirichlet boundary conditions imposed on a PDE solution. Rather, it refers to known simulator input–output relationships at the boundary of an input domain, where the output is a scalar summary of the solution of a PDE model and the inputs are parameters of the PDE model (geometric parameters, source term parameters, parameters in boundary and initial conditions, and parameters in the PDE). The methods by Gulian et al. [16] and Lange-Hegermann [23] for imposing boundary conditions on a GP are limited to input domains of low (one to three) dimensions since the authors suggest using the solution to a PDE with given boundary conditions as the mean function of the GP and computational methods to solve PDEs are well developed for PDEs defined on one- to three-dimensional spatial domains only. Similarly, [15] gives an abstract method to build a GP model that satisfies boundary conditions but only explicitly consider a two-dimensional domain. Extension of this method to higher dimensional domains seems difficult. Unfortunately, simulators often have high-dimensional inputs. In contrast to the methods by Graepel [15], Gulian et al. [16], and Lange-Hegermann [23], the proposed BMGP model is easily made to satisfy multiple continuous boundary constraints for a domain of dimension higher than three. Nonetheless, we shall compare Graepel's model for a two-dimensional domain with the BMGP model in Sect. 4.1.

The remainder of this chapter is organized as follows. In Sect. 2, we state the type of boundary information for a two-dimensional experiment region that can be exploited by the BMGP model, and we give a motivating example with such boundary information. In Sect. 3, we review the standard GP emulator and then present the BMGP emulator. Section 4 applies the BMGP model to two examples and compares it with several other alternative GP emulators. Section 5 concludes the article.

2 A Motivating Example

In this section, we introduce the form of boundary information for a unit square experiment region/domain that can be exploited by the proposed BMGP model. Then, we introduce a practical example that involves a simulator with such boundary information. More general forms of boundary information that can be incorporated into the BMGP model are given in Sect. 3.3.

Suppose that $\{y(\mathbf{x}) \in \mathbb{R} : \mathbf{x} \in [0, 1]^2\}$ denotes the continuous functional relationship represented by the simulator, where we assume that $[0, 1]^2$ is the experiment domain for simplicity. In practice, there is often boundary information/constraints given by one or more of the following expressions:

$$y(0, x_2) = b_{01}(x_2)\ \forall x_2 \in [0, 1], \tag{1}$$

$$y(1, x_2) = b_{11}(x_2)\ \forall x_2 \in [0, 1], \tag{2}$$

$$y(x_1, 0) = b_{02}(x_1)\ \forall x_1 \in [0, 1], \tag{3}$$

$$y(x_1, 1) = b_{12}(x_1)\ \forall x_1 \in [0, 1]. \tag{4}$$

Note that the subscript ij in $b_{ij}(\cdot)$, which we call a boundary function, indicates that the boundary function is for the edge defined by $x_j = i$. These boundary functions may be obtained through mathematical analysis or physical considerations or cheap-to-compute low-fidelity simulators. Thus, if any of Eqs. (1)–(4) holds, the simulator output at one or more of the edges of the experiment domain is known, as illustrated in Fig. 1. Hence, if one or more of (1)–(4) are known, the goal of an emulator is to approximate the simulator well in the interior $(0, 1)^2$ of the experiment domain and the edges of the unit square where the boundary functions are unknown. The BMGP emulator discussed in this chapter can be made to satisfy any combination of (1)–(4), which helps improve predictions, especially near the boundary.

As an example, we consider a simulator implemented with Matlab PDE Toolbox [25] that predicts the steady-state temperature T (a function of spatial coordinates) in a cylindrical rod of length $l = 0.9$ m and diameter $\delta = 0.15$ m (Fig. 2) by solving

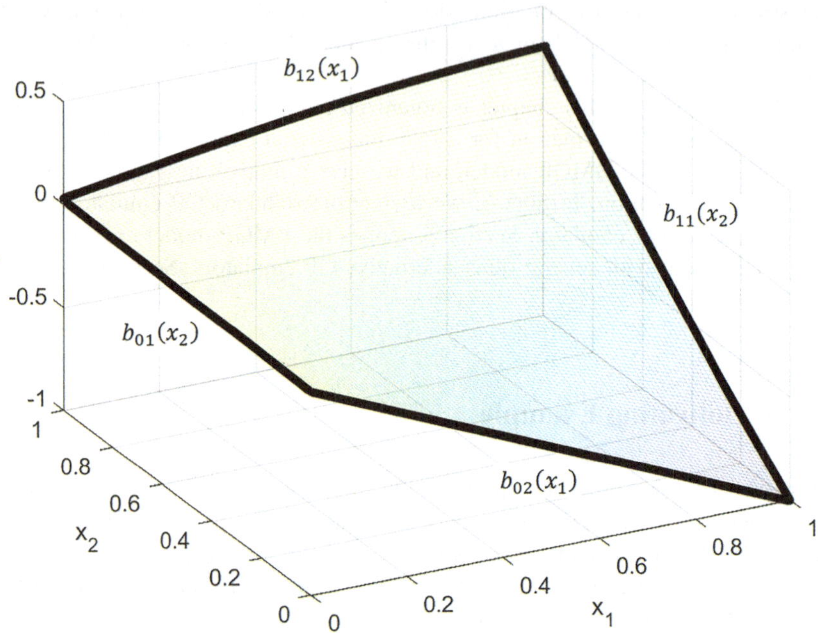

Fig. 1 Illustration of boundary functions

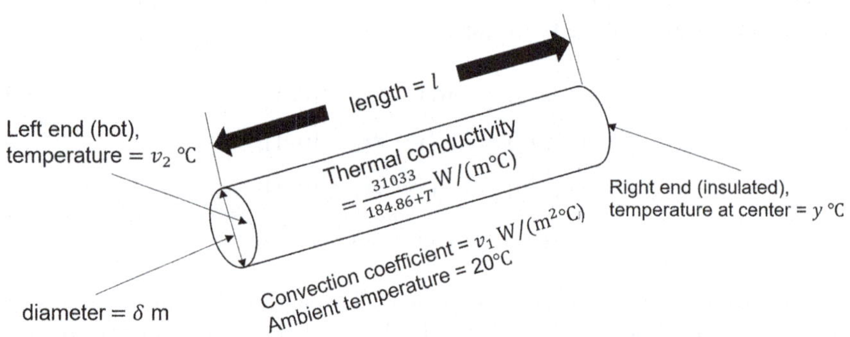

Fig. 2 Schematic diagram of rod temperature problem

the steady-state heat equation with boundary conditions. Convective heat transfer between the cylindrical surface and the ambient air at temperature 20 °C occurs with a convection coefficient of v_1W/(m^2°C), while the right end of the rod is insulated. The left end of the rod is uniformly at a fixed temperature of v_2 °C. The rod is made of silicon with thermal conductivity $31033/(184.86 + T)$W/(m°C). This equation is obtained by fitting the model $\kappa(T) = \alpha_1/(\alpha_2 + T)$ to the thermal conductivity $\kappa(T)$ versus temperature T (in °C) data in the temperature range of 250K to 1000K from [10], which gives a good fit. Since the thermal conductivity depends on T, the steady-state heat equation is a nonlinear PDE. Both v_1 and v_2 are inputs to the simulator. The range of v_1 of interest is 0 to $30\mathrm{W}/(\mathrm{m}^2\,^\circ\mathrm{C})$, while the range of v_2 of interest is 20 °C to 720 °C. The output of interest y is the temperature (in °C) at the center of the right end of the rod.

Suppose we standardize v_1 (by dividing by 30) to give the standardized input x_1 with range [0, 1] and we standardize v_2 to give the standardized input x_2 with range [0, 1]. It is known that (1) holds with $b_{01}(x_2) = 700x_2 + 20$ since if $x_1 = 0$, there will be no heat transfer between the cylindrical surface of the rod and the ambient air, which makes the temperature of the entire rod equal to the temperature at the left end $v_2 = 700x_2 + 20$. It is also known that (3) holds with $b_{02}(x_1) = 20$ since if the temperature at the left end of the rod is equal to the ambient temperature of 20°C ($x_2 = 0$ yields $v_2 = 20$°C), the temperature in the entire rod will be equal to 20°C also. This constant temperature is easily verified to be the PDE solution at $v_2 = 20$°C. Thus, boundary information of the form (1) and (3) is known to us.

3 Gaussian Process Modeling with Boundary Information

This section first provides a review of the standard GP emulator. Then, the proposed BMGP emulator that exploits boundary information is described.

3.1 Review of Standard GP Emulator

In this section, we discuss the standard GP emulator for a simulator with d inputs that each takes values in [0, 1]. The *continuous* functional relationship represented by the simulator is denoted as $\{y(\mathbf{x}) \in \mathbb{R} : \mathbf{x} \in [0, 1]^d\}$. The commonly employed GP emulator assumes a stationary real-valued GP prior $\{Y(\mathbf{x}) : \mathbf{x} \in [0, 1]^d\}$ for $\{y(\mathbf{x}) \in \mathbb{R} : \mathbf{x} \in [0, 1]^d\}$. The prior mean and prior variance of $Y(\cdot)$ (which is shorthand for $\{Y(\mathbf{x}) : \mathbf{x} \in [0, 1]^d\}$) are constants denoted by μ and σ^2, respectively. The prior correlation function, denoted by $\rho(\mathbf{x}, \mathbf{x}'|\boldsymbol{\theta})$, $\mathbf{x}, \mathbf{x}' \in [0, 1]^d$, where $\boldsymbol{\theta}$ is a vector of correlation parameters, is usually chosen so that the GP is one or more times mean square differentiable. A common choice, which we use in this chapter, is

$$\rho(\mathbf{x}, \mathbf{x}'|\boldsymbol{\theta}) = \prod_{i=1}^{d} \left[\exp\left(-\frac{|x_i - x_i'|}{\theta_i}\right)\right]\left(\frac{|x_i - x_i'|}{\theta_i} + 1\right), \tag{5}$$

where $\mathbf{x} = (x_1, \ldots, x_d)$, $\mathbf{x}' = (x_1', \ldots, x_d')$, and $\boldsymbol{\theta} = (\theta_1, \ldots, \theta_d) \in (0, \infty)^d$. The correlation function (5) is a member of the product Matérn correlation function family ([34], page 39). It gives a once mean square differentiable GP ([32], page 85). Other choices can be used as well.

The prior GP is updated with the output data $\mathbf{y} = \left(y(\mathbf{x}^1), \ldots, y(\mathbf{x}^n)\right)^T$ obtained by evaluating the simulator at the points in the design $\mathcal{D} = \{\mathbf{x}^1, \ldots, \mathbf{x}^n\}$, which is typically an LHD or other space-filling designs [34]. This gives the posterior GP with posterior mean and covariance functions that are easily obtained. In particular, for fixed μ, σ^2, and $\boldsymbol{\theta}$, the posterior GP is [8]

$$Y(\cdot)|(\mathbf{y}, \mu, \sigma^2, \boldsymbol{\theta}) \sim \mathrm{GP}\left(M(\cdot|\mu, \boldsymbol{\theta}), C(\cdot, \cdot|\sigma^2, \boldsymbol{\theta})\right), \tag{6}$$

i.e., a GP with mean function $M(\cdot|\mu, \boldsymbol{\theta})$ and covariance function $C(\cdot, \cdot|\sigma^2, \boldsymbol{\theta})$. The posterior mean function is

$$M(\mathbf{x}|\mu, \boldsymbol{\theta}) = \mu + \mathbf{r}(\mathbf{x})^T \mathbf{R}^{-1}(\mathbf{y} - \mu \mathbf{1}_n), \tag{7}$$

where $\mathbf{r}(\mathbf{x}) = \left(\rho(\mathbf{x}, \mathbf{x}^1|\boldsymbol{\theta}), \ldots, \rho(\mathbf{x}, \mathbf{x}^n|\boldsymbol{\theta})\right)^T$, $\mathbf{R} = \left(\rho(\mathbf{x}^i, \mathbf{x}^j|\boldsymbol{\theta})\right)_{1 \le i,j \le n}$ (an $n \times n$ matrix with $\rho(\mathbf{x}^i, \mathbf{x}^j|\boldsymbol{\theta})$ in the i-th row and j-th column), and $\mathbf{1}_n$ is an $n \times 1$ vector of $1'$s. The posterior covariance function is

$$C(\mathbf{x}, \mathbf{x}'|\sigma^2, \boldsymbol{\theta}) = \sigma^2 \left[\rho(\mathbf{x}, \mathbf{x}'|\boldsymbol{\theta}) - \mathbf{r}(\mathbf{x})^T \mathbf{R}^{-1} \mathbf{r}(\mathbf{x}')\right]. \tag{8}$$

In practice, the parameters μ, σ^2, and $\boldsymbol{\theta}$ are estimated by maximizing the likelihood (see [34]), and these estimates are plugged into the formulas for the posterior mean and covariance functions (7)–(8).

3.2 Boundary Modified GP Emulator

Let us now consider a simulator with two inputs x_1 and x_2, and experiment region $[0, 1]^2$. Clearly, the stationary GP prior described in Sect. 3.1 does not satisfy the boundary constraints (1)–(4), i.e., it ignores the prior information given by (1)–(4). Our proposed BMGP emulator employs a nonstationary GP prior $\{\mathcal{Y}(\mathbf{x}) : \mathbf{x} \in [0, 1]^2\}$ that *satisfies given boundary constraints* (at a boundary edge with a given boundary function, the prior GP equals the boundary function at each point with probability one). To illustrate, suppose (1) and (3) are known, but (2) and (4) are not known, and assume that $y(\cdot)$ is continuous (so that $b_{01}(\cdot)$ and $b_{02}(\cdot)$ are also continuous). Then, we choose the prior mean function $m(\cdot)$ to be a function that is continuously differentiable in $(0, 1)^2$ and continuous on $[0, 1]^2$ such that

$$m(0, x_2) = b_{01}(x_2)\ \forall x_2 \in [0, 1], \tag{9}$$

$$m(x_1, 0) = b_{02}(x_1)\ \forall x_1 \in [0, 1]. \tag{10}$$

We choose the prior variance function $v(\cdot)$ to be a function that is continuously differentiable in $(0, 1)^2$ and continuous on $[0, 1]^2$ such that

$$v(0, x_2) = 0 \; \forall x_2 \in [0, 1], \tag{11}$$

$$v(x_1, 0) = 0 \; \forall x_1 \in [0, 1], \tag{12}$$

and $v(\mathbf{x}) > 0 \; \forall \mathbf{x} \in (0, 1]^2$. The act of selecting a nonconstant variance function is called vertical rescaling in [32]. The correlation function for our BMGP model is (5). Given these choices, we see that

$$\lim_{x_1 \to 0} E\left\{\left[\mathcal{Y}(x_1, x_2) - b_{01}(x_2)\right]^2\right\} = 0 \; \forall x_2 \in [0, 1], \tag{13}$$

$$\lim_{x_2 \to 0} E\left\{\left[\mathcal{Y}(x_1, x_2) - b_{02}(x_1)\right]^2\right\} = 0 \; \forall x_1 \in [0, 1], \tag{14}$$

which implies that the BMGP prior converges in mean square to $b_{01}(x_2)$ as $x_1 \to 0$ and to $b_{02}(x_1)$ as $x_2 \to 0$. The mean square convergence properties (13)–(14) provide a stronger justification of the BMGP model than the simple statement that the BMGP prior satisfies the boundary constraint (1) and (3) as it is the properties of the BMGP prior near the boundary (rather than at the boundary) that is of concern. The BMGP prior is mean square differentiable in $(0, 1)^2$ as $m(\cdot)$ is differentiable, and the correlation function (5) and variance function $v(\cdot)$ give a covariance function $c(\mathbf{x}, \mathbf{x}') = \rho(\mathbf{x}, \mathbf{x}'|\boldsymbol{\theta})\sqrt{v(\mathbf{x})}\sqrt{v(\mathbf{x}')}, \mathbf{x}, \mathbf{x}' \in (0, 1)^2$ whose mixed partial derivative with respect to x_i and x_i' exists at all $\mathbf{x}, \mathbf{x}' \in (0, 1)^2$ ([1], page 27). We assume that $m(\cdot)$ and $v(\cdot)$ are continuously differentiable because continuous but non-differentiable $m(\cdot)$ and $v(\cdot)$ in the interior $(0, 1)^2$ of the experiment domain yield a GP prior that is too rough for many applications. Note that a GP prior that is continuous but non-differentiable (in mean square sense) often yields a posterior mean with poor prediction accuracy and very wide prediction intervals. Furthermore, the input–output relationship of a simulator is usually at least continuously differentiable.

There are many possible specific choices for the mean function that satisfy (9)–(10). However, for practical purposes, we would like a parameterized function that is not only sufficiently flexible to model nonlinear relationships but also analytically and computationally tractable. The existence of parameters in the mean function allows the function to be adjusted based on data. Our proposed prior mean function is

$$m(\mathbf{x}) = \lambda_0(\mathbf{x})\mu + \lambda_1(\mathbf{x})b_{01}(x_2) + \lambda_2(\mathbf{x})b_{02}(x_1), \tag{15}$$

$$\lambda_0(\mathbf{x}) = \left(\sum_{j=1}^{2} d^2(x_j)\right) \Big/ \left[\sum_{j=1}^{2} d^2(x_j) + \sum_{j=1}^{2} \frac{\alpha}{d^2(x_j)}\right],$$

$$\lambda_i(\mathbf{x}) = \frac{\alpha}{d^2(x_i)} \Big/ \left[\sum_{j=1}^{2} d^2(x_j) + \sum_{j=1}^{2} \frac{\alpha}{d^2(x_j)} \right], i = 1, 2,$$

$$d^2(x_i) = |0.5/(0.5 + x_i) - 1|^2, i = 1, 2,$$

where $d(x_i)$ is a measure of distance between x_i and 0 that is obtained from the distance metric $d(x, x') = \left|0.5/(0.5 + x) - 0.5/(0.5 + x')\right|$, and μ and $\alpha > 0$ are parameters. One should extend (15) by continuity when evaluating it at $(x_1, 0)$ or $(0, x_2)$, which gives (9)–(10). The prior mean function has the intuitively appealing property that it is a convex combination of μ, $b_{01}(x_2)$, and $b_{02}(x_1)$. The parameter μ may be viewed as the prior mean far away from the boundaries given by $x_1 = 0$ and $x_2 = 0$ as $\lambda_0(\mathbf{x})$ increases with x_1 and x_2. We also see that as x_i decreases to zero, more weight is assigned to $b_{0i}(\cdot)$. Our proposed prior variance function is

$$v(\mathbf{x}) = s^2 \prod_{i=1}^{2} d^{2\eta}(x_i), \tag{16}$$

where s^2 and η are positive parameters. It can be seen from (16) that as the distance from the point $\mathbf{x}$ to the boundary $x_i = 0$ increases, the prior variance increases, i.e., prior uncertainty about the simulator output increases, which is a sensible property.

When other combinations of (1)–(4) hold, it is simple to modify the above approach to construct $m(\cdot)$ and $v(\cdot)$ so that the boundary constraints are satisfied. The appropriate distance measure to use is given by (24) in Sect. 3.3, and one can refer to [39] for a generalization of the formulas (15)–(16) to accommodate any number of boundary constraints. There are other ways to build $m(\cdot)$ and the prior covariance function $c(\cdot, \cdot)$ (which gives $v(\cdot)$) so that some combination of (1)–(4) is satisfied (e.g., see [9] for alternative choices of $m(\cdot)$ and $c(\cdot, \cdot)$, and Sect. 4.1). However, we have found that the choices (15)–(16) for the case where (1) and (3) hold give good and reliable results.

As in Sect. 3.1, the prior GP is updated with the output data $\mathbf{y} = \left(y(\mathbf{x}^1), \ldots, y(\mathbf{x}^n)\right)^T$ obtained by evaluating the simulator at the design points $\mathcal{D} = \{\mathbf{x}^1, \ldots, \mathbf{x}^n\}$. This gives the posterior GP with mean and covariance functions that are easily obtained. In particular, the posterior GP is

$$\mathcal{Y}(\cdot)|(\mathbf{y}, \mu, \alpha, s^2, \eta, \boldsymbol{\theta}) \sim \text{GP}\left(M'(\cdot|\mu, \alpha, \eta, \boldsymbol{\theta}), C'(\cdot, \cdot|s^2, \eta, \boldsymbol{\theta})\right), \tag{17}$$

i.e., a GP with mean function $M'(\cdot|\mu, \alpha, \eta, \boldsymbol{\theta})$ and covariance function $C'(\cdot, \cdot|s^2, \eta, \boldsymbol{\theta})$ (note that the mean function depends on $\mu, \alpha, \eta, \boldsymbol{\theta}$, while the covariance function depends on s^2, η, $\boldsymbol{\theta}$). In particular, the posterior mean function is

$$M'(\mathbf{x}|\mu, \alpha, \eta, \boldsymbol{\theta}) = m(\mathbf{x}) + \mathbf{q}(\mathbf{x})^T \mathbf{Q}^{-1}(\mathbf{y} - \mathbf{m}), \tag{18}$$

where $\mathbf{Q} = \left(\sqrt{v(\mathbf{x}^i)v(\mathbf{x}^j)}\rho(\mathbf{x}^i, \mathbf{x}^j|\boldsymbol{\theta})\right)_{1\leq i,j\leq n}$, $\mathbf{m} = \left(m(\mathbf{x}^1), \ldots, m(\mathbf{x}^n)\right)^T$, and $\mathbf{q}(\mathbf{x}) = \sqrt{v(\mathbf{x})}\left(\sqrt{v(\mathbf{x}^1)}\rho(\mathbf{x}, \mathbf{x}^1|\boldsymbol{\theta}), \ldots, \sqrt{v(\mathbf{x}^n)}\rho(\mathbf{x}, \mathbf{x}^n|\boldsymbol{\theta})\right)^T$. The posterior covariance function is

$$C'(\mathbf{x}, \mathbf{x}'|s^2, \eta, \boldsymbol{\theta}) = \sqrt{v(\mathbf{x})v(\mathbf{x}')}\rho(\mathbf{x}, \mathbf{x}'|\boldsymbol{\theta}) - \mathbf{q}(\mathbf{x})^T\mathbf{Q}^{-1}\mathbf{q}(\mathbf{x}'). \tag{19}$$

Note that the boundary constraints (1) and (3) are satisfied with probability one a posteriori. This is clear from (18)–(19), as $m(\mathbf{x})$ equals the known boundary function and $v(\mathbf{x}) = 0$ when $\mathbf{x} = (0, x_2)$ or $\mathbf{x} = (x_1, 0)$. Moreover, as the posterior mean and posterior variance functions are continuous, the BMGP posterior converges in mean square to $b_{01}(x_2)$ as $x_1 \to 0$ and to $b_{02}(x_1)$ as $x_2 \to 0$. We estimate the parameters μ, α, s^2, η, and $\boldsymbol{\theta}$ based on data and then plug the estimates into the posterior mean and covariance functions (18)–(19). The mean parameters μ and α are estimated by the least squares method, i.e., by minimizing the sum of squared differences between the components of $\mathbf{y}$ and $\mathbf{m}$, which yields $\hat{\mu}$ and $\hat{\alpha}$. Then, given $(\mu, \alpha) = (\hat{\mu}, \hat{\alpha})$, we estimate the covariance parameters $(s^2, \eta, \boldsymbol{\theta})$ by maximizing the likelihood. For fixed $(\eta, \boldsymbol{\theta})$, $s^2 = \hat{s}^2 = (\mathbf{y} - \mathbf{m})^T\mathbf{P}^{-1}(\mathbf{y} - \mathbf{m})/n$ maximize the likelihood, where $\mathbf{P} = s^{-2}\mathbf{Q}$ (which does not depend on s^2). Thus, the maximum likelihood estimate $(\hat{\eta}, \hat{\boldsymbol{\theta}})$ of $(\eta, \boldsymbol{\theta})$ is obtained by minimizing

$$n\log(\hat{s}^2) + \log(|\mathbf{P}|). \tag{20}$$

The maximum likelihood estimate of s^2 is $\hat{s}^2$ evaluated at $(\eta, \boldsymbol{\theta}) = (\hat{\eta}, \hat{\boldsymbol{\theta}})$. Note that [39] uses the maximum likelihood method to estimate all parameters (including μ and α), but we found that this can yield prediction intervals with somewhat low coverage. The method for parameter estimation proposed above tends to give superior coverage and equally good prediction accuracy, possibly because the least squares method gives better estimates of mean function parameters, as shown in [30]. As with the standard GP model in Sect. 3.1, the covariance matrix $\mathbf{Q}$ of the BMGP model can become ill conditioned when n is large. To deal with this issue, one can add a nugget $s^2\xi$ to the diagonal elements of $\mathbf{Q}$ [29], where ξ is a small positive number.

In general, to make sure that the BMGP model works well, the model should be validated. One can refer to [4] for a few methods that can be applied to validate the BMGP model if a test set of modest size that is not used in training the BMGP model is available. A more economical way to check the adequacy of the BMGP model is leave-one-out cross validation [26].

We have found that space-filling designs such as LHDs and quasi-Monte Carlo sequences [11, 34] that contain *no boundary design points with known outputs* (which are zero prior variance points) are good choices for the design $\mathcal{D}$ as they yield BMGP emulators that predict well. However, the result obtained by fitting a BMGP emulator with an initial space-filling design can be improved by employing a sequential design approach, which adds design points one by one at the point of maximum prediction variance. The details are given in [39].

3.3 BMGP Model for More General Cases

In Sect. 3.2, we propose the BMGP emulator for a two-dimensional experiment region $[0, 1]^2$ where boundary information of the form (1) and (3) is available. It is straightforward to generalize this method to a d-dimensional experiment region $[0, 1]^d$ for a continuous function $y(\cdot)$ (computed by a simulator) with boundary information on a $(d-k)$-face boundary of the form

$$\lim_{\mathbf{x}^{k-}\to\mathbf{c}^{k-}} y(\mathbf{x}) = y(\mathbf{c}^{k-}, \mathbf{x}^{k+}) = b(\mathbf{x}^{k+})\ \forall \mathbf{x}^{k+} \in [0, 1]^{d-k}, \tag{21}$$

where $\mathbf{x}^{k-} = (x_1, \ldots, x_k)$, $\mathbf{c}^{k-} = (c_1, \ldots, c_k)$, and $\mathbf{x}^{k+} = (x_{k+1}, \ldots, x_d)$. Note that $c_i \in \{0, 1\}\ \forall i = 1, \ldots, k$, and $b(\cdot)$ is continuous as $y(\cdot)$ is continuous. Clearly, if $d = 2$, $c_1 = 0$, and $k = 1$, (21) reduces to (1). The prior mean function $m(\cdot)$ and prior variance function $v(\cdot)$ of the BMGP model are chosen as functions that are continuous on $[0, 1]^d$, continuously differentiable in $(0, 1)^d$, and respectively satisfy

$$\lim_{\mathbf{x}^{k-}\to\mathbf{c}^{k-}} m(\mathbf{x}) = m(\mathbf{c}^{k-}, \mathbf{x}^{k+}) = b(\mathbf{x}^{k+})\ \forall \mathbf{x}^{k+} \in [0, 1]^{d-k}, \tag{22}$$

and

$$\lim_{\mathbf{x}^{k-}\to\mathbf{c}^{k-}} v(\mathbf{x}) = v(\mathbf{c}^{k-}, \mathbf{x}^{k+}) = 0\ \forall \mathbf{x}^{k+} \in [0, 1]^{d-k}. \tag{23}$$

The prior correlation function $\rho(\cdot, \cdot|\boldsymbol{\theta})$ is given by (5). After updating the BMGP prior with data $\mathbf{y} = \left(y(\mathbf{x}^1), \ldots, y(\mathbf{x}^n)\right)^T$, a posterior GP with mean function $M'(\cdot)$ and covariance function $C'(\cdot, \cdot)$ given by expressions identical to the right-hand sides of (18) and (19), respectively, is obtained. Given the above assumptions on $m(\cdot)$, $v(\cdot)$, and $\rho(\cdot, \cdot|\boldsymbol{\theta})$, we have the following result:

Proposition 1 *The BMGP prior $\mathcal{Y}(\mathbf{x})$ and the BMGP posterior $\mathcal{Y}(\mathbf{x})|\mathbf{y}$ converge to $b(\mathbf{x}^{k+})$ in mean square as $\mathbf{x}^{k-} \to \mathbf{c}^{k-}$, i.e., $\lim_{\mathbf{x}^{k-}\to\mathbf{c}^{k-}} \mathcal{Y}(\mathbf{x}) = b(\mathbf{x}^{k+})$ and $\lim_{\mathbf{x}^{k-}\to\mathbf{c}^{k-}} \mathcal{Y}(\mathbf{x})|\mathbf{y} = b(\mathbf{x}^{k+})$ hold in mean square. The first limit also holds with probability one, i.e., $\mathcal{Y}(\mathbf{x})$ converges to $b(\mathbf{x}^{k+})$ almost surely as $\mathbf{x}^{k-} \to \mathbf{c}^{k-}$.*

Proof We first prove the limits hold in mean square. An immediate consequence of (22)–(23) is that $\lim_{\mathbf{x}^{k-}\to\mathbf{c}^{k-}} \mathcal{Y}(\mathbf{x}) = b(\mathbf{x}^{k+})$ holds in mean square. Since $m(\cdot)$, $v(\cdot)$, and $\rho(\cdot, \cdot|\boldsymbol{\theta})$ are continuous, the posterior mean function $M'(\cdot)$ and the posterior covariance function $C'(\cdot, \cdot)$ are continuous. This fact and (22)–(23)

give $\lim_{\mathbf{x}^{k-}\to\mathbf{c}^{k-}} M'(\mathbf{x}) = b(\mathbf{x}^{k+})$ and $\lim_{\mathbf{x}^{k-}\to\mathbf{c}^{k-}} C'(\mathbf{x},\mathbf{x}) = 0$, which imply that $\lim_{\mathbf{x}^{k-}\to\mathbf{c}^{k-}} \mathcal{Y}(\mathbf{x})|\mathbf{y} = b(\mathbf{x}^{k+})$ holds in mean square. □

Due to continuity of $m(\cdot)$ and $v(\cdot)$, and the choice of $\rho(\cdot,\cdot|\boldsymbol{\theta})$ in (5), the prior GP $\mathcal{Y}(\cdot)$ is sample path continuous almost surely. This is because we can write $\mathcal{Y}(\mathbf{x}) = m(\mathbf{x})+\sqrt{v(\mathbf{x})}\tilde{\mathcal{Y}}(\mathbf{x})$, where $\tilde{\mathcal{Y}}(\cdot) \sim \mathrm{GP}(0, \rho(\cdot,\cdot|\boldsymbol{\theta}))$ is sample path continuous almost surely ([28], page 47). As $\mathcal{Y}(\cdot)$ is continuous with probability one, $\mathcal{Y}(\mathbf{x})$ converges to $b(\mathbf{x}^{k+})$ almost surely as $\mathbf{x}^{k-} \to \mathbf{c}^{k-}$.

Remark 1 Proposition 1 applies to cases with multiple constraints of the form (21).

Remark 2 Mean square convergence of the BMGP posterior and almost sure convergence of its prior to $b(\mathbf{x}^{k+})$ as $\mathbf{x}^{k-} \to \mathbf{c}^{k-}$ are new results stated in this paper.

We now give our recommendation for $m(\cdot)$ and $v(\cdot)$. First, we generalize the measure of distance, i.e., $\mathcal{d}(\cdot)$, defined in Sect. 3.2. We define $\mathcal{d}^2_{\mathbf{c}^{k-}}(\mathbf{x}^{k-}) = k^{-1}\sum_{l=1}^{k}\varphi^2_{c_l}(x_l)$ as the squared distance between $\mathbf{x}^{k-}$ and $\mathbf{c}^{k-}$, where

$$\varphi_c(x) = \begin{cases} \left|\dfrac{0.5}{0.5+x} - \dfrac{0.5}{0.5+c}\right|, & c = 0 \\[2ex] \left|\dfrac{0.5}{0.5+1-x} - \dfrac{0.5}{0.5+1-c}\right|, & c = 1. \end{cases} \tag{24}$$

Note that the quantity $\varphi_c(x)$ can be written as $\varphi_c(x) = \varphi_c(x, c)$, where $\varphi_0(x, x') = \left|\frac{0.5}{0.5+x} - \frac{0.5}{0.5+x'}\right|$, and $\varphi_1(x, x') = \left|\frac{0.5}{0.5+1-x} - \frac{0.5}{0.5+1-x'}\right|$. It is easy to show that $\varphi_0(x, x')$ and $\varphi_1(x, x')$ are valid distance metrics if viewed as functions of (x, x'). Thus, $\mathcal{d}_{\mathbf{z}^{k-}}(\mathbf{x}^{k-}) = \left[k^{-1}\sum_{l=1}^{k}\varphi^2_{z_l}(x_l)\right]^{1/2}$ is a valid distance metric also if viewed as a function of $(\mathbf{x}^{k-}, \mathbf{z}^{k-})$, where $\mathbf{z}^{k-} = (z_1, \ldots, z_k)$. We plot (24) (recommended distance metric) in Fig. 3 for the case where $c = 0$. The Euclidean distance between x and $c = 0$, which is a straight line, is also plotted in Fig. 3. It is seen that (24) has a concave shape when $c = 0$. The reason that we choose $\varphi_c(x)$ so that $\varphi_1(x) = \varphi_0(1-x)$ is for the sake of symmetry, i.e., the function $\{\varphi_1(x) : x \in [0, 1]\}$ is a reflection of $\{\varphi_0(x) : x \in [0, 1]\}$ about $x = 0.5$, as we should obtain the same distance between x and $c = 0$ and between $1 - x$ and $c = 1$. Note that instead of (24), [39] sets $\varphi_c(x) = \left|\frac{u}{u+x} - \frac{u}{u+c}\right|$ for some $u > 0$, which is appropriate for the input domain $[0, \infty)^d$.

Analogous to (15), our suggested prior mean function is $m(\mathbf{x}) = \lambda_0(\mathbf{x})\mu + \lambda_1(\mathbf{x})b(\mathbf{x}^{k+})$, where $\lambda_0(\mathbf{x}) = \mathcal{d}^2_{\mathbf{c}^{k-}}(\mathbf{x}^{k-})/[\mathcal{d}^2_{\mathbf{c}^{k-}}(\mathbf{x}^{k-}) + \alpha/\mathcal{d}^2_{\mathbf{c}^{k-}}(\mathbf{x}^{k-})]$, $\lambda_1(\mathbf{x}) = [\alpha/\mathcal{d}^2_{\mathbf{c}^{k-}}(\mathbf{x}^{k-})]/[\mathcal{d}^2_{\mathbf{c}^{k-}}(\mathbf{x}^{k-}) + \alpha/\mathcal{d}^2_{\mathbf{c}^{k-}}(\mathbf{x}^{k-})]$, and μ and $\alpha > 0$ are parameters. Similarly, our suggested variance function is $v(\mathbf{x}) = s^2[\mathcal{d}^2_{\mathbf{c}^{k-}}(\mathbf{x}^{k-})]^{\eta}$. It is easily seen that these choices satisfy (22)–(23). The reason we recommend the distance measure $\mathcal{d}_{\mathbf{c}^{k-}}(\mathbf{x}^{k-})$ constructed from (24) is because we found empirically that concave shaped distance measures like $\varphi_c(x)$ in (24) give a GP emulator with better coverage than the Euclidean distance measure. This is because the distance between a point near 0 and the boundary 0 is a larger fraction of the maximum distance to the boundary 0 achieved at $x = 1$ when the recommended distance measure is used instead of the Euclidean distance. This makes the prior variance at a point near 0 a larger fraction of the maximum prior variance within the experiment region.

Another kind of boundary information is obtained when an input goes to infinity, which gives rise to asymptotes in graphs. Consider a simulator with a scalar input x that produces output $y(x)$ given input x. Suppose it is known that $\lim_{x\to\infty} y(x) = b$, where $x = \infty$ can be viewed as a boundary of the input domain where the simulator can be run. Note that this is not the boundary of the experiment region, which needs to be specified as a compact set. For this problem, we recommend

$$m(x) = \frac{\mu d^2(x,\infty)}{d^2(x,\infty) + \alpha/d^2(x,\infty)} + \frac{b\alpha/d^2(x,\infty)}{d^2(x,\infty) + \alpha/d^2(x,\infty)}, \tag{25}$$

where $\alpha > 0$, and μ are parameters to be estimated, $d(x,\infty) = \left|\frac{0.5}{0.5+x}\right|$, and μ can be viewed as the GP mean far away from ∞. Note that $d(x,x') = \left|\frac{0.5}{0.5+x} - \frac{0.5}{0.5+x'}\right|$ is a distance metric on $[0,\infty]$ when viewed as a function of (x,x'). We suggest a variance function of the form

$$v(x) = s^2\left[d^2(x,\infty)\right]^\eta. \tag{26}$$

Clearly, a prior GP $\{\mathcal{Y}(x) : x \in [0,\infty)\}$ with these choices of mean function and variance function will give $\lim_{x\to\infty} E\left\{[\mathcal{Y}(x) - b]^2\right\} = 0$, i.e., the boundary information is satisfied in mean square sense.

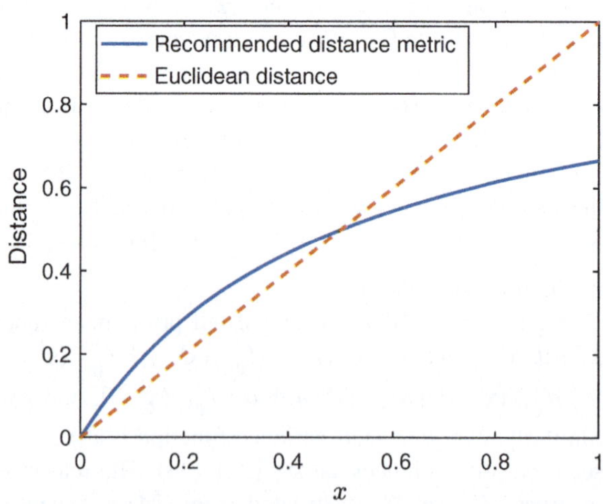

Fig. 3 Distance between x and 0 as given by recommended distance metric $\varphi_0(x)$ and usual Euclidean metric

4 Numerical Examples

In this section, we give two simple but realistic examples to compare the proposed BMGP model with some alternative GP emulators. Matlab codes for reproducing the results in this section can be found in GitHub, https://github.com/ustclzh/Gaussian-Process-With-Boundary-Information.git. The codes were run in Matlab® R2020a (MathWorks, Inc., Natick, MA, USA).

4.1 Example 1

We revisit the rod temperature simulator (based on a nonlinear PDE model) described in Sect. 2. We compare the BMGP emulator with four alternative GP emulators to demonstrate the improvements that can be gained by forcing an emulator to satisfy known boundary constraints and to show that the prior mean and variance functions (15)–(16) tend to be better than alternative choices. The first alternative GP emulator is the standard GP emulator given in Sect. 3.1. It does not satisfy the known boundary constraints (1) and (3). The second, third, and fourth alternative GP emulators satisfy the boundary constraints (1) and (3) (a priori and a posteriori), as with the BMGP emulator. In fact, these alternative emulators also give GP priors/posteriors that converge in mean square to $b_{01}(x_2)$ as $x_1 \to 0$ and to $b_{02}(x_1)$ as $x_2 \to 0$, as with the BMGP emulator. The second alternative emulator is obtained from the BMGP emulator by replacing the covariance function $c(\mathbf{x}, \mathbf{x}') = \sqrt{v(\mathbf{x})}\sqrt{v(\mathbf{x}')}\rho(\mathbf{x}, \mathbf{x}'|\boldsymbol{\theta})$, $\mathbf{x}, \mathbf{x}' \in [0, 1]^2$ of the BMGP emulator with a product of fractional Brownian covariance functions, which we call the fractional Brownian GP (FBGP) emulator. The product fractional Brownian covariance function is given by

$$C_{\mathrm{FB}}(\mathbf{x}, \mathbf{x}') = s^2 \prod_{i=1}^{2} \left(|x_i|^{p_i} + |x_i'|^{p_i} - |x_i - x_i'|^{p_i}\right)/2, \tag{27}$$

where $0 < p_i < 2, i = 1, 2$, and $s^2 > 0$ are parameters. The fractional Brownian covariance function is well known (see [24], page 196) and it is used in [45] for emulator construction. For the FBGP model, we employ (15) as its prior mean function, as with the BMGP model. The FBGP model satisfies the boundary constraints (1) and (3) as $C_{\mathrm{FB}}((x_1, 0), (x_1, 0)) = C_{\mathrm{FB}}((0, x_2), (0, x_2)) = 0$. Comparing the BMGP and FBGP models helps justify our proposed choice of $c(\cdot, \cdot)$. The third alternative emulator is the BdryGP emulator [9], which uses a so-called BdryMatérn covariance function

$$C_{\mathrm{Bdry}}(\mathbf{x}, \mathbf{x}') = s^2 \prod_{i=1}^{2} \sinh\left[\theta_i \min\{x_i, x_i'\}\right] \exp\left[-\theta_i \max\{x_i, x_i'\}\right]. \tag{28}$$

For the BdryGP emulator, the prior mean function is a radial basis function interpolator of boundary data points (see [9]). Its value at $\mathbf{x} = (x_1, x_2)$ is obtained by interpolating the output values at boundary points $(x_1, 0)$ and $(0, x_2)$ using the radial basis kernel $\max\left\{1 - \left\|\mathbf{x} - \mathbf{x}'\right\|_2, 0\right\}^{\nu}$, where a kernel with $\nu = 3$ is used as it gives good results. The fourth alternative GP emulator is constructed according to the model proposed by Graepel [15], who models the solution $T(\cdot)$ to a linear PDE with a Dirichlet boundary condition at the edges of $[0, 1]^2$ as $b(\cdot) + a(\cdot)Z(\cdot)$, where $a(\cdot)$ is zero at the boundary and $b(\cdot)$ is equal to the Dirichlet boundary condition at the edges. Using this idea and the $a(\cdot)$ and $b(\cdot)$ proposed in [15] to build a GP model that satisfies (1) and (3) gives us the prior GP

$$\mathcal{Z}(\mathbf{x}) = (1-x_1)b_{01}(x_2)+(1-x_2)\left[b_{02}(x_1) - (1 - x_1)b_{02}(0)\right]+x_1x_2Z(\mathbf{x}), \quad (29)$$

where $Z(\cdot) \sim \mathrm{GP}\left(0, \sigma^2\rho(\cdot, \cdot|\boldsymbol{\theta})\right), \sigma^2 > 0$, and $\rho(\cdot, \cdot|\boldsymbol{\theta})$ is defined in (5). We call the emulator given by this prior GP as the Graepel GP (GGP) emulator. All parameters in the FBGP, BdryGP, and GGP models (including the mean function parameters of the FBGP emulator) are estimated by the maximum likelihood method.

We generate a sliced LHD with three slices for two inputs, where each slice is an LHD with 20 points, using the R package SLHD [3]. The first slice is used as the design to build the BMGP emulator (Sect. 3.2) and all four alternative GP emulators described above. The union of the second and third slices, which has an empty intersection with the first slice, is taken as the test input set. The simulator is run at all points in the design and test input set, and the predictions at the test input sites given by the BMGP emulator and the four alternative GP emulators are computed. This process is repeated 60 times, i.e., 60 designs and 60 test input sets are generated.

For each pair of design and test input set, we compute the mean absolute prediction error (MAE), root mean square prediction error (RMSE), and average length (ALPI) and empirical coverage (Coverage) of 98% prediction intervals for the standard GP, FBGP, BdryGP, GGP, and BMGP emulators. Table 1 gives the sample means (averages) of the 60 values of MAE, RMSE, ALPI, and Coverage and their standard errors. We see that the average MAE and average RMSE for the BMGP emulator are the smallest among all GP emulators, and the size of the standard error suggests that the average MAE/RMSE of the BMGP model is

Table 1 Sample mean and its standard error (in parentheses) of the MAE, RMSE, ALPI, and Coverage given by the standard GP, FBGP, BdryGP, BMGP, and GGP emulators for 60 pairs of design and test set

Emulator	MAE	RMSE	ALPI	Coverage
Standard GP	7.31 (0.21)	15.33 (0.68)	43.32 (1.16)	0.934 (0.005)
FBGP	7.41 (0.34)	12.61 (0.54)	58.22 (1.49)	0.928 (0.006)
BdryGP	6.23 (0.20)	11.78 (0.49)	90.95 (1.04)	0.978 (0.003)
BMGP	4.22 (0.10)	8.23 (0.24)	25.38 (0.82)	0.938 (0.004)
GGP	6.50 (0.37)	10.36 (0.67)	93.68 (3.12)	0.928 (0.005)

significantly smaller than that of all other models. We see that forcing an emulator to satisfy known boundary constraints helps to improve prediction accuracy as all four emulators that satisfy the boundary constraints (the FBGP, BdryGP, GGP, and BMGP models) have smaller average RMSE than the standard GP emulator, and all these emulators except the FBGP emulator have smaller average MAE than the standard GP emulator. The BMGP model also gives significantly smaller ALPI on average compared to all other emulators. The wide prediction intervals of the BdryGP emulator give it an average coverage closest to the nominal value of 0.98. However, the BMGP emulator gives an average coverage of 0.938, which is not far from the nominal 0.98, and this is achieved with far shorter prediction intervals than the BdryGP emulator. Overall, the BMGP emulator is superior to the other four GP emulators as it gives the smallest sample means for the MAE, RMSE, and ALPI, and it gives an average coverage that is close to nominal. It outperforms the standard GP emulator and three other GP emulators (all with different prior covariance functions and two with different prior mean functions) that satisfy the known boundary constraints. This justifies the choice of the prior mean function (15) and prior variance function (16) for the BMGP emulator.

For the standard GP model and the BMGP model, the maximums of the condition numbers of the 60 prior covariance matrices for the experiment data ($\mathbf{R}\sigma^2$ for the standard GP model and $\mathbf{Q}$ for the BMGP model) with covariance parameters fixed at the maximum likelihood estimates are 188,390 and 35,450, respectively. Thus, in this example, the BMGP model gives a prior covariance matrix that is better conditioned than the standard GP model in the worst-case setting and neither model suffers from serious ill conditioning problems as their maximum condition numbers are not large. However, if there are design points very close to boundary edges where the BMGP model gives a prior variance of zero, the prior covariance matrix $\mathbf{Q}$ can be ill conditioned. One way to overcome this problem is to simply avoid using a design with points very close to the boundary. Nonetheless, space-filling designs like LHDs, which have zero probability of placing design points at the boundary, usually do not yield an ill-conditioned prior covariance matrix for the data in our experience.

4.2 Example 2

The second example considers the deflection of a $10\,\text{m} \times 10\,\text{m}$ square plate clamped on all edges and deformed by a uniform load in the downward direction, which is normal to the plate surface (see [25], page 3–7 and page 3–8, for details). The simulator for this problem is developed from the Matlab codes given in pages 3–8 to 3–11 of [25]. The plate thickness (in meters) is the input x, while the other parameters are fixed at the values given in [25]. The output y is the deflection (displacement) at the midpoint of the plate, which is negative as the deflection is downward. It is known that when the thickness x goes to infinity, the midpoint deflection $y(x)$ will converge to 0, i.e., $\lim_{x\to\infty} y(x) = 0$.

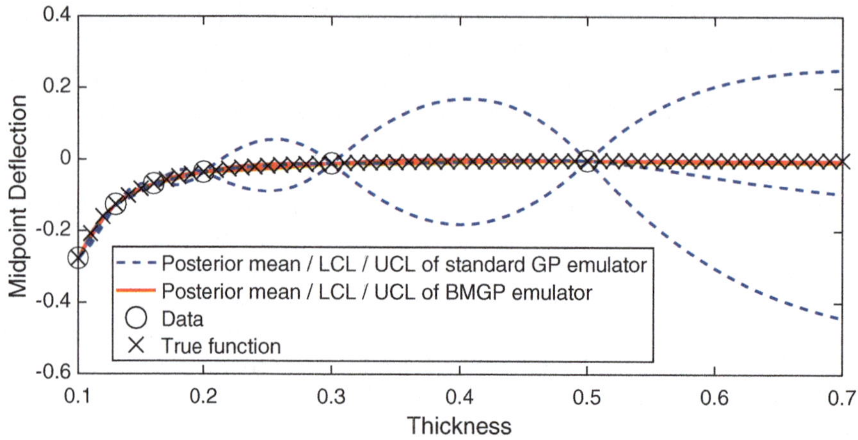

Fig. 4 Plot of posterior means and 98% credible intervals of prediction [LCL, UCL] given by the BMGP and standard GP emulators constructed with a six-point design for the plate simulator. The data and the true simulator output, i.e., midpoint deflection of the plate, are also plotted

We employ the design $\mathcal{D} = \{0.1, 0.13, 0.16, 0.2, 0.3, 0.5\}$ to fit both BMGP and standard GP emulators. Figure 4 plots the true output and posterior means and 98% credible intervals of prediction for the BMGP and standard GP models, where the BMGP model has prior mean and variance functions given by (25)–(26), respectively. The BMGP model gives point and interval predictions that are nearly identical to the true output, and it performs better than the standard GP model in both the experiment region $[0.1, 0.5]$ and the extrapolation region $(0.5, 0.7]$. In addition, the prediction interval length of the BMGP model, which is near zero everywhere, is far smaller than that of the standard GP model for points in $[0.2, 0.7]$ that are away from design points. It can be shown that the posterior variance of the standard GP emulator converges to its maximum, which is the prior variance, as x goes to infinity. Thus, the prediction interval length of the standard GP emulator will increase as x increases above 0.5 and eventually converges to a maximum. This tendency can be seen from Fig. 4. Moreover, it is seen from Fig. 4 that the posterior mean of the BMGP model gives better prediction accuracy than the posterior mean of the standard GP model in the interval $(0.5, 0.7]$. The tendency of the posterior mean of the standard GP emulator to revert toward the prior mean as x goes to infinity can also be seen from Fig. 4.

We now consider another scenario with the same simulator where the goal is to predict within the experiment domain $[0.1, 1]$, and we assume the output at $x = 1$ is known, i.e., boundary information is given by $y(1) = -2.754 \times 10^{-4}$. We use the design $\mathcal{D} = \{0.1, 0.13, 0.16, 0.2, 0.3, 0.5\}$ to fit the BMGP model, which employs the prior mean and variance functions given by (25)–(26) with $b = y(1)$ and $d(x, \infty)$ replaced by the distance measure (24) with $c = 1$, i.e., $\varphi_1(x)$. For this one-dimensional problem, the standard GP model can be made to satisfy the boundary constraint at $x = 1$ by simply including the boundary data point $(1, y(1))$

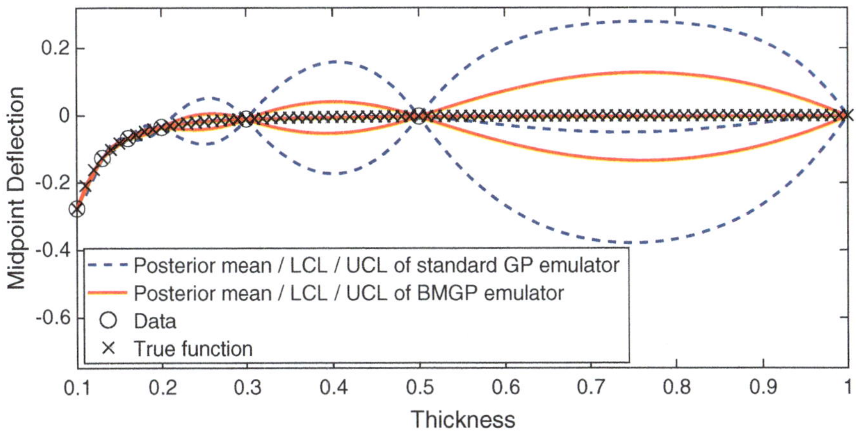

Fig. 5 Plot of posterior means and 98% credible intervals of prediction [LCL, UCL] given by the BMGP emulator constructed with a six-point design $\mathcal{D} = \{0.1, 0.13, 0.16, 0.2, 0.3, 0.5\}$ and the standard GP emulator constructed with a seven-point design $\mathcal{D}_1 = \mathcal{D} \cup \{1\}$ for the plate simulator. The data and the true simulator output, i.e., midpoint deflection of the plate, are also plotted

in the dataset, i.e., by using the design $\mathcal{D}_1 = \mathcal{D} \cup \{1\}$ to fit the model, which is the method for taking into account boundary information studied in [41]. Figure 5 plots the point and interval predictions of the BMGP model fitted with data from $\mathcal{D}$ and the standard GP model fitted with data from $\mathcal{D}_1$. The true output is also plotted in the figure. The BMGP model gives far shorter prediction intervals in [0.2, 1] at points far from $\mathcal{D}_1$ (distance to the closest point in $\mathcal{D}_1$ is large). This is because the standard GP model employs a stationary GP prior (with constant variance), while the BMGP model uses a nonstationary GP prior such that as x approaches one, its prior variance gets smaller. It is also seen that the BMGP model gives more accurate point predictions for $x \in [0.6, 0.9]$ than the standard GP model. This is because the standard GP model employs a constant prior mean μ, which causes its posterior mean to be pulled toward $\mu = -0.101$ (the maximum likelihood estimate) at points away from $\mathcal{D}_1$. In contrast, as the BMGP model uses a nonconstant prior mean function that converges to $y(1)$ as $x \to 1$, its posterior mean is not pulled toward a value different from $y(1)$ between [0.5, 1], which makes its predictions more accurate.

5 Conclusions

In this article, we consider the problem of using known simulator input–output relationships at the boundary of the input domain to construct a GP emulator with improved prediction performance. These relationships can often be discovered by physical considerations, mathematical analysis, or the use of cheap-to-compute low-fidelity simulators, as the boundary faces of an input domain frequently correspond to simplified physical processes. We propose to use judiciously chosen nonstationary mean and variance functions to build a GP emulator so that the emulator converges to the known boundary functions in mean square sense. The proposed GP emulator, called the BMGP emulator, is numerically compared with the standard stationary GP emulator and alternative GP emulators that satisfy the known boundary constraints in two realistic examples. Three such alternative GP emulators are considered in the first example, i.e., a GP model with the product fractional Brownian covariance function and the same prior mean function as the BMGP model, the BdryGP model proposed by Ding et al. [9], and a GP model adapted from the GP model proposed by Graepel [15] for inferring the solution of a linear PDE with Dirichlet boundary conditions. The BMGP emulator outperforms the standard stationary GP emulator and all three alternative GP emulators that satisfy the boundary constraints in terms of prediction accuracy and prediction interval length, and the prediction intervals of the BMGP emulator have close-to-nominal coverage. In the second example, the BMGP emulator is compared with the approach of adding boundary points to the standard stationary GP emulator. The BMGP model is also shown to predict better in this case.

While we only consider the case where the boundary function on one face of the d-dimensional experiment region $[0, 1]^d$ is known in Sect. 3.3, it is straightforward to extend our approach to problems where the boundary functions on multiple faces are known. Some problems need further research. First, the proposed BMGP model needs to be modified when dealing with problems with nonrectangular input domains and boundaries that are not planar faces or boundaries that are planar faces but not perpendicular to any Cartesian axis. Second, theoretical properties of the BMGP model aside from the property that it converges to given boundary functions need further study. Third, it is of interest to use the BMGP emulator for parameter calibration and study the improvements in calibration achieved with this emulator over the standard GP emulator as many PDE models contain parameters that need to be calibrated.

Acknowledgments The work described in this chapter was supported by three grants from the Research Grants Council of the Hong Kong Special Administrative Region, China (General Research Fund Project Nos.: CityU 11226716, CityU 11201117, and CityU 11205118). We also thank two reviewers for comments that helped improve the article.

References

1. Adler, R.J.: The Geometry of Random Fields. SIAM, Philadelphia (2010)
2. Alvarez, M.A., Luengo, D., Lawrence, N.D.: Linear latent force models using gaussian processes. IEEE Trans. Pattern Anal. Mach. Intell. **35**(11), 2693–2705 (2013). https://doi.org/10.1109/TPAMI.2013.86
3. Ba, S., Myers, W.R., Brenneman, W.A.: Optimal sliced Latin hypercube designs. Technometrics **57**(4), 479–487 (2015). https://doi.org/10.1080/00401706.2014.957867

4. Bastos, L.S., O'Hagan, A.: Diagnostics for gaussian process emulators. Technometrics **51**(4), 425–438 (2009). https://doi.org/10.1198/TECH.2009.08019
5. Carpinteri, A.: Structural Mechanics: A Unified Approach. CRC Press, Boca Raton (2017)
6. Chen, J., Chen, Z., Zhang, C., Wu, C.F.J.: APIK: Active physics-informed Kriging model with partial differential equations (2020). Preprint. arXiv:201211798
7. Coleman, M.P.: An Introduction to Partial Differential Equations with MATLAB. CRC Press, Boca Raton (2013)
8. Currin, C., Mitchell, T., Morris, M., Ylvisaker, D.: Bayesian prediction of deterministic functions, with applications to the design and analysis of computer experiments. J. Am. Stat. Assoc. **86**(416), 953–963 (1991). https://doi.org/10.2307/2290511
9. Ding, L., Mak, S., Wu, CFJ.: BdryGP: a new gaussian process model for incorporating boundary information (2019). Preprint. arXiv:190808868
10. eFunda Inc: Thermal conductivity: Silicon (2021). https://www.efunda.com/materials/elements/TC_Table.cfm?Element_ID=Si, on April 29, 2021
11. Fang, K.T., Li, R., Sudjianto, A.: Design and Modeling for Computer Experiments. CRC Press, Boca Raton (2005)
12. Farlow, S.J.: Partial Differential Equations for Scientists and Engineers. Wiley, New York (1982)
13. Gockenbach, M.S.: Partial Differential Equations: Analytical and Numerical Methods, vol. 122. SIAM, Philadelphia (2011)
14. Golchi, S., Bingham, D.R., Chipman, H., Campbell, D.A.: Monotone emulation of computer experiments. SIAM/ASA J. Uncertain. Quantif. **3**(1), 370–392 (2015). https://doi.org/10.1137/140976741
15. Graepel, T.: Solving noisy linear operator equations by gaussian processes: application to ordinary and partial differential equations. In: Proceedings of the 20th International Conference on Machine Learning, vol. 3, pp. 234–241 (2003)
16. Gulian, M., Frankel, A., Swiler, L.: Gaussian process regression constrained by boundary value problems (2020). Preprint. arXiv:201211857
17. Hahn, D.W., Özisik, M.N.: Heat Conduction. John Wiley & Sons, New York (2012)
18. Jackson, S.E., Vernon, I.: Efficient emulation of computer models utilising multiple known boundaries of differing dimensions (2019). Preprint. arXiv:191008846
19. Jidling, C., Wahlström, N., Wills, A., Schön, T.B.: Linearly constrained gaussian processes. In: Proceedings of the 31st International Conference on Neural Information Processing Systems, pp. 1215–1224 (2017)
20. Jidling, C., Hendriks, J., Wahlström, N., Gregg, A., Schön, T.B., Wensrich, C., Wills, A.: Probabilistic modelling and reconstruction of strain. Nucl. Instrum. Methods Phys. Res., Sect. B Beam Interactions Mat. Atoms **436**, 141–155 (2018). https://doi.org/10.1016/j.nimb.2018.08.051
21. Joseph, V.R., Hung, Y.: Orthogonal-maximin latin hypercube designs. Stat. Sin. **18**(1), 171–186 (2008). https://www.jstor.org/stable/24308251
22. Lange-Hegermann, M.: Algorithmic linearly constrained gaussian processes. In: Proceedings of the 32nd International Conference on Neural Information Processing Systems, pp. 2141–2152 (2018)
23. Lange-Hegermann, M.: Linearly constrained gaussian processes with boundary conditions. In: Proceedings of the 24th International Conference on Artificial Intelligence and Statistics, PMLR, pp. 1090–1098 (2021)
24. Lord, G.J., Powell, C.E., Shardlow, T.: An Introduction to Computational Stochastic PDEs, vol. 50. Cambridge University Press, Cambridge (2014)
25. MathWorks: Partial differential equation toolbox: user's guide (r2019b) (2019). https://www.mathworks.com/help/pdf_doc/pde/pde.pdf, on September 20, 2019
26. Mitchell, T.J., Morris, M.D.: Bayesian design and analysis of computer experiments: two examples. Stat. Sin. **2**(2), 359–379 (1992). https://www.jstor.org/stable/24304865

27. Ndlovu, P.L., Moitsheki, R.J.: Application of the two-dimensional differential transform method to heat conduction problem for heat transfer in longitudinal rectangular and convex parabolic fins. Commun. Nonlinear Sci. Numer. Simul. **18**(10), 2689–2698 (2013). https://doi.org/10.1016/j.cnsns.2013.02.019
28. Paciorek, C.J.: Nonstationary gaussian processes for regression and spatial modelling. PhD Thesis. Carnegie Mellon University (2003)
29. Peng, C.Y., Wu, C.F.J.: On the choice of nugget in kriging modeling for deterministic computer experiments. J. Comput. Graph. Stat. **23**(1), 151–168 (2014). https://doi.org/10.1080/10618600.2012.738961
30. Plumlee, M., Joseph, V.R.: Orthogonal gaussian process models. Stat. Sin. **28**(2), 601–619 (2018). https://www.jstor.org/stable/44841917
31. Raissi, M., Perdikaris, P., Karniadakis, G.E.: Machine learning of linear differential equations using Gaussian processes. J. Comput. Phys. **348**, 683–693 (2017). https://doi.org/10.1016/j.jcp.2017.07.050
32. Rasmussen, C.E., Williams, C.K.: Gaussian Processes for Machine Learning, vol. 2. MIT Press, Cambridge (2006)
33. Sacks, J., Welch, W.J., Mitchell, T.J., Wynn, H.P.: Design and analysis of computer experiments. Stat. Sci. **4**(4), 409–423 (1989). https://doi.org/10.1214/ss/1177012413
34. Santner, T.J., Williams, B.J., Notz, W.I.: The Design and Analysis of Computer Experiments. Springer Science + Business Media, New York (2018)
35. Solin, A., Kok, M.: Know your boundaries: constraining gaussian processes by variational harmonic features. In: The 22nd International Conference on Artificial Intelligence and Statistics, PMLR, pp. 2193–2202 (2019)
36. Solin, A., Kok, M., Wahlström, N., Schön, T.B., Särkkä, S.: Modeling and interpolation of the ambient magnetic field by gaussian processes. IEEE Trans. Robot. **34**(4), 1112–1127 (2018). https://doi.org/10.1109/TRO.2018.2830326
37. Tan, M.H.Y.: Monotonic metamodels for deterministic computer experiments. Technometrics **59**(1), 1–10 (2017). https://doi.org/10.1080/00401706.2015.1105759
38. Tan, M.H.Y.: Gaussian process modeling of a functional output with information from boundary and initial conditions and analytical approximations. Technometrics **60**(2), 209–221 (2018). https://doi.org/10.1080/00401706.2017.1345702
39. Tan, M.H.Y.: Gaussian process modeling with boundary information. Stat. Sin. **28**(2), 621–648 (2018). https://www.jstor.org/stable/44841918
40. Tan, M.H.Y.: Gaussian process modeling of finite element models with functional inputs. SIAM/ASA J. Uncertain. Quantif. **7**(4), 1133–1161 (2019). https://doi.org/10.1137/17M1112942
41. Vernon, I., Jackson, S.E., Cumming, J.A.: Known boundary emulation of complex computer models. SIAM/ASA J. Uncertain. Quantif. **7**(3), 838–876 (2019). https://doi.org/10.1137/18M1164457
42. Wang, X., Berger, J.O.: Estimating shape constrained functions using Gaussian processes. SIAM/ASA J. Uncertain. Quantif. **4**(1), 1–25 (2016). https://doi.org/10.1137/140955033
43. Wang, C.M., Reddy, J.N., Lee, K.H.: Shear Deformable Beams and Plates: Relationships with Classical Solutions. Elsevier Science, Oxford (2000)
44. Wheeler, M.W., Dunson, D.B., Pandalai, S.P., Baker, B.A., Herring, A.H.: Mechanistic hierarchical Gaussian processes. J. Am. Stat. Assoc. **109**(507), 894–904 (2014). https://doi.org/10.1080/01621459.2014.899234
45. Zhang, N., Apley, D.W.: Fractional Brownian fields for response surface metamodeling. J. Qual. Technol. **46**(4), 285–301 (2014). https://doi.org/10.1080/00224065.2014.11917972

Part II
Challenges and Solutions in Applications

An Overview and General Framework for Spatiotemporal Modeling and Applications in Transportation and Public Health

Lishuai Li, Kwok-Leung Tsui, and Yang Zhao

Abstract Spatiotemporal modeling and forecasting is an essential task for many real-world problems, especially in the field of transportation and public health. The complex and dynamic patterns with dual attributes of time and space create unique challenges for effective modeling and forecasting. With the advancement of data collection, storage, and sharing technologies, the amount of data and the types of data available for spatiotemporal modeling research in transportation and public health are rapidly increasing. Some traditional spatiotemporal methods become obsolete. There is a need to review existing methods and propose new ones to harness the power of newly available data. Therefore, in this chapter, we conduct a comprehensive survey of methods and algorithms for spatiotemporal monitoring and forecasting, focusing on applications in transportation and public health. Then, we propose a systematic framework to incorporate three different approaches: statistical methods, machine learning methods, and mechanistic simulation methods. The proposed framework is expected to help researchers in the field to better formulate spatiotemporal problems, construct appropriate models, and facilitate new developments that combine the strengths of mechanistic approaches and data-driven ones. The proposed general framework is illustrated via examples of spatiotemporal methods developed in transportation and public health.

L. Li (✉)
Air Transport and Operations, Faculty of Aerospace Engineering, Delft University of Technology, Delft, Netherlands

School of Data Science, City University of Hong Kong, Hong Kong, China
e-mail: lishuai.li@cityu.edu.hk

K.-L. Tsui
Grado Department of Industrial and Systems Engineering, Virginia Polytechnic Institute and State University, Blacksburg, VA, USA
e-mail: kltsui@vt.edu

Y. Zhao
School of Public Health (Shenzhen), Sun Yat-sen University, Shenzhen, China
e-mail: zhaoy393@mail.sysu.edu.cn

A. Steland, K.-L. Tsui (eds.), *Artificial Intelligence, Big Data and Data Science in Statistics*, https://doi.org/10.1007/978-3-031-07155-3_8

Keywords Spatiotemporal modeling · Transporation · Public health · Statistical learning · Machine learning · Simulation

1 Introduction

A wide variety of events in the real-world problems are featured by spatiotemporal dynamics, such as traffic flows, population migration, infectious diseases transmission, diffusion of air pollutants, power grids failure, etc. The ubiquitous events with dual attributes of time and space have created challenges for effective forecasting methods to predict future changes in a timely manner. The objective of spatiotemporal monitoring and forecasting is to analyze such patterns in events at both temporal and spatial dimensions and to understand and predict current and future developments. A large number of approaches under various applications have been developed in the literature, including statistical methods, machine learning/deep learning methods, and simulation models. The methods and algorithms related to spatiotemporal modeling are overabundant, yet there is no systematic approach for researchers and practitioners to choose what to use, why and when the method(s) works for what kind of real-world problems.

Developing a generic framework for any spatiotemporal problems is overly ambitious; in this chapter, we focus on spatiotemporal modeling in two specific applications: transportation and public health, particularly infectious disease transmission. Both have significant societal and economic impact, and there are common characteristics in transportation flow and infectious disease spreading in terms of dynamics and modeling methods used. In transportation systems, congestion and delays feature typical spatiotemporal patterns and cause significant economic and environment costs. For road traffic, congestion cost the US economy nearly 87 billion dollars in 2018 [29]. For air traffic, domestic flight delays were found to cost the US economy 33 billion dollars in 2019 [27]. In public health, the outbreaks and prevalence of infectious disease can be highly life-threatening. Up to July in year 2021, the COVID-19 pandemic has resulted in around 4.2 million deaths worldwide. In year 2009, H1N1 pandemic resulted in between 151,700 and 575,400 deaths worldwide. It is crucial to efficiently and accurately detect and forecast the event occurrence patterns in these problems.

With the deployment of wireless sensors, decreasing cost of data transmission and storage, the data available for studying these spatiotemporal problems become increasingly large and diverse. This brings both opportunities and challenges for spatiotemporal modeling in transportation and public health. Existing spatiotemporal methods are not readily available to harness the power of emerging data for real-life situations. Existing studies generally adopt three approaches: (1) statistical models that consider mathematical description of the physical processes of related variable, (2) machine learning/deep learning approaches that utilize a complex model structure to estimate the future event occurrences, and (3) mechanistic/simulation approaches that built upon domain knowledge. For the statistical approaches, many models are limited to strong probability and distribution assumptions, which are not always valid for data from wireless sensors, text data, or data generated from human behaviors. They also have limited capability to model the spatiotemporal structure of multivariate data and are lack of flexibility to incorporate external factors that influence the spatiotemporal patterns. These limitations have prohibited broad applications of statistical methods involving high-dimensional variables, especially in complex socio-technical problems. For machine learning and deep learning methods, experience-based model construction, feature engineering, and parameter tuning have made it difficult for researchers and practitioners to generalize and implement for a wide range of applications. In addition, model interpretability is another major shortcoming of deep learning methods. The lack of a clear understanding of the model and the meaning of its results limits its deployment and impact in real world. At last, the mechanistic/simulation approaches are strong in explaining the underlying mechanism of the spatiotemporal dynamics, yet few of the traditional mechanistic/simulation approaches can harness the power of big data. To our best knowledge, few papers address both mechanistic and data-driven approaches.

Motivated by the above challenges, we review methods and algorithms for spatiotemporal monitoring and forecasting in transportation and public health applications. We propose a systematic framework to incorporate statistical methods, machine learning methods, and mechanistic simulation methods. The framework is not meant to solve any specific spatiotemporal problems, but rather to structure the problems, construct appropriate spatiotemporal models, and facilitate new developments that combine the strengths of mechanistic approaches and data-driven ones.

More specifically, we plan to achieve the following objectives in this chapter.

- Summarize existing spatiotemporal models and compare statistical learning, machine learning, and simulation methods. The goal is to understand what methods have been developed for what kind of problems, infer how they can complement each other, and suggest what new models can be developed to address problems in transportation and public health.
- Propose a systematic framework that integrates statistical models, machine learning methods, and simulation approaches for modeling spatiotemporal problems, focusing on applications in transportation and public health. The focus is on how to handle typical categories of spatiotemporal problems and what key steps are involved in spatiotemporal modeling.
- Illustrate the proposed method and strategy with examples in transportation and public health applications. A number of example methods are shown with different focuses and application areas.

This chapter is expected to be useful to researchers and practitioners to meet the increasing demand and challenges for spatiotemporal monitoring and forecasting in transportation and public health applications. The comprehensive analysis of spatiotemporal methods helps us to understand which methods work for what problems in real life. It also contributes to the development of robust methods with meaningful interpretability for spatiotemporal monitoring and forecasting problems in transportation and public health. New methods proposed in this framework are expected to incorporate multiple data sources with various data structure, reveal the inherent evolution of target event occurrences, and deliver accurate predictions of future changes for transportation and public health problems.

2 Literature Review

We review existing general methods for spatiotemporal modeling and forecasting and conduct an in-depth survey of methods to address spatiotemporal problems in the field of transportation and public health. Most of the existing methods for spatiotemporal modeling in general take the data-driven approach, including statistical approaches and machine learning/deep learning-based approaches. The survey of methods used in transportation and public health provided us new perspectives, particularity on the value of the mechanistic simulation approaches on spatiotemporal modeling.

2.1 Statistical Approaches to Spatiotemporal Modeling

Time series forecasting has fundamental importance to various practical domains [26, 64, 88]. Autoregressive integrated moving average (ARIMA) and its family is the most general class of models for forecasting time series. Hamed et al. [35] applied ARIMA model for short-term forecasting by using traffic volume data of urban arterials. Williams and Hoel [118] presented the theoretical basis for modeling univariate data streams as seasonal autoregressive integrated moving average (SARIMA) processes. ARIMA and SARIMA work well in specific conditions, but they are limited by the assumption of "stationary" data. Motivated by the superior capability to cast the regression problem of a Kalman Filter (KF) [44, 45], numerous KF-based prediction studies began to emerge [34, 76, 122]. However, traditional time series analysis methods cannot consider the spatiotemporal correlation.

Statistical spatiotemporal models are often constructed by combining time series models with variogram-based models. Popular time series approaches include autoregressive moving average models [11] for stationary data and state-space models [117] for non-stationary components. In the spatial setting, the early work often involves kriging-based models. Spatial processes consider the correlation depends on location as well as distance. For example, the STARMA [84] and STARMAX [98] models were constructed by adding spatial covariance matrices to standard vector autoregressive moving average models. However, they are limited to stationary temporal processes. Stroud et al. [99] developed a statistical model for non-stationary spatiotemporal data. The model is cast in a Gaussian state-space framework and can include temporal components such as trends, seasonal effects, and autoregressions. Some other methods [25, 56] use the vector autoregressive models (VARs) for spatiotemporal data, in which the Y variable is a vector of observations at difference sites at time t, and the coefficient matrices are carefully constructed to model the spatiotemporal autoregressive relationships.

More generally, regression analysis is widely used for prediction and forecasting [1, 6, 47, 70, 73, 96]. The major advantage of regression models is that they can be used to capture important relationships between the forecast variable of interest and the predictor variables. By modeling the spatial as well as the temporal dependence of the errors, [79] applied spatial–temporal regression with 14 variables to forecast real estate prices. Yang et al. [128] improved the accuracy of flu activity predictions by establishing an autoregressive model of Google search data as an external explanatory variable. Lu et al. [66] adapted a multivariate dynamic regression method integrating Google searches, Twitter posts, electronic health records, and a crowd-sourced influenza reporting system to forecast influenza activity. To estimate regional activity, [74] proposed a two-step augmented regression model that efficiently combines publicly available Google search data at different resolutions (national and regional) and spatial dependence of influenza transmission.

Probabilistic graphical models (PGMs) are a powerful framework that bring together graph theory and probability theory. Considering the uncertainty of the noisy data and simplifying the complexity of the real world are the main advantages of PGMs, [61] proposed dynamic cost predictions for a trip planner by using a spatiotemporal Markov random field (STMRF). Hoang et al. [40] proposed a Gaussian Markov random field-based model to forecast citywide crowd flows based on traffic big data. However, PGM is highly computationally complex at the training stage of the algorithm, making it very difficult to retrain the model when newer data become available.

A related topic to spatiotemporal modeling is change detection of spatiotemporal processes, yet it is not the focus of this chapter. Many methods have been proposed for change detection in monitoring industrial processes [85, 86], remote sensing with digital aerial imagery [2, 9], and other applications [92].

2.2 *Machine Learning/Deep Learning-Based Approaches to Spatiotemporal Modeling*

Machine learning/deep learning methods have advanced in many application fields over the past years, especially under the big data environment. Compared with conventional statistical methods, machine learning and deep learning methods have more flexibility in handling data with complex structure, such as graphs and networks. The well-known machine learning methods for forecasting include k-nearest neighbors (KNN) algorithm [14, 67, 136], support vector machine (SVM) [41, 135], and neural networks (NNs) [94, 112]. To evaluate a wider range of machine learning methods, [42] implemented a network of stacked sparse auto-encoders to detect and predict event occurrence. Besides, some researchers developed various extensions of SVM [21, 91, 114, 127] on short-term forecasting problem. In addition to these machine learning approaches, deep learning has attracted many attention and shown superior performance in spatiotemporal modeling.

Convolutional Neural Networks (CNNs) CNNs have been widely used in mining spatiotemporal data because it is effective in capturing the spatial correlations in the data [57]. Especially, for data types of spatial maps and spatial rasters, which can be represented as a two-dimensional matrix, CNN is well suited to learn the spatial features [17, 24, 51, 68, 69, 75]. Spatiotemporal data are sometimes represented as a tensor or a sequence of tensors, and three-dimensional CNNs can be used to learn the complex spatial and temporal dependencies of the data [16, 53].

Recurrent Neural Networks (RNNs) RNNs have been well recognized for sequence learning tasks [103]. Incorporating long short-term memory (LSTM) or gated recurrent unit (GRU) enables RNNs to capture the long-term temporal dependency of time series. RNN and LSTM are increasingly used in time series prediction. For example, RNNs are applied for future weather forecasting where the weather variables are modeled as time series [18]. Volkova et al. [113] evaluated the predictive power of neural network architectures based on LSTM units and demonstrate its capability of nowcasting and forecasting ILI dynamics. However, these algorithms cannot capture the spatial features.

Hybrid Models of CNN+RNN New hybrid models that combine CNN and RNN have been proposed to extract spatiotemporal dependencies simultaneously in spatiotemporal forecasting models. The basic idea is to structure the input as a sequence of image-like matrices, and then a hybrid model that combines CNN and RNN can be used, where CNN extracts the spatial relationships embedded in the matrices and RNN learns the temporal pattern from the sequences. For example, [121, 131] proposed the structures with the combination of CNN and LSTM for spatiotemporal forecasting. Instead of simply stacking the architectures of CNN and RNN, by extending the fully connected LSTM (FC-LSTM) to have convolutional structures in both the input-to-state and state-to-state transitions, [93] proposed the convolutional LSTM (ConvLSTM) model for the precipitation nowcasting problem. ConvLSTM was then used in the spatiotemporal forecasting on transportation applications [3, 48]. The hybrid architectures show good performance on extracting spatiotemporal dependencies and correlations in forecasting. But the training procedure may become time consuming as the size of dataset increases because the complexity of RNNs is determined by the size of data sequences. Furthermore, [137] proposed a spatiotemporal residual network for forecasting crowd flow in each regular region of a city, yet it cannot be adapted to deal with irregular regions.

Graph Neural Networks (GNNs) and the Hybrid Models of GNNs+RNNs CNNs are commonly applied for dealing with Euclidean data such as images, regular grids, etc. However, spatial features based on the topological structure of a network or a graph have strong effects on modeling graph-structured data. Graph Convolutional Networks (GCNs) were widely used to capture network-based spatial dependencies as GCNs can handle arbitrary graph-structured data. Zhao et al. [138] developed a spatiotemporal neural network named Temporal Graph Convolutional Network for forecasting problem, which combines the GCN with GRU. However, the experimental results show that it has difficulty to capture the sudden changes of events occurrence. Sun et al. [100] proposed a novel multi-view deep learning model, named Multi-View Graph Convolutional Network, to predict the inflow and outflow in each irregular region of a city. Besides, another graph neural network, Diffusion Convolutional Neural Networks (DCNNs) were also developed for graph-structured data [5]. Later on, [60] proposed the diffusion convolutional recurrent neural network (DCRNN) to model the traffic flow as a diffusion process on a directed graph and incorporate both spatial and temporal dependency in the traffic flow for traffic forecasting. More recently, attention mechanism was widely applied into temporal and graph-structured spatial dependencies' extractions. Spatiotemporal graph attention models were proposed for spatiotemporal forecasts of traffic states [80, 116].

2.3 *Spatiotemporal Modeling in Transportation*

In the field of transportation, spatiotemporal modeling is often used for the modeling and forecasting of (1) **traffic conditions** (including flow speed, volume, congestion level, etc.) and (2) **travel demand**.

Spatiotemporal modeling and forecasting for **traffic conditions** is a fast evolving field in recent years. Many papers have been published using machine learning or deep learning-based approach to do traffic condition forecasting. Ermagun and Levinson [25] provides a comprehensive literature review. The output of these models includes traffic flow [142], traffic speed [31], travel time [89], relative velocity [46], etc. The forecast time horizon is normally short term, e.g., a few minutes to 1 h. The modeling techniques are primarily data driven, ranging from statistical approaches [20] to machine learning/deep learning methods [130]. Many efforts have been made on developing effective methods to model the spatial dependency, the temporal dependency, and their dynamics for traffic condition forecasting. Regarding the spatial dependency in traffic condition forecasting, it can be coded as regions or positions in Euclidean coordinates without network structures. For example, [137] developed a deep learning-based approach, called ST-ResNet, to collectively forecast two types of crowd flows (i.e., inflow and outflow) in each and every region of a city. Yu et al. [131] proposed a network grid representation method for traffic speed forecasting on a transportation network. The network-wide traffic speeds are transformed into a series of images and fed into a deep architecture combining both convolutional neural networks (DCNNs) and long short-term memory (LSTM) neural networks for traffic forecasting. Meanwhile, networks are very common in transportation, such as highways, railroads, and airways. Some recent developments in traffic condition forecasting utilize the graph/network structure in the deep learning framework and show significant improvement in terms of forecasting accuracy [22, 140]. He et al. [36] focused on exploring the influencing factors on forecasting urban rail transit (URT) ridership. In this chapter, the authors proposed an approach based on spatial models considering spatial autocorrelation of variables, which outperform the traditional global regression model, OLS, in terms of model fitting and spatial explanatory power. He et al. [37] made an effort to incorporate multiple factors, including spatial factors (distance and network topology), temporal factors (e.g., period and trend), and external factors (e.g., land use and socioeconomics) to estimate metro ridership based on general estimating equation (GEE) models. A following study investigated local model selection in ridership prediction [38]. In this study, an adapted geographically weighted LASSO (Ada-GWL) framework is proposed for modeling subway ridership, which involves regression coefficient shrinkage and local model selection. It takes subway network layout into account and adopts network-based distance instead of Euclidean-based distance [38].

Similarly, in air transportation, many methods have been developed to monitor and forecast traffic flow, travel time, congestion level, and delay time. In the past, statistical methods or probabilistic approaches are used to analyze factors that influenced flight delays and estimate delay distributions [111, 123]. Several machine learning methods have been used to predict delays, e.g., k-nearest neighbors, neural networks, support vector machine, fuzzy logic, and tree-based methods, yet did not explicitly utilize the spatiotemporal patterns [19, 83, 87, 97]. With the vast volume of commercial aviation system data being collected, classic methods are not sufficient to incorporate the complexity involved in the real-world operational data collected from multiple sources. For instance, tracking of aircraft position becomes available for many areas in the world, which contains rich spatiotemporal information needed for delay predictions. However, how to use and model this kind of raw position data, as well as incorporating external data sources (e.g., weather, airline records), is still an open research question [133]. More recently, Kim et al. proposed a recurrent neural network (RNN) to predict the flight delays of an individual airport with day-to-day sequences [50].

Different from the data-driven approaches, various simulation models have been developed and commonly used for traffic condition forecasting in research as well as in practice. The simulation models are built upon system-level abstractions of real world (e.g., queue theory), component-level abstractions (agent-based), or a combination of both (e.g., delay propagation). The purpose of using simulation-based approaches for spatiotemporal modeling is to understand the "physics" of transportation systems, design for "optimality," and manage operations in real time. Some simulation models are based on representations of driver behavior, e.g., car following, gap acceptance, and lane choice. These are considered as microscopic models or agent-based simulations. Another type of simulation models is based on macroscopic traffic flow theory developed during the 1950s [62, 90]. The objective of these models is to represent temporal congestion phenomena over a road network based on system-level equations instead of individual vehicle level.

Regarding the **travel demand** modeling and forecasting, the conventional focus is on aggregate forecasts for transportation facilities, districts, cities, and regions, and it is normally for long-term forecasts, such as forecasting the travel demand for the next few year(s) [23, 78]. The basic principle used to construct travel demand models is similar to the one used in standard microeconomics, which is utility maximization. The travel activity is a reflection of explicit preferences and limitations [78]. Individuals are defined by socioeconomic variables. In the conventional models, the time is discretized into intervals and space into zones, the depiction of travel patterns can be trip-based, tour-based, or activity schedule-based, and the forecasting methods are regression-based [10]. More recently, there are a lot of developments in simulation-based approaches for travel demand modeling and forecasting [52, 129, 134].

2.4 Spatiotemporal Modeling in Public Health

In public health, considerable attention has been paid to spatiotemporal modeling approaches for real-time tracking, timely detection and early intervention of disease outbreaks and other public health-related events. Correspondingly, related papers can be broadly grouped into three categories: forecasting, surveillance, and simulation.

Forecasting The outbreaks and prevalence of infectious disease can be highly life-threatening. Therefore, the focus of papers in this category is on predictive modeling and forecasting of the spread of infectious disease and other public health-related events. One of the earliest key studies in this category is often referred to as "Google Flu" [33]. The authors proposed the first method to use Google search queries to track influenza-like illness and detect influenza epidemics in areas with a large population of web search users. Later, [128] proposed an influenza tracking model, ARGO (AutoRegression with GOogle search data), with significantly improved performance of Google search-based real-time tracking than other existing models for influenza epidemics at the national level of the USA, including Google Flu Trends. This first ARGO model only considers the temporal trends of influenza epidemics. To overcome this limitation, the ARGO2 (2-step Augmented Regression with GOogle data) was proposed to model both spatial and temporal trends [74]. ARGO2 can efficiently combine publicly available Google search data at different resolutions (national and regional) with traditional influenza surveillance data from the Centers for Disease Control and Prevention (CDC) for accurate, real-time regional tracking of influenza [74]. Along this line of research, there are many recent research studies, such as forecasting influenza in Hong Kong with Google search queries and statistical model fusion [64, 125], using Baidu index to nowcast hand–foot–mouth disease in China [139], a hybrid autoregressive integrated moving average–linear regression (ARIMA–LR) approach for forecasting patient visits in emergency department [124], and personalized health monitoring of elderly wellness at community level [132].

Surveillance The objective of public health surveillance is to systematically collect, analyze, and interpret public health data (chronic or infectious diseases) in order to understand trends and detect changes in disease incidence and death rates and for planning, implementation, and evaluation of public health practices. The focus is on the accurate detection of the time or/and location(s) of changes in the occurrence rate as soon as possible. Several review papers examined existing methods for public health surveillance and discussed research opportunities and challenges [107–110, 120]. The most common existing disease spread monitoring methods can be categorized into temporal, spatial, and spatiotemporal surveillance techniques. Most basic methods such as SPC, regression, time series, and forecast-based methods were originally developed as temporal approaches. On the other hand, popular public health surveillance methods such as scan statistics were originally developed as spatial approaches [54] and later extended as temporal and spatiotemporal approaches [55]. Most spatial surveillance techniques rely on existing statistical clustering methods. Many techniques have been developed to expand those models to spatiotemporal methods that also search for clusters in time. Despite many independent implementations of surveillance systems have been deployed across different disciplines, such as ESSENCE, Google Flu Trends, and Global Microbial, the ability to accurately detect infectious disease outbreaks and pandemics is still in its nascent stages. Current surveillance systems lack the means to integrate disparate data sources, although recently proposed methods for multivariate surveillance hold promise for deployment in future systems to provide accurate prediction for infectious disease outbreaks and spreading trends.

Simulation The third category includes studies that use simulation models or mechanistic models to describe the spread of infectious diseases. Mechanistic models are built with structures that make explicit hypotheses about the biological mechanisms that drive infection dynamics. Such hypotheses include the dynamics of disease process among individuals (e.g., susceptible, infected, immune) and social interactions of people in an entire country or even the world [59]. As early as 1930s, mechanical epidemic simulators are used as research tools as well as teaching tools for epidemic theory [58]. Since that time, modeling has become an integral part of epidemiology and public health. The history and typical methods of mechanistic models of infectious disease are reviewed in [58, 59]. The most commonly used mechanistic approach is the Susceptible-Exposed-Infectious-Recovered (SEIR) compartmental model [49]. Smieszek [95] presented a mechanistic model that considers the different duration and intensity of contacts. Balcan et al. [7] presented the Global Epidemic and Mobility (GLEaM) model to simulate the spread of epidemics at the worldwide scale. The model used in a spatially structured stochastic disease approach to integrate sociodemographic and population mobility data [7]. Andradóttir et al. [4] developed a stochastic, individual-level simulation model of influenza spread within a structured population to investigate reactive strategies for containing developing outbreaks of pandemic influenza [4].

The three categories of research, forecasting, surveillance and simulation tackle different aspects of the spatiotemporal problems in public health, therefore using different underlying methods. In the group of public health forecasting, the key question to answer is how would the trend develop in time and space for a particular kind of infectious diseases? Statistical or machine learning-based spatiotemporal methods are used in these studies. For the public heath surveillance problems, the focus is on detection, detecting the time and (or) location(s) of significant changes in disease incidence and death rates. Therefore, statistical process control (SPC) and related process monitoring methods are often used. Regarding the simulation models, the goal is to understand the dynamic nature of the process, to evaluate the impact of policy change over time, to develop better intervention strategies for epidemics. Therefore, the development of models that reveal the nature of infection dynamics is more important.

2.5 Challenges and Opportunities

To summarize, many methods for spatiotemporal forecasting have been applied in various contexts, and they generally adopt three approaches: (1) statistical models that consider mathematical description of the spatiotemporal process of related variables, (2) machine learning/deep learning approaches that use the complex model structures to learn spatiotemporal patterns and forecast future event occurrences, and (3) simulation models or mechanistic models built with structures describing the physical/biological mechanisms that drive dynamics. In the two specific applications, transportation and public health, classic statistical methods and simulation models have been used for traffic prediction, infectious disease forecasting, and other spatiotemporal modeling. Recent efforts have been made to utilize the power of artificial intelligence for spatiotemporal modeling and forecasting. Most papers focus on either data-driven approaches or mechanistic approaches. Few papers address both mechanistic and data-driven approaches as well as their interaction in feature engineering.

Each of the three approaches has its own limitations. For the statistical approaches, many models are limited to strong probability and distribution assumptions, which are not always valid for data from wireless sensors, text data, or data generated from human behaviors. They also have limited capability to model the spatiotemporal structure of multivariate data and are lack of flexibility to incorporate external factors that influence the spatiotemporal patterns. These limitations have prohibited broad applications of statistical methods involving high-dimensional variables, especially in complex socio-technical problems. For machine learning and deep learning methods, different methods/models work for different situations. Experience-based model construction, feature engineering, and parameter tuning have made it difficult for researchers and practitioners to generalize and implement for a wide range of applications. In addition, model interpretability is another major shortcoming of deep learning methods. The lack of a clear understanding of the model and the meaning of its results limits its deployment and impact in real world. The simulation models or mechanistic models require specific knowledge of particular problems, which is difficult to be scaled up to deal with different applications. To fill this gap, there is an urgent need to examine existing spatiotemporal monitoring and forecasting methods and develop improved solutions for practice and complex problems.

3 A General Framework for Spatiotemporal Modeling

With existing methods and algorithms, we propose a systematic framework for modeling spatiotemporal problems focusing on applications in transportation and public health. The framework covers statistical models, machine learning methods, and mechanistic simulations for spatiotemporal modeling. It introduces a unified way for effective and efficient mining and modeling of spatial and temporal dependences among diversified data sources while integrating domain knowledge and various forecasting methods. The proposed framework can serve as a guideline for researchers and practitioners to understand and structure the spatiotemporal problems they are facing and configure the modeling steps, i.e., feature engineering, model selection and fusion, parameter tuning, performance evaluation, and results interpretation. The proposed systematic framework for modeling spatiotemporal problems is shown in Fig. 1. The details of individual modules are explained in the following subsections.

3.1 Mechanistic/Simulation Approach

The mechanistic/simulation approach models the spatiotemporal dynamics based on explicitly theories or hypotheses about the traffic flow or the infectious diseases transmission mechanisms. Depending on whether the fundamental unit models the dynamics among individuals or the dynamics at a system level, agent-based simulations or system-level simulations are used. Many classic simulation models have been developed in the field of transportation as well as the field of public health, and new developments of mechanistic models and simulation are still evolving.

For the system (compartment)-level simulations, there are SEIR/SIR models [32] and stochastic Markov chain models [30, 77] in public health. Ghaffarzadegan [32] is an example of the system-level simulation model. The model analyzes the spread of COVID-19 in universities that can be used to conduct a what-if analysis and estimate infection cases under different policies. In transportation, the classic system-level simulations are LWR models [62, 90, 119] and traffic flow simulations [13, 82].

For agent-based (component-level) simulations, examples in public health include the FluTE model [15, 65], the EpiSimdemics simulations [8], GSAM [81], and Andradáttir's model [4]. Andradáttir et al. [4] is an example of the agent-based simulation model. It simulates the transmission of pandemic influenza with

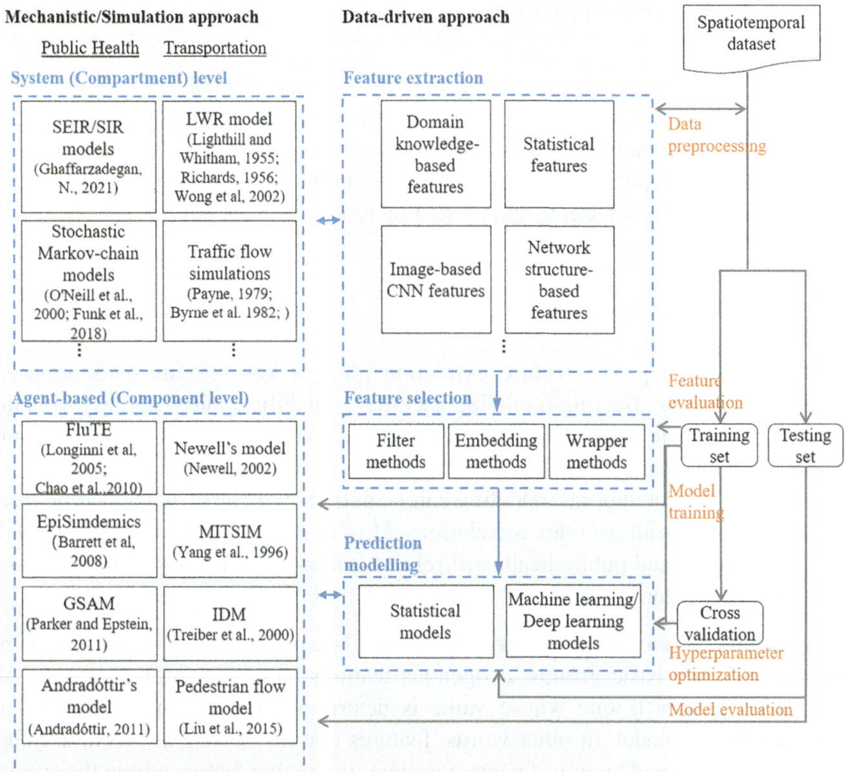

Fig. 1 A general framework for spatiotemporal modeling

the purpose of examining reactive strategies and concluded that in reaction to developing outbreaks combination strategies of reactive vaccination and limited antiviral use can be substantially more effective than vaccination alone in terms of controlling outbreaks and economic cost. In transportation, the well-known agent-based simulations include Newell's model [72], MITSIM [126], and IDM [106]. Liu et al. [63] is a new development that takes the mechanistic/simulation approach. The authors developed an agent-based simulation to model movement direction choice and collision avoidance for pedestrian flow. The results reveal the joint effect of several physical, psychological, and sociological factors dominating the real-world pedestrian walking behaviors.

Empirical data are important for the development of the mechanistic/simulation approach. The data are used to estimate and calibrate the chosen parameters in the model, as well as to validate the output accuracy. In the process of calibration and validation of mechanistic/simulation models, the focus is on the mechanism of interest to investigate and the intermediate steps, while the data-driven approach strives for the final step accuracy.

3.2 Data-Driven Approach

The data-driven approach focuses on the performance of final predictions, rather than the intermediate steps/processes involved in the spatiotemporal modeling. It does not require domain knowledge, yet its performance can be enhanced by domain knowledge. The typical steps involved in a data-driven approach include feature engineering, feature selection, and prediction model.

3.2.1 Feature Engineering

Feature engineering and feature extraction plays a key role in spatiotemporal modeling, directly affecting modeling accuracy, reliability, and generality. It is the process to generate informative features from the existing raw data by discovering hidden patterns inside them. For real-life problems, the types and characteristics of raw data for spatiotemporal modeling can be extremely diverse as they come from multiple sources with complex correlations. Most of existing spatiotemporal models in transportation and public heath still rely on a trial-and-error approach to choose and construct features.

Exogeneous Versus Endogenous Features The spatiotemporal features can be divided into two basic groups: exogenous features and endogenous features. An exogenous feature is one whose value is determined outside the model and is imposed on the model, in other words, features that affect a model without being affected by it. An endogenous feature correlates with other factors within the system being studied. It is changed or determined by its relationship with other features within the model.

For transportation applications, the information of weather, holiday, and calendar, as well as social media information containing the big events, etc. is the commonly used exogenous features for assisting the spatiotemporal modeling in transportation [48, 104, 137]. Sun et al. [102] is an example of using exogenous features, Baidu Index and Google Index, in the forecasting model. The Internet search data were integrated into extreme learning machine (KELM) models, and the forecasting performance was significantly improved in terms of both forecasting accuracy and robustness analysis.

For public health applications, the feature engineering part focuses on identifying exogenous features for improving the prediction accuracy, rather than incorporating complex network structures as in transportation models. Taking the forecasts of infectious disease as an example, Internet search index is the exogenous feature for improving the prediction accuracy of ILI rate [74, 128, 139].

Temporal Versus Spatial Within endogenous features, they can be further classified into (1) temporal features, which refers to any feature that is associated with or changes over time, and (2) spatial correlation features, which refers to any feature that is associated with or changes over space. Temporal features are typically classified as three different components: correlation with previous temporal measurements, upward or downward trends, and cyclical or seasonal pattern. Some spatial features (Euclidean-based) are better represented by Euclidean-based distance measures or an image-like matrix. Other spatial features (network-based) are better captured by the topological structure of a network or a graph. In addition, the spatial structure may change over time. Thus, special feature extraction and modeling techniques are needed to deal with the dynamic spatial structure.

Spatial Versus Network Features When choosing between image-based CNN features and network structure-based features, it mainly depends on how to characterize different types of spatial dependency in the forecasting problem. There are mainly two classes, regular or network spatiotemporal problems, which require different sets of methods. Most existing spatiotemporal models for disease transmission forecasting do not consider the network features, while for transportation systems, the network effect is naturally embedded, e.g., road network, rail network, air traffic network. How to model the network structure also differs depending on the mode of transportation. Road traffic prediction focuses on traffic condition on links, while forecasting problems in rail operations and air transport pay more attention to the delays at stations and airports.

Casual Versus Correlated Features The input features of a data-driven approach model could be correlated with or (and) casual factors to the output variable(s). Correlation and causation can both exist at the same time. However, correlation does not imply causation. Causation means that one event or action causes another event to happen. Correlation simply means there is a relationship between two events or two variables. Most models in the data-driven approach are only assuming and checking correlations between input features and the output variable(s). In contrast, the mechanistic/simulation approach has a better chance to identify casual features because a physical or mechanistic process is explicitly modeled and causal relationships are normally used in these models.

Typical Ways to Extract Features from Raw Data There are four typical ways to construct these endogenous features from raw data: domain knowledge-based features, statistical features, image-based CNN features, and network structure-based features. Domain knowledge-based features are constructed using a set of variables (usually have physical meanings), rules, or mechanism based on the knowledge and experience of the specific system or problem. For example, [28] extracted 42 parameters as features for battery lifetime prediction based on domain knowledge. These parameters can effectively reflect the aging dynamics of lithium-ion batteries, such as the voltage, capacity, temperature, etc. This is an effective way to construct features as it utilizes the domain knowledge accumulated for many years. However, it requires deep understanding of the domain, which could be expensive and time consuming for less popular applications. Statistical features are commonly used in various applications, including mean, standard deviation, variance, skewness, and correlation coefficients, autoregressive coefficients. The advantage of using these statistical features is that they are not limited to a particular domain or system. However, the downside is that these statistical features may not be able to capture the hidden patterns and the fundamental structure of the system. For examples, in traffic forecasting, the raw data collected are normally flow-rates at major road segments. Statistical features of flow-rates are not enough to build a traffic forecasting model because flow-rate is not sufficient to determine the traffic condition—a small flow-rate value may correspond to either a very light traffic or a congested traffic [71].

3.2.2 Feature Selection

Feature selection is the method to reduce a large set of features to a small number of features. The reduced feature set size makes it computationally feasible and easier to interpret when using certain algorithms. It may also lead to better results by reducing overfitting.

Three typical methods are used for feature selection: (1) filter methods, (2) wrapper methods, and (3) embedded methods, as shown in Fig. 2. Filter methods assess feature importance based on some ranking criteria. Typical filter methods are ANOVA, Pearson correlation, variance threshold, and information gain. Wrapper methods evaluate and select feature subsets based on model performance. The most commonly used wrapper methods include forward selection, backward elimination, and bidirectional elimination (Stepwise Selection).

Embedding methods take all available features as input and perform feature selection in model training as part of the model construction process, e.g., LASSO, elastic net, decision tree, deep learning methods, etc.

In addition, dimension reduction techniques can also be broadly categorized under feature selection methods, such as PCA, SVD, autoencoders, etc. These methods transform the original features into other variables via parametric or nonparametric projection.

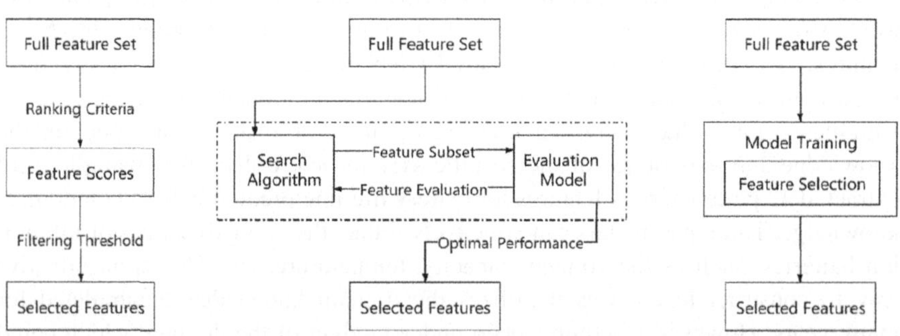

(a) (b) (c)

Fig. 2 Three categories of feature selection methods [28]. (**a**) Filter method. (**b**) Wrapper method. (**c**) Embedded method

3.2.3 Prediction Modeling

Prediction models take the input of selected features or transformed features, model the patterns and relationships among the features and their influence on the output variables (sometimes they are the input features with a different time window) in the training data, and predict the output variables. Both statistical methods and machine learning methods have been developed for prediction models.

Most statistical forecasting methods are developed based on autoregressive moving average models. However, these statistical forecasting methods are limited by the assumptions that they rely on, such as stationary of time series, known statistical distributions of features, etc. Such kind of methods have difficulty to incorporate complex spatiotemporal correlation into modeling.

Deep learning methods have advanced in many application fields over the past years, especially under the big data environment. Compared with conventional statistical methods, deep learning methods have more flexibility in handling data with complex structure, such as spatial data like maps, rasters, graphs and spatiotemporal data like sequence of spatial data, and 3D tensors. However, the training procedure may become time consuming as the size of dataset becomes very large. Besides, existing methods may have difficulty in state forecasting in irregular regions, directed network flow forecasting, or graph-structured data forecasting in non-Euclidian spaces. Many ongoing research efforts are made to develop better learning-based prediction models to overcome these difficulties.

3.3 Combining the Mechanistic and the Data-Driven Approach

A hybrid approach that involves both the mechanistic/simulation approach and the data-driven approach is expected to combine strengthens of both. For example, the mechanism and knowledge used in the mechanistic/simulation approach can be used to generate better domain knowledge-based features, as well as better structured features, such as network structure-based features. The formulated dynamics used in the mechanistic/simulation approach can also be used to design better prediction models in the data-driven approach, such as the structure of the statistical models, the architecture of the deep learning models, etc.

3.4 Evaluation Metrics and Methods

The standard way of evaluating the spatiotemporal forecasting models is to test the prediction accuracy on testing dataset. In the data preprocessing part, the original spatiotemporal dataset is divided into training set and testing set. Training set is used for feature evaluation, model training, and hyper-parameter optimization, while testing set is used for the model evaluation.

The model performance is typically evaluated based on three evaluation metrics: RMSE (root-mean-squared error), MAPE (mean absolute percentage error), and R2 (coefficient of determination). The RMSE describes the absolute error measured by variations in data errors, and the MAPE indicates the relative error, in terms of percentages of predicted values. Smaller values of RMSE and MAPE indicate better prediction accuracy. R2 measures how much of the variance in the predicted variables can be explained by the model. A larger value of R2 represents better performance. To evaluate the spatiotemporal forecasting models comprehensively, the models can also be assessed from other several aspects in addition to the prediction accuracy and prediction robustness, including computational efficiency, model interpretability, etc.

One important aspect of spatiotemporal forecasting models is about the prediction performance across different forecast horizons. The comparison of short-term and long-term predictions can help provide a better understanding of the robustness and adaptability of forecasting methods. For instance, an influenza forecast model might have different performance when forecast influenza activity 1 week ahead, 1 month ahead, and 1 year ahead. However, the natural difference in the structures of statistical models, deep learning models, and the mechanistic simulation models may cause unfair comparisons in sliding window evaluation. Most statistical models are naturally structured for sliding window evaluation. Many of these models are also capable of doing one-step or multi-step ahead forecasting with parameter re-estimation. Model parameters can be adjusted based on newly available data after each window sliding is performed. However, conventional machine learning/deep learning-based models reuse data at time t-1, t-2, . . . to predict output values at time t. The hyper-parameters of the model are not refitted as the prediction time shifts in sliding window evaluations unless the training process is explicitly re-performed on the new training set. Moreover, the comparison of mechanistic forecasting model may be different from the other types of forecasting models. In the simulation-based forecasting, the changes in policy and behavior could be factored in the model, while the data-driven approach does not have the natural structure to do this. Therefore, the comparison among different methods needs to be carefully designed to make it fair.

4 Examples of the General Framework for Spatiotemporal Modeling

Many real-world data sets are available for research in transportation and public health applications in recent years, including ILI data at different region levels, inter-city passenger travel data, the smart card records data collected from metro Automatic Fare Collection (AFC) system, and datasets for external factors, such as Google search and meteorological data. We performed a number of studies following proposed systematic framework for spatiotemporal modeling to achieve better prediction results, understand the spatiotemporal patterns better, and generate application insights. In this section, we illustrate the systematic framework for spatiotemporal modeling via several examples. The methods or models developed in these examples are different and they have unique characteristics best suited for a particular kind of application problem, since the framework is not meant to solve any specific spatiotemporal problems, but rather to structure the problems and construct appropriate spatiotemporal models.

4.1 Spatiotemporal Modeling for Road Traffic

This is a fast evolving field. Many papers have been published using machine learning or deep learning-based approaches to forecast traffic conditions on road network in recent years. A number of open datasets provide traffic speed and traffic flow over major road segments or intersections measured by sensors installed on the road or real-time information provided by the vehicles on the road. [43] provides a summary of open data and big data tools used for traffic estimation and prediction.

How to structure the problem, including the selection of the output variables, feature extraction, and the design of model structures, is critical for spatiotemporal modeling in the field of road transportation. Regarding the design and selection of the output variables, the output variable is normally flow, speed, congestion level, relative velocity, and other traffic condition measures. Yet it can be categorical or continuous. The selection of output variables depends on data availability and model specification.

In terms of feature extraction, most of the recent deep learning-based approaches to forecast traffic conditions on road network have been focused on how to extract and model spatial features in the road network. In addition to these endogenous features, [102] is an example of using exogenous features, Baidu Index and Google Index, in the forecasting model. The Internet search data were integrated into extreme learning machine (KELM) models, and the forecasting performance was significantly improved in terms of both forecasting accuracy and robustness analysis.

One of our ongoing work is the development of data-driven approaches to predict traffic condition on a city road network [115]. In this study, we are using a dataset provided by Baidu, named MapBJ, which provides the traffic conditions categorized into four levels (unblocked, slow, congested, extreme congested) over major roads in Beijing [18], and another dataset provided by DiDi Chuxing for a similar set of traffic condition measures in Xi'an. Following the proposed general framework, the challenges of developing a data-driven approach to predict traffic condition on a city road network come from how to structure the model so that it can capture the temporal dependency, the spatial dependency, and the changes of spatial dependency overt time over a road network. A number of deep learning model architectures are being examined, and below is an architecture that we proposed. The proposed architecture is named, periodic spatial–temporal deep neural network (PSTN) as

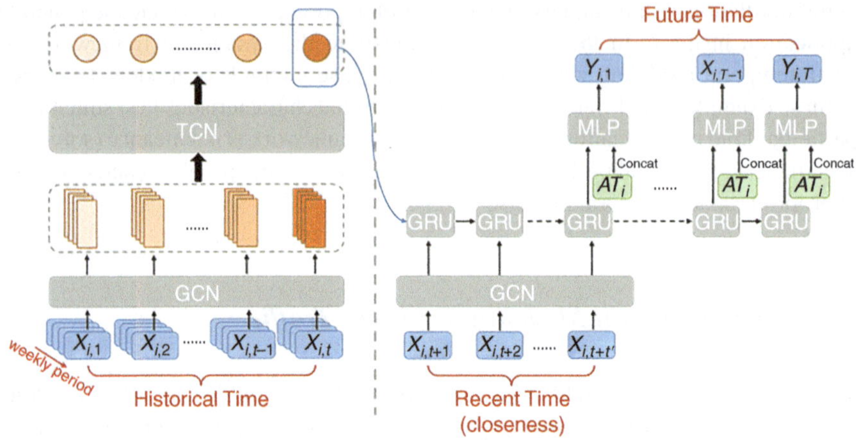

Fig. 3 Illustration of PSTN architecture of spatiotemporal model for road traffic. From [115], ©Tiange Wang, Zijun Zhang, Kwok-Leung Tsui, 2021, used under the Creative Commons Attribution 4.0 International License: https://creativecommons.org/licenses/by/4.0/

shown in Fig. 3. The basic idea is to have three sequentially parts: (1) graph convolutional networks (GCNs) to capture the topological structure of road network, (2) the temporal convolutional networks (TCNs) and the gated recurrent units (GRUs) to capture periodic temporal dependency and local temporal dependency, respectively, and (3) the multi-layer perceptron (MLP) to combine road attributes and make the final prediction.

4.2 Spatiotemporal Modeling for Transit Passenger Flow

Forecasting short-term passenger flow on urban metro networks is an essential task for proactive traffic management in cities, which monitors real-time traffic conditions and forecast the condition in the immediate future. The challenges are mainly driven by complex spatiotemporal characteristics in metro passenger flow data and other external influence factors such as weather effect. We performed a number of studies following proposed systematic framework for spatiotemporal modeling to achieve better prediction results, understand the spatiotemporal patterns better, and generate application insights [36, 63, 104, 105].

Liu et al. [63] is an example of taking the mechanistic/simulation approach. The authors developed an agent-based simulation to model movement direction choice and collision avoidance for pedestrian flow. This developed microscopic simulation of pedestrian flow could be used for studying problems related to pedestrian traffic and evacuation dynamics.

Tang et al. [104] is an example of taking the statistical approach to forecast the short-term passenger flow on Shenzhen metro. In the proposed framework, there are three modules: traffic data profiling (feature engineering), feature extraction, and predictive modeling. In the feature engineering and feature extraction part, three types of features were comprehensively investigated in this study. They are (1) temporal features from passenger flow time series data, (2) spatial features based on origin-destination (OD) patterns, and (3) external weather factors. In the prediction model part, this study employed a number of forecasting models to evaluate the performance of the proposed framework, i.e., the time series model autoregressive integrated moving average, linear regression, and support vector regression. Moreover, the evaluation of this framework pays special attention to forecasting steps and horizons. The results suggest that smaller forecasting step predicts better for longer forecasting horizon, while larger forecasting step performs well for t + 1 prediction yet the prediction performance degrades when forecasting horizon grows.

Focusing on how to construct the features and extract the complex spatiotemporal relationships, we have studied both the statistical approach and the deep learning approach. He et al. [36] used a statistical approach, focusing on the travel demand forecasting and exploring the influencing factors on urban rail transit (URT) ridership. In this paper, the authors proposed an approach based on spatial models considering spatial autocorrelation of variables, which outperform the traditional global regression model, OLS, in terms of model fitting and spatial explanatory power. A following study investigated local model selection in ridership prediction [38]. In this study, an adapted geographically weighted LASSO (Ada-GWL) framework was proposed for modeling subway ridership, which involves regression coefficient shrinkage and local model selection. It takes subway network layout into account and adopts network-based distance instead of Euclidean-based distance. In addition, [37] made an effort to incorporate multiple factors, including spatial factors (distance and network topology), temporal factors (e.g., period and trend), and external factors (e.g., land use and socioeconomics) to estimate metro ridership based on general estimating equation (GEE) models.

He et al. [39] is an example of taking the deep learning approach for short-term passenger flow forecasting on Shenzhen metro. This work focused on investigating how to encode the network-based spatial features and other heterogeneous inter-station correlations in the model. The solution proposed in this work is a multi-graph convolutional recurrent neural network (MGC-RNN) (shown in Fig. 4) to generate multiple graphs that each represents a type of network structure and then to employ multiple parallel graph convolutional operators on multigraphs in the prediction model. This work illustrates that feature engineering and feature selection are both embedded in the deep learning-based prediction model. Specifically, by incorporating various types of inter-station correlations, temporal dependencies, and exogenous factors, the framework exhibits a possibility for multi-source heterogeneous data fusion in a big data environment.

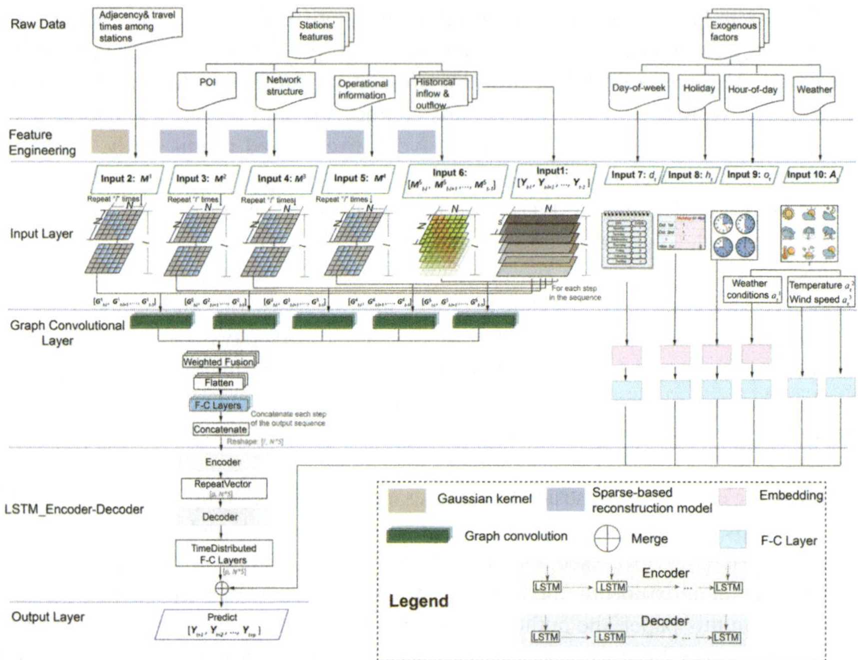

Fig. 4 The architecture of MGC-RNN [39]

4.3 Spatiotemporal Modeling for Air Traffic

Flight time prediction and, a related topic, flight delay prediction have been studied for years in the field of aviation. Many prior works employ statistical methods or probabilistic approaches. However, the accuracy of these models is not sufficient for the individual flight predictions. With the increasing amount of aviation system data being collected and available, such as Automatic Dependent Surveillance—Broadcast (ADS-B) data, aviation meteorological data, it is possible to utilize machine learning methods to learn the patterns of aircraft movement on a national air traffic network and predict individual flight time.

Sun et al. [101] is an example of combing statistical approach and deep learning approach to forecast air passenger flows. The proposed model incudes nonlinear vector autoregression and neural network. The results show that it outperforms single models and other hybrid approaches in terms of level forecasting accuracy, directional forecasting accuracy, and robustness analysis.

Zhu and Li [141] developed a novel spatial weighted recurrent neural network (SWRNN) model to provide flight time predictions for individual flights at a scale of national air traffic network, as shown in Fig. 5. Following the systematic framework for spatiotemporal modeling, the feature engineering part is a combination of domain knowledge-based and imagine-based CNN features. Based on domain

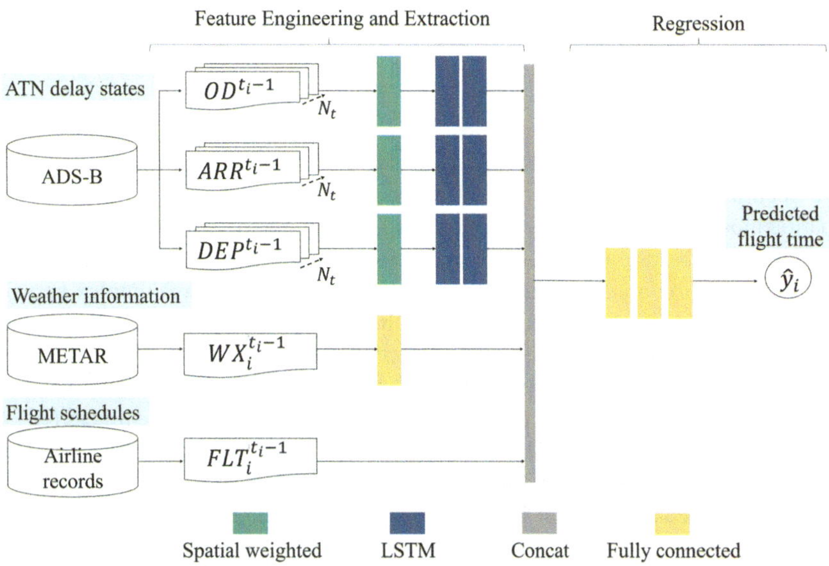

Fig. 5 Framework of SWRNN model [141]

knowledge, the network delay state features are extracted from the aircraft position tracking data, ADS-B, manually, including the average flight delay of each origin–destination (OD) pair, the average flight delay at each arrival airport, and the average flight delay at each departure airport for a specific time interval. Then, these network delay state matrices are sequenced based on time and fed into the spatial weighed layer to extract the spatial dependency and reduce the dimensionality for the network delay state features. The learnable weights of the spatial weighted layer show the importance of different OD pair/airports to the sample flight. Then, long short-term memory (LSTM) networks are used after the spatial weighted layer to extract the temporal dependency of network delay states. Therefore, the feature selection is an embedded method in this work. Finally, features from delays, weather, and flight schedules are fed into a fully connected neural network to predict the flight time of a particular flight. Evaluation of the SWRNN model was conducted using 1 year of historical data from an airline's real operations. Results show that the SWRNN model can provide a more accurate flight time predictions than baseline methods, especially for flights with extreme delays. In the paper, the authors also demonstrated that fuel loading can be optimized with the improved flight time prediction and resulting reduced fuel consumption by 0.016%–1.915% without increasing the fuel depletion risk for airlines.

4.4 Spatiotemporal Modeling for Infectious Disease Transmission

Taking the forecast of infectious disease transmission as an example, the systematic framework for spatiotemporal modeling can also be followed. Tsui et al. [108] provided a comprehensive review of research and developments in temporal and spatiotemporal surveillance for public health. Compared with the transportation problems, the feature engineering part focuses on identifying exogenous features for improving the prediction accuracy, rather than incorporating complex network structures. Several studies have shown that Google search data are effective exogenous features for improving the prediction accuracy of ILI rate [74, 128, 139].

Following the statistical approach, [125] studied the value of using online social media and web search queries to forecast new cases of influenza-like illness (ILI) in general outpatient clinics (GOPC) in Hong Kong. The study tested four individual models to forecast ILI-GOPC both 1 week and 2 weeks in advance, which are generalized linear model (GLM), least absolute shrinkage and selection operator (LASSO), autoregressive integrated moving average (ARIMA), and deep learning (DL) with feedforward neural networks (FNNs). Furthermore, the authors also proposed a statistical fusion model using Bayesian model averaging (BMA) to integrate multiple forecast scenarios.

Regarding the machine learning approach, [64] used a deep learning method to forecast influenza epidemics in Hong Kong, which also uses Google search queries. In this method, the innovative parts are mainly feature engineering on the output data. Variational mode decomposition (VMD), a signal decomposition method, is used to decompose the influenza data (the output data) into modes with different frequencies. Then, each mode extracted by VMD is forecasted by artificial neural networks (ANNs) and then these forecasts of each mode are added to generate the final forecasting results.

Zhao et al. [139] is an example of combining both the traditional statistical approach and the machine learning approach for spatiotemporal modeling of infectious disease transmissions. A meta learning framework (shown in Fig. 6) is proposed to select appropriate predictive model based on the statistical and time series meta features to nowcast the monthly hand, foot, and mouth disease (HFMD). In addition, the feature engineering part incorporated search engine index. The proposed meta learning method significantly improves the HFMD prediction accuracy, demonstrating that (1) the Internet-based information offers the possibility for effective HFMD nowcasts and (2) the meta learning approach is capable of adapting to a wide variety of data and enables selecting appropriate method for improving the nowcasting accuracy.

More recently, a study evaluated thirteen different methods for short-term forecasting of COVID-19 in Germany and Poland for 10 weeks, 12 October–19 December 2020, [12]. The study found that these forecasts from thirteen different teams are heterogeneous in terms of both point predictions and forecast spread. The performance of ensemble forecasts was relatively better on coverage, but ensemble

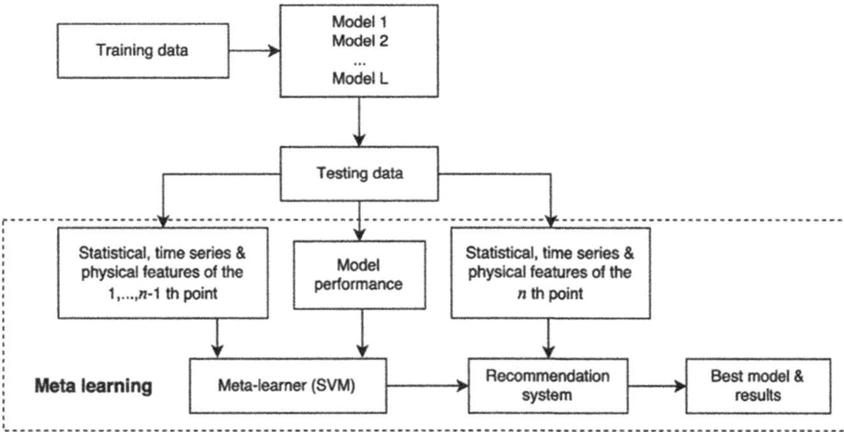

Fig. 6 Meta learning framework. From [139], ©The Author(s), 2018, used under the Creative Commons Attribution 4.0 International License (http://creativecommons.org/licenses/by/4.0/)

forecasts did not clearly dominate single-model predictions. The findings are consistent with the increasingly trend of acknowledgment that combining multiple models can improve the reliability of outputs.

5 Conclusion

There is an increasing demand for spatiotemporal monitoring and forecasting under various applications. This work reviews recent research and developments of spatiotemporal modeling in transportation and public health. Current spatiotemporal modeling methods are designed for specific applications, and various techniques and algorithms are proposed at different stages involved in the spatiotemporal modeling. This chapter proposes a systematic framework for developing spatiotemporal modeling, covering mechanistic/simulation approaches, statistical methods, and machine learning/deep learning methods. The proposed framework is illustrated via a few examples of spatiotemporal modeling in transportation and public health. The proposed framework will be useful to help researchers and practitioners formulate and structure the spatiotemporal modeling and forecasting problems, develop effective and accurate models, and improve the effectiveness of spatiotemporal modeling in solving real-life problems.

References

1. Abuella, M., Chowdhury, B.: Solar power probabilistic forecasting by using multiple linear regression analysis. In: SoutheastCon 2015, pp. 1–5. IEEE, New York (2015)
2. Agouris, P., Mountrakis, G., Stefanidis, A.: Automated spatiotemporal change detection in digital aerial imagery. In: Roper, W.E., Hamilton, M.K. (eds.) Automated Geo-Spatial Image and Data Exploitation, vol. 4054, pp. 2–12. International Society for Optics and Photonics, SPIE, New York (2000)
3. Ai, Y., Li, Z., Gan, M., Zhang, Y., Yu, D., Chen, W., Ju, Y.: A deep learning approach on short-term spatiotemporal distribution forecasting of dockless bike-sharing system. Neural Comput. Applic. **31**(5), 1665–1677 (2019)
4. Andradáttir, S., Chiu, W., Goldsman, D., Lee, M.L., Tsui, K.L., Sander, B., Fisman, D.N., Nizam, A.: Reactive strategies for containing developing outbreaks of pandemic influenza. BMC Public Health **11**(SUPPL. 1), 1–15 (2011)
5. Atwood, J., Towsley, D.: Diffusion-Convolutional neural networks. In: Advances in Neural Information Processing Systems (2016)
6. Austin, M.P., Nicholls, A.O., Margules, C.R.: Measurement of the realized qualitative niche: environmental niches of five Eucalyptus species. Ecol. Monogr. **60**(2), 161–177 (1990)
7. Balcan, D., Gonçalves, B., Hu, H., Ramasco, J.J., Colizza, V., Vespignani, A.: Modeling the spatial spread of infectious diseases: The global epidemic and mobility computational model. J. Comput. Sci. **1**(3), 132–145 (2010)
8. Barrett, C.L., Bisset, K.R., Eubank, S.G., Feng, X., Marathe, M.V.: EpiSimdemics: An efficient algorithm for simulating the spread of infectious disease over large realistic social networks. In 2008 SC—International Conference for High Performance Computing, Networking, Storage and Analysis, SC 2008 (2008)
9. Boulila, W., Farah, I.R., Ettabaa, K.S., Solaiman, B., Ghézala, H.B.: Improving spatiotemporal change detection: A high level fusion approach for discovering uncertain knowledge from satellite image databases. In: Icdm, vol. 9, pp. 222–227. Citeseer, New York (2009)
10. Bowman, J.L., Ben-Akiva, M.E.: Activity-based disaggregate travel demand model system with activity schedules. Transp. Res. A Policy Pract. **35**(1), 1–28 (2001)
11. Box, G., Jenkins, G.M., Reinsel, G.C.: Time Series Analysis: Forecasting and Control, 3rd ed.. Prentice-Hall, Englewood Cliffs (1994)
12. Bracher, J., Wolffram, D., Deuschel, J., Görgen, K., Ketterer, J., Ullrich, A., Abbott, S., Barbarossa, M., Bertsimas, D., Bhatia, S., et al.: A pre-registered short-term forecasting study of COVID-19 in Germany and Poland during the second wave. Nat. Commun. **12**(1), 1–16 (2021)
13. Byrne, A.S., Courage, K.G., Wallace, C.E.: Handbook of computer models for traffic operations analysis. Technical report, Technology Sharing Report FHWA-TS-82-213, Washington, D.C. (1982)
14. Cai, P., Wang, Y., Lu, G., Chen, P., Ding, C., Sun, J.: A spatiotemporal correlative k-nearest neighbor model for short-term traffic multistep forecasting. Transportation Research Part C: Emerging Technologies **62**, 21–34 (2016)
15. Chao, D.L., Halloran, M.E., Obenchain, V.J., Longini, I.M.: FluTE, a publicly available stochastic influenza epidemic simulation model. PLoS Comput. Biol. **6**(1), e1000656 (2010)
16. Chen, C., Li, K., Teo, S.G., Chen, G., Zou, X., Yang, X., Vijay, R.C., Feng, J., Zeng, Z.: Exploiting spatio-temporal correlations with multiple 3D convolutional neural networks for citywide vehicle flow prediction. Proceedings—IEEE International Conference on Data Mining, ICDM **2018-Novem**(61661146006), 893–898 (2018)
17. Chen, M., Yu, X., Liu, Y.: PCNN: Deep Convolutional Networks for Short-term Traffic Congestion Prediction. IEEE Trans. Intell. Transp. Syst. **19**(11), 3550–3559 (2020)
18. Cheng, X., Zhang, R., Zhou, J., Xu, W.: DeepTransport: learning spatial-temporal dependency for traffic condition forecasting. In: Proceedings of the International Joint Conference on Neural Networks, vol. 2018-July. Institute of Electrical and Electronics Engineers Inc. (2018)

19. Choi, S., Kim, Y.J., Briceno, S., Mavris, D.: Prediction of weather-induced airline delays based on machine learning algorithms. In: 2016 IEEE/AIAA 35th Digital Avionics Systems Conference (DASC), pp. 1–6. IEEE, New York (2016)
20. Clark Dougherty, S.D., M.S., Kirby, H.R.: The Use of Neural Network and Time Series Modes for Short Term Forecasting: A Comparative Study (1993)
21. Cong, Y., Wang, J., Li, X.: Traffic flow forecasting by a least squares support vector machine with a fruit fly optimization algorithm. Procedia Engineering **137**, 59–68 (2016)
22. Cui, Z., Henrickson, K., Ke, R., Wang, Y.: Traffic Graph Convolutional Recurrent Neural Network: A Deep Learning Framework for Network-Scale Traffic Learning and Forecasting. IEEE Trans. Intell. Transp. Syst. **21**(11), 4883–4894 (2020)
23. Davidson, W., Donnelly, R., Vovsha, P., Freedman, J., Ruegg, S., Hicks, J., Castiglione, J., Picado, R.: Synthesis of first practices and operational research approaches in activity-based travel demand modeling. Transp. Res. A Policy Pract. **41**(5), 464–488 (2007)
24. Duan, L., Hu, T., Cheng, E., Zhu, J., Gao, C.: Deep convolutional neural networks for spatiotemporal crime prediction. In: Proceedings of the International Conference on Information and Knowledge Engineering (IKE), pp. 61–67 (2017)
25. Ermagun, A., Levinson, D.: Spatiotemporal traffic forecasting: review and proposed directions. Transp. Rev. **38**(6), 786–814 (2018)
26. Fan, K., Li, C., Heller, K.: A unifying variational inference framework for Hierarchical Graph-Coupled HMM with an application to influenza infection. In: Thirtieth AAAI Conference on Artificial Intelligence (2016)
27. Federal Aviation Administration: Cost of Delay Estimates. Technical report, Federal Aviation Administration, Washington (2020)
28. Fei, Z., Yang, F., Tsui, K.L., Li, L., Zhang, Z.: Early prediction of battery lifetime via a machine learning based framework. Energy **225**, 120205 (2021)
29. Fleming, S. Traffic Congestion Cost the US Economy Nearly $87 Billion in 2018. Future of the environment, World Economic Forum (2019)
30. Funk, S., Camacho, A., Kucharski, A.J., Eggo, R.M., Edmunds, W.J.: Real-time forecasting of infectious disease dynamics with a stochastic semi-mechanistic model. Epidemics **22**, 56–61 (2018)
31. Fusco, G., Colombaroni, C., Isaenko, N.: Comparative analysis of implicit models for real-time short-term traffic predictions. IET Intell. Transp. Syst. **10**(4), 270–278 (2016)
32. Ghaffarzadegan, N.: Simulation-based what-if analysis for controlling the spread of Covid-19 in universities. PLoS One **16**(2 February), 1–24 (2021)
33. Ginsberg, J., Mohebbi, M.H., Patel, R.S., Brammer, L., Smolinski, M.S., Brilliant, L.: Detecting influenza epidemics using search engine query data. Nature **457**(7232), 1012–1014 (2009)
34. Guo, J., Huang, W., Williams, B.M.: Adaptive Kalman filter approach for stochastic short-term traffic flow rate prediction and uncertainty quantification. Transportation Research Part C: Emerging Technologies **43**, 50–64 (2014)
35. Hamed, M.M., Al-Masaeid, H.R., Said, Z.M.B.: Short-term prediction of traffic volume in urban arterials. J. Transp. Eng. **121**(3), 249–254 (1995)
36. He, Y., Zhao, Y., Tsui, K.L.: Exploring influencing factors on transit ridership from a local perspective. Smart and Resilient Transport **1**(1), 2–16 (2019)
37. He, Y., Zhao, Y., Tsui, K.L.: Modeling and analyzing modeling and analyzing impact factors of metro station ridership: an approach based on a general estimating equation factors influencing metro station ridership: an approach based on general estimating equation. IEEE Intell. Transp. Syst. Mag. **12**(4), 195–207 (2020)
38. He, Y., Zhao, Y., Tsui, K.L.: An adapted geographically weighted LASSO (Ada-GWL) model for predicting subway ridership. Transportation **48**(3), 1185–1216 (2021)
39. He, Y., Li, L., Zhu, X., Tsui, K.L.: Multi-Graph Convolutional-Recurrent Neural Network (MGC-RNN) for Short-Term Forecasting of Transit Passenger Flow, arXiv:2107.13226, (2021)

40. Hoang, M.X., Zheng, Y., Singh, A.K.: FCCF: forecasting citywide crowd flows based on big data. In: Proceedings of the 24th ACM SIGSPATIAL International Conference on Advances in Geographic Information Systems, p. 6. ACM, New York (2016)
41. Hong, W.-C.: Traffic flow forecasting by seasonal SVR with chaotic simulated annealing algorithm. Neurocomputing **74**(12–13), 2096–2107 (2011)
42. Jacobson, O., Dalianis, H.: Applying deep learning on electronic health records in Swedish to predict healthcare-associated infections. In: Proceedings of the 15th workshop on biomedical natural language processing, pp. 191–195 (2016)
43. Jiang, W., Jiayun L.: Big data for traffic estimation and prediction: a survey of data and tools. Appl. Syst. Innov. **5**(1) 23, (2022)
44. Kalman, R.E.: A new approach to linear filtering and prediction problems. J. Basic Eng. **82**(1), 35–45 (1960)
45. Kalman, R.E., Bucy, R.S.: New results in linear filtering and prediction theory. J. Basic Eng. **83**(1), 95–108 (1961)
46. Kamarianakis, Y., Prastacos, P.: Forecasting traffic flow conditions in an Urban network: comparison of multivariate and univariate approaches. Transp. Res. Rec. **1**(1857), 74–84 (2003)
47. Kaytez, F., Taplamacioglu, M.C., Cam, E., Hardalac, F.: Forecasting electricity consumption: A comparison of regression analysis, neural networks and least squares support vector machines. Int. J. Electr. Power Energy Syst. **67**, 431–438 (2015)
48. Ke, J., Zheng, H., Yang, H., Chen, X.M.: Short-term forecasting of passenger demand under on-demand ride services: A spatio-temporal deep learning approach. Transportation Research Part C: Emerging Technologies **85**(October), 591–608 (2017)
49. Kermack, W., Mckendrick, A.G.: A contribution to the mathematical theory of epidemics. Proceedings of the Royal Society of London. Series A, Containing Papers of a Mathematical and Physical Character **115**(772), 700–721 (1927)
50. Kim, Y.J., Choi, S., Briceno, S., Mavris, D.: A deep learning approach to flight delay prediction. In: AIAA/IEEE Digital Avionics Systems Conference—Proceedings, 2016-Decem:1–6 (2016)
51. Kira, Z., Li, W., Allen, R., Wagner, A.R., Georgia, A.: Leveraging deep learning for spatio-temporal understanding of everyday environments. In: IJCAI Workshop on Deep Learning and Artificial Intelligence (2016)
52. Kitamura, R., Chen, C., Pendyala, R.M., Narayanan, R.: Micro-simulation of daily activity-travel patterns for travel demand forecasting. Transportation **27**(1), 25–51 (2000)
53. Kleesiek, J., Urban, G., Hubert, A., Schwarz, D., Maier-Hein, K., Bendszus, M., Biller, A.: Deep MRI brain extraction: A 3D convolutional neural network for skull stripping. NeuroImage **129**, 460–469 (2016)
54. Kulldorff, M.: A spatial scan statistic. Communications in Statistics—Theory and Methods **26**(6), 1481–1496 (1997)
55. Kulldorff, M.: Prospective time periodic geographical disease surveillance using a scan statistic. J. R. Stat. Soc. Ser. A Stat. Soc. **164**(1), 61–72 (2001)
56. Lamberti, A., Naccarato, A.: VAR models for spatio-temporal structures: An application to environmental data. In: Studies in Classification, Data Analysis, and Knowledge Organization (2005)
57. LeCun, Y., Boser, B., Denker, J.S., Henderson, D., Howard, R.E., Hubbard, W., Jackel, L.D.: Backpropagation applied to digit recognition. Neural Comput. **1**(4), 541–551 (1989)
58. Lessler, J., Cummings, D.A.: Mechanistic models of infectious disease and their impact on public health. Am. J. Epidemiol. **183**(5), 415–422 (2016)
59. Lessler, J., Azman, A.S., Grabowski, M.K., Salje, H., Rodriguez-Barraquer, I.: Trends in the mechanistic and dynamic modeling of infectious diseases. Current Epidemiology Reports **3**(3), 212–222 (2016)
60. Li, Y., Yu, R., Shahabi, C., Liu, Y.: Diffusion convolutional recurrent neural network: Data-driven traffic forecasting. arXiv preprint arXiv:1707.01926, pp. 1–16 (2017)

61. Liebig, T., Piatkowski, N., Bockermann, C., Morik, K.: Predictive trip planning-smart routing in smart cities. In: EDBT/ICDT Workshops, pp. 331–338 (2014)
62. Lighthill, M., Whitham, G.: On kinematic waves II: A theory of traffic flow on long crowded roads. Proc. R. Soc. Lond. A Math. Phys. Sci. **229**(1178), 317–345 (1955)
63. Liu, S.B., Lo, S.M., Tsui, K.L., Wang, W.L.: Modeling movement direction choice and collision avoidance in agent-based model for pedestrian flow. J. Transp. Eng. **141**(6), 04015001 (2015)
64. Liu, Y., Feng, G., Tsui, K.-L., Sun, S.: Forecasting influenza epidemics in Hong Kong using Google search queries data: a new integrated approach. Expert Systems with Applications **185**, 115604 (2021)
65. Longini, I.M., Nizam, A., Xu, S., Ungchusak, K., Hanshaoworakul, W., Cummings, D.A., Halloran, M.E.: Containing pandemic influenza at the source. Science **309**(5737), 1083–1087 (2005)
66. Lu, F.S., Hou, S., Baltrusaitis, K., Shah, M., Leskovec, J., Hawkins, J., Brownstein, J., Conidi, G., Gunn, J., Gray, J.: Accurate influenza monitoring and forecasting using novel internet data streams: a case study in the Boston Metropolis. JMIR Public Health Surveill. **4**(1), e4 (2018)
67. Luo, X., Li, D., Yang, Y., Zhang, S.: Spatiotemporal traffic flow prediction with KNN and LSTM. J. Adv. Transp. 0197–6729 (2019)
68. Lv, J., Li, Q., Sun, Q., Wang, X.: T-CONV: A convolutional neural network for multi-scale taxi trajectory prediction. In: *Proceedings—2018 IEEE International Conference on Big Data and Smart Computing, BigComp 2018*, pp. 82–89 (2018)
69. Ma, X., Dai, Z., He, Z., Ma, J., Wang, Y., Wang, Y.: Learning traffic as images: a deep convolutional neural network for large-scale transportation network speed prediction. Sensors **17**(4), 818 (2017)
70. Mao, L.-f., Jiang, Y.-c.: Medium-and long-term load forecasting based on partial least squares regression analysis. Power System Technology **32**(19), 71–77 (2008)
71. McCrea, J., Moutari, S.: A hybrid macroscopic-based model for traffic flow in road networks. Eur. J. Oper. Res. **207**(2), 676–684 (2010)
72. Newell, G.F.: A simplified car-following theory: A lower order model. Transp. Res. B Methodol. **36**(3), 195–205 (2002)
73. Ng, S.T., Cheung, S.O., Skitmore, M., Wong, T.C.Y.: An integrated regression analysis and time series model for construction tender price index forecasting. Constr. Manag. Econ. **22**(5), 483–493 (2004)
74. Ning, S., Yang, S., Kou, S.C.: Accurate regional influenza epidemics tracking using Internet search data. Sci. Rep. **9**(1), 1–8 (2019)
75. Niu, X., Zhu, Y., Zhang, X.: DeepSense: A novel learning mechanism for traffic prediction with taxi GPS traces. In: 2014 IEEE Global Communications Conference, GLOBECOM 2014, pp. 2745–2750 (2014)
76. Okutani, I., Stephanedes, Y.J.: Dynamic prediction of traffic volume through Kalman filtering theory. Transp. Res. B Methodol. **18**(1), 1–11 (1984)
77. O'Neill, P.D., Balding, D.J., Becker, N.G., Eerola, M., Mollison, D.: Analyses of infectious disease data from household outbreaks by Markov chain Monte Carlo methods. J. R. Stat. Soc.: Ser. C: Appl. Stat. **49**(4), 517–542 (2000)
78. Oppenheim, N.: Urban Travel Demand Modeling: From Individual Choices to General Equilibrium. Wiley, New York (1995)
79. Pace, R.K., Barry, R., Gilley, O.W., Sirmans, C.F.: A method for spatial–temporal forecasting with an application to real estate prices. Int. J. Forecast. **16**(2), 229–246 (2000)
80. Park, C., Lee, C., Bahng, H., Tae, Y., Jin, S., Kim, K., Ko, S., Choo, J.: ST-GRAT: A Novel Spatio-temporal graph attention networks for accurately forecasting dynamically changing road speed. In: Proceedings of theInternational Conference on Information and Knowledge Management, pp. 1215–1224 (2020)
81. Parker, J., Epstein, J.M.: A distributed platform for global-scale agent-based models of disease transmission. ACM Transactions on Modeling and Computer Simulation: A Publication of the Association for Computing Machinery **22**(1), 2 (2011)

82. Payne, H.J.: FREFLO: A macroscopic simulation model of freeway traffic. Transp. Res. Rec. **722**, 68–77 (1979)
83. Pérez–Rodríguez, J.V., Pérez–Sánchez, J.M., Gómez–Déniz, E.: Modelling the asymmetric probabilistic delay of aircraft arrival. J. Air Transp. Manag. **62**, 90–98 (2017)
84. Pfeifer, P.E., Deutsch, S.J.: A STARIMA model-building procedure with application to description and regional forecasting. Trans. Inst. Brit. Geogr. **5**(3), 330–349 (1980)
85. Prause, A., Steland, A.: Detecting changes in spatial-temporal image data based on quadratic forms. In: Steland, A., Rafajłowicz, E., Szajowski, K. (eds.) Stochastic Models, Statistics and Their Applications, pp. 139–147. Springer International Publishing, Cham (2015)
86. Rafajłowicz, E.: Detection of essential changes in spatio-temporal processes with applications to camera based quality control. In: Steland, A., Rafajłowicz, E., Szajowski, K. (eds.) Stochastic Models, Statistics and Their Applications, pp. 433–440. Springer International Publishing, Cham (2015)
87. Rebollo, J.J., Balakrishnan, H.: Characterization and prediction of air traffic delays. Transportation Research Part C: Emerging Technologies **44**, 231–241 (2014)
88. Reis, B.Y., Mandl, K.D.: Time series modeling for syndromic surveillance. BMC Med. Inform. Decis. Mak. **3**(1), 2 (2003)
89. Reza, R.M.Z., Pulugurtha, S.S., Duddu, V.R.: ARIMA Model for Forecasting Short-Term Travel Time due to Incidents in Spatio-Temporal Context. In: 94th Annual Meeting of the Transportation Research Board, vol. 257 (2015)
90. Richards, P.I.: Shock waves on the highway. Oper. Res. **4**(1), 42–51 (1956)
91. Santillana, M., Nguyen, A.T., Dredze, M., Paul, M.J., Nsoesie, E.O., Brownstein, J.S.: Combining search, social media, traditional data sources to improve influenza surveillance. PLoS Comput. Biol. **11**(10), e1004513 (2015)
92. Shekhar, S., Jiang, Z., Ali, R.Y., Eftelioglu, E., Tang, X., Gunturi, V., Zhou, X.: Spatiotemporal data mining: A computational perspective. ISPRS Int. J. Geo Inf. **4**(4), 2306–2338 (2015)
93. Shi, X., Chen, Z., Wang, H., Yeung, D.-Y.: Convolutional LSTM network: a machine learning approach for precipitation nowcasting. Adv. Neural Inf. Proces. Syst. **28**, 802–810 (2015)
94. Sirvio, K., Hollmén, J.: Spatio-temporal road condition forecasting with Markov chains and artificial neural networks. In: International Workshop on Hybrid Artificial Intelligence Systems, pp. 204–211. Springer, Berlin (2008)
95. Smieszek, T.: A mechanistic model of infection: Why duration and intensity of contacts should be included in models of disease spread. Theor. Biol. Med. Model. **6**(1), 1–10 (2009)
96. Song, K.-B., Baek, Y.-S., Hong, D.H., Jang, G.: Short-term load forecasting for the holidays using fuzzy linear regression method. IEEE Trans. Power Syst. **20**(1), 96–101 (2005)
97. Sternberg, A., Soares, J., Carvalho, D., Ogasawara, E.: A review on flight delay prediction. arXiv preprint arXiv:1703.06118, pp. 1–21 (2017)
98. Stoffer, D.S.: Estimation and identification of space-time ARMAX models in the presence of missing data. J. Am. Stat. Assoc. **81**(395), 762–772 (1986)
99. Stroud, J.R., Müller, P., Sansö, B.: Dynamic models for spatiotemporal data. J. R. Stat. Soc. Ser. B Stat Methodol. **63**(4), 673–689 (2001)
100. Sun, J., Zhang, J., Li, Q., Yi, X., Zheng, Y.: Predicting Citywide Crowd Flows in Irregular Regions Using Multi-View Graph Convolutional Networks. arXiv preprint arXiv:1903.07789 (2019)
101. Sun, S., Lu, H., Tsui, K.L., Wang, S.: Nonlinear vector auto-regression neural network for forecasting air passenger flow. J. Air Transp. Manag. **78**(September 2018), 54–62 (2019)
102. Sun, S., Wei, Y., Tsui, K.L., Wang, S.: Forecasting tourist arrivals with machine learning and internet search index. Tour. Manag. **70**(July 2018), 1–10 (2019)
103. Sutskever, I., Vinyals, O., Le, Q.V.: Sequence to sequence learning with neural networks. In: Advances in neural information processing systems, pp. 3104–3112 (2014)
104. Tang, L., Zhao, Y., Cabrera, J., Ma, J., Tsui, K.L.: Forecasting short-term passenger flow: an empirical study on Shenzhen Metro. IEEE Trans. Intell. Transp. Syst. **20**(10), 3613–3622 (2019)

105. Tang, L., Zhao, Y., Tsui, K.L., He, Y., Pan, L.: A clustering refinement approach for revealing urban spatial structure from smart card data. Applied Sciences (Switzerland) **10**(16), 5606 (2020)
106. Treiber, M., Hennecke, A., Helbing, D.: Congested traffic states in empirical observations and microscopic simulations. Phys. Rev. E Stat. Phys. Plasmas Fluids Relat. Interdiscip. Topics **62**(2), 1805–1824 (2000)
107. Tsui, K.L., Chiu, W., Gierlich, P., Goldsman, D., Liu, X., Maschek, T.: A review of healthcare, public health, and syndromic surveillance. Qual. Eng. **20**(4), 435–450 (2008)
108. Tsui, K.L., Wong, S.Y., Jiang, W., Lin, C.J.: Recent research and developments in temporal and spatiotemporal surveillance for public health. IEEE Trans. Reliab. **60**(1), 49–58 (2011)
109. Tsui, K.L., Han, S.W., Jiang, W., Woodall, W.H.: A review and comparison of likelihood-based charting methods. IIE Transactions (Institute of Industrial Engineers) **44**(9), 724–743 (2012)
110. Tsui, K.L., Wong, Z.S.Y., Goldsman, D., Edesess, M.: Tracking infectious disease spread for global pandemic containment. IEEE Intell. Syst. **28**(6), 60–64 (2013)
111. Tu, Y., Ball, M.O., Jank, W.S.: Estimating flight departure delay distributions—A statistical approach with long-term trend and short-term pattern. J. Am. Stat. Assoc. **103**(481), 112–125 (2008)
112. Vlahogianni, E.I., Karlaftis, M.G., Golias, J.C.: Spatio-temporal short-term urban traffic volume forecasting using genetically optimized modular networks. Comput. Aided Civ. Inf. Eng. **22**(5), 317–325 (2007)
113. Volkova, S., Ayton, E., Porterfield, K., Corley, C.D.: Forecasting influenza-like illness dynamics for military populations using neural networks and social media. PLoS One **12**(12), e0188941 (2017)
114. Wang, J., Shi, Q.: Short-term traffic speed forecasting hybrid model based on chaos–wavelet analysis-support vector machine theory. Transportation Research Part C: Emerging Technologies **27**, 219–232 (2013)
115. Wang, T., Zhang, Z., Tsui, K.-L.: PSTN: Periodic Spatial-Temporal Deep Neural Network for Traffic Condition Prediction, arXiv:2108.02424, (2021)
116. Wei, C., Sheng, J.: Spatial-temporal graph attention networks for traffic flow forecasting. IOP Conference Series: Earth and Environmental Science **587**(1), 1853–1862 (2020)
117. West, M., Harrison, J.: Bayesian Forecasting and Dynamic Models, 2nd edn. Springer, New York (1997)
118. Williams, B.M., Hoel, L.A.: Modeling and forecasting vehicular traffic flow as a seasonal ARIMA process: Theoretical basis and empirical results. J. Transp. Eng. **129**(6), 664–672 (2003)
119. Wong, G.C., Wong, S.C.: A multi-class traffic flow model—An extension of LWR model with heterogeneous drivers. Transp. Res. A Policy Pract. **36**(9), 827–841 (2002)
120. Woodall, W.H., Tsui, K.L.: Comments on 'Some methodological issues in biosurveillance'. Stat. Med. **30**(5), 430–433 (2011)
121. Wu, Y., Tan, H.: Short-term traffic flow forecasting with spatial-temporal correlation in a hybrid deep learning framework. arXiv preprint arXiv:1612.01022 (2016)
122. Xie, Y., Zhang, Y., Ye, Z.: Short-term traffic volume forecasting using Kalman filter with discrete wavelet decomposition. Comput. Aided Civ. Inf. Eng. **22**(5), 326–334 (2007)
123. Xu, N., Donohue, G., Laskey, K.B., Chen, C.H.: Estimation of delay propagation in the national aviation system using Bayesian networks. In: Proceedings of the 6th USA/Europe Air Traffic Management Research and Development Seminar, ATM 2005, pp. 353–363 (2005)
124. Xu, Q., Tsui, K.L., Jiang, W., Guo, H.: A Hybrid approach for forecasting patient visits in emergency department. Qual. Reliab. Eng. Int. **32**(8), 2751–2759 (2016)
125. Xu, Q., Gel, Y.R., Ramirez, L.L.R., Nezafati, K., Zhang, Q., Tsui, K.L.: Forecasting influenza in Hong Kong with Google search queries and statistical model fusion. PLoS One **12**(5), 1–17 (2017)

126. Yang, Q., Koutsopoulos, H.N.: A microscopic traffic simulator for evaluation of dynamic traffic management systems. Transportation Research Part C: Emerging Technologies **4**(3 PART C), 113–129 (1996)
127. Yang, Z., Mei, D., Yang, Q., Zhou, H., Li, X.: Traffic flow prediction model for large-scale road network based on cloud computing. Math. Probl. Eng. 926251 (2014)
128. Yang, S., Santillana, M., Kou, S.C.: Accurate estimation of influenza epidemics using Google search data via ARGO. Proc. Natl. Acad. Sci. U. S. A. **112**(47), 14473–14478 (2015)
129. Yin, W., Murray-Tuite, P., Ukkusuri, S.V., Gladwin, H.: An agent-based modeling system for travel demand simulation for hurricane evacuation. Transportation Research Part C: Emerging Technologies **42**, 44–59 (2014)
130. Yu, B., Song, X., Guan, F., Yang, Z., Yao, B.: k-nearest neighbor model for multiple-time-step prediction of short-term traffic condition. J. Transp. Eng. **142**(6), 04016018 (2016)
131. Yu, H., Wu, Z., Wang, S., Wang, Y., Ma, X.: Spatiotemporal recurrent convolutional networks for traffic prediction in transportation networks. Sensors (Switzerland) **17**(7), 1501 (2017)
132. Yu, L., Chan, W.M., Zhao, Y., Tsui, K.L.: Personalized health monitoring system of elderly wellness at the community level in Hong Kong. IEEE Access **6**, 35558–35567 (2018)
133. Yu, B., Guo, Z., Asian, S., Wang, H., Chen, G.: Flight delay prediction for commercial air transport: a deep learning approach. Transportation Research Part E: Logistics and Transportation Review **125**(March), 203–221 (2019)
134. Zhang, L., Levinson, D.: Agent-based approach to travel demand modeling exploratory analysis. Transp. Res. Rec. **1898**, 28–36 (2004)
135. Zhang, Y., Xie, Y.: Forecasting of short-term freeway volume with v-support vector machines. Transp. Res. Rec. **2024**(1), 92–99 (2007)
136. Zhang, L., Liu, Q., Yang, W., Wei, N., Dong, D.: An improved k-nearest neighbor model for short-term traffic flow prediction. Procedia. Soc. Behav. Sci. **96**, 653–662 (2013)
137. Zhang, J., Zheng, Y., Qi, D., Li, R., Yi, X., Li, T.: Predicting citywide crowd flows using deep spatio-temporal residual networks. Artif. Intell. **259**, 147–166 (2018)
138. Zhao, L., Song, Y., Deng, M., Li, H.: Temporal graph convolutional network for urban traffic flow prediction method. arXiv preprint arXiv:1811.05320 (2018)
139. Zhao, Y., Xu, Q., Chen, Y., Tsui, K.L.: Using Baidu index to nowcast hand-foot-mouth disease in China: A meta learning approach. BMC Infect. Dis. **18**(1), 1–11 (2018)
140. Zhao, L., Song, Y., Zhang, C., Liu, Y., Wang, P., Lin, T., Deng, M., Li, H.: T-GCN: A Temporal Graph Convolutional Network for Traffic Prediction. IEEE Trans. Intell. Transp. Syst. **21**(9), 3848–3858 (2020)
141. Zhu, X., Li, L.: Flight time prediction for fuel loading decisions with a deep learning approach. Transportation Research Part C: Emerging Technologies **128**, 103179 (2021)
142. Zhu, Z., Peng, B., Xiong, C., Zhang, L.: Short-term traffic flow prediction with linear conditional Gaussian Bayesian network. J. Adv. Transp. **50**(6), 1111–1123 (2016)

Introduction to Wafer Tomography: Likelihood-Based Prediction of Integrated-Circuit Yield

Michael Baron, Emmanuel Yashchin, and Asya Takken

Abstract A concept of wafer tomography is introduced referring to a detailed reconstruction of hidden information on integrated circuits given incomplete and sparse layer-by-layer data that are usually available. Proposed tools associate chip failures with all observed, partially observed, and unobserved defects on a chip via a cause-and-effect relationship to predict the final yield at any time during the production process. The method also allows to determine the most probable causes of failures, the most dangerous defects, the most vulnerable layers, the most influential factors, and their combinations.

Keywords Diagnostics · EM algorithm · Kill ratio · Tomography · Wafer inspection · Yield prediction

1 Introduction

Over the last three decades, a variety of stochastic models have been proposed that explain the patterns of defective chips in semiconductor manufacturing. Earlier works focus on the total number of failed chips [11, 21, 29], while recent papers deal with their spatial dependence [9, 10, 14, 20, 23, 28], effect of the critical area [13, chap. 4], [27, 31], and defect types [5, 8, 12], emphasizing more and more often the large volumes of integrated-circuit test data and the consequent computational

M. Baron (✉)
American University, Washington, DC, USA
e-mail: baron@american.edu

E. Yashchin
IBM Research Division, Yorktown Heights, NY, USA
e-mail: yashchi@us.ibm.com

A. Takken
Alliant Cooperative Data, Brewster, NY, USA
e-mail: asya@stanfordalumni.org

A. Steland, K.-L. Tsui (eds.), *Artificial Intelligence, Big Data and Data Science in Statistics*, https://doi.org/10.1007/978-3-031-07155-3_9

challenges [2–4, 17, 22], and others. For a more complete list of classical references on semiconductor yield modeling and prediction, see [1].

As recognized in the literature, the probability for a chip to fail is affected by the number of defects, their size, type, and location. These factors are accounted in the model proposed in [1] that relates chip failures to all the measurable and available characteristics of defects. In reality, for the sake of cost economy, many defects are observed but not classified, and a majority of layers are not inspected at all. Thus, a multitude of substantial information is missing, whereas the fatal defects causing chip failures are often unobserved or unclassified. Nevertheless, it is possible to use all the pieces of available information effectively to reconstruct relationships between defects and chip failures and to predict failures with a reasonably high probability. We refer to such a process as *wafer tomography*, by analogy with the tomographic image reconstruction techniques via mathematical superposition of information obtained from sectional images, absorbed neutrons, photoacoustic signals, or wave travel time measurements.

In this chapter, we supply the proposed methodology with a theoretical treatment, explain mathematics supporting wafer tomography, assess its benefits and limitations for semiconductor yield prediction, and propose new computational tools for mining large volumes of detailed multidimensional integrated-circuit defect data.

Incorporating all the observed and unobserved information for the explanation and prediction of chip failures, the proposed model inevitably contains *a very large number of parameters*. It has to account for effects of tens of defect types occurring on tens of different layers of the chip, hundreds of interactions between defect types and layers, effects of other causes, distribution of defect sizes of each type, as well as thousands of frequencies of defects of different types on different layers in different lots. Direct optimization over such a large number of parameters is computationally problematic. At the same time, inclusion of unobserved and unclassified chips into the model is rather crucial. Our analysis of over 1000 lots and millions of chips showed that failure of any chip is more likely to be caused by one or several unobserved or unclassified defects than by an observed and classified defect. In addition, a chip may fail due to reasons specific to a lot but other than observable defects.

In the next section, the detailed tomographic model of the cause-and-effect relationship between the defects and yield is developed. The value and limitations of this model are discussed in a theoretical context of the accuracy of *any* model predicting the yield. Computational aspects of wafer tomography are studied in Sect. 3. Proposed modifications of the EM algorithm enhance the success and fast convergence of the multidimensional estimation routine. Section 4 discusses the goodness-of-fit criteria for the tomographic yield predicting models and their limitations for the correct prediction of yield. The proofs are given in Appendix.

2 Tomographic Models

In this section, we derive a stochastic model explaining a cause-and-effect relationship between (a) defects of different types and sizes, occurring at different layers of a wafer, and (b) the yield of each chip.

This model was originally introduced in [1]. Here, we provide theoretical justification for each of its components and discuss possible variations. We then assess limitations of the model and present a theoretical result on the prediction power of *any* yield predicting model.

We start by modeling the probability for a chip to survive a single observed and classified defect of a given size, type, and location. Then, using basic rules of probability, we extend the formula to unclassified defects and unobserved defects occurring on uninspected layers. Combining all the components, we obtain the probability for a chip to survive given all its classified and unclassified observed defects as well as all the uninspected layers and lot-specific other causes.

The model is based on the following general assumptions:

(1) Failure of a chip can be caused by defects occurring on it as well as other factors that are specific to a lot.
(2) Any defect can be fatal, independently of other defects, with a probability that is a function of the defect type, its size, and the layer on which it occurs.
(3) The number of defects of each type on each layer follows a Poisson distribution with a parameter determined by the defect type, layer, and lot.

2.1 Notation

Let us introduce the following notation. Throughout the paper, index k represents a defect, j is a defect type, s is a defect size, x is the transformed defect size, l is a layer, i is a chip, w is a wafer, and m is a lot. Thus, j_k is the type of the k-th defect, l_k is a layer on which it occurred, etc. The numbers of chips, wafers, layers, etc., are denoted by the corresponding capital letters, I for the number of chips, W for the number of wafers, L for the number of lots, and so on.

Next, C and U will denote the sets of classified and unclassified defects, respectively. Likewise, C_{lw} is the set of classified defects on layer l of wafer w. The number of classified defects (of type j on layer l) is denoted by d (d_{jl}), whereas the number of unclassified defects (on layer l) is u (u_l). The total number of defects on layer l is then $N_l = u_l + \sum_j d_{jl}$.

Also, let ξ_i be a binary variable representing the quality of chip i; $\xi_i = 1$ if the chip is good, 0 otherwise. At the same time, φ_i will be the probability for chip i to survive, given all its defects and other causes. Thus, each ξ_i is a Bernoulli random variable with parameter φ_i.

Finally, let L_w be a set of layers that were inspected on wafer w. For any layer $l \in L_w$, all the defects are counted and measured although only a small portion of them is classified. No information is available about defects on the remaining, uninspected layers $l \notin L_w$. Layers are inspected by wafer; hence, under normal circumstances, each layer is either inspected on all chips $i \in w$ or left uninspected on all chips.

2.2 Parameters

The proposed model is parameterized by the following five groups of parameters, where $r(j)$ for $j = 1, \ldots, J$ is the effect of defect type j; $a(l)$ for $l = 1, \ldots, L$ is the effect of layer or operation l; $b(m)$ for $m = 1, \ldots, M$ is the effect of other causes for lot m; $\lambda(j, l, m)$ is the frequency of defects of type j on layer l of lot m, i.e., the expected number of such defects per chip; and $\pi_j(x) = \pi(x \mid \text{defect type } j)$ is the density function of transformed sizes x of defects of type j. The whole parameter vectors will be denoted by $\boldsymbol{r} = (r(1), \ldots, r(J))$, $\boldsymbol{a} = (a(1), \ldots, a(L))$, $\boldsymbol{b} = (b(1), \ldots, b(M))$, and $\boldsymbol{\lambda} = \{\lambda(j, l, m)\}$.

The final group of parameters are *interactions* between defect types and layers, which is found significant for the prediction of chip failures. Essentially, defects of the same type can have different effects on the yield if they occur on different layers of a wafer. Without losing prediction power, the number of interaction parameters can be reduced by considering *groups of layers* such as light metal layers, darkfield, brightfield, and so on. Then, by a defect type j, we consider a given type of defects occurring within a given group of layers. For example, let j_1 be a scratch on darkfield, j_2 be a puddle on lightfield, j_3 be a scratch on lightfield, etc. In other words, j is understood as a pair of a defect type and a group of layers where it occurred.

When the overall number of parameters is too large for the available computing resources, one may also consider grouping defect types. For example, surface defects include scratches and improperly etched traces; processing defects including poorly connected wires, inadvertent shorts, and improperly drilled vias; misalignment of layers, and so on. Grouping reduces the number of parameters, especially interactions and defect frequencies, simplifying computations and accelerating the data analysis.

2.3 Effect of Classified Defects

We start building the likelihood from a *single defect*. Suppose a defect of type j and size s occurred on layer l of chip i. What is the probability $\mathbf{P}\{\xi_i = 1\}$ that the chip survives this defect? A number of competing models can be proposed, such as:

(1) $\mathbf{P}\{\text{survival}\} = \exp\{-r(j)a(l)g(s)\}$ (multiplicative model)
(1a) $\mathbf{P}\{\text{survival}\} = \exp\{-[r(j) + a(l)]\, g(s)\}$ (additive model)
(1b) $\mathbf{P}\{\text{survival}\} = \exp\{-r(j)g(s) - a(l)\}$ (simplified additive model)

and others. Effects of categorical factors such as defect types and layers are given in the most general form by parameters $r(j)$ and $a(l)$, whereas the effect of the quantitative variable size, s, is represented by a function $g(s)$.

According to our experiments, model (1), calibrated on training data, provided the best fit in terms of the highest prediction accuracy on test data. Therefore, the rest of this chapter is based on the multiplicative model. Also, among other functions of size, the *logarithmic* transformation

$$x = g(s) = \log(1 + s) \tag{2}$$

was found to give the best fit to the actual data.

No theory can guarantee that the multiplicative model will continue to dominate for future chip designs. Thus, it is natural to keep *a bank of plausible models* that can be compared for each new mode of production, so that the model with the best fit can always be chosen. Also, another form of the function $g(s)$ may perform well for certain types of chips. A possible approach is to specify this function up to unknown parameters that will then be estimated among many other parameters of the model.

Estimation, prediction, and diagnostics methods proposed here and in [1] are not tied to any specific form of $g(s)$, and they can be readily applied to any model in (1)–(1b).

2.4 *Effect of Unclassified Defects on Inspected Layers*

From our experience, only a small portion of defects gets classified. A majority of defects remain unclassified, which means that the defect types are unknown. The observed information on these defects is confined to their size s and layer l on which they occur.

The portion of the likelihood that corresponds to unclassified defects can be derived from basic laws of probability. A chip will survive an unclassified defect k with probability

$$\mathbf{P}\{\xi = 1 \mid x_k\} = \sum_{j=1}^{J} \mathbf{P}\{j_k = j \mid x_k\}\, \mathbf{P}\{\xi = 1 \mid j_k = j, x_k\}, \tag{3}$$

according to the formula of total probability. Essentially, the expectation is taken with respect to an unknown defect type j.

Components of (3) are computed as follows. Conditional survival probabilities $\mathbf{P}\{\xi = 1 \mid j_k = j, x_k\}$, given the defect type, are obtained from (1), the formula that assumes known defect types. Next, the marginal probabilities of defect types, given their sizes, follow from the Bayes rule,

$$\mathbf{P}\{j_k = j \mid x_k = x\} = \frac{\mathbf{P}\{j_k = j\}\pi(x \mid j_k = j)}{\sum_{j'} \mathbf{P}\{j_k = j'\}\pi(x \mid j_k = j')}.$$

Finally, the overall marginal probability of each defect type is the proportion of defects of the given type among all the defects, which is the ratio of the corresponding defect frequencies. Hence, $\mathbf{P}(j_k = j) = \lambda(j, l_k, m_k)/\lambda(l_k, m_k)$, where $\lambda(l, m) = \sum_j \lambda(j, l, m)$ is a cumulative frequency of defects per chip on layer l of lot m, regardless of the defect type.

Substituting the obtained expressions into (3), we obtain the likelihood function for the probability for a chip to survive an unclassified defect k of a given size x_k,

$$\mathbf{P}\{\xi = 1 \mid x_k\} = \frac{\sum_j \lambda(j, l_k, m_k)\pi_j(x_k)\exp\{-r(j)a(l_k)x_k\}}{\sum_j \lambda(j, l_k, m_k)\pi_j(x_k)}. \tag{4}$$

2.5 Effect of Uninspected Layers

Equations (1) and (4) express the probabilities for a chip to survive all its *observed* defects, classified and unclassified. According to them, the probability of surviving all the inspected layers can be evaluated.

Often, however, inspection of all the layers is expensive and time consuming. Then, a majority of layers remain uninspected, and therefore, they represent the main source of potential fatal defects that cause chip failures. Among tens of layers, only a few wafers had more than 75% of their layers inspected, whereas the majority had only one or two inspected layers.

Any uninspected layer provides no observed information. The numbers of defects, their types, and sizes remain unknown. Nevertheless, the effect of uninspected layers can be included into the likelihood function. The probability for a chip to survive uninspected layers can be evaluated by taking expectations over all the unobserved and unknown pieces of information. Again, the law of total probability will be the main tool in this derivation.

Let N_{ijlm} be the (unobserved) number of type j defects on layer l of chip i in lot m. Chip i survives the entire layer l if all the defects of all types on this layer appear not fatal. According to our main assumptions, N_{ijlm} has Poisson distribution with the frequency parameter $\lambda(j, l, m)$, and effects of all the defects on the chip's failure are independent.

Then, following the equation (4) of [24],

$$
\begin{aligned}
&\mathbf{P}\{\text{chip } i \text{ survives layer } l\} \\
&= \prod_{j=1}^{J} \left(\sum_{n=0}^{\infty} \mathbf{P}\left\{N_{ijlm} = n\right\} \mathbf{P}\left\{ \begin{matrix} \text{chip } i \text{ survives } n \text{ defects} \\ \text{of type } j \text{ on layer } l \end{matrix} \right\} \right) \\
&= \prod_{j=1}^{J} \mathbf{E}\left(\mathbf{P}\{\text{chip } i \text{ survives a type } j \text{ defect on layer } l\}\right)^{N_{ijlm}} \\
&= \prod_{j=1}^{J} \mathbf{E}\,\psi_{jl}^{N_{ijlm}} = \prod_{j=1}^{J} \exp\left\{-\lambda(j,l,m)\left(1-\psi_{jl}\right)\right\},
\end{aligned} \tag{5}
$$

where ψ_{jl} is the probability that a type j defect is not fatal. Applying the law of total probability once again, integrating over the (transformed) size of the defect, this probability is evaluated as

$$
\begin{aligned}
\psi_{jl} &= \mathbf{P}\{\text{a type } j \text{ defect is not fatal}\} \\
&= \mathbf{E}_j^x\ \mathbf{P}\{\text{a type } j \text{ defect of size } x \text{ is not fatal}\} \\
&= \mathbf{E}_j^x e^{-r(j)a(l)x} = \int e^{-r(j)a(l)x} \pi_j(x)dx.
\end{aligned} \tag{6}
$$

Thus, the survival probability ψ_{jl} is the value of the moment generating function of transformed defect sizes of type j defects, computed at $(-r(j)a(l))$. For its estimation, see Sect. 6 and, in particular, Eq. (15).

2.6 Survival of All the Observed and Unobserved Defects and Other Causes

Components of the likelihood derived in the previous three subsections will now be combined to obtain the overall likelihood function of defects and chip failures.

The chip's survival of a collection of defects is the intersection of events representing the chip's survival of all the individual defects. Studies show [24, 26] that these events are mutually independent and also independent of the effect of other causes, specific to the lot. Therefore, the survival probabilities given by (1), (4), and (5) can be multiplied over all the defects on a chip.

Also, conditioned on all the defects, the chip failures are conditionally independent. This means that the chip failure can be caused by its own defects only, and not by defects occurring on another chip. With chip failures denoted by Bernoulli (indicator) variables ξ_i, we obtain the likelihood function

$$\mathcal{L}(\boldsymbol{a}, \boldsymbol{b}, \boldsymbol{r}, \boldsymbol{\lambda}) \tag{7}$$
$$= \prod_{m} \prod_{w \in m} \prod_{l \in L_w} \prod_{i \in w} e^{-\lambda(l,m)} \frac{\lambda^{N_{il}}(l, m)}{N_{il}!} \prod_{j=1}^{J} \left(\frac{\lambda(j, l, m)}{\lambda(l, m)} \right)^{d_{ijl}} \varphi_i^{\xi_i} (1 - \varphi_i)^{1-\xi_i},$$

in which the products are taken over all lots m in the dataset, wafers w in these lots, chips i and inspected layers $l \in L_w$ on these wafers, and defect types j. In this formula,

$$p_{lm}(N_{il}) = e^{-\lambda(l,m)} \frac{\lambda^{N_{il}}(l, m)}{N_{il}!}$$

is the Poisson probability of observing N_{il} defects (of all types) on inspected layer l of chip i;

$$p_{jlm} = \lambda(j, l, m)/\lambda(l, m)$$

is the probability that the observed defect is of type j, and it is a part of the likelihood for all d_{ijl} *classified* type j defects on layer l of chip i.

The last portion of the likelihood function (7) represents the probability mass function of the survival indicator ξ_i, a Bernoulli variable with parameter

$$\varphi_i = \mathbf{P}\{\xi_i = 1\} = \mathbf{P}\{\text{ chip } i \text{ is good }\}.$$

This parameter, the probability that chip i survives all its defects and lot-specific other causes, equals

$$\varphi_i = e^{-b(m)} \prod_{l \in L_i} \left\{ \prod_{k \in C_{il}} e^{-r(j_k)a(l)x_k} \prod_{k \in U_{il}} \frac{\sum_j \lambda(j, l, m)\pi_j(x_k)e^{-r(j)a(l)x_k}}{\sum_j \lambda(j, l, m)\pi_j(x_k)} \right\}$$
$$\times \prod_{l \notin L_i} \prod_{j=1}^{J} e^{-\lambda(j,l,m)(1-\psi_{jl})}, \tag{8}$$

according to expressions (1), (4), and (5) derived above for the effects of classified, unclassified, and unobserved defects. The products here are taken over all the layers $l \in L_i$ inspected on chip i and all the layers $l \notin L_i$ uninspected on it. The defects k on inspected layers are further divided into classified $k \in C_{il}$ and unclassified $k \in U_{il} = \bar{C}_{il}$.

The likelihood function (7) of all the observed defects and chip failures, derived from the model assumptions, is now ready for further analysis.

3 Computational Aspects and Optimization

The cause-and-effect model and the likelihood function derived in the previous section contain a rather large number of parameters consisting of effects of L layers, J defect types on each layer type, M lot-specific other causes, and $L \times J \times M$ defect frequencies. The latter group is the largest, sometimes reaching 100,000 parameters to be estimated. However, besides explaining the chip failures and predicting the yield, estimation of defect frequencies is a problem of separate importance in semiconductor manufacturing.

The large number of parameters presents the main challenge in their estimation, making most of the classical methods (maximum likelihood, least squares, Bayesian) computationally difficult if not completely infeasible.

To deal with the large dimension of the problem, [1] use the *expectation–maximization (EM) algorithm* [6, 7, 15, 18]. According to it, thousands of defect frequencies are estimated during the *E-step*, by computing the expected number of defects of each type on each layer of each lot. All the other parameters representing effects of defects and other causes on chip failures are estimated during the M-step by maximizing the likelihood function (7), given the defect frequencies $\lambda(j, l, m)$ obtained during the preceding E-step.

In our experience, direct application of the EM algorithm for model (7) is not always successful. The algorithm converges rather slowly and often ends in a local extremum of the log-likelihood function rather than the global maximum likelihood estimator. This is due to the complicated multidimensional structure of the likelihood function and non-existence of a tractable stochastic model that provides a *perfect* fit to the chip and defect data.

Two modifications of the algorithm are proposed here. First, we propose an efficient data-driven initialization of the algorithm that is essentially a quick and sketchy estimation of all the model parameters. Second, we propose to introduce the third *directional* step ("D-step") during each iteration cycle, where we compute the most recently found direction of the growth of log likelihood and search for the optimal parameter vector along that direction. This often allows to make a strong step into the direction of higher log likelihood, preventing the algorithm from staying in the neighborhood of a local maximum and converging to it eventually.

Initialization of the parameter estimates and the analysis necessary for the E-step, M-step, and D-step are given below. Longer derivations are in the appendix.

3.1 Initialization of Parameter Estimates

A meaningful initial point in the iterative numerical routine may accelerate the entire scheme and prevent it from converging to a local but not global extremum. Here we propose simple choices for the initial values of parameter estimates, $\boldsymbol{a}^{(0)}$, $\boldsymbol{b}^{(0)}$, $\boldsymbol{r}^{(0)}$, and $\boldsymbol{\lambda}^{(0)}$, that follow from the observed data by a quick, sketchy computation.

It is natural to set the initial frequencies of defects of different types proportional to the corresponding observed counts of classified defects, i.e.,

$$\lambda^{(0)}(j,l,m)/\lambda^{(0)}(l,m) = d_{jlm}/d_{lm}.$$

Denominator $\lambda^{(0)}(l,m)$ in this expression is the total frequency of all defects on layer l of lot m. It can be quickly estimated by the sample frequency of defects on all the chips where layer l is inspected,

$$\lambda^{(0)}(l,m) = \frac{d_{lm} + u_{lm}}{\sum_{i \in m} I\{l \in L_i\}}.$$

Then, all frequencies are initialized as $\lambda^{(0)}(j,l,m) = \left(d_{jlm}/d_{lm}\right)\lambda^{(0)}(l,m)$.

Next, without any additional information at the initial step, suppose that $r(j) \equiv r$, $b(m) \equiv b$, and $a(l) \equiv 1$ (we notice that a_l are multipliers in model (1); hence, they are determined only up to a constant coefficient). Replacing, for a rough approximation, transformed defect sizes x_k by their sample average $\bar{x}$, we obtain from (8) that for any chip i of lot m,

$$\varphi_i \approx e^{-b} \prod_{j=1}^{J} \prod_{l=1}^{L} \left(e^{-r(j)a(l)\bar{x}}\right)^{\lambda(j,l,m)} = e^{-b-r\bar{x}\sum_j \sum_l \lambda(j,l,m)} = e^{-b-r\bar{x}JL\bar{\lambda}(m)},$$

where $\bar{\lambda}(m) = \sum_j \sum_l \lambda(j,l,m)/(JL) \approx \bar{\lambda}$ is the average number of defects per chip, defect type, and layer for lot m, which, for the initial approximation, we assume equal for all the lots.

Equating, by the method of moments, the expected and the actual yield, $\sum \varphi_i$ and $\sum \xi_i$, one obtains

$$r^{(0)} = -\frac{\log(\sum \xi_i/I) + b^{(0)}}{JL\bar{x}\bar{\lambda}^{(0)}} = -M\frac{\log(\sum \xi_i/I) + b^{(0)}}{\bar{x}\sum\sum\sum \lambda^{(0)}(j,l,m)}, \tag{9}$$

an equation connecting the initial choice of the averaged effect of a defect and the averaged effect of other causes, where M is the number of lots and I is the number of chips, so that $\sum \xi_i/I = \bar{\xi}$ is the observed yield per chip.

It remains to choose a reasonable initial approximation for the effect of other causes b. It can be obtained, for example, from the average yield $\bar{\xi}^*$ of all defect-free chips if they are available. Failures of defect-free chips can only be attributed to other causes, i.e., $\varphi_i = e^{-b}$, and by the introduced version of the method of moments, $b^{(0)} = -\log(\bar{\xi}^*)$.

3.2 M-Step

Available optimization routines can handle computations arising during the M-step if they involve a moderate number of parameters. In the proposed estimation scheme, the likelihood function given by (7) and (8) is maximized over all the effects $a(l)$, $b(m)$, and $r(j)$, whereas the frequencies $\lambda(j, l, m)$ are updated during the E-step.

Then, at this step, only a part of the likelihood function containing $\boldsymbol{a}$, $\boldsymbol{b}$, and $\boldsymbol{r}$ needs to be maximized, which is equivalent to maximizing

$$\sum_i \{\xi_i \log \varphi_i + (1 - \xi_i) \log(1 - \varphi_i)\}, \tag{10}$$

with $\log \varphi_i$ given in (8). The speed and accuracy of the algorithm depend on the chosen optimization routine and convergence criteria. Also, the following remarks allow to reduce the number of operations significantly.

Remark 1 Since only a few layers are inspected on each wafer, there is a large number, often a vast majority, of chips without a single detected defect. Within each wafer, such chips share the same value of φ_i. Thus the corresponding terms of (10) can be computed only once for each wafer (but distinguish between good chips with no defects and bad chips with no defects).

Remark 2 It is not difficult to compute and supply the analytic gradient of (10); for its explicit derivation, see Sect. 6. Thus, it is recommended to include it into the routine instead of forcing its estimation by finite differences. Supplying the Hessian would be efficient too although it is cumbersome.

Remark 3 It is beneficial to terminate the computationally intensive M-step under rather *mild convergence criterion* and proceed to other steps, since high-precision optimization is inefficient (and not really necessary) in intermediate stages.

3.3 E-Step

An *unbiased* estimator of defect frequencies $\lambda(j, l, m)$ is

$$\hat{\lambda}(j, l, m) = I_m^{-1} \, \mathbf{E} \left\{ N_{jlm} \mid \boldsymbol{d}, \boldsymbol{u}, \boldsymbol{x}, \boldsymbol{\xi}, \boldsymbol{a}, \boldsymbol{b}, \boldsymbol{r}, \boldsymbol{\lambda} \right\}, \tag{11}$$

where I_m is the total number of chips in lot m; N_{jlm} is the number of defects of type j on layer l of lot m; and vectors $\boldsymbol{d}$, $\boldsymbol{u}$, $\boldsymbol{x}$, $\boldsymbol{\xi}$, $\boldsymbol{a}$, $\boldsymbol{b}$, $\boldsymbol{r}$, and $\boldsymbol{\lambda}$ represent, respectively, the number of classified and unclassified defects, their transformed sizes, yield of each chip, and the current refined values of all estimated parameters. This frequency estimator is computed as the conditional expectation of the number of type j, layer l defects per chip, given the number and sizes of classified and unclassified defects (u_i, d_i, x_k), the quality of chips (ξ_i), and the *true* values of parameters.

Then, at each iteration n, the set $\boldsymbol{\lambda}$ of defect frequencies will be updated as

$$\boldsymbol{\lambda}^{(n+1)} = I_m^{-1}\,\mathbf{E}\left\{\boldsymbol{N} \mid \boldsymbol{d}, \boldsymbol{u}, \boldsymbol{x}, \boldsymbol{\xi}, \boldsymbol{a}^{(n)}, \boldsymbol{b}^{(n)}, \boldsymbol{r}^{(n)}, \boldsymbol{\lambda}^{(n)}\right\},$$

which results in the following refining equation (for details, see Sect. 6)

$$\lambda^{(n+1)}(j,l,m) = \left\{ d_{jlm} + \sum_{i:l\in L_i} \left(\frac{1-\xi_i}{1-\varphi_i^{(n)}} \sum_{k\in U_{il}} \frac{v_{jkm}^{(n)}}{v_{km}^{(n)}} + \frac{\xi_i - \varphi_i^{(n)}}{1-\varphi_i^{(n)}} \sum_{k\in U_{il}} \frac{w_{jkm}^{(n)}}{w_{km}^{(n)}} \right) \right.$$
$$\left. + \lambda^{(n)}(j,l,m) \sum_{i:l\notin L_i} \left(\frac{1-\xi_i}{1-\varphi_i^{(n)}} + \frac{\xi_i - \varphi_i^{(n)}}{1-\varphi_i^{(n)}} \psi_{jlm}^{(n)} \right) \right\} / I, \quad (12)$$

where, for each defect k,

$$v_{jkm}^{(n)} = \lambda^{(n)}(j, l_k, m)\pi_j(x_k), \quad v_{km}^{(n)} = \sum_j v_{jkm}^{(n)}, \quad (13)$$

$$w_{jkm}^{(n)} = v_{jkm}^{(n)} e^{-r^{(n)}(j)a^{(n)}(l)x_k} = \lambda^{(n)}(j, l_k, m)\pi_j(x_k)e^{-r^{(n)}(j)a^{(n)}(l)x_k},$$

and

$$w_{km}^{(n)} = \sum_j w_{jkm}^{(n)}.$$

During each iteration, the functions of parameters φ_i, v_{jkm}, and w_{jkm} are re-estimated with the use of updated parameter estimates.

During the E-step, each frequency is recomputed once, and no iterations are involved. Therefore, the E-step is much faster and computationally cheaper than the M-step, where a numerical optimization routine is used to maximize the likelihood under fixed $\boldsymbol{\lambda}$.

3.4 *Modification and Directional D-Step*

The EM algorithm possesses a number of appealing properties [15, 30]; however, in a wide range of practical problems (specifically, those dealing with a large number of parameters and large datasets), its performance can typically be improved via suitable modifications [15, chap. 4], [16]. The problem described in this chapter is not an exception.

Here we introduce an additional step that enables one to achieve a sizeable improvement in the speed of convergence and prevent the algorithm to get caught near a local but not global extremum. This step essentially tries to guess the correct search *direction* for the maximum of the likelihood function $\mathcal{L}(\boldsymbol{a}, \boldsymbol{b}, \boldsymbol{r}, \boldsymbol{\lambda})$. When it succeeds, it starts making increasingly larger steps in that direction preventing the routine from too many iterations in the area where the likelihood is increasing slowly. If it fails to find the direction of improvement, we skip the step for the current cycle and proceed with the standard EM algorithm until the next iteration.

The step is introduced as follows. It starts by analyzing results of the latest E-step and M-step. Let $\boldsymbol{\theta}_0$ be the vector of parameter estimates $(\hat{\boldsymbol{a}}, \hat{\boldsymbol{b}}, \hat{\boldsymbol{r}}, \hat{\boldsymbol{\lambda}})$ obtained as a result of the previous cycle, and $\boldsymbol{\theta}_1$ be the refined vector. That is, the latest E-step and M-step transformed $\boldsymbol{\theta}_0$ into $\boldsymbol{\theta}_1$. If the chosen global convergence criterion is met, then the last cycle failed to improve the value of $\mathcal{L}(\boldsymbol{\theta})$ by more than ε, and the entire estimation routine stops. In all other cases, we obtain that $\mathcal{L}(\boldsymbol{\theta}_1) > \mathcal{L}(\boldsymbol{\theta}_0) + \varepsilon$; hence, the likelihood function is seemingly increasing in the direction of

$$\Delta\boldsymbol{\theta} = \boldsymbol{\theta}_1 - \boldsymbol{\theta}_0.$$

We now follow this direction and check if the likelihood continues to increase. Also, each time we increase the step, therefore "shaking" the system and preventing it from convergence to a local extremum. That is, we consider a sequence of vectors $\{\boldsymbol{\theta}_n\}$ defined recursively as

$$\boldsymbol{\theta}_n = \boldsymbol{\theta}_{n-1} + \gamma(n)\Delta\boldsymbol{\theta}, \quad n \geq 2,$$

where $\gamma(n)$ is a chosen increasing function of n (polynomial or even exponential) that controls the rate of search. The value of $\mathcal{L}(\boldsymbol{\theta}_n)$ is calculated for each $n = 2, 3, \ldots$, and the algorithm proceeds only if this value is improved. The D-step stops at the time

$$T = \min\left\{n \geq 2 : \ \mathcal{L}(\boldsymbol{\theta}_n) < \mathcal{L}(\boldsymbol{\theta}_{n-1})\right\}.$$

Then, $\boldsymbol{\theta}_{T-1}$, the best set of parameter estimates obtained so far, serves as an initial point of the next EM iteration.

Clearly, this step is activated only if it leads to larger values of $\mathcal{L}(\boldsymbol{\theta})$. Otherwise, it is skipped, and the routine proceeds to the next EM iteration. Thus, it will generally result in an equal or a higher value of the likelihood. And above all, it is computationally the cheapest of all three steps, requiring only computation of the likelihood function, but no optimization or gradient evaluation.

Based on our experience, insertion of this step into the EM algorithm always resulted in the same or, even more often, higher value of the maximum likelihood. It always accelerated the EM algorithm in the beginning by making aggressive steps and saving a considerable number of EM iterations. During the late iterations, it rarely went beyond $\boldsymbol{\theta}_2$.

3.5 Scalability and Computational Complexity

Due to the increasing complexity of integrated circuits, one may wonder how scalable the proposed modified EM algorithm is in modern manufacturing standards. In general, its computational complexity depends on the sample size, the number of parameters, and the number of iterations [15, 19]. A majority of estimated parameters are frequencies $\lambda(j, l, m)$, which are updated during the E-step and are completely excluded from the M-step. As a result, the time complexity of the algorithm largely depends on $(J + L + M)$ parameters that are being estimated during the most expensive M-step.

The study described in [1, section 5], involved 25 lots with 33,634 chips and 284,796 observed defects of 65 types on 24 layers. Reportedly, each iteration took approximately 30 min of pure CPU time on a 2.4 mhz computer.

The introduced D-step and the use of previous parameters estimates as initial values reduce the number of iterations to about 7–10, according to our experience. On the other hand, increasing the sample size affects the computational complexity nearly proportionally, which may slow the algorithm. When it becomes a concern, grouping of layers or defect types can be considered along the lines of Sect. 2.2 above.

4 Goodness of Fit and Its Theoretical Limitations

Two general goodness-of-fit criteria are proposed for the introduced tomographic model.

One method measures the closeness between the yield *predicted* by the model for each lot m

$$\hat{Y}_m = \text{predicted yield} = \sum_{i \in m} \mathbf{E}(\xi_i) = \sum_{i \in m} \varphi_i$$

and the actually *observed* yield

$$Y_m = \text{actual yield} = \sum_{i \in m} \xi_i .$$

One looks for a small difference between $\{Y_m\}$ and $\{\hat{Y}_m\}$ or a high correlation coefficient between Y_m and $\hat{Y}_m$.

The second method compares the *proportional predicted yield among good chips*

$$\hat{y}_g = \hat{\mathbf{P}}\{\text{ predicted good } \mid \text{ actually good }\} = \frac{\sum_i \varphi_i \xi_i}{\sum_i \xi_i}$$

with the *proportional predicted yield among bad chips*

$$\hat{y}_b = \hat{\mathbf{P}}\{\text{ predicted good } \mid \text{ actually bad }\} = \frac{\sum_i \varphi_i(1-\xi_i)}{\sum_i(1-\xi_i)}.$$

Here, we look for the best separation between $\hat{y}_g$ and $\hat{y}_b$. Of course, we would like to see a predicted yield of 100% among good chips and 0% among bad chips. However, it is not possible due to the uncertainty and randomness involved in chip failures, even in the extreme and unrealistic case when all the layers are inspected, all the defects are classified, and therefore, the yield prediction is based on all the occurred defects.

Indeed, it is not unusual to see a failed chip that was only exposed to the non-defect causes and no detected defects and a good chip that survived the same non-defect causes in addition to many observed defects.

Then, how well can the model separate $\hat{y}_g$ and $\hat{y}_b$, and what difference between them should be considered satisfactory, or a good fit? The following theoretical result answers these questions.

Lemma 1 *Let $\{\varphi_i,\ i = 1, \ldots, I\}$ be mutually independent, identically distributed random variables with the distribution $F(\varphi)$. For each i, consider ξ_i, a Bernoulli variable with parameter φ_i, and let $\xi_1, \ldots, \xi_I$ be conditionally independent, given $\varphi_1, \ldots, \varphi_I$. Let $\hat{y}_g = \sum_i \varphi_i\xi_i / \sum_i \xi_i$ and $\hat{y}_b = \sum_i \varphi_i(1-\xi_i)/\sum_i(1-\xi_i)$.*

Then the strong law of large numbers holds for $\hat{y}_g$ and $\hat{y}_b$. Namely,

$$\lim_{I\to\infty} \hat{y}_g = \frac{\boldsymbol{E}_F\varphi^2}{\boldsymbol{E}_F\varphi} \quad \textit{and} \quad \lim_{I\to\infty} \hat{y}_b = \frac{\boldsymbol{E}_F\varphi - \boldsymbol{E}_F\varphi^2}{1-\boldsymbol{E}_F\varphi},$$

with probability one, where $\boldsymbol{E}_F$ represents the expectation with respect to the distribution F.

The proof of Lemma 1 is given in the Appendix.

Applying this lemma to defects and failures, we define φ_i as the probability that chip i is good, given its defects, and $\xi_i \sim \text{Bernoulli}(\varphi_i)$ as the binary variable that equals 1 if the chip is good. For each i, the value of φ_i is a function of the number, types, locations, and sizes of defects occurring on chip i, as in (8). In turn, all these factors are random variables that collectively determine the distribution F of φ_i.

Corollary 1 *Under the conditions of Lemma 1,*

$$\lim_{I\to\infty}\left(\hat{y}_g - \hat{y}_b\right) = \frac{\mathrm{Var}_F(\varphi)}{\boldsymbol{E}_F\varphi(1-\boldsymbol{E}_F\varphi)} \tag{14}$$

with probability one.

This corollary along with Lemma 1 establishes theoretical limitations of yield prediction and possible separation of predicted proportional yield among good and bad chips. Under no circumstances can we achieve $\hat{y}_g \approx 100\%$ and $\hat{y}_b \approx 0\%$. As follows from (14), the difference between $\hat{y}_g$ and $\hat{y}_b$ is controlled by the variation among φ_i, the probabilities of good chips, which, in turn, depend on a number of factors considered in Sect. 2.

As an extreme situation, suppose that all the defects are getting eliminated by constant modification and improvement of the manufacturing line. Then $\varphi \uparrow 1$, and both $\hat{y}_g$ and $\hat{y}_b$ approach 1. At the other extreme, if $\varphi \downarrow 0$, then both $\hat{y}_g$ and $\hat{y}_b$ approach 0, and in both cases, the difference between them vanishes.

5 Performance of the Algorithm and the Case Study

The algorithm proposed in this chapter has been used in the IBM Microelectronics Division for parameter estimation and yield prediction on many lots of different grades. Besides forecasting the yield, parameter estimates allowed to compare effects of different factors, evaluate kill ratios, determine the most critical defect-layer combinations and the most influential layers, identify the most probable fatal defects, and as a result, suggest business decisions and optimal yield enhancement strategies.

This section focuses on three aspects of the algorithm performance discussed in our paper—effect of uninspected layers on the prediction power, efficiency of the additional D-step for the acceleration of the EM algorithm, and the spread between predicted yield on the good and bad chips.

Based on the data analyzed at IBM, the accuracy and power of wafer tomography strongly depend on the amount of available information, i.e., the number of inspected layers within a lot. This is clearly seen in Fig. 1. In this figure, two sets of lots differ in the number of inspected layers.

In Fig. 1a, each lot has at least 7 out of 11 layers inspected. As a result, there is a noticeable variation in the predicted yield on different lots. Although a few predictions appear totally inaccurate, the predicted yield is strongly correlated to the actual yield on the majority of lots.

Each lot in Fig. 1b has only one or two layers inspected. As a result, the predicted yield is close to the overall average yield, with some variation caused by these few inspected layers. For the other layers, survival probability is computed by (5), based on the general parameter estimates instead of the actual defects, and it is approximately the same for all the lots.

Figure 1 shows the real data; the scale is removed because of commercial confidentiality.

Efficiency of each step of the proposed accelerated EM algorithm can be traced in Table 1. For this data analysis, it took the algorithm 9 iterations to converge, and only during the first iteration, the directional D-step resulted in the increase of the overall joint likelihood. However, it was a large step in the correct general direction of parameter estimates. It resulted in a very significant change of the negative log-likelihood function from 363,927 to 222,530. After this step, the likelihood did not experience substantial changes, and the algorithm converged quickly.

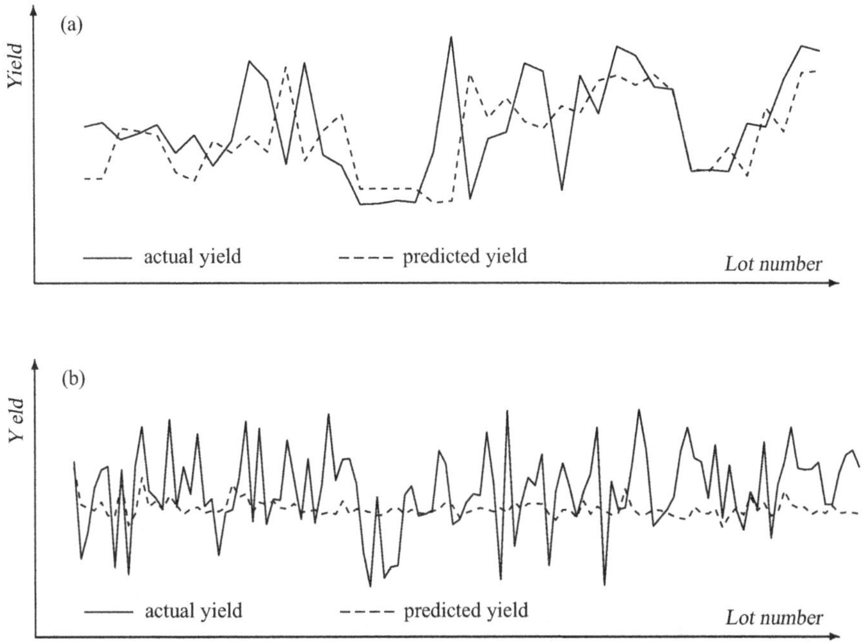

Fig. 1 Yield prediction on 41 lots with many inspected layers (**a**) and on 115 lots with only a few inspected layers (**b**)

Application of the D-step saved the algorithm from slow convergence to a local but possibly not global maximum of the likelihood. After Step 1, no D-step yielded any likelihood increase, so it was not activated. Moreover, after Step 2, the E-step corrected the estimated defect frequencies but also did not yield a larger likelihood. Only M-steps were able to further reduce the current smallest negative log-likelihood function $-\log \mathcal{L}(\hat{\boldsymbol{a}}, \hat{\boldsymbol{b}}, \hat{\boldsymbol{r}}, \hat{\boldsymbol{\lambda}})$ shown in column 3 of the table.

Besides the likelihood function, two other indicators of a good fit are in the last two columns of Table 1. Column 4 contains

$$\text{relative prediction error} = \frac{|\text{predicted yield} - \text{actual yield}|}{\text{actual yield}} = \frac{|\hat{Y} - Y|}{Y}.$$

The last column contains

$$\text{prediction ratio} = \frac{\text{predicted yield among good chips}}{\text{predicted yield among bad chips}} = \frac{\hat{y}_g}{\hat{y}_b}.$$

Ideally, we would like to see a very low prediction error and a very high prediction ratio; however, we know the limitations established by Lemma 1.

Looking for the maximum likelihood estimators of all the parameters, the algorithm continued as long as $(-\log \mathcal{L})$ reduced.

Table 1 Steps, convergence, and performance of the accelerated EM algorithm

Iteration	Step	Min. negative log likelihood	Relative prediction error	Prediction ratio
0	Initiation	406,254	0.1227	1.0590
1	E-step	404,988	0.0189	1.2947
	M-step	363,927	0.3261	1.5298
	D-step	222,530	0.2750	1.5292
2	E-step	221,074	0.0370	1.3752
	M-step	220,638	0.0244	1.3706
3	E-step	220,638	0.0244	1.3706
	M-step	220,453	0.0224	1.3714
4	E-step	220,385	0.0146	1.3552
	M-step	220,453	0.0224	1.3714
5	E-step	220,453	0.0224	1.3714
	M-step	220,360	0.0059	1.3645
6	E-step	220,360	0.0059	1.3645
	M-step	220,355	0.0043	1.3614
7	E-step	220,355	0.0043	1.3614
	M-step	220,343	0.0063	1.3617
8	E-step	220,343	0.0063	1.3617
	M-step	220,340	0.0039	1.3649
9	E-step	220,340	0.0039	1.3649
	M-step	220,336	0.0039	1.3649

At the same time, it can be noticed, in line with the standard statistical practice, that lower values of the negative log likelihood are not necessarily associated with the improvement of predictive performance in terms of the relative prediction error and the prediction ratio. For example, a low relative prediction error of 0.0189 is observed early during the EM algorithm (the first iteration, E-step), although the likelihood is far from its maximum at this stage; hence, the model fit is still problematic. Despite an imperfect fit, the subsequent M-step yields a relatively high prediction ratio of 1.5298, although the relative prediction error is high, indicating that this value should be generally evaluated only in the context of the overall performance of the estimation procedure.

6 Summary and Conclusions

Despite a large number of parameters to be estimated, many unclassified defects, a majority of layers left uninspected, and the overall complexity of the model, the proposed computational modifications make wafer tomography feasible and reasonably fast. As a result, the yield on a wafer, a lot, or a series of lots can be predicted at any time during the manufacturing process based on the observed defects. All the information about the defects, their location, size, and type, is included in the model, whether it is observed or not. Accuracy of wafer tomography depends on the proportion of inspected layers and classified defects and the overall quality of data, in particular, our ability to identify and handle outliers.

As noted, the number of estimated parameters can be high, including possibly tens of thousands of defect frequencies λ. Due to a very large amount of processed training data, there was no need for regularization. However, in the cases when the number of parameters is compatible with the sample size, one may turn to shrinkage methods and augment our derived likelihood function with a penalty term resulting in a penalized likelihood and enhanced prediction accuracy [25].

A good measure of prediction power is the discrepancy between the predicted yield on functioning chips and the predicted yield on failed chips. Although it may seem that it is a must for an ideal statistical procedure to predict a 100% yield on good chips and a 0% yield on bad chips, such a result is theoretically possible only if yield is a deterministic (non-random) function of defects. Accounting for the uncertainty of failures, prediction limitations on good and failing chips are stated by Lemma 1.

Accurate yield prediction carries several important benefits. Besides planning, business development, and process control considerations, a reliable yield forecast obtained during lot manufacturing can be used to decide whether to continue processing the lot, to scrap it, or to rework the most recent layers.

In addition to its prediction capability, the algorithm results in parameter estimates such as the effects of defects, layers, and other causes and frequency of defects of each type on each layer in each lot. These statistics can be further used for determining the most critical defect-layer combinations, the most influential layers, the most probable fatal defects, etc. One can also predict, given the causal nature of the model, the yield improvement that elimination or reduction of certain defect types can cause, suggesting the optimal yield enhancing strategies.

Appendix

Gradient of the Log Likelihood for the M-Step

Here we derive analytic expressions for $\nabla \log \mathcal{L}$ that are used by the optimization routine during the M-step. It is seen from (7) and (10) that for any parameter $\theta \in \{a(1), \ldots, a(L); b(1), \ldots, b(M); r(1), \ldots, r(J)\}$,

$$\frac{\partial \log \mathcal{L}}{\partial \theta} = \sum_i \left(\frac{\xi_i - \varphi_i}{1 - \varphi_i} \right) \frac{\partial \log \varphi_i}{\partial \theta}.$$

Thus, it remains to compute the partial derivatives of φ_i for each chip i. One has

$$\frac{\partial \log \varphi_i}{\partial b(m)} = \begin{cases} -1 & \text{if chip } i \text{ belongs to lot } m \\ 0 & \text{otherwise;} \end{cases}$$

$$\frac{\partial \log \varphi_i}{\partial a(l)} = -\sum_{k\in C_{il}} r(j_k)x_k - \sum_{k\in U_{il}} x_k \sum_j r(j)w_{jk}/w_k$$

for any layer l *inspected* on the chip i (C_{il} and U_{il} represent the sets of classified and unclassified defects, respectively, j_k is the type of classified defect k, and the quantities w_{jk} and w_k are defined in (14));

$$\frac{\partial \log \varphi_i}{\partial a(l)} = \sum_j \lambda(j, l, m_i) \frac{\partial \psi_{jl}}{\partial a(l)}$$

for all layers l that are *not inspected* on chip i; and

$$\frac{\partial \log \varphi_i}{\partial r(j)} = -\sum_{k\in C_{ij}} a(l_k)x_k - \sum_{k\in U_{il}} a(l_k)x_k w_{jk}/w_k + \sum_{l\in L_i} \lambda(j, l, m) \frac{\partial \psi_{jl}}{\partial r(j)}.$$

The last two expressions contain partial derivatives of the moment generating function (6) of the distribution of defect sizes. They will certainly depend on the model used for this distribution. However, simple *nonparametric estimators* for these partial derivatives are available, as well as for ψ_{jl} itself.

Indeed, since $\psi_{jl} = \mathbf{E}_j^x e^{-a(l)r(j)x}$, it has partial derivatives

$$\frac{\partial \psi_{jl}}{\partial a(l)} = -r(j)\,\mathbf{E}_j^x x e^{-a(l)r(j)x} \quad \text{and} \quad \frac{\partial \psi_{jl}}{\partial r(j)} = -a(l)\,\mathbf{E}_j^x x e^{-a(l)r(j)x}.$$

All three sets of quantities can be estimated by the method of moments from the classified defects,

$$\widehat{\psi_{jl}} = \frac{1}{d_{jl}} \sum_{k\in C_{jl}} e^{-a(l)r(j)x_k}; \tag{15}$$

$$\widehat{\frac{\partial \psi_{jl}}{\partial a(l)}} = -\frac{r(j)}{d_{jl}} \sum_{k\in C_{jl}} x_k e^{-a(l)r(j)x_k}; \quad \widehat{\frac{\partial \psi_{jl}}{\partial r(j)}} = -\frac{a(l)}{d_{jl}} \sum_{k\in C_{jl}} x_k e^{-a(l)r(j)x_k}.$$

This completes the computation of the analytic gradient that is supplied to the optimization routine that maximizes the log-likelihood function with respect to $\boldsymbol{a}$, $\boldsymbol{b}$, and $\boldsymbol{r}$ during the M-step.

Derivation of the E-Step

In this section, we derive recursive estimators of $\lambda(j, l, m)$ for the E-step and prove Eq. (12).

The expected number of (j, l)-defects in (11) consists of three parts: (1) all the detected defects classified to type j on layer l, (2) a suitable portion of unclassified defects that "should" be attributed to type j, and (3) a portion of (j, l)-defects that is expected on wafers where layer l is uninspected.

That is,

$$\hat{\lambda}(j, l, m) = I_m^{-1}\left(d_{jlm} + \sum_{k\in U_{lm}} \mathbf{P}\{j_k = j \mid x_k, \xi_i\} + \sum_{i:l\notin L_i} \mathbf{E}\left\{N_{ijl} \mid \xi_i\right\}\right), \tag{16}$$

where I_m is the total number of chips in lot m.

We compute these three terms separately. The first term is simply the observed number of classified type j defects observed on layer l. For the second term, consider two cases, when the chip containing defect k is good and when it is bad.

Suppose for a moment that an unclassified defect k is in fact of type j and consider the conditional probability

$$\varphi_i(j, k) = \mathbf{P}\{\xi = 1 \mid j_k = j, x_k\}. \tag{17}$$

The only difference between $\log \varphi_i(j, k)$ and $\log \varphi_i$ in (8) is caused by this defect appearing in the set C_{ijl} instead of U_{il}. Hence,

$$\log \varphi_i(j, k) - \log \varphi_i = -r(j)a(l_k)x_k - \log \frac{\sum_{j'} \lambda(j', l, m)\pi_{j'}(x_k)e^{-r(j')a(l_k)x_k}}{\sum_{j'} \lambda(j', l, m)\pi_{j'}(x_k)},$$

so that

$$\varphi_i(j, k) = e^{-r(j)a(l_k)x_k}\varphi_i/\rho_{km},$$

where

$$\rho_{km} = \frac{w_{km}}{v_{km}} = \frac{\sum_{j'} w_{j'km}}{\sum_{j'} v_{j'km}}$$

is the probability for a chip to survive an unclassified defect k, which is independent of the (unknown) defect type j, and v_{jkm}, v_{km}, w_{jkm}, and w_{km} are defined in (13) and (14).

Then, for an *unclassified* defect $k \in U_l$ occurring on a *good* chip i,

$$\begin{aligned}
&\mathbf{P}\{j_k = j \mid x_k, \xi_i = 1\} \\
&= \frac{\mathbf{P}\{j_k = j\}\,\pi(x_k \mid j_k = j)\,\mathbf{P}\{\xi_i = 1 \mid j_k = j, x_k\}}{\sum_{j'} \mathbf{P}\{j_k = j'\}\,\pi(x_k \mid j_k = j')\,\mathbf{P}\{\xi_i = 1 \mid j_k = j', x_k\}} \\
&= \frac{[\lambda(j, l_k, m)/\lambda(l_k, m)]\,\pi_j(x_k)\varphi_i(j,k)}{\sum_{j'} \left[\lambda(j', l_k, m)/\lambda(l_k, m)\right]\pi_{j'}(x_k)\varphi_i(j',k)} \\
&= \frac{v_{jkm} e^{-r(j)a(l_k)x_k}\varphi_i/\rho_{km}}{\sum_{j'} v_{j'km} e^{-r(j')a(l_k)x_k}\varphi_i/\rho_{km}} = \frac{w_{jkm}}{w_{km}}.
\end{aligned}$$

Similarly, for an *unclassified* defect k occurring on a *bad* chip i,

$$\begin{aligned}
&\mathbf{P}\{j_k = j \mid x_k, \xi_i = 0\} \\
&= \frac{[\lambda(j, l_k, m)/\lambda(l_k, m)]\,\pi_j(x_k)\,[1-\varphi_i(j,k)]}{\sum_{j'} \left[\lambda(j', l_k, m)/\lambda(l_k, m)\right]\pi_{j'}(x_k)\left[1-\varphi_i(j',k)\right]} \\
&= \frac{v_{jkm}\left[1 - e^{-r(j)a(l_k)x_k}\varphi_i/\rho_{km}\right]}{\sum_{j'} v_{j'km}\left[1 - e^{-r(j')a(l_k)x_k}\varphi_i/\rho_{k}\right]} = \frac{v_{jkm} - w_{jkm}\varphi_i/\rho_{km}}{v_{km} - w_{km}\varphi_i/\rho_{km}} \\
&= \frac{v_{jkm} - w_{jkm}\varphi_i/\rho_{km}}{v_{km}(1-\varphi_i)} = \frac{v_{jkm}/v_{km} - \varphi_i w_{jkm}/w_{km}}{1-\varphi_i}.
\end{aligned}$$

Hence, the second term of (16), the expected number of type j defects among unclassified defects on layer l, equals

$$\begin{aligned}
\sum_{k \in U_l} \mathbf{P}\{j_k = j | \boldsymbol{x}, \boldsymbol{\xi}\} &= \sum_{i: l \in L_i} \sum_{k \in U_{il}} \left\{ \xi_i \frac{w_{jkm}}{w_{km}} + (1-\xi_i)\frac{v_{jkm}/v_{km} - \varphi_i w_{jkm}/w_{km}}{1-\varphi_i} \right\} \\
&= \sum_{i: l \in L_i} \left\{ \frac{1-\xi_i}{1-\varphi_i} \sum_{k \in U_{il}} \frac{v_{jk}}{v_{km}} + \frac{\xi_i - \varphi_i}{1-\varphi_i} \sum_{k \in U_{il}} \frac{w_{jk}}{w_{km}} \right\}. \qquad (18)
\end{aligned}$$

Finally, we compute the expected number of type j defects on an *uninspected layer* l. This expectation is not just the ratio of corresponding defect frequencies. Although the defect situation on an uninspected layer is hidden, the quality of a chip (ξ_i) is still known, and it should be used in our computation.

Similarly to (17), we define $\varphi_{in}(j,l)$ to be the probability for chip i to be good, despite of its n defects of type j on an uninspected layer l. Sizes of these defects are hidden and thus replaced by the corresponding expectation as in (6). Then, $\log \varphi_{in}(j,l)$ can be obtained from $\log \varphi_i$ by moving the effect of all n defects of type j from the set $\{k \in i,\ l \notin L_i\}$ of defects on uninspected layers to the set C_{ijl} of classified defects, replacing, by the formula of total probability, their missing sizes by expectations ψ_{jl}. That is,

$$\log \varphi_{in}(j,l) = \log \varphi_i + \lambda(j,l)(1-\psi_{jl}) + n \log \psi_{jl}.$$

Then, the expected number of type j defects on an *uninspected layer* of a *good* chip i equals

$$\begin{aligned}
\mathbf{E}\left\{N_{ijl} \mid \xi_i = 1\right\} &= \sum_{n=0}^{\infty} n\, \mathbf{P}\left\{N_{ijl} = n \mid \xi_i = 1\right\} \\
&= \sum_{n} n \frac{\varphi_{in}(j,l) e^{-\lambda(j,l,m)} \lambda^n(j,l,m)/n!}{\varphi_i} \\
&= \sum_{n=0}^{\infty} n \left(e^{\lambda(j,l,m)(1-\psi_{jl})} \psi_{jl}^n\right)\left(e^{-\lambda(j,l,m)} \lambda^n(j,l,m)/n!\right) \\
&= \sum_{n=0}^{\infty} n e^{-\lambda(j,l,m)\psi_{jl}} \left(\lambda(j,l,m)\psi_{jl}\right)^n / n! = \lambda(j,l,m)\psi_{jl}.
\end{aligned} \tag{19}$$

Similarly, for a bad chip i of lot m,

$$\begin{aligned}
\mathbf{E}\left\{N_{ijl} \mid \xi_i = 0\right\} &= \sum_{n=0}^{\infty} n\, \mathbf{P}\left\{N_{ijl} = n \mid \xi_i = 0\right\} \\
&= \sum_{n} n \frac{\left[1-\varphi_{in}(j,l)\right] e^{-\lambda(j,l,m)} \lambda^n(j,l,m)/n!}{1-\varphi_i} \\
&= \frac{1}{1-\varphi_i} \sum_{n=0}^{\infty} \left(1 - \varphi_i e^{\lambda(j,l,m)(1-\psi_{jl})} \psi_{jl}^n\right) e^{-\lambda(j,l,m)} \lambda^n(j,l,m)/(n-1)! \\
&= \frac{1}{1-\varphi_i} \left(\sum_{n=0}^{\infty} e^{-\lambda(j,l,m)} \frac{\lambda^n(j,l,m)}{(n-1)!} - \sum_{n=0}^{\infty} \varphi_i e^{-\lambda(j,l,m)\psi_{jl}} \frac{(\lambda(j,l,m)\psi_{jl})^n}{(n-1)!}\right) \\
&= \frac{1}{1-\varphi_i} \left(\lambda(j,l,m) - \varphi_i \lambda(j,l,m)\psi_{jl}\right) = \frac{\lambda(j,l,m)(1-\varphi_i\psi_{jl})}{1-\varphi_i}.
\end{aligned} \tag{20}$$

Combining (19) and (20), we obtain the third term of (16),

$$\sum_{i:l\notin L_i} \mathbf{E}\{N_{ijl}\,|\,\boldsymbol{\xi}\} = \sum_{i:l\notin L_i} \left\{\xi_i\lambda(j,l,m)\psi_{jl} + \frac{1-\xi_i}{1-\varphi_i}\lambda(j,l,m)(1-\varphi_i\psi_{jl})\right\}$$

$$= \lambda(j,l,m)\sum_{i:l\notin L_i}\left(\frac{\xi_i-\varphi_i}{1-\varphi_i}\psi_{jl} + \frac{1-\xi_i}{1-\varphi_i}\right). \tag{21}$$

Finally, using (18) and (21) in (16) and adding the classified type j defects, we obtain the expression for the refined frequency estimator,

$$\lambda^{(n+1)}(j,l,m) = \left\{d_{jl} + \sum_{i:l\in L_i}\left(\frac{1-\xi_i}{1-\varphi_i}\sum_{k\in U_{il}}\frac{v_{jkm}}{v_{km}} + \frac{\xi_i-\varphi_i}{1-\varphi_i}\sum_{k\in U_{il}}\frac{w_{jkm}}{w_{km}}\right)\right.$$

$$\left. +\lambda^{(n)}(j,l,m)\sum_{i:l\notin L_i}\left(\frac{1-\xi_i}{1-\varphi_i} + \frac{\xi_i-\varphi_i}{1-\varphi_i}\psi_{jl}\right)\right\}/I.$$

Such a refinement of $\lambda^{(n)}(j,l,m)$ for all defect types, layers, and lots completes the E-step.

Prediction on Good and Failed Chips: Proof of Lemma 1

Since $\xi_i \sim \text{Bernoulli}(\varphi_i)$, we have $\mathbf{E}\{\xi\,|\,\varphi\} = \varphi$. Unconditionally, ξ_i are i.i.d. random variables with the *compound* distribution,

$$\mathbf{P}\{\xi=1\} = \int \varphi dF(\varphi), \quad \mathbf{P}\{\xi=0\} = \int (1-\varphi)dF(\varphi).$$

By the strong law of large numbers,

$$\hat{y}_g = \frac{\sum_{i=1}^I \varphi_i\xi_i}{\sum_{i=1}^I \xi_i} \to \frac{\mathbf{E}(\varphi\xi)}{\mathbf{E}(\xi)} = \frac{\mathbf{E}(\varphi;\ \xi=1)}{\mathbf{P}\{\xi=1\}} = \mathbf{E}\{\varphi\,|\,\xi=1\}. \tag{22}$$

By the Bayes formula,

$$dF(\varphi\,|\,\xi=1) = \frac{\mathbf{P}\{\xi=1\,|\,\varphi\}dF(\varphi)}{\int \mathbf{P}\{\xi=1\,|\,\varphi\}dF(\varphi)} = \frac{\varphi dF(\varphi)}{\mathbf{E}_F(\varphi)}. \tag{23}$$

Objectively, we deal with a Bayesian model, where $F(\varphi)$ is a *prior distribution* of φ_i. Then, taking expectation over the *posterior* distribution of φ and combining it with (22), we obtain

$$\lim_{I\to\infty} \hat{y}_g = \mathbf{E}\{\varphi \mid \xi = 1\} = \int \varphi \frac{\varphi}{\mathbf{E}_F(\varphi)} dF(\varphi) = \frac{\mathbf{E}_F(\varphi^2)}{\mathbf{E}_F(\varphi)}.$$

The posterior distribution of φ_i given a bad chip, $\xi = 0$, is considered similarly. In this case, we have

$$F(\varphi \mid \xi = 0) = \frac{\mathbf{P}\{\xi = 0 \mid \varphi\}\ F(\varphi)}{\int \mathbf{P}\{\xi = 0 \mid \varphi\} dF(\varphi)} = \frac{(1-\varphi)F(\varphi)}{1 - \mathbf{E}_F(\varphi)},$$

so that

$$\lim_{I\to\infty} \hat{y}_b = \mathbf{E}\{\varphi \mid \xi = 0\} = \int \varphi \frac{(1-\varphi)}{1 - \mathbf{E}_F(\varphi)} dF(\varphi) = \frac{\mathbf{E}_F(\varphi) - \mathbf{E}_F(\varphi^2)}{1 - \mathbf{E}_F(\varphi)}.$$

Acknowledgments This project resulted from a 1-year academic visit of the first author to the IBM T. J. Watson Research Center and the joint work of IBM Research and Microelectronics divisions. We are thankful to our colleagues at IBM who made this collaboration possible, for their support and encouragement.

References

1. Baron, M., Takken, A., Yashchin, E., Lanzerotti, M.: Modeling and forecasting of defect-limited yield in semiconductor manufacturing. IEEE Trans. Semicond. Manuf. **21**(4), 614–624 (2008)
2. Butte, S., Patil, S.: Big data and predictive analytics methods for modeling and analysis of semiconductor manufacturing processes. In: Proceedings of the 2016 IEEE Workshop on Microelectronics and Electron Devices (WMED), pp. 1–5. IEEE, New York (2016)
3. Chien, C.-F., Chang, K.-H., Wang, W.-C.: An empirical study of design-of-experiment data mining for yield-loss diagnosis for semiconductor manufacturing. J. Intell. Manuf. **25**(5), 961–972 (2014)
4. Chien, C.-F., Liu, C.-W., Chuang, S.-C.: Analysing semiconductor manufacturing big data for root cause detection of excursion for yield enhancement. Int. J. Prod. Res. **55**(17), 5095–5107 (2017)
5. Chien, C.-F., Chen, Y.-H., Lo, M.-F.: Advanced quality control (AQC) of silicon wafer specifications for yield enhancement for smart manufacturing. IEEE Trans. Semicond. Manuf. **33**(4), 569–577 (2020)
6. Dempster, A., Laird, N., Rubin, D.: Maximum likelihood from incomplete data via the EM algorithm. J. Royal Statist. Soc. B **39**(1), 1–38 (1977)
7. Hastie, T., Tibshirani, R., Friedman, J.: The Elements of Statistical Learning: Data Mining, Inference, and Prediction. Springer, New York (2009)
8. Hessinger, U., Chan, W.K., Schafman, B.T.: Data mining for significance in yield-defect correlation analysis. IEEE Trans. Semicond. Manuf. 27(3), 347–356 (2014)
9. Huang, K., Kupp, N., Carulli Jr, J.M., Makris, Y.: Handling discontinuous effects in modeling spatial correlation of wafer-level analog/RF tests. In: Proceedings of the Conference on Design, Automation and Test in Europe, pp. 553–558. EDA Consortium, New York (2013)

10. Hwang, J.Y., Kuo, W., Ha, C.: Modeling of integrated circuit yield using a spatial nonhomogeneous Poisson process. IEEE Trans. Semicond. Manuf. **24**(3), 377–384 (2011)
11. Ketchen, M.B.: Point defect yield model for wafer scale integration. IEEE Circuits and Devices **1**, 24–34 (1985)
12. Krueger, D.C., Montgomery, D.C., Mastrangelo, C.M.: Application of generalized linear models to predict semiconductor yield data using defect metrology data. IEEE Trans. Semicond. Manuf. **24**, 44–58 (2012)
13. Kuo, W., Chien, W.-T.K., Kim, T.: Reliability, Yield, and Stress Burn-in: A Unified Approach for Microelectronics Systems Manufacturing and Software Development. Springer, Berlin (2013)
14. Lin, J.: Constructing a yield model for integrated circuits based on a novel fuzzy variable of clustered defect pattern. Expert Systems with Applications **39**, 2856–2864 (2012)
15. McLachlan, G.J. Krishnan, T.: The EM Algorithm and Extensions. Wiley, New York (2008)
16. Minami, M.: Convergence speed and acceleration of the EM algorithm. The EM Algorithm and Related Statistical Models **170**, 85–94 (2004)
17. Moyne, J., Samantaray, J., Armacost, M.: Big data emergence in semiconductor manufacturing advanced process control. In: Proceedings of the 2015 26th Annual SEMI Advanced Semiconductor Manufacturing Conference (ASMC), pp. 130–135. IEEE, New York (2015)
18. Ng, S.K., Krishnan, T., McLachlan, G.J.: The EM algorithm. In: Handbook of Computational Statistics, pp. 139–172. Springer, Berlin (2012)
19. Parra, L., Barrett, H.H.: List-mode likelihood: EM algorithm and image quality estimation demonstrated on 2-D pet. IEEE Trans. Med. Imaging **17**(2), 228–235 (1998)
20. Singh, E.: Impact of radial defect clustering on 3D stacked IC yield from wafer to wafer stacking. In: 2012 IEEE International Test Conference, pp. 1–7. IEEE, New York (2012)
21. Stapper, C.H.: On yield, fault distributions, and clustering of particles. IBM J. Res. Develop. **30**, 326–338 (1986)
22. Tam, W.C.J., Blanton, R.D.S.: LASIC: Layout analysis for systematic IC-defect identification using clustering. IEEE Trans. Comput. Aided Des. Integr. Circuits Syst. **34**(8), 1278–1290 (2015)
23. Tikkanen, J., Siatkowski, S., Sumikawa, N., Wang, L.-C., Abadir, M.S.: Yield optimization using advanced statistical correlation methods. In: 2014 International Test Conference, pp. 1–10. IEEE, New York (2014)
24. Tirkel, I., Rabinowitz, G.: Modeling cost benefit analysis of inspection in a production line. Int. J. Prod. Econ. **147**, 38–45 (2014)
25. Van Houwelingen, J.C.: Shrinkage and penalized likelihood as methods to improve predictive accuracy. Statistica Neerlandica **55**(1), 17–34 (2001)
26. Venkataraman, A., Koren, I.: Determination of yield bounds prior to routing. In: Proceedings of the 1999 IEEE International Symposium on Defect and Fault Tolerance in VLSI Systems, pp. 4–13 (1999)
27. Wang, J., Cao, H.: Range selection of critical area optimization. In: Proceedings of the 2013 International Conference on Anti-Counterfeiting, Security and Identification (ASID), pp. 1–4. IEEE, New York (2013)
28. Wang, H., Li, B., Tong, S.H., Chang, I.-K., Wang, K.: A discrete spatial model for wafer yield prediction. Qual. Eng. **30**(2), 169–182 (2018)
29. Warner, R.M.: Applying a composite model to the IC yield problem. IEEE J. Solid-State Circuits SC-9, 86–95 (1974)
30. Watanabe, M., Yamaguchi, K.: The EM Algorithm and Related Statistical Models. Marcel Dekker, New York (2003)
31. Zhu, J.-J., Luo, X.-H., Chen, L.-S., Ye, Y., Yan, X.-l.: A fast method for extracting and optimizing critical area to improve yield. Journal of Circuits and Systems, **2**, 371–375 (2013)

Uncertainty Quantification Based on Bayesian Neural Networks for Predictive Quality

Simon Cramer, Meike Huber, and Robert H. Schmitt

Abstract In the context of production metrology, the field *Predictive Quality* develops methods based on statistics and machine learning to predict quality characteristics from process data. In prior work, conventional machine learning methods such as feed-forward neural networks have been successfully applied. Yet, an uncertainty quantification for the prediction is not provided. Therefore, it is not possible to prove the suitability of the applied predictive quality methods for quality inspections. However, we can estimate the uncertainty by taking a Bayesian perspective and utilizing suitable algorithms.
Here we define *Prediction of Quality Characteristics* (PQC), which is the foundation for every Predictive Quality application. We extend our definition of PQC into a general Bayesian framework to interpret predicted quality characteristics. As an example, we show how Bayesian neural networks are applied to PQC to estimate the uncertainty of every prediction. We interpret the results in the industrial context and determine the suitability of the PQC method.
Our results demonstrate that the application of Bayesian methods is highly promising to get Predictive Quality recognized in industry as an accredited method for quality inspections.

Keywords Predictive quality · Prediction of quality characteristics · Bayesian neural networks · Uncertainty quantification

---

S. Cramer (✉) · M. Huber
WZL, RWTH Aachen University, Aachen, Germany
e-mail: s.cramer@wzl.rwth-aachen.de; m.huber@wzl.rwth-aachen.de

R. H. Schmitt
WZL, RWTH Aachen University and Fraunhofer IPT, Aachen, Germany

Fraunhofer IPT, Aachen, Germany
e-mail: r.schmitt@wzl.rwth-aachen.de

A. Steland, K.-L. Tsui (eds.), *Artificial Intelligence, Big Data and Data Science in Statistics*, https://doi.org/10.1007/978-3-031-07155-3_10

1 Prediction of Quality Characteristics

As Industry 4.0 strategies are rolled out progressively, process data is becoming accessible in large amounts. The available data offers engineers and scientist innumerable opportunities to analyze and improve production processes. Some exemplary applications are predictive maintenance and process mining [23]. The research field *Predictive Quality* describes the user's ability to optimize product and process-related quality characteristics by using data-driven forecasts as a basis for actions to be taken [5]. The foundation for all predictive quality applications is the *prediction of quality characteristics* (PQC).

The prediction those characteristics can be regarded as a virtual inspection process, as it replaces a physical inspection.

In conventional physical inspection processes for determining product quality, a specific operation (e.g., measuring or gauging) is used to decide whether a quality characteristic meets a pre-defined requirement. In order to make this decision, it is checked whether the considered quality characteristic lies within previously defined specification limits.

Since every inspection process is subject to uncertainties (e.g., due to the uncertainty of the underlying measurement process), the decision whether the characteristic meets the requirement is also uncertain. Due to the uncertainty of inspection results, an erroneous decision is possible. Characteristics that are within the specification limits are rejected (α-error), and characteristics that are outside the specification limits are accepted (β-error). Both errors entail technical, economic, and legal consequences. To reduce the risk of a wrong decisions, the limits of conformity are narrower than the specification limits to account for the uncertainty of the inspection process (e.g., the measurement uncertainty). To guarantee a product within the specification limits, the process variance, the variance of the test process, and the specification limits must be aligned according to DIN EN ISO 14253-1 (see Fig. 1) [41].

In order to consider an inspection process as *suitable*, it must be ensured that the quotient of uncertainty of the inspection process U and tolerance of the considered quality characteristic T does not exceed a certain threshold. This threshold value is defined differently in various standards and guidelines (see MSA [18], VDA5 [40], ISO 22514-7 [42]). As a rule of thumb, the *golden rule of metrology* states that the ratio U/T should not be greater than one-tenth to one-fifth [28, 39]. To deploy PQC in industry, the suitability of the (virtual) inspection process must be guaranteed. Hence, the uncertainty of the underlying model must be quantified. The determination of the uncertainty of a model is a typical example from the mathematical field of Uncertainty Quantification [37].

Uncertainty Quantification (UQ) focuses on the quantitative characterization of uncertainties in both real and computer-based applications. UQ methods are used to quantify the probability of certain results if some or all input variables are uncertain. A mathematical model is used to describe the system's behavior extracted from the measured data. UQ problems are divided into two classes: forward uncertainty

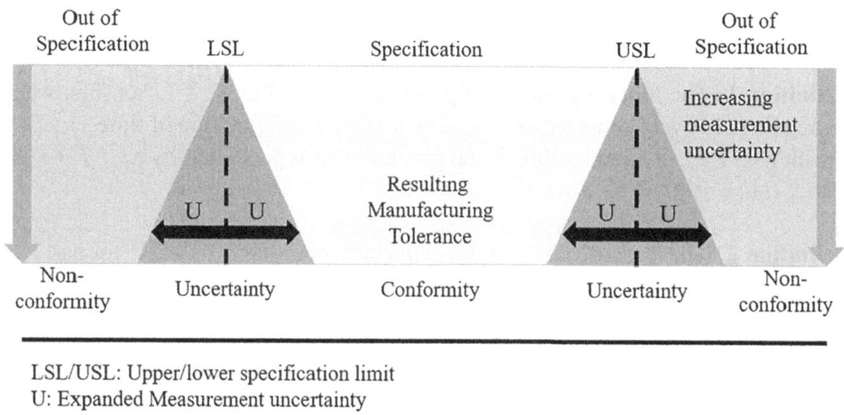

Fig. 1 Limitation of the specification range due to measurement uncertainty according to ISO 14253-1 (see [28, 41])

propagation and inverse uncertainty quantification. Forward uncertainty propagation aims to estimate the different sources of uncertainty, acting on a model to predict an overall uncertainty of the system response. Inverse uncertainty quantification involves estimating the so-called bias correction (i.e., the discrepancy between the measured value and the model) and unknown parameters of the model [6, 37].

In PQC, we estimate the parameters for a given model structure from data. The data used for parameter estimation are usually measurement data and, therefore, affected by uncertainty [28]. For a given model structure and some data, the objective is to minimize the model prediction's uncertainty by setting the parameters appropriately. The determination of uncertainty in the field of predictive quality can, therefore, be considered an inverse uncertainty quantification problem by definition [37].

2 Definition of Prediction of Quality Characteristics

We first define PQC in a deterministic way before introducing a Bayesian perspective. The definition is provided for a single product in discrete manufacturing. Thus, the index $i \in \mathbb{N}$ identifies a unique part of one product type. With minor modifications, the definition of PQC can be extended to the process industry. The foundation for any machine learning (ML) application is a sufficient database. In the case of PQC it contains the *quality characteristics* and the *process data* on a per-part basis. PQC is an inverse problem, as we want to infer a function H from some infinite-dimensional function space predicting the quality characteristics from process data [37].

We define process data and quality characteristics before constructing a database and deriving the resulting inverse problem.

Definition 1 The *process data* x_i for part i is generated by $m \in \mathbb{N}$ sensors, where the readings of every sensor s_j $0 \leq j < m$ are given as a function of time $s_j : T \rightarrow S$ with $t \in T \subset \mathbb{R}^+$. Accordingly the process data is modelled by $x_i : T \rightarrow S^m$ with $x_i(t) := [s_0(t), \ldots, s_m(t)]^T$.

Definition 2 The measurements of the *quality characteristics* $y_i \in \mathbb{R}^n$ for part i are given by $n \in \mathbb{N}$ measurements, where every measurement v_l $0 \leq l < n$ is a fixed value $y_i := [v_l]^T$.

In comparison to the process data x_i we assume that the quality characteristics are time-invariant—or measured only once. Based on Definitions 1 and 2 the data for a unique part i is given by the tuple (x_i, y_i). Hence, we denote $\mathcal{D} := \{(x_i, y_i)\}$ $(0 \leq i < k)$ the database for a given PQC application with $k \in \mathbb{N}$ entries.

Given the database $\mathcal{D}$ we want to determine the parameters $\mathbf{w} \in \mathbf{W}$ of the mapping $H_\mathbf{w}$ with

$$y_i = H_\mathbf{w}(x_i) \quad \forall (x_i, y_i) \in \mathcal{D}. \tag{1}$$

Thus, the inverse problem has become a *parameter estimation* problem, which is usually ill-posed [37]. A common approach is the computation of a least-squares solution:

$$\arg\min_{\mathbf{w}} ||y_i - H_\mathbf{w}(x_i)||^2_{\mathcal{D}}. \tag{2}$$

Note here that some kind of regularization usually improves the solution as noise in the data is considered [37]. The presence of noise in the data motivates the expansion of this deterministic interpretation of the parameter estimation using a Bayesian perspective.

The measurement of a quality characteristic is subject to measurement uncertainty; thus, it is better represented by a random variable. All sensor readings are also subject to measurement uncertainty and hence – to preserve the time dependency – interpreted as a stochastic process, which we define as follows:

Definition 3 Let $u(t, \omega) : T \times \Omega \rightarrow S$ be a *stochastic process*, where $t \in T \subset \mathbb{R}^+$ and $\omega \in \Omega$. Here Ω is the sample space of the probability space $(\Omega, \mathcal{F}, P)$ with $\mathcal{F}$ being a σ-algebra and P a probability measure.

Accordingly we give the definitions of process data and quality characteristics in the Bayesian sense:

Definition 4 The *process data* X is generated by $m \in \mathbb{N}$ sensors, where the sensor readings u_j $0 \leq j < m$ are given by a stochastic process. Accordingly the process data is modelled by $X : T \times \Omega^m \to S^m$ with $X(t, \bar{\omega}) := [u_j(t, \omega_j)]^T$ where $\bar{\omega} := [\omega_j]^T$.

Definition 5 The measurements of the *quality characteristics* $Y : \Omega^n \to \mathbb{R}^n$ are given by $n \in \mathbb{N}$ measurements, where every measurement v_l $0 \le l < n$ is a random variable $Y(\bar{\omega}) := [v_l(\omega_l)]^T$ where $\bar{\omega} := [\omega_l]^T$.

Based on Definition 4 and 5 the data of a single part i is given by (x_i, y_i), where $(x_i = X(\cdot, \bar{\omega}_i), y_i = Y(\cdot, \bar{\omega}_i))$ is a realization of (X, Y). Taking a Bayesian point of view, Eq. (1) introduces the conditioned random variable $Y|X, \mathbf{w}$ and the solution to the inverse problem is the conditioned random variable $\mathbf{w}|\mathcal{D}$ [37]. The parameters can be determined with maximum likelihood estimation (MLE) as

$$\mathbf{w}^{MLE} = \arg\max_{\mathbf{w}} \log P(\mathcal{D}|\mathbf{w}) = \arg\max_{\mathbf{w}} \sum_i \log P(y_i|x_i, \mathbf{w}) \tag{3}$$

or by introducing a prior $P(\mathbf{w})$ on the parameters and finding the maximum a posteriori (MAP) parameters

$$\mathbf{w}^{MAP} = \arg\max_{\mathbf{w}} \log P(\mathbf{w}|\mathcal{D}) = \arg\max_{\mathbf{w}} \log P(\mathcal{D}|\mathbf{w}) + \log P(\mathbf{w}). \tag{4}$$

Example 1 Let the product have $n = 2$ quality characteristics, and the total amount of sensors on the involved machinery be $m = 3$. Then the database $\mathcal{D}$ is constructed from Table 1. For sensor $j = 0$ there are two readings, for sensor $j = 1$ there is one reading and for sensor $j = 2$ there are three readings. We append all sensor readings into a single vector $x \in \mathbb{R}^6$. The same procedure applies to the quality characteristics, which form the vector $y \in \mathbb{R}^2$.

Assume that $H_{\mathbf{w}}(x) := \mathbf{w}x = y$ is a linear operator with $\mathbf{w} \in \mathbb{R}^{2\times 6}$, then the least-squares solution $\hat{\mathbf{w}}$ according to Eq. (2) is

$$\hat{\mathbf{w}} \approx \begin{pmatrix} -12.92 & -89.1 & 1.69 & 3.35 & -3.17 & 24.28 \\ 0.93 & 0.12 & -0.002 & 0.04 & 0.002 & -0.16 \end{pmatrix}. \tag{5}$$

Table 1 Database entries for an exemplary predictive quality application

	x_i			y_i	
Part i	0	1	2	0	1
0	[0.49, 0.4]	[29]	[3.7, 3.7, 3.8]	100.1	0.02
1	[0.52, 0.39]	[27]	[3.6, 3.8, 3.9]	98.99	0.03
2	[0.5, 0.42]	[30]	[3.7, 3.6, 3.8]	100.2	0.03
3	[0.49, 0.37]	[29]	[3.6, 3.7, 3.7]	100.01	0.028
4	[0.52, 0.4]	[27]	[3.6, 3.2, 3.9]	100	0.03
5	[0.51, 0.42]	[30]	[3.5, 3.6, 3.8]	99.4	0.031

3 State of Uncertainty Quantification for Predictive Quality

The formal proof of suitability requires a determination of the measurement uncertainty. We present the results of our literature review regarding the prediction of quality characteristics and on uncertainty quantification in (deep) machine learning. The UQ methods are designated keystones to provide a measurement uncertainty for PQC applications.

Current State of Predictive Quality The quality of product depends on the interaction of the individual production steps and the condition of the components/machinery and material characteristics. Due to the increasing complexity in production processes, the number of interactions between individual processes is rising. Further, the increasing individualization of products leads to a significant increase in process variance [5].

To improve the understanding of products and processes in production engineering, data analytics methods are used to extract information from data and derive actions based on this information [15, 35]. In this sense, data analytics describes the steps of data investigation, data understanding, and knowledge acquisition, which aim to uncover new relationships within the production process [11]. There are many different methods for the implementation of this decision support, starting with statistical methods up to complex machine learning models, which differ in their application and depend on various factors such as purpose, expertise, and available resources. Data analytics methods can be categorized as descriptive analytics, diagnostic analytics, predictive analytics, or prescriptive analytics. The categories can be seen as steps in the data analysis, which partly rely on each other [26].

Considering the categories, PQ focuses on the application of *predictive analytics* to determine product quality based on process data [5]. Besides considering data from different process steps, existing information on intermediates and the individual assembly can also be taken into account. This enables a comprehensive optimization of the production process.By including data from product usage, the fulfillment of customer requirements can be increased [16, 36].

In recent years, the use of ML algorithms for PQC has been investigated in a manifold of applications. Especially the use of neural networks has shown potential for predicting quality characteristics, as they are capable of mapping and detecting complex cause-effect dependencies while the user is not required to contribute a high amount of expert knowledge [28, 34]. For example, Chen et al. used a back-propagation neural network algorithm and the Taguchi method for quality prediction in plasma-enhanced chemical vapor deposition for semiconductor manufacturing already in 2007 [12]. Ogordnyk et al. introduce a neural networks approach for PQ in the injection molding process. The task here was to classify the product quality based on 18 machine and process parameters [30]. Baturynska et al. describe a prediction model for selective laser sintering. They use neural networks to predict the deviation of manufactured parts in three dimensions depending on their orientation and positioning in the 3D printer [3].

The examples have in common that a model is set up to predict quality characteristics without quantifying the model's uncertainty. Thus, no proof of suitability is obtained, making the use as an inspection tool in an industrial environment challenging. There are, however, machine learning methods which can be used to quantify the uncertainty of the model. These are introduced in the following.

Uncertainty Quantification in (Deep) Machine Learning In the rise of (deep) machine learning since the 2010s, the importance of UQ has been underestimated in the scientific community. As adoption of ML progresses in industrial and consumer applications, safety and security regulations make some types of UQ necessary: verification, robustness, and interpretability [13]. Verification of a ML system provides formal guarantees about its behavior [8, 33, 44]. The robustness (i.e., the reaction to novel/noisy data) is highly relevant for industrial applications, as self-learning robots, and consumer applications, as autonomous vehicles [10, 27, 32]. Interpretability is another active field, where researchers try to understand why an ML system behaves a certain way [31]. We argue that verification and robustness are a form of UQ and that at least a subset of interpretability can be classified as UQ. In all cases, uncertainty in the model or the data are investigated.

Uncertainty in the data and the model are studied using Bayesian approaches since 1989. Early examples of Bayesian learning and Bayesian approaches to neural networks are [25] and [22]. In the 1980s, data sets were significantly smaller than today, and computational power was expensive. Since, the definition of UQ has been significantly expanded. Sullivan et al. consider the treatment of all uncertainties in real and computer-based applications [37]. Especially in the simulation community, where finite element and finite volume methods and their variants are commonly used, UQ did not gain traction until the early 2000s [43]. This was mainly due to the curse of dimensionality and the lack of computational power to perform the simulations for all parameter sets to be investigated [4]. The development of improved methods (e.g., sparse collocation) opened novel possibilities to overcome the curse of dimensionality and explore large parameter spaces efficiently [37].

In deep learning, there are three main movements for UQ [9]. There is Concrete Dropout [14]. The dropout rate becomes a learnable parameter, and nodes are dropped during the evaluation. Thus, a sample from a posterior distribution is generated from a single neural network by randomly omitting a certain percentage of neurons in each layer at each evaluation. This method is an extension to Dropout, which is used as a regularization method to prevent overfitting during model training [19]. Secondly, Deep Ensembles, as introduced in [24], are more sophisticated than Concrete Dropout. Depending on the algorithm's variant, multiple neural networks are trained with different initializations and on different data subsets. At the evaluation, the outputs of all the neural networks are interpreted as samples from a posterior distribution. If we expand the number of models to infinity, we converge to Bayesian Neural Networks (BNN). For a BNN, the weights of each layer are represented by a probability distributions [17]. These networks are evaluated by sampling multiple times from the posterior distributions. In [20] a different classification is discussed, which takes other approaches into account that do not apply to PQC.

BNN are capable of representing aleatoric uncertainty (e.g., variability in the data) and epistemic uncertainty (e.g., model neglecting effects or missing data) via the posterior distribution [7]. This is a crucial feature for PQC applications as by Definitions 4 and 5 we have (commonly) unknown uncertainty in our data and no indication whether an employed model structure is sufficiently expressive. Even though we have seen successful applications of neural networks to PQC (cmp. [3, 12, 30] and more), assumptions regarding the structure or the hyperparameters of the models may be inherently flawed. BNN are successfully applied to various disciplines as physics [38], civil engineering [1], and others [2, 21, 45]. The BNN have shown excellent results, not only on theoretical toy problems (cmp. [7]) but in real world applications. Thus, we focus on BNN given their benefits and apply them to production engineering, and in particular to PQC. We demonstrate briefly how we apply BNNs to PQC, when predicting a quality characteristic $\hat{y}$ from process data $\hat{x}$.

The (posterior) predictive distribution of the unknown value $\hat{y}$ for the test item $\hat{x}$ is given by $P(\hat{y}|\hat{x}) = \mathbb{E}_{P(\mathbf{w}|\mathcal{D})}\left[P(\hat{y}|\hat{y}, \mathbf{w})\right]$. The unknown distribution $P(\mathbf{w}|\mathcal{D})$ can be rewritten using Bayes' theorem:

$$P(\mathbf{w}|\mathcal{D}) = \frac{P(\mathcal{D}|\mathbf{w})P(\mathbf{w})}{P(\mathcal{D})}, \tag{6}$$

where $P(\mathbf{w})$ is the prior on the weights, $P(\mathcal{D})$ is a normalizing constant, and $P(\mathcal{D}|\mathbf{w})$ is the likelihood of observation. To enable PQC in industrial settings, the predicted distribution $P(\hat{y}|\hat{x})$ requires a small variance σ^2. However, this is not a specific goal of training a BNN since this method aims to approximate the distribution based on the given data. Hence the ambitions of quality engineers and mathematicians are not necessarily aligned.

There is not yet a consensus on how to quantify the quality of uncertainty quantification. Standard measures for a good fit of the posterior are the average marginal-log-likelihood, the prediction interval coverage probability, or the mean prediction interval width. However, Yao et al. show that these measures depend on the inference method used to determine the posterior distribution; we refer to [46] for a discussion of this matter.

Interim Conclusion As detailed above, ML algorithms are successfully applied to PQC applications. In special use cases, we even see deployments in industrial applications even though uncertainties are not considered. Further, we established that UQ essential part for PQC and almost all other ML applications outside of laboratories.

To accomplish the overall goal to certify PQC methods as an inspections process, the application of UQ on PQC methods is imminent. We focus our upcoming research on BNN, as we see them as the most comprehensive and expressive method.

4 Application of Bayesian Neural Networks to the Prediction of Quality Characteristics

We apply a BNN to an injection molding process of a thin-walled thermoplastic part. In expert interviews, 14 process parameters (e.g., tool temperature, cycle time, pressure) were identified, each of which is recorded with one sensor. Hence, the machine provides $m = 14$ sensors for process data. We focus on $n = 1$ quality characteristic, i.e., a length of the exemplary part with a nominal value of 72.6 mm. The database $\mathcal{D}$ was generated using a full-factorial design of experiments (DoE), where machine settings are explicitly varied, with $k = 600$ experiments. The measurements of the quality characteristic were performed on a coordinate-measuring machine, whose suitability was proven by a Gage R&R Study (MSA) in advance [29].

The data quality is excellent, as it was manually verified during the recording and before model training. All sensors and the quality characteristic are scaled to the interval [0, 1] to facilitate efficient model training. The original scaling is used for the interpretation in the industrial context in Sect. 4.1.

We use a feed-forward neural network with two hidden layers and leaky ReLU activation functions. The first hidden layer has four nodes, while the second hidden layer has two nodes. The second layer's output is used to parametrize a normal distribution $\mathcal{N}(\mu, \sigma)$: the first node is interpreted as the mean μ, while the second node is understood as the variance σ.

Comparably to [7], we use a prior $P(\mathbf{w})$ on our weights $\mathbf{w}$ and fit a posterior $P(\mathbf{w}|\mathcal{D})$. A prior is placed on the weights $P_t(\mathbf{w}) = \prod_j \mathcal{N}(\mathbf{w}_j|t_j, \sigma_p)$ where $\mathcal{N}(x|\mu_p, \sigma_p)$ is the Gaussian density evaluated at x with mean μ_p and variance σ_p. The prior is learnable as the means t_j are fitted during training, while $\sigma_p = 1$ is fixed. We use a Gaussian variational posterior with trainable mean and variance.

The network is trained for 1250 epochs with a learning rate of 0.001 using the Adam optimizer. The other hyperparameters of the optimizer are the default values.[1] For the loss L we use the sum of the Kullback–Leibler divergence from both hidden layers and add the negative log-likelihood:

$$L = KL_1 + KL_2 + \mathbb{E}_{q_1(\mathbf{w}_1|\theta_1), q_2(\mathbf{w}_2|\theta_2)}\left[log P(\mathbf{w}|\mathcal{D})\right]. \tag{7}$$

Here $KL_i = KL\left[q_i(\mathbf{w}_i|\theta_i)||P(\mathbf{w}_i)\right]$ where $i = 1, 2$ indexes the hidden layers and θ_i are the parameters of a distribution on the weights. We keep the notation according to [7] and refer the interested reader for details. The loss L over the 1250 epochs is given in Fig. 2. After plateauing for about 1000 epochs, a final drop occurs over another 200 epochs before optimal performance is reached.

We train the BNN on 540 data points ($\approx 90\%$) and randomly select 60 ($\approx 10\%$) points for the evaluation. We sample the trained BNN 5000 times for each evaluation

[1] https://www.tensorflow.org/api_docs/python/tf/keras/optimizers/Adam.

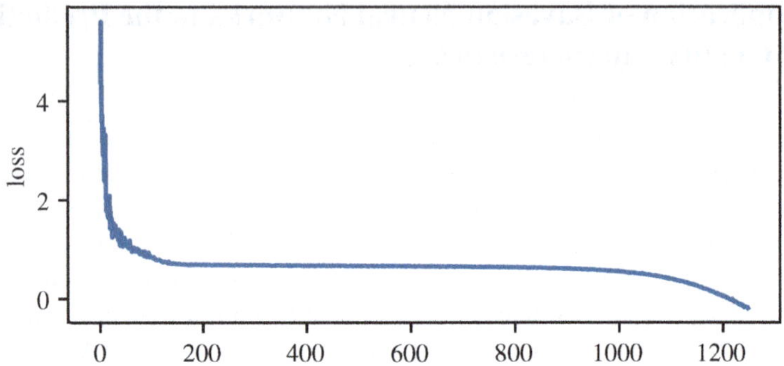

Fig. 2 Loss L during the training with 1250 epochs

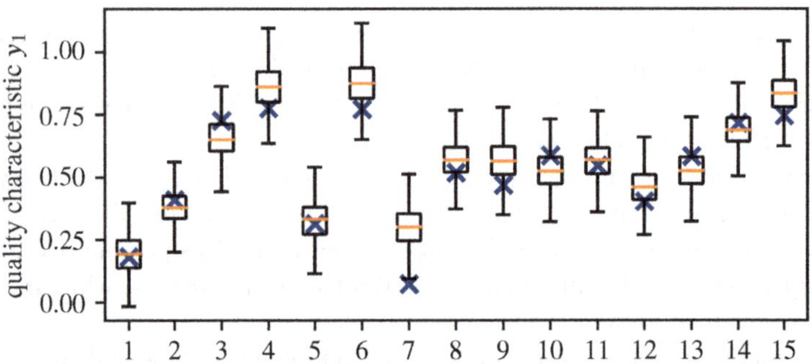

Fig. 3 Quality characteristic y_1 (blue) with a box plot of the prediction $H_{\mathbf{W}}(x)$ based on 5000 model evaluations for 15 samples from the test data set

point to generate as many pairs (μ_i, σ_i) for the parametrized normal distribution. Figure 3 depicts the means μ_i in a box plot for the first 15 evaluation points and give the results for the first 10 as tabular data in Table 2. The actual quality characteristics y_1 are given in blue in the box plot for comparison. The mean absolute error (MAE) between the mean of means $\frac{1}{5000}\sum_{i=0}^{5000}\mu_i$ and the actual value y_1 is ≈ 0.1814. In relation to the size of the data set, this is a reasonably low MAE. In Fig. 3 only sample $i = 7$ is an outlier regarding the mean of means. A more extensive data set would allow more rigorous training of the BNN and yield a better MAE. We provide code and the scaled data set in our GitHub repository.[2]

[2] https://github.com/predictive-quality/bnn-example.

Table 2 Quality characteristic y_1 with the mean prediction $\mathbb{E}[H_{\mathbf{W}}(x)]$ and the variance of the prediction $\mathbb{V}[H_{\mathbf{W}}(x)]$ after the training

y_1	$\mathbb{E}[H_{\mathbf{W}}(x)]$	$\mathbb{V}[H_{\mathbf{W}}(x)]$
0.30239546	0.2830744	0.00976317
0.48470376	0.43315677	0.00700909
0.30063383	0.27861683	0.00965479
0.15363564	0.1343955	0.00571805
0.49191045	0.5456889	0.00547123
0.35253981	0.356259	0.00638528
0.30983144	0.296187	0.00978199
0.32386012	0.26562449	0.00686061
0.51326475	0.51090966	0.00551332
0.35963779	0.36319379	0.0046631

Table 3 Quality characteristic y_1 with the mean prediction $\mathbb{E}[H_\theta(x)]$ and the variance of the prediction $\mathbb{V}[H_\theta(x)]$ after transformation to the original scale

y_1	$\mathbb{E}[H_\theta(x)]$	$\mathbb{V}[H_\theta(x)]$
72.15644055	72.1800606	0.01459118
72.33991666	72.402933	0.01047518
72.15099116	72.177907	0.0144292
71.97468018	71.9982013	0.0085457
72.47748751	72.4117432	0.00817683
72.24590892	72.2413622	0.00954288
72.17247074	72.1891511	0.01461932
72.13510798	72.2063012	0.01025328
72.43496978	72.4378489	0.00823973
72.25438673	72.2500395	0.00696906

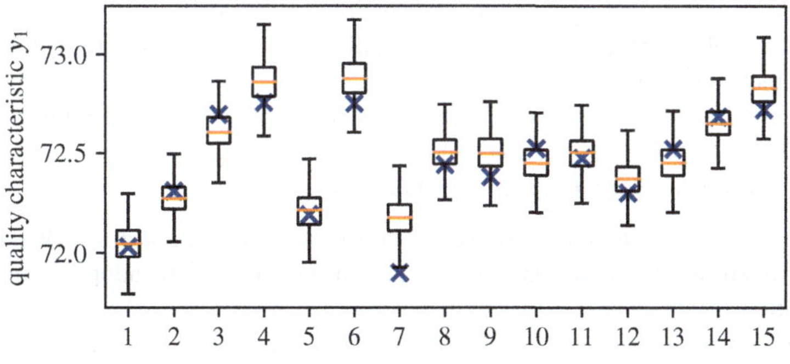

Fig. 4 Quality characteristic y_1 (blue) in its original scale with a box plot of the prediction $H_{\mathbf{W}}(x)$ based on 5000 model evaluations for 15 samples from the test data set

4.1 Interpretation in the Industrial Context

For the industrial practitioner, the raw results of the BNN need further interpretation. Primarily, we have to restore the original scaling to evaluate the PQC in context. In Table 3 and Fig. 4 the predicted values are restored to their original scaling. It is notable how the variance decreases after the rescaling. This does not indicate a better model performance but is rather due to the dependency of variance on the mean. Similarly, the MAE decreases to ≈ 0.0641.

To prove the suitability for this virtual inspections process, the *golden rule of metrology* according to which the ratio $\frac{U}{T}$ of the uncertainty of measurements U to tolerance T shall not be greater to one-tenth to one-fifth [39]. For our example, we can interpret the 2σ-interval γ of $H_{\mathbf{W}}(x)$ as the uncertainty of measurement. Then with $\mathbb{V}[H_{\mathbf{W}}(x)] < 0.0167$:

$$\gamma = 2\sqrt{\mathbb{V}[H_{\mathbf{W}}(x)]} \leq 0.258. \tag{8}$$

Given $T = 0.6$ and choosing $U = \lceil \gamma \rceil$, we derive

$$\frac{U}{T} \leq \frac{0.258}{0.6} = 0.43 \overset{!}{\leq} 0.2. \tag{9}$$

Thus, based on this conservative estimate of the uncertainty of measurement, this BNN is not suitable as an inspection process. However, the following aspects need further consideration:

- Using a more advanced inference method (e.g., Hamiltonian Monte Carlo) can better approximate the posterior and generate more favorable results regarding the suitability.
- As the database was generated by a DOE, the process variation is deliberately high. This is in stark contrast to a real production environment, where the variation is usually low, and process capability is ensured.
- The size of the database is relatively small compared to the number of trainable parameters (≈ 210) in the BNN.
- The hyperparameters have a significant influence on the performance of the BNN. Deliberate, application-specific manual tuning or the use of AutoML-methods could guarantee proof of suitability.

Overall, we are certain that BNN are a well-suited method for PQC, but we openly acknowledge that more research is necessary before adopting industrial applications.

Furthermore, for a formal evaluation of the suitability, the measurement uncertainty must be determined by an approved procedure as the GUM or the VDA 5 (see [39] for details). However, none of these procedures considers algorithms based on process data. Many aspects from physical inspection procedures are transferable to PQC, yet some error sources (e.g., numerical concerns) are not addressed. As the adoption and development of PQC methods progress, the process to determine suitability will be extended as well.

5 Concluding Remarks

We identified the prediction of quality characteristics as the fundamental foundation of every predictive quality method. To give a framework for future research, we provided a formal definition of *prediction of quality characteristics*. Further, we established PQC as a virtual inspection process, which can complement and/or reduce costly physical inspections. For every inspection process, a proof of suitability is necessary, which requires the determination of the measurement uncertainty of the underlying method. Hence we added a Bayesian perspective to our definition to PQC, to consider model- and data-inherent uncertainties.

Based on our literature review, we reason that existing machine learning methods, as BNN, can provide an adequate uncertainty estimation. The uncertainty estimates are a decisive keystone to establish PQC as a virtual inspection process and permit proof of suitability. As a showcase, we applied a BNN to an injection molding process and give several hints on how to improve the uncertainty estimate for future applications. To facilitate adoption in industry, we advocate for a revision of standards as the VDA5 or the ISO 22514-7 to accommodate for virtual inspection processes.

Acknowledgments Funded by the Deutsche Forschungsgemeinschaft (DFG, German Research Foundation) under Germany's Excellence Strategy—EXC-2023 Internet of Production—390621612.

The data set was created in cooperation with Festo Polymer GmbH.

References

1. Arangio, S., Beck, J.L.: Bayesian neural networks for bridge integrity assessment. Struct. Control. Health Monit. **19**(1), 3–21 (2012). https://doi.org/10.1002/stc.420
2. Auld, T., Moore, A.W., Gull, S.F.: Bayesian neural networks for internet traffic classification. IEEE Trans. Neural Netw. **18**(1), 223–239 (2007). https://doi.org/10.1109/TNN.2006.883010
3. Baturynska, I., Semeniuta, O., Wang, K.: Application of machine learning methods to improve dimensional accuracy in additive manufacturing. In: Wang, K., Wang, Y., Strandhagen, J.O., Yu, T. (eds.) Advanced Manufacturing and Automation VIII, Lecture Notes in Electrical Engineering, pp. 245–252. Springer, Singapore (2019). https://doi.org/10.1007/978-981-13-2375-1_31
4. Bellman, R.: Dynamic Programming. Princeton University Press, Princeton, NJ (1984)
5. Bergs, Thomas: Internet of Production—Turning Data into Value (2020). https://doi.org/10.24406/IPT-N-589615
6. Biegler, L.T. (ed.): Large-scale inverse problems and quantification of uncertainty. In: Wiley series in computational statistics. Wiley, Chichester, West Sussex (2011)
7. Blundell, C., Cornebise, J., Kavukcuoglu, K., Wierstra, D.: Weight Uncertainty in Neural Networks (2015). ArXiv: 1505.05424
8. Borg, M., Englund, C., Wnuk, K., Duran, B., Levandowski, C., Gao, S., Tan, Y., Kaijser, H., Lönn, H., Törnqvist, J.: Safely Entering the Deep: A Review of Verification and Validation for Machine Learning and a Challenge Elicitation in the Automotive Industry (2018). ArXiv: 1812.05389

9. Caldeira, J., Nord, B.: Deeply uncertain: comparing methods of uncertainty quantification in deep learning algorithms. Machine Learning: Science and Technology **2**(1), 015002 (2020). https://doi.org/10.1088/2632-2153/aba6f3. ArXiv: 2004.10710
10. Carlini, N., Wagner, D.: Towards Evaluating the Robustness of Neural Networks. arXiv:1608.04644 [cs] (2017). ArXiv: 1608.04644
11. Cattaneo, L., Fumagalli, L., Macchi, M., Negri, E.: Clarifying data analytics concepts for industrial engineering. IFAC-PapersOnLine **51**(11), 820–825 (2018). https://doi.org/10.1016/j.ifacol.2018.08.440
12. Chen, W.C., Lee, A.H.I., Deng, W.J., Liu, K.Y.: The implementation of neural network for semiconductor PECVD process. Expert Systems with Applications **32**(4), 1148–1153 (2007). https://doi.org/10.1016/j.eswa.2006.02.013
13. Döbel, I., Leis, M., Molina Vogelsang, M., Welz, J., Neustroev, D., Petzka, H., Riemer, A., Püping, S., Voss, A., Wegele, M.: Maschinelles Lernen. Eine Analyse zu Kompetenzen, Forschung und Anwendung. Study, Fraunhofer-Gesellschaft, München (2018)
14. Gal, Y., Hron, J., Kendall, A.: Concrete dropout. In: Guyon, I., Luxburg, U.V., Bengio, S., Wallach, H., Fergus, R., Vishwanathan, S., Garnett, R. (eds.) Advances in Neural Information Processing Systems, vol. 30, pp. 3581–3590. Curran Associates, Inc., Red Hook (2017)
15. Ge, Z., Song, Z., Ding, S.X., Huang, B.: Data mining and analytics in the process industry: the role of machine learning. IEEE Access **5**, 20590–20616 (2017). https://doi.org/10.1109/ACCESS.2017.2756872
16. GQW-Jahrestagung: Qualitätsmanagement 4.0—Status quo! Quo vadis? Bericht zur GQW-Jahrestagung 2016 in Kassel. No. Band 6 in Kasseler Schriftenreihe Qualitätsmanagement. Kassel University Press, Kassel (2016)
17. Graves, A.: Practical variational inference for neural networks. In: Shawe-Taylor, J., Zemel, R.S., Bartlett, P.L., Pereira, F., Weinberger, K.Q. (eds.) Advances in neural information processing systems 24, pp. 2348–2356. Curran Associates, Inc., Red Hook (2011)
18. Group, A.I.A.: Measurement systems analysis: [MSA] ; reference manual, 4th edn. Automotive Industry Action Group, Southfield, Mich (2010)
19. Hinton, G.E., Srivastava, N., Krizhevsky, A., Sutskever, I., Salakhutdinov, R.R.: Improving neural networks by preventing co-adaptation of feature detectors (2012). ArXiv: 1207.0580
20. Hüllermeier, E., Waegeman, W.: Aleatoric and Epistemic Uncertainty in Machine Learning: An Introduction to Concepts and Methods (2020). ArXiv: 1910.09457
21. Khan, M.S., Coulibaly, P.: Bayesian neural network for rainfall-runoff modeling. Water Resour. Res. **42**(7) (2006). https://doi.org/10.1029/2005WR003971
22. Kononenko, I.: Bayesian neural networks. Biol. Cybern. **61**(5), 361–370 (1989). https://doi.org/10.1007/BF00200801
23. Krauß, J., Dorißen, J., Mende, H., Frye, M., Schmitt, R.H.: Machine learning and artificial intelligence in production: application areas and publicly available data sets. In: Wulfsberg, J.P., Hintze, W., Behrens, B.A. (eds.) Production at the leading edge of technology, pp. 493–501. Springer, Berlin (2019). https://doi.org/10.1007/978-3-662-60417-5_49
24. Lakshminarayanan, B., Pritzel, A., Blundell, C.: Simple and Scalable Predictive Uncertainty Estimation using Deep Ensembles (2017). ArXiv: 1612.01474
25. Lansner, A., Ekeberg, O.: A one-layer feedback artificial neural network with a Bayesian learning rule. Int. J. Neural Syst. **01**(01), 77–87 (1989). https://doi.org/10.1142/S0129065789000499
26. Lin, N.: Applied Business Analytics: Integrating Business Process, Big Data, and Advanced Analytics. Pearson Education, Upper Saddle River (2014)
27. Mangal, R., Nori, A.V., Orso, A.: Robustness of Neural Networks: A Probabilistic and Practical Approach (2019). ArXiv: 1902.05983
28. Mueller, T., Huber, M., Schmitt, R.: Modelling complex measurement processes for measurement uncertainty determination. International Journal of Quality and Reliability Management **37**(3), 494–516 (2020). https://doi.org/10.1108/IJQRM-07-2019-0232

29. Mueller, T., Kiesel, R., Schmitt, R.H.: Automated and predictive risk assessment in modern manufacturing based on machine learning. In: Advances in Production Research, pp. 91–100. Springer, Cham (2018). https://doi.org/10.1007/978-3-030-03451-1_10
30. Ogorodnyk, O., Lyngstad, O.V., Larsen, M., Wang, K., Martinsen, K.: Application of machine learning methods for prediction of parts quality in thermoplastics injection molding. In: Wang, K., Wang, Y., Strandhagen, J.O., Yu, T. (eds.) Advanced Manufacturing and Automation VIII, Lecture Notes in Electrical Engineering, pp. 237–244. Springer, Singapore (2019). https://doi.org/10.1007/978-981-13-2375-1_30
31. Otte, C.: Safe and interpretable machine learning: a methodological review. In: Moewes, C., Nürnberger, A. (eds.) Computational Intelligence in Intelligent Data Analysis, vol. 445, pp. 111–122. Springer, Berlin (2013). https://doi.org/10.1007/978-3-642-32378-2_8
32. Patel, D., Hazan, H., Saunders, D.J., Siegelmann, H.T., Kozma, R.: Improved robustness of reinforcement learning policies upon conversion to spiking neuronal network platforms applied to Atari Breakout game. Neural Netw. **120**, 108–115 (2019). https://doi.org/10.1016/j.neunet.2019.08.009
33. Pei, K., Cao, Y., Yang, J., Jana, S.: Towards Practical Verification of Machine Learning: The Case of Computer Vision Systems (2017). ArXiv: 1712.01785
34. Schmitt, J., Böning, J., Borggräfe, T., Beitinger, G., Deuse, J.: Predictive model-based quality inspection using Machine Learning and Edge Cloud Computing. Adv. Eng. Inform. **45**, 101101 (2020). https://doi.org/10.1016/j.aei.2020.101101
35. Schmitt, R.H., Ellerich, M., Schlegel, P., Ngo, Q.H., Emonts, D., Montavon, B., Buschmann, D., Lauther, R.: Datenbasiertes Qualitätsmanagement im Internet of Production. In: Frenz, W. (ed.) Handbuch Industrie 4.0: Recht, Technik, Gesellschaft, pp. 489–516. Springer, Berlin (2020). https://doi.org/10.1007/978-3-662-58474-3_25
36. Schuh, G., Riesener, M., Prote, J.P., Dölle, C., Molitor, M., Schloesser, S., Liu, Y., Tittel, J.: Industrie 4.0: Agile Entwicklung und Produktion im Internet of Production. In: Frenz, W. (ed.) Handbuch Industrie 4.0: Recht, Technik, Gesellschaft, pp. 467–488. Springer, Berlin (2020). https://doi.org/10.1007/978-3-662-58474-3_24
37. Sullivan, T.: Introduction to uncertainty quantification. Texts in Applied Mathematics, vol. 63. Springer, Cham (2015). https://doi.org/10.1007/978-3-319-23395-6
38. Utama, R., Piekarewicz, J.: Refining mass formulas for astrophysical applications: A Bayesian neural network approach. Phys. Rev. C **96**(4), 044308 (2017). https://doi.org/10.1103/PhysRevC.96.044308. Publisher: American Physical Society
39. e. V., V.D.I.: VDI/VDE-Richtline 2600 Blatt 1: 2013—10 Prüfprozessmanagement—Identifizierung, Klassifizierung und Eignungsnachweise von Prüfprozessen (VDI/VDE-Guideline 2600 Part 1: 2013—10 Inspection process management—Identification, classification and proof of suitability for inspection processes). (2013)
40. VDA (ed.): VDA 5 -Prüfprozesseignung, Eignung von Messsystemen, Mess- und Prüfprozessen, Erweiterte Messunsicherheit, Konformitätsbewertung, 2 edn., vol. 5 (2011)
41. Verlag, B.: Geometrical product specifications (GPS)—Inspection by measurement of workpieces and measuring equipment—Part 1: Decision rules for verifying conformity or nonconformity with specifications (ISO 14253-1:2017); German version EN ISO 14253-1:2017. Tech. rep., Beuth Verlag GmbH (2017). https://doi.org/10.31030/2693140
42. Verlag, B.: DIN ISO 22514-7:2020-06, Statistische Verfahren im Prozessmanagement_-Fähigkeit und Leistung_- Teil_7: Fähigkeit von Messprozessen (ISO/DIS_22514-7:2020); Text Deutsch und Englisch. Tech. rep., Beuth Verlag GmbH (2020). https://doi.org/10.31030/3160215
43. Wojtkiewicz, S., Eldred, M., Field Jr., R., Urbina, A., Red-Horse, J.: Uncertainty quantification in large computational engineering models. In: 19th AIAA Applied Aerodynamics Conference. American Institute of Aeronautics and Astronautics, Anaheim (2001). https://doi.org/10.2514/6.2001-1455
44. Xiang, W., Musau, P., Wild, A.A., Lopez, D.M., Hamilton, N., Yang, X., Rosenfeld, J., Johnson, T.T.: Verification for Machine Learning, Autonomy, and Neural Networks Survey (2018). ArXiv: 1810.01989

45. Xie, Y., Lord, D., Zhang, Y.: Predicting motor vehicle collisions using Bayesian neural network models: An empirical analysis. Accid. Anal. Prev. **39**(5), 922–933 (2007). https://doi.org/10.1016/j.aap.2006.12.014
46. Yao, J., Pan, W., Ghosh, S., Doshi-Velez, F.: Quality of Uncertainty Quantification for Bayesian Neural Network Inference (2019). ArXiv: 1906.09686

Two Statistical Degradation Models of Batteries Under Different Operating Conditions

Jin-Zhen Kong and Dong Wang

Abstract The commercialization of electric vehicles (EVs) demands higher performances of rechargeable batteries. Accurate assessments of state of health (SOH) and remaining useful life (RUL) of batteries are important to indicate battery status and ensure EVs safety. However, the accuracies of existing battery capacity degradation models are not sufficient to describe battery states under the complicated impacts of usage environments. Various operating conditions will make degradation modeling more challenging and difficult, for instance, different discharge rates and discontinuous charge and discharge can influence the capacity degradation tendencies of batteries. To address the above issues, two statistical degradation models are respectively proposed to implement battery prognostics in different usage conditions based on the knowledge of big data and data science. Results show that the proposed methods outperform many existing works.

Keywords Statistical degradation model · Remaining useful life · Batteries · Data science

1 Introduction

Rechargeable batteries are widely applied to provide power for equipment such as smartphones, unmanned aerial vehicles, and EVs. They are getting more attention because of great power density and long lifetime [1]. The health status of batteries has a key impact on the safety of devices so that a reliable battery management system (BMS) is much necessary. To guarantee the performances of batteries in actual scenarios, accurate prognostics and health management (PHM) of batteries are essential, including SOH and RUL [2]. However, since operating conditions in reality are complicated and changeable, PHMs of batteries contain notable errors,

J.-Z. Kong · D. Wang (✉)
The State Key Laboratory of Mechanical Systems and Vibration, Shanghai Jiao Tong University, Shanghai, P.R. China
e-mail: kongjinzhen@sjtu.edu.cn; dongwang4-c@sjtu.edu.cn

A. Steland, K.-L. Tsui (eds.), *Artificial Intelligence, Big Data and Data Science in Statistics*, https://doi.org/10.1007/978-3-031-07155-3_11

and inaccurate estimations of SOH and RUL may cause safety hazards [3]. Most estimation methods assume that operating conditions are continuously changeless, ignoring the effects of operating conditions on batteries to simplify the modeling process, which may cause these methods do not satisfy the actual usage of battery PHM assessment [4].

Battery PHM has attracted much attention. The existing battery PHM approaches can be classified into data-driven methods and mechanism-based model. Data-driven methods consist of machine learning (ML) approaches and statistical models. ML approaches have already got much attention due to their flexibility and predictive accuracy [5]. Severson et al. [6] extracted features from discharge voltage curves and used a ML approach to predict the lifetime of batteries under fast-charging conditions. Li et al. [7] proposed an online battery capacity estimation method based on random forest, one of the ML approaches. The advantage of this method is that online data can be used as input features without pre-processing so that online assessment can be convenient. Besides, neural networks are appropriate tools in battery PHM. Sbarufatti et al. [8] developed an algorithm using radial basis function neural networks and achieved adaptive prognosis of batteries. Ma et al. [9] combined a hybrid neural network and false nearest neighbors to predict the remaining useful life of batteries.

In addition, approaches using statistics and probability knowledge are also attractive. This kind of methods can provide prognostics results and uncertainty simultaneously, making up for the disadvantages of ML methods [10]. Statistics and probability modeling has been widely used in battery PHM assessment, and it shows the superiority of uncertainty measurement, which is vital in this field. Tang et al. [11] put forward a migration-based method combining Bayesian Monte Carlo to predict the battery aging trajectory. Also, He et al. presented a RUL prediction method of batteries using Dempster–Shafer theory and the Bayesian Monte Carlo method. Cripps [12] proposed a Bayesian nonlinear random-effect model to identify defective batteries and used it for statistical analysis. Wang et al. [13] considered heterogeneous noise variances into consideration and utilized a state-space model to conduct prognostics of batteries.

PHM assessments of batteries are influenced by some operation conditions, such as discharge rates and discontinuous charge–discharge that will cause capacity recovery. To the best of our knowledge, there have remained research gaps that seldom researchers regard how to model the battery capacity at different discharge rates and deal with capacity recovery phenomenon (CRP) while degradation modeling.

This chapter aims at analyzing battery capacity degradation and constructing battery estimation models at different operating conditions, including the degradation with CRP because of discontinuous charge and discharge, and degradation under different discharge rates. We develop two statistical degradation models of batteries separately. The first one models battery degradation considering CRP, while the second one considers different discharge rates of batteries in degradation modeling. The major contributions of this work are summarized as follows. First, a piecewise degradation model considering the CRP of batteries is constructed. To verify the effect of the model, NASA battery data is utilized, and the prediction error compared with the true RUL is less than 1%. Besides, according to the battery degradation data under different discharge rates, a coupling capacity degradation model considering the coupling effect of cycle number and discharge rate is presented. A prognostic surface under various discharge rates can be established, and the PHM of batteries can be obtained. Battery degradation data from a four-cycle rotation fading strategy-based cycle testing were generated to validate the effectiveness of the model.

The rest of this chapter is organized as follows. Two statistical degradation modeling ideas of batteries are introduced respectively in Sects. 2 and 3. Section 2 presents a piecewise degradation model for batteries in view of CRP. In Sect. 3, a capacity degradation model for batteries under different discharge rates is presented. Finally, Sect. 4 draws conclusions of this work.

2 Piecewise Degradation Model for Batteries in View of Capacity Recovery Phenomenon

For batteries, discontinuous charging and discharging behaviors can cause the phenomenon of battery capacity recovery. The existence of CRP will make degradation modeling more difficult. In this section, considering CRP, a piecewise degradation model is proposed based on statistical domain knowledge.

2.1 Theory of Piecewise Degradation Modeling

As shown in Fig. 1, CRPs occur in battery capacity fade curves due to discontinuous charge–discharge of batteries. The battery capacity was measured during discharge processing of each cycle number. Current research usually ignores CRPs and

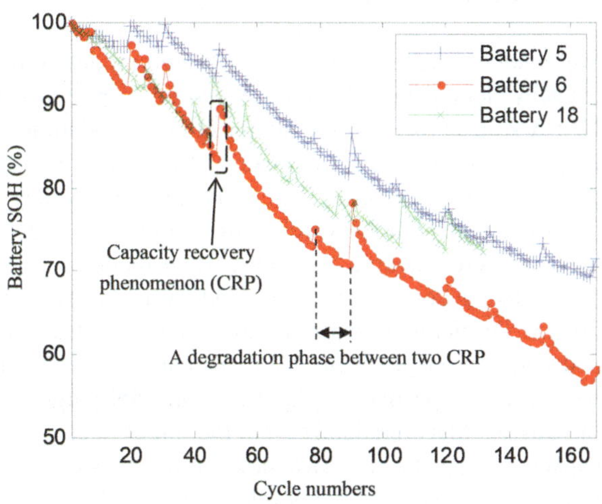

Fig. 1 NASA battery degradation data of three batteries with CRP

constructs global degradation models. However, CRPs will hinder the accurate estimation of battery PHM because degradation modeling for global degradation process cannot capture the detailed information during capacity fade. A piecewise degradation model is constructed considering locations of CRPs and local fade information in this work, expressed as Eq. (1).

$$X_k = \begin{cases} p_1(k, \gamma_1, \Theta_1), k < \gamma_1 \\ p_i(k - \gamma_i, \gamma_i, \Theta_i), \gamma_i \leq k < \gamma_{i+1} \end{cases}. \quad (1)$$

In Eq. (1), k is the charging and discharging cycle number; X_k is a stochastic process representing battery fade data in cycle k; i ($i > 0, i \in N$) denotes the series number of CRP; the phases are divided by i; γ_i represents a random variable representing the location of ith CRP, and it splits the whole process into $i+1$ phases; p_i denotes the general piecewise degradation model in ith phase, whose parameters can be marked as Θ_i.

The key of the model is determining the patterns of p_i and γ_i. The p_i can be selected or designed according to the actual degradation trajectory, and γ_i should be detected by some approaches. In the training stage, it assumes that locations of CRPs are known using historical data so that we can directly obtain the parameters of piecewise degradation models. However, in the testing stage, the new data are updated online, locations of CRPs should be detected probabilistically using the probability approach such as expectation–maximization (EM) algorithm, and parameters should be updated as new data are available. After having detected the locations of CRPs, remaining useful cycles can be predicted resorting to the extrapolation of the proposed piecewise model.

2.2 *Case Study A*

2.2.1 Dataset Description

We conduct a case study to describe the modeling process in detail. The battery capacity degradation dataset we choose is provided by the Prognostics CoE at NASA Ames [14]; here we demonstrate degradation data of batteries 5, 6, and 18; in the data exists the phenomenon of capacity recovery in the whole degradation process, shown as Fig. 1. The CRPs split battery degradation process into several phases. The experiments were carried out at room temperature. Charging process was conducted under a constant current mode of 1.5 A first, then voltage reached 4.2 V, and a constant voltage mode was utilized until the current dropped to 20 mA. Discharging process was conducted at a constant current mode of 2 A, and cut-off voltages were 2.7 V for battery 5, 2.5 V for battery 6, and 2.5 V for battery 18, respectively. As charge and discharge cycles increased, the capacity of batteries faded gradually from the initial rated capacity 2 Ahr. While the actual capacity dropped to 1.4 Ahr, the experiments were stopped. Note that in our study, normalized capacities are regarded as battery SOH.

2.2.2 Piecewise Model Construction and Monitoring Locations of CRPs

For constructing a piecewise degradation model for the NASA dataset based on Eq. (1), we analyze the degradation rules of NASA batteries and general fade model pattern. Exponential model [15, 16] and polynomial model [17] are reasonable functions to describe different kinds of batteries. Finally, we determine the exponential model as the pattern of sub-models in our work. The piecewise degradation model we first built is shown as follows:

$$X_k = \begin{cases} \theta_1 \exp(\beta k) \exp\left(\varepsilon(k - \frac{\sigma^2}{2})\right), k < \gamma_1 \\ \theta_1 \exp(\beta(k - \gamma_i)) \exp\left(\varepsilon(k - \gamma_i) - \frac{\sigma^2}{2}\right), \gamma_i \leq k < \gamma_{i+1} \end{cases}. \tag{2}$$

In contrast to Eq. (1), in this case study, the sub-model p_i is defined as $p_i = \theta_1 \exp(\beta(k - \gamma_i)) \exp\left(\varepsilon(k - \gamma_i) - \frac{\sigma^2}{2}\right)$; the parameter set Θ_i is $\Theta_i = \{\beta, \theta_i, \varepsilon\}$. Specifically, in Eq. (2), θ_i denotes a logarithm random variable that represents initial capacity, and its mean and variance are μ_{θ_1}, $\sigma^2_{\theta_1}$, respectively, that is $\theta_i \sim N(\mu_{\theta_1}, \sigma^2_{\theta_1})$. β reflects the fade rate of batteries in different phases, with a mean μ_β and a variance σ^2_β, that is $\beta \sim N(\mu_\beta, \sigma^2_\beta)$. $\varepsilon(k)$ is an error term having a mean 0 and variance σ^2, that is $\varepsilon(k) \sim N(0, \sigma^2)$; and it can be easily obtained that the expectation of $E\left(\exp\left(\varepsilon(k) - \frac{\sigma^2}{2}\right)\right) = 1$ according to the property of log-normal distribution; in Eq. (2), each phase shows an exponential trend $\theta_1 \exp(\beta(k - \gamma_i))$. After applying logarithm operation to Eq. (2), the piecewise degradation model can be expressed as follows:

$$\ln X_k = \begin{cases} \ln \theta + \beta k + \varepsilon(k), k < \gamma_1 \\ \ln \theta_1 + \beta(k - \gamma_i) + \varepsilon(k - \gamma_i), \gamma_i \leq k < \gamma_{i+1} \end{cases}. \tag{3}$$

In Eq. (3), $\ln \theta = \ln \theta_1 - \sigma^2/2$ with a mean μ_θ and variance σ^2_θ. As a result, a prognostic model for NASA batteries considering CRP into consideration is constructed. Three main parts of battery prognostics in view of CRP consist of battery degradation modeling, detection of the location γ_i of the ith CRP, and RUL prediction of batteries.

In the training stage, capacity data of battery 6 are regarded as historical data and used to obtain distributions of $\ln \theta_1$ and β. The whole degradation process includes 11 phases split by CRPs, and for each phase, an exponential sub-model $\theta_1 \exp(\beta k)$ is utilized to fit the battery degradation data. With exponential sub-models, the fitting results of exponential sub-models and actual SOH of battery 6 are illustrated in Fig. 2a. To investigate the goodness of fitting, a goodness-of-fit statistic R-square is used, and the boxplots of R-square values are plotted in Fig. 2b. With the help of the toolbox cftool in Matlab, the R-square values can be obtained. R-square values

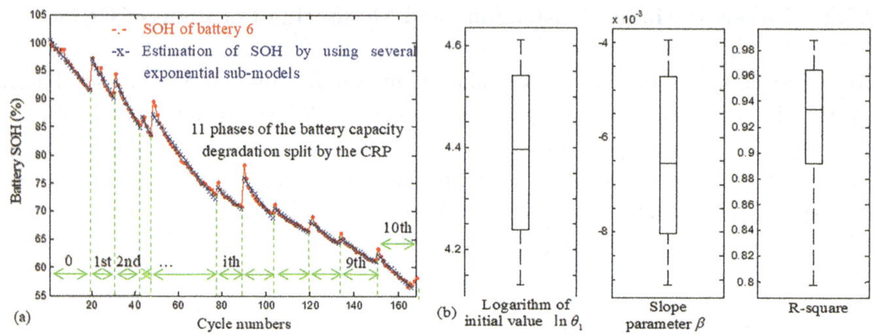

Fig. 2 Fitting results in the training stage of NASA battery degradation data: (**a**) comparison between fitting of exponential sub-models and actual SOH of battery 6; (**b**) the logarithm of the initial values and slope parameter of fitted exponential sub-models and their R-square values

are around 0.94 so that the fitting effect is good. Besides, the logarithm of the initial values and slope parameter of fitted exponential sub-models are given in Fig. 2b; it can be obvious that $\ln\theta_1$ and β are subject to normal distributions, respectively. After testing by Lilliefors test [18], the normality of $\ln\theta_1$ and β is verified. The mean and variance can be derived. Note that the command "lillietest" in Matlab can be easily used to conduct Lilliefors test. The results indicate that Lilliefors test does not reject the null hypothesis that the data comes from a normal distribution at the 5 % significance level.

Then, we are going to monitor locations of CRP for the whole degradation data. Given the information in ith phase, including observations $x_1^i, \ldots, x_l^i$ of cycle $c_1, \ldots, c_l$, and location γ_i of the ith CRP. If the $i+1$th CRP does not exist, battery capacity degradation data in the ith phase will lie within a band around $E(\ln\theta + \beta t_j) = \mu_{\theta'} + \mu_{\beta'} t_j$, where $\mu_{\theta'}$, $\mu_{\beta'}$ denote the posterior means of $\ln\theta$ and β; otherwise, if $i+1$th CRP occurs, battery capacity degradation data at γ_i are supposed to lie within a band around the mean of $\ln\theta$. To estimate the latent variable γ_{i+1}, the expectation–maximization (EM) algorithm can be employed. It is deduced that the probability of $c_j < (\gamma_{i+1} - \gamma_i)$ is proportional to $\exp(-(\ln x_j - \mu_{\theta'} - \mu_{\beta'} c_j)^2 / 2\sigma_{l,m}^2)$, where $\sigma_{l,m}^2$ denotes the covariance at cycle c_l and iteration m. The above relation can be expressed as follows:

$$p(c_j < (\gamma_{i+1} - \gamma_i) | \sigma_{l,m}^2, \gamma_i, \mu_{\theta'}, \mu_{\beta'}, \mu_\theta, \sigma_\theta^2) = \frac{\frac{1}{\sqrt{2\pi\sigma_{l,m}^2}} \exp\left(-\frac{(\ln x_j - \mu_{\theta'} - \mu_{\beta'} c_j)^2}{2\sigma_{l,m}^2}\right)}{\frac{1}{\sqrt{2\pi\sigma_{l,m}^2}} \exp\left(-\frac{(\ln x_j - \mu_{\theta'} - \mu_{\beta'} c_j)^2}{2\sigma_{l,m}^2}\right) + \frac{1}{\sqrt{2\pi\sigma_\theta^2}} \exp\left(-\frac{(\ln x_j - \mu_\theta)^2}{2\sigma_\theta^2}\right)}. \tag{4}$$

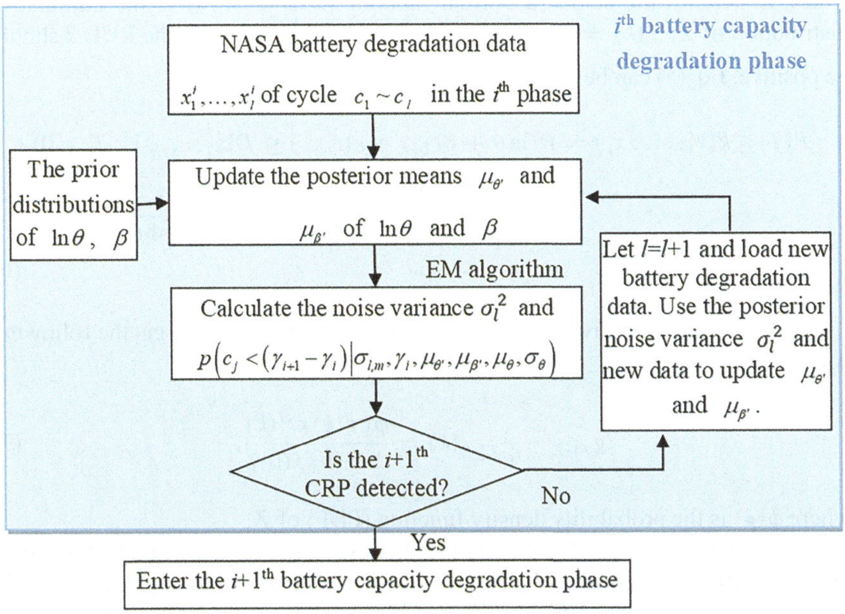

Fig. 3 Flowchart of the proposed method for monitoring locations of CRP

By using Bayesian updating thought to update posterior estimations of parameters and EM algorithm to estimate latent variables, the locations of CRP can be obtained automatically. The flowchart of the proposed method for monitoring locations of CRP is illustrated in Fig. 3.

2.2.3 RUL Prediction for Batteries

Using battery 6 as historical data and detecting the locations of CRP of battery 5, the probabilities of CRP and SOH estimation can be derived. Given observations $x_1^i, \ldots, x_l^i$ of cycle $c_1, \ldots, c_l$ and location γ_i of the ith CRP, we can get that $\ln\theta + \beta c_{l+k} + \varepsilon(c_{l+k})$ is the extrapolated capacity data of battery at cycle c_{l+k} that follows a normal distribution. And the RUL can be predicted by setting a failure threshold D (D is defined as 75 % in our study):

$$\begin{aligned} P(T \leq k | x_1^i, \ldots, x_l^i) &= P(\ln\theta + \beta c_{l+k} + \varepsilon(c_{l+k}) \leq D | x_1^i, \ldots, x_l^i) \\ &= P\left(Z \leq \frac{D - E(\ln\theta + \beta c_{l+k} + \varepsilon(c_{l+k}))}{\sqrt{var(\ln\theta + \beta c_{l+k} + \varepsilon(c_{l+k}))}}\right) = \Phi(f(k)), \end{aligned} \tag{5}$$

where Z denotes the standard normal random variable; $\Phi(\bullet)$ is the cumulative distribution of Z; $f(k) = Z \leq \frac{D-E(\ln\theta+\beta c_{l+k}+\varepsilon(c_{l+k}))}{\sqrt{var(\ln\theta+\beta c_{l+k}+\varepsilon(c_{l+k}))}}$. Because the RUL Tshould be positive, Eq. (5) can be revised as

$$\begin{aligned} P(T \leq k|x_1^i, \ldots, x_l^i) &= P(\ln\theta + \beta c_{l+k} + \varepsilon(c_{l+k}) \leq D|x_1^i, \ldots, x_l^i, T \geq 0) \\ &= \frac{P(0 \geq T \leq k|x_1^i, \ldots, x_l^i)}{P(T \geq 0|x_1^i, \ldots, x_l^i)} = \frac{\Phi(f(k)) - \Phi(f(0))}{1 - \Phi(f(0))} \cdot \end{aligned} \tag{6}$$

Implementing the derivative of Eq. (6) with respect to k, we can get the following equation:

$$g_{T|x_1^i,\ldots,x_l^i,T\geq 0}(k) = \frac{\phi(f(k))f'(k)}{1-\Phi(f(0))}, \tag{7}$$

where $\phi(\bullet)$ is the probability density function (PDF) of Z.

Then we can get the RUL prediction results shown in Fig. 4, excluding the impact of cycle shifts resulting from the future unknown CRP at a certain prediction time. The results show that the 5th, 50th, and 95th percentiles of the PDF of the RUL are 55, 61, and 69 cycles, respectively. The 50th percentile of the PDF of the RUL is regarded as the predicted RUL in our work, and the prediction error compared with the true RUL is less than 1%, which is more accurate than the existing literature [16]. The detailed results and descriptions are demonstrated in the literature [19].

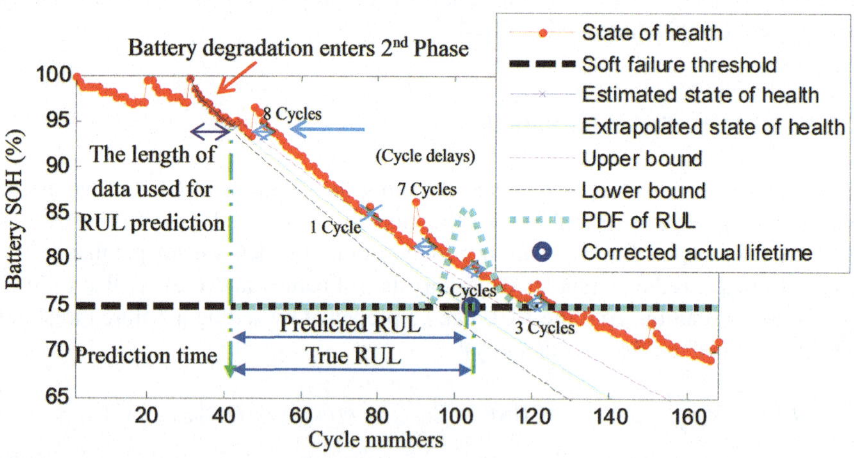

Fig. 4 RUL prediction at cycle 43 of battery 5 using the proposed method

3 State-Space-Based Capacity Degradation Model for Batteries Under Different Discharge Rates

Various operation conditions such as different discharge rates will influence the degradation of batteries; however, current research seldom considers it while modeling. In this section, a state-space-based capacity degradation model for batteries under different discharge rates is presented to improve the accuracy of battery PHM.

3.1 Theory of Degradation Modeling Under Different Discharge Rates

Battery discharge capacity degradations are remarkably affected by several environmental factors, and discharge rate is one of the key factors. In actual scenarios, batteries may work under different discharge rates because users have various usage habits. The velocities of chemical reactions inside the battery are diverse so that the capacity fades vary from various discharge rates.

Most existing studies pay attention to battery capacity estimation at constant working conditions, assuming the discharge environments are changeless [20]. To investigate the influences of discharge rates on battery capacity degradation and estimate SOH of batteries under corresponding discharge rates, we design an experiment and construct a SOH estimation model.

Considering the discharge rates variable while estimating SOH, we construct the following framework of SOH estimation, expressed as Eq. (8).

$$Q(k, R) = H \cdot \Gamma + \varepsilon(k). \tag{8}$$

In Eq. (8), Q is the discharge capacity of battery; R represents the discharge rate; k is the cycle number; H denotes the related variables; here we can define $H = [1, k, R, kR]$, where the term kR reflects the coupling effect of cycle number and discharge rate on battery capacity degradation; Γ is the coefficient variable of above four factors, which can be marked as $\Gamma = [\lambda_1, \lambda_2, \lambda_3, \lambda_4]'$.

3.2 Case Study B

3.2.1 Experimental Dataset Description and Capacity Model Construction

The experiments we conducted are continuous battery cycle testing at several discharge rates. Due to the battery degrades slowly at low discharge rates, a four-cycle rotation fading strategy-based cycle testing was designed, aiming at accelerating degradation process and saving experimental time. In this experiment, each battery went through four continuous charge–discharge cycles. The battery was charged under a constant current of 1 C (C-rate, a measure of the rate at which a battery is discharged relative to its maximum capacity) until the voltage reached 3.6 V, and then a constant voltage mode was conducted until the current dropped to C/20. And in the discharge stage, a 0.5 C constant current was utilized until the voltage reached 2 V. Then the battery was charged again and discharged at 1 C constant current, and experimental conditions of 3 C and 5 C discharge rates are similar. The current and voltage curves in one cycle are shown in Fig. 5. The capacity dataset obtained from the designed experiments includes capacity fade at 0.5 C, 1 C, 3 C, and 5 C, respectively, shown in Fig. 6a. We can deduce that the dataset agrees with the linear framework of capacity estimation in Eq. (8). According to Eq. (8), the concrete model is reformulated as follows:

$$Q(k, R) = \lambda_1 + \lambda_2 \cdot k + \lambda_3 \cdot R + \lambda_4 \cdot k \cdot R + \varepsilon(k), \tag{9}$$

where the error term follows Gaussian distribution with a zero mean and variance respect to cycle numbers, marked as $\varepsilon(k) \sim N(0, \sigma^2)$.

Fitting results in Fig. 6b show that the above model in Eq. (9) can match the capacity degradation at different discharge rates well because R-square reaches 0.9778. For sake of letting the model adapt unit-to-unit variances and use online operating data of a certain battery to update model-related parameters posteriorly, the model in Eq. (9) is transformed into a state-space-based model:

$$\begin{cases} \Gamma_k = \Gamma_{k-1} + \omega_k \\ Q_k = H_k \cdot \Gamma_k + \varepsilon(k), \end{cases} \tag{10}$$

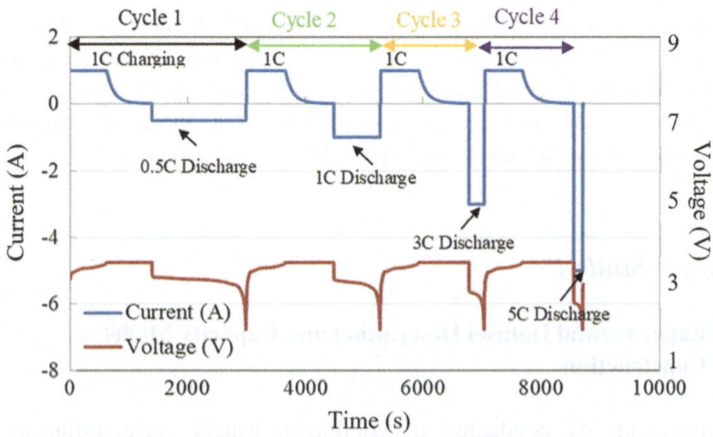

Fig. 5 The experimental setting of current and voltage curves in one cycle

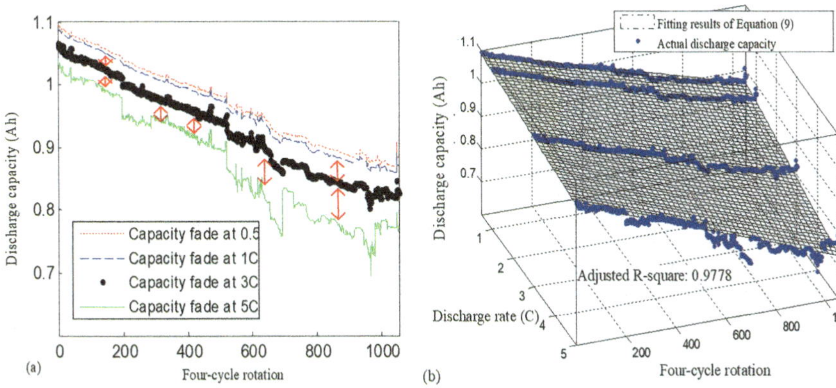

Fig. 6 Capacity data and model results: (**a**) Discharge capacity degradation at different discharge rates; (**b**) fitting of capacity and results of the proposed model at different discharge rates

where ω_k is the multivariate Gaussian noises with a zero mean and a covariance σ_ω^2; other parameters have the same meanings as above. The models in Eq. (10) have linear forms so that in this work, traditional Kalman filter was used, including predicting step and updating step. More information about Kalman filter can be found in Ref. [21].

3.2.2 RUL Prediction for Batteries at Different Discharge Rates

RUL prediction at different discharge rates can be divided into offline stage and online stage. The flowchart of the proposed method to predict RUL of an operating battery at different discharge rates is illustrated in Fig. 7.

In the offline stage, using historical capacity data at different discharge rates, the initial discharge capacity model is established. Then, transforming the proposed model to a state-space-based model and obtain prior distributions of parameters.

Secondly, in the online stage, the online capacity data of a certain operating battery at different discharge rates are fed into the state-space-based model and posterior distributions of the state-space-based model can be derived using Kalman filter. In addition, RUL prediction process is similar to Sect. 2.2.3 (Eqs. (6) and (7) can be references.) so that we will not introduce it in detail here.

Define the predicted time as a percent of actual battery lifetime and set 80% of actual battery lifetime as failure threshold; we can get the RUL prediction and uncertainty measurement by extrapolations of the proposed method, according to the procedures in Fig. 7. The results of RUL prediction and PDF of RUL for batteries at four different discharge rates are shown in Fig. 8. We can obtain that MSEs of prediction are less than 1.7×10^{-6}. The proposed method has good performance on RUL prediction at different discharge rates. The detailed results and descriptions are explained in the literature [22].

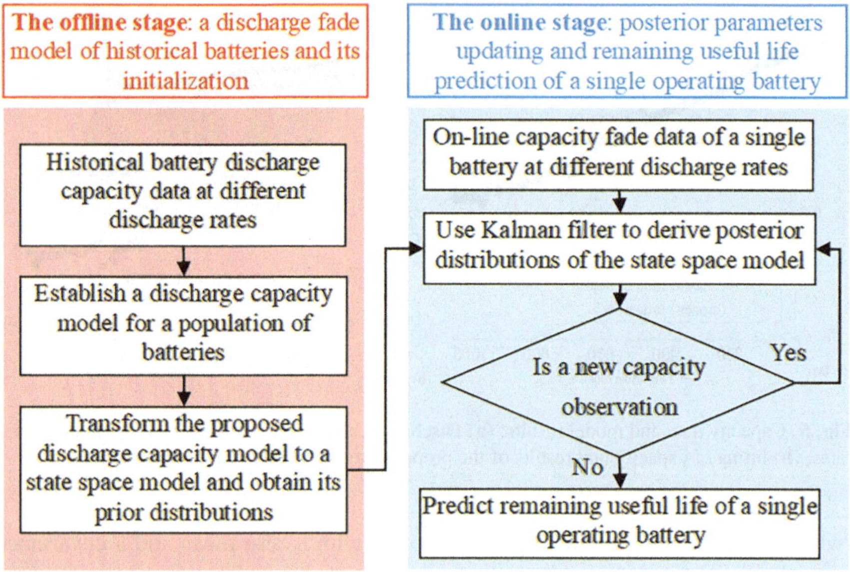

Fig. 7 Flowchart of the proposed method to predict RUL of an operating battery at different discharge rates

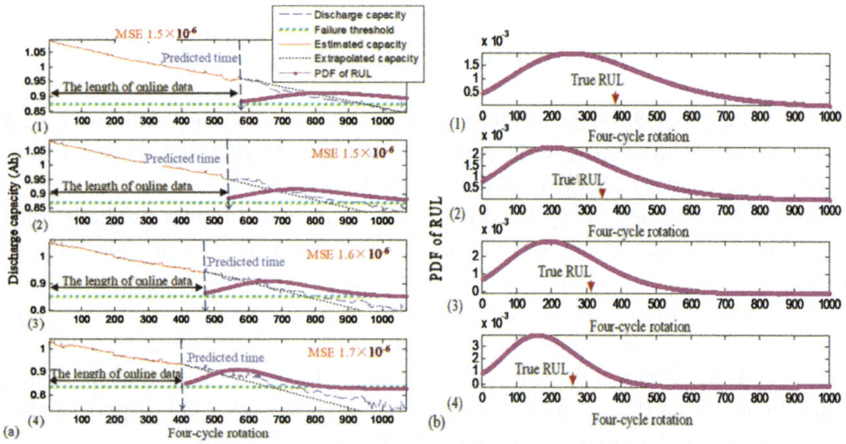

Fig. 8 RUL prediction results at different discharge rates using the proposed method: (**a**) RUL prediction results of batteries at four different discharge rates; (**b**) comparison of PDF of RUL and true RUL

The SOH estimations at different temperatures are similar to this work, but the model form should be modified, according to the analysis, capacity fade tends to be exponential, and then the linear model is not suitable.

4 Conclusions

PHM of batteries is an attractive and enabling research domain as EVs become more commercialized. In this chapter, we explored two actual issues in battery capacity degradation modeling and provided two modeling ideas by statistical and probabilistic-related approaches. Considering CRP of batteries caused by discontinuous charge and discharge, the piecewise degradation model was proposed by considering CRP, and then RUL of batteries was predicted based on some specific degradation phases. Next, battery degradation modeling under different discharge rates was studied. The state-space-based capacity degradation model for the batteries under different discharge rates was presented. The battery degradation data at different discharge rates could be described by a surface. These two works achieved good prediction effects, which exhibited the ability of statistical approaches for data modeling and prediction.

In future work, more operating conditions will be considered such as depth of discharge, environmental factors, etc. More modeling ideas are needed for improving the accuracy of battery PHM with the help of statistical approaches and big data, which include huge amounts of battery historical data and online monitoring data.

References

1. Lucu, M., Martinez-Laserna, E., Gandiaga, I., Camblong, H.: A critical review on self-adaptive Li-ion battery ageing models. J. Power Sources **401**, 85–101 (2018)
2. Meng, H., Li, Y.-F.: A review on prognostics and health management (PHM) methods of lithium-ion batteries. Renew. Sust. Energ. Rev. **116**, 109405 (2019)
3. Xiong, R., Li, L., Tian, J.: Towards a smarter battery management system: A critical review on battery state of health monitoring methods. J. Power Sources **405**, 18–29 (2018)
4. Si, X.-S., Wang, W., Hu, C.-H., Zhou, D.-H.: Remaining useful life estimation A review on the statistical data driven approaches. Eur. J. Oper. Res. **213**(1), 1–14 (2011)
5. Tan, P., Jiang, H.R., Zhu, X.B., An, L., Jung, C.Y., Wu, M.C., et al.: Advances and challenges in lithium-air batteries. Appl. Energy **204**, 780–806 (2017)
6. Severson, K.A., Attia, P.M., Jin, N., Perkins, N., Jiang, B., Yang, Z., et al.: Data-driven prediction of battery cycle life before capacity degradation. Nat. Energy **4**(5), 383–91 (2019)
7. Li, Y., Zou, C., Berecibar, M., Nanini-Maury, E., Chan, J.C.W., van den Bossche, P., et al.: Random forest regression for online capacity estimation of lithium-ion batteries. Appl. Energy **232**, 197–210 (2018)
8. Sbarufatti, C., Corbetta, M., Giglio, M., Cadini, F.: Adaptive prognosis of lithium-ion batteries based on the combination of particle filters and radial basis function neural networks. J. Power Sources **344**, 128–40 (2017)

9. Ma, G., Zhang, Y., Cheng, C., Zhou, B., Hu, P., Yuan, Y.: Remaining useful life prediction of lithium-ion batteries based on false nearest neighbors and a hybrid neural network. Appl. Energy **253**, 113626 (2019)
10. Peng, W., Ye, Z.-S., Chen, N.: Bayesian deep learning based health prognostics towards prognostics uncertainty. IEEE Trans. Ind. Electron. **67**(3), 2283–2293 (2019)
11. Tang, X., Zou, C., Yao, K., Lu, J., Xia, Y., Gao, F.: Aging trajectory prediction for lithium-ion batteries via model migration and Bayesian Monte Carlo method. Appl. Energy **254**, 113591 (2019)
12. Cripps, E., Pecht, M.: A Bayesian nonlinear random effects model for identification of defective batteries from lot samples. J. Power Sources **342**, 342–350 (2017)
13. Wang, D., Yang, F., Zhao, Y., Tsui, K.-L.: Prognostics of Lithium-ion batteries based on state space modeling with heterogeneous noise variances. Microelectron. Reliab. **75**, 1–8 (2017)
14. B. Saha KG.: Battery Data Set. In: (http://ti.arc.nasa.gov/project/prognostic-data-repository) NAPDR, editor. NASA Ames Research Center, Moffett Field, CA2007
15. He, W., Williard, N., Osterman, M., Pecht, M.: Prognostics of lithium-ion batteries based on Dempster Shafer theory and the Bayesian Monte Carlo method. J. Power Sources **196**(23), 10314–10321 (2011)
16. Gebraeel, N.Z., Lawley, M.A., Li, R., Ryan, J.K.: Residual-life distributions from component degradation signals: A Bayesian approach. IIE Trans. **37**(6), 543–557 (2005)
17. Micea, M.V., Ungurean, L., Cârstoiu, G.N., Groza, V.: Online state-of-health assessment for battery management systems. IEEE Trans. Instrum. Meas. **60**(6), 1997–2006 (2011)
18. Dallal, G.E., Wilkinson, L.: An analytic approximation to the distribution of Lilliefors' test statistic for normality. Am. Stat. **40**(4), 294–296 (1986)
19. Wang, D., Kong, J.Z., Zhao, Y., Tsui, K.L.: Piecewise model based intelligent prognostics for state of health prediction of rechargeable batteries with capacity regeneration phenomena. Measurement. **147**, 106836 (2019)
20. Ng, S.S.Y., Xing, Y., Tsui, K.L.: A naive Bayes model for robust remaining useful life prediction of lithium-ion battery. Appl. Energy **118**, 114–23 (2014)
21. Welch, G., Bishop, G.: An Introduction to the Kalman Filter: University of North Carolina, Chapel Hill (1995)
22. Wang, D., Kong, J.-z., Yang, F., Zhao, Y., Tsui, K.-L.: Battery prognostics at different operating conditions. Measurement **151**, 107182 (2020)

Detecting Diamond Breakouts of Diamond Impregnated Tools for Core Drilling of Concrete by Force Measurements

Christine H. Müller, Hendrik Dohme, Dennis Malcherczyk, Dirk Biermann, and Wolfgang Tillmann

Abstract Diamond impregnated tools for core drilling consist of segments in which synthetical diamonds are bounded in a metal matrix. The wear of these tools depends on the time points when active diamonds breakout and new diamonds from deeper layers of the metal matrix become active. Up to now, these time points were measured only by visual inspection at given inspection time points, a measurement which is very error-prone and labor-intensive. Hence the aim is to use the automatic force measurements during the drilling process for detecting the breakouts of the diamonds. These force measurements consist of three time series observed over about 75 min, each minute with over 300,000 measurements. At first, we present here an approach of an analysis of these time series in three steps: identification of the time periods of active drilling, identification of the rotation periods, and determination of differences between successive rotations. Based on the detected rotation periods, 147 features for classification of minutes with and without diamond breakout are created. Some of these features are based on the differences between successive rotations and some on p-values for testing the independence of the detected rotation lengths. After a feature selection step, random forest and logistic regression are applied. This leads at least for one of two considered series of experiments to a classification error which is smaller than the trivial classification error.

C. H. Müller (✉) · H. Dohme · D. Malcherczyk
Department of Statistics, TU University Dortmund, Dortmund, Germany
e-mail: cmueller@statistik.tu-dortmund.de; hendrik.dohme@tu-dortmund.de; dennis.malcherczyk@tu-dortmund.de

D. Biermann
Institute of Machining Technology, TU University Dortmund, Dortmund, Germany
e-mail: biermann@isf.de

W. Tillmann
Institute of Materials Engineering, TU University Dortmund, Dortmund, Germany
e-mail: wolfgang.tillmann@udo.edu

A. Steland, K.-L. Tsui (eds.), *Artificial Intelligence, Big Data and Data Science in Statistics*, https://doi.org/10.1007/978-3-031-07155-3_12

Keywords Classification · LASSO · Logistic regression · Random forest · Dynamic time warping

1 Introduction

The wear of diamond impregnated tools for core drilling of concrete depends on the wear of active diamonds visible on the surface of the segments of the tool. The segments consist of synthetical diamonds bounded in a metal matrix. As soon as an active diamond breaks out, usually a new diamond embedded in a deeper layer of the metal matrix becomes active. This so-called self-sharpening property means that new sharp diamonds are exposed at the tool surface at any time of the process.

Several authors already identified the main wear mechanisms of diamond impregnated tools, e.g., [8, 21]. The authors as [3, 12, 17] considered also some statistical analysis. However, these approaches mainly concern diamond impregnated tools for sawing applications of rock. Only a few authors are dealing with the diamond core drilling process, see, e.g., [1, 9, 10, 15]. In particular, [15] showed how the size of the diamonds and the used concrete influence the lifetime of the active diamonds. This analysis was complicated by the fact that the breakout times of the active diamonds and the appearance times of new active diamonds were only measured by visual inspections at given inspection times. Thereby, the number of visible and active diamonds on the tool surface was determined by microscopical inspections of the tool at the given points in time. This leads to the so-called doubly interval-censored data. Moreover, the intervals between the inspection times lasted always 1 min which is not the best choice as indicated by Malevich and Müller [14]. However, more grave is the fact that the visual inspections are very labor-intensive and so error-prone that different inspectors provided different results.

Hence an important aim is to detect automatically the time points of the breakout of active diamonds and the appearance of new diamonds. The automatic measurements of the process forces during the drilling process are especially appropriate for this task. Since in the given experimental setup, the force measurements are given by the intervals between the visual inspections, we consider the task to identify the intervals with and without diamond breakout via the force measurements. Each interval consists of three time series in x, y, and z direction of drilling, each with about 300,000 observations. The first attempt of classifying these intervals with and without diamond breakout was done in [9] by using simple features like classical and robust measures of location and scale of the force measurements in the intervals. Additionally, the number of bivariate change points was used by applying the method of [7] to two of the three time series. In particular, the number of change points in the intervals looked promising for the classification problem in the first series of experiments with 25 intervals. However, this result could not be confirmed using further 25 intervals, see [10].

The change point analysis suffers from the fact that there are additional oscillations within each rotation, see Fig. 6. These oscillations vary over time. Hence, we consider here the approach to identify at first the rotations and then to measure the differences of the oscillations between successive rotations. However, in the first step, the time periods of active drilling must be identified automatically. Although this was done already in [9], even this task is challenging. Further challenges appear by identifying the rotations and by calculating the differences between the rotations. In particular, to test the quality of the identification of the rotations, we test for independence of the detected rotation lengths with the runs test of [20] (see [5] pp. 78–86) and a new test based on the recently proposed generalized sign test of [13]. The differences between rotations are calculated by applying the method of dynamic time warping of [6].

We create 147 features from the identified rotations, the differences between rotations, the p-values of the independence tests, and additionally from more simple quantities per interval. After a feature selection step, we use the random forest and the logistic regression for classifying minutes with and without diamond breakout. We apply this for two series of experiments with two types of concrete. The experiments at the more homogeneous concrete provide a low breakout rate of 0.173 so that the trivial classification method which classifies all minutes as "no breakout" could not be beaten. However, the experiments at the inhomogeneous concrete show a breakout rate of 0.342 while the leave-one-out misclassification rate was 0.260 for the logistic regression and 0.329 for the random forest.

The paper is organized as follows. The experimental setup is given in Sect. 2. Section 3 deals with identification of periods of active drilling and Sect. 4 with the identification of the rotations. Section 5 shows how time warping can be used to calculate differences between rotations. In Sect. 6, the 147 features are given and used for the classification problem. Finally, a discussion of the results is given in Sect. 7.

2 Experimental Setup and the Data

Four sequences of experiments, each with 75 sequential drilling phases of (approximately) equal length, were conducted. In each drilling experiment, automatic force measurements were obtained in time intervals of length of 61 up to 83 seconds where each time interval should contain active drilling of about 1 min length. During each interval, the process forces F_z, F_x, F_y in z, x, and y direction were measured with measurement frequency of about 5000 Hz so that each process time series consists in average about 300 000 measurements per time interval of active drilling. The circumferential speed was 3.225 m/s leading to ca. 616 rotations per minute with ca. 487 observations per rotation. After each experiment, the number of diamonds which have been broken out and which newly appeared were recorded by visual inspections of photos obtained by a microscope. For more details of the experiments, see [9].

The four sequences of experiments differ by the size of diamonds (small diamonds from grid size of $d_k = 40/50$ mesh and large diamonds from grid size of

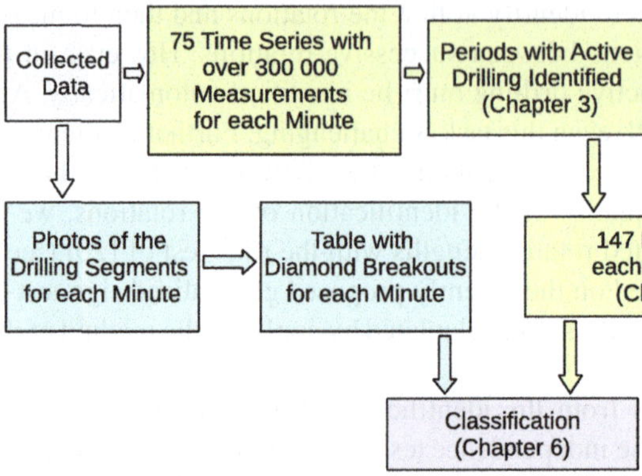

Fig. 1 Flowchart of data collection, preprocessing, and analyzing

$d_k = 20/30$ mesh) and two types of concrete (conventional concrete with compressive strength of C20/25 and homogeneous concrete with high strength of C100/115). However, only the sequences of experiments with the small diamonds provided enough intervals with diamond breakouts. These two sequences of experiments are called B28 and B29, where B28 concerns the drilling in the C20/25 concrete and B29 the drilling in the C100/115 concrete. In the sequence B28, 22 diamonds were visible at the beginning and a diamond breakout was observed in 25 intervals while 35 diamonds were visible in the beginning of the series B29 and here 13 intervals showed a diamond breakout.

Figure 1 shows a flowchart describing how the data were collected, preprocessed, and analyzed. The parts displayed in yellow concern the preprocessing of the time series of the force measurements as well as the feature generation based on these measurements as described in this paper. The blue parts provide the preprocessing steps to get the classification of minutes with and without diamond breakout by visual inspection. Details concerning this procedure are given in [15]. The final aim of this paper, which is the classification of minutes with and without diamond breakout by force measurements, is given in the purple part. All calculations were done in R [18] and all cited packages are R packages available on the website given by R Development Core Team [18].

3 Identification of Periods of Active Drilling

Each single drilling experiment in the sequence of experiments should last about 1 min before it is interrupted for the visual inspection. Because of these interruptions, there are phases of no active drilling at the beginning and at the end of each of the 75 time series so that the time intervals for each experiment are longer than 1 min. Hence, identifying the phase of active drilling is a necessary first step. For this, the examination of the feed forces F_z in the time intervals is the most appropriate.

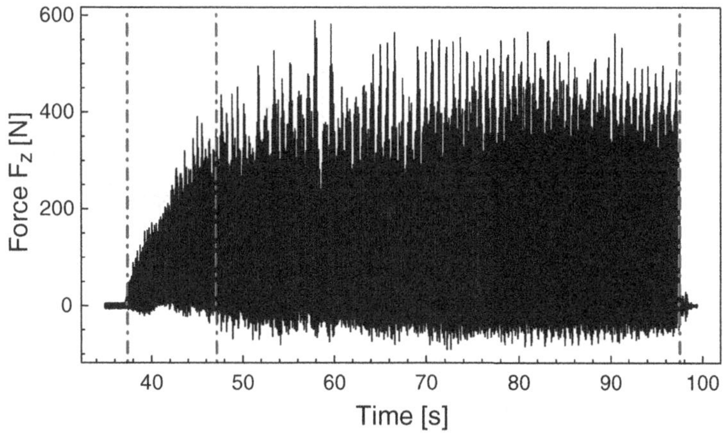

Fig. 2 F_z time series of the 25th min of the B28 data with different phases separated by red lines

Figure 2 depicts the different phases of F_z in a time slot of 66 s, which are separated by the dashed, red lines. In the beginning of the drilling no forces are acting. As soon as the drilling tool comes into contact with the concrete workpiece, the feed forces start to rise until a stationary main phase is reached. At the end of the drilling, the segment is drawn back and the acting forces decrease rapidly.

The beginning and end phase of each drilling are not relevant for the analysis of diamond breakouts and should not be considered for statistical analysis. Nevertheless, breakouts might possibly occur in the phase of rising forces. But since the length and form of this phase varies a lot over the different experiments (especially in B28), including this phase would lead to a distortion of statistical properties of the drilling. Hence these phases also were excluded from further analysis, concentrating merely on the stationary phase.

An automatic detection of the phase of interest for each experiment was achieved by calculating the standard deviation in running windows of 100 observations throughout the time series of the feed forces F_z (see Fig. 3) using the function `rollapply` from the `zoo`-package [22]. The mean of these windowed standard deviations was defined as μ_w and their standard deviation as σ_w. At first, the end of the stationary phase was detected by simply identifying the point in time when the windowed standard deviation of the feed forces takes a value above a threshold T_{End} for the last time. Here, T_{End} was set as $T_{End} = \mu_w + 0.15 \cdot \sigma_w$, where the factor 0.15 appeared to be most appropriate for the experiments B28 and B29.

Next, a threshold for the beginning of the stationary phase was defined. For this purpose, the standard deviation of 10,000 preceding observations before reaching the detected endpoint of the stationary phase was calculated and defined as σ_{main}. This represents the standard deviation during the main phase of the drilling. The threshold T_{Start} was set as $T_{Start} = \sigma_{main} + 0.5 \cdot \sigma_w$. Defining the beginning of this phase as the first point in time when the windowed standard deviations exceed this threshold leads to satisfactory results for each experiment in the B29

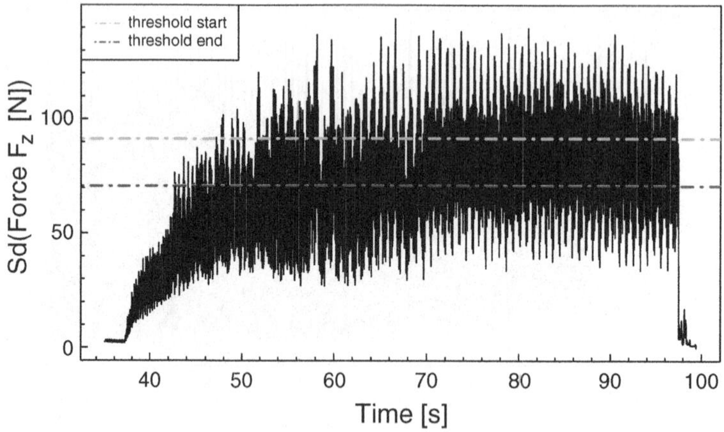

Fig. 3 Windowed standard deviations of F_z time series of Fig. 2 with thresholds for the beginning and the end of the stationary phase

as well as in the B28 data. However, some minutes of the B28 data provide very bad results. An ensuing inspection of those minutes shows that those minutes were badly affected by measuring disturbances so that they were removed ending up with 73 min.

In other drilling experiments the thresholds might require some adjusting which can be achieved by modifying the factors of the standard deviations σ_w and σ_{main}. In choosing the thresholds, it is particularly difficult to classify the beginning of the stationary phase since it varies a lot over the different experiments and minutes. Thus, it was important that the corresponding threshold is a function of the actual variance of this phase (here σ_{main}). Furthermore, if the threshold was defined too small some drilling periods contained big parts of the phase of the rising forces leading to distortion of statistical properties. Too large thresholds, on the other hand, resulted in later onset points and a loss of information due to unnecessary short active drilling periods that were to be considered for further analysis. In extreme cases, no onset points can be detected.

4 Identification of the Rotations

Since statistical methods will be applied for the single rotations of the drilling tool (e.g., dynamic time warping), the second step is to identify the starting and endpoints of each rotation. To accomplish this, a local polynomial regression (loess) (see [2]) was run on the time series of the tangential forces F_x of the experiments since the periodic structure, which can be attributed to the rotations, is clearly visible in these time series (see Fig. 4). For reasons of symmetry, using F_y would lead to similar results.

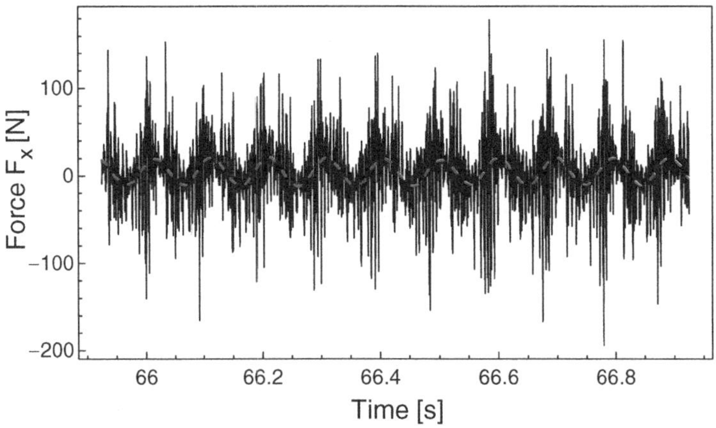

Fig. 4 Close up of the time series of tangential force F_x together with the smoothed time series by `loess` (red dashed line)

The loess method is used to smooth a time series by fitting a polynomial function to a neighborhood $N(x_0, h) = [x_0 - h(x_0), x_0 + h(x_0)]$ of an observation x_0 from a time series $(x_t)_{t \in T}$, where h is a span function. The function `loess` in R uses a so-called `span` parameter α, which defines the relative amount $p = \lfloor \alpha n \rfloor$ of n observations in the neighborhood $N(x_0, h)$ of x_0. The degree of the polynomial can be specified using the parameter `degree` and fitting is accomplished by using weighted least squares with a tricube weight function, see [2].

Here, `loess` was used to smooth F_x by using quadratic polynomials and a final `span` parameter of 0.0015 for the B28 experiments and 0.00125 for B29. The starting and endpoints of each rotation were then calculated by applying the differences operator Δ on the smoothed time series. Then the sign changes from negative to positive of these differences represent the minima of the smoothed curve. These minima are used as the onset points of a new rotation.

Note that too small span parameters lead to unrealistic short rotations which result in strong distortions of the computed features for classification. Too large parameters, on the other hand, result in alternating lengths of rotations. In particular, if a rotation length is determined as slightly too long by a too large parameter then the following identified rotation length is too short and vice versa. This leads to the chain structure of rotation lengths shown in Fig. 5 and means that the rotation lengths are negatively correlated. Hence in choosing appropriate span parameters, it is necessary to compromise between correlated rotation lengths and the amount of unrealistic short rotations.

To test for correlation, we applied the runs test of [20] (see [5] pp. 78–86) and the generalized sign test of [13]. This generalized sign test is based on the 3-sign

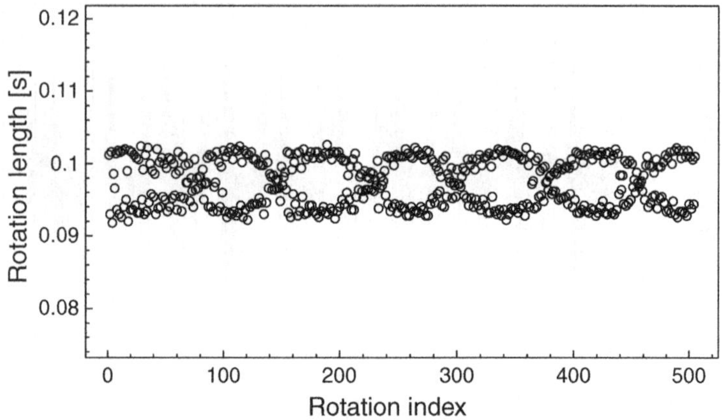

Fig. 5 Detected rotation lengths in the first minute of Experiment B29 using the span parameter $\alpha = 0.002$

depth (3-depth) of residuals $r_1, \ldots, r_N$ defined by

$$d_3(r_1, \ldots, r_N) := \frac{1}{\binom{N}{3}} \sum_{1 \le n_1 < n_2 < n_3 \le N} \Big(\mathbb{1}\{r_{n_1} > 0, r_{n_2} < 0, r_{n_3} > 0\} + \mathbb{1}\{r_{n_1} < 0, r_{n_2} > 0, r_{n_3} < 0\} \Big)$$

where $\mathbb{1}\{\cdot\}$ denotes the indicator function. Hence, the 3-depth is the relative number of 3-tuples with alternating residuals. Here, the residuals are the deviations of the rotation lengths from their median in the considered minute. The null hypothesis

$$H_0 : \text{The residuals are independent}$$

is rejected if the 3-depth is too small or too large. A too small 3-depth indicates positive correlation while a too large 3-depth indicates negative correlation.

Table 1 provides the rejection rates of H_0 by the 3-sign depth tests (3-depth tests) and the runs tests, as well as the 1%-quantiles of the rotation length, the number of unrealistic short rotations and the minimal rotation length over all minutes using different values for the span parameter α. A rotation is regarded as too short, when its shorter than the median of rotation lengths minus 3 times their IQR (interquartile range). Table 1 shows that the runs test always rejects the independence assumption in more than 50% of the minutes. The rejection rates of the 3-depth test are smaller but also increases with growing span parameter. These high rejection rates also occur when two successive rotations were put together. Hence, these rejection rates indicate to choose the span parameter as small as possible as soon as the number of short rotations is small enough. Based on Table 1, the choice of the span parameters (0.0015 for B28 and 0.00125 for B29) is comprehensible. Note that the number of

Table 1 Rejection rates of the 3-depth test and the runs test of H_0, 1%-quantiles of rotation lengths, number of short rotations and minimal rotation length for the experiments B28 and B29 over all minutes using α as `span` parameter for detecting the rotations

Span parameter α	Rejection rate 3-depth test	Rejection rate runs test	1%-quantile of rotation lengths	Minimal rotation length	Number of short rotations
	B28				
0.00075	0.0000	0.8082	0.0048	0.0004	0
0.001	0.0274	0.7945	0.0042	0.0008	7912
0.00125	0.0685	0.6712	0.0918	0.0024	92
0.0015	0.1096	0.6986	0.0918	0.0836	1
0.002	0.4110	0.7397	0.0904	0.0854	0
	B29				
0.00075	0.0267	0.6267	0.0046	0.0008	5817
0.001	0.1200	0.5067	0.0052	0.0006	930
0.00125	0.2933	0.7600	0.0934	0.0890	0
0.0015	0.3067	0.7867	0.0928	0.0890	0
0.002	0.4933	0.8000	0.0908	0.0888	0

short rotations for the B28 experiment and span parameter 0.00075 is zero, because almost all rotations are predicted as too short. Furthermore for $\alpha = 0.0015$, only one rotation is considered as too short (with a rotation length of 0.0836). Nevertheless, this circumstance is negligible since its length is not unrealistic like it is the case for smaller span parameters.

An explanation for the different optimal values of the span parameters may be the different kind of concrete used. Here, the time series of the more inhomogenous concrete (B28) might require more smoothing. This results in a higher optimal span parameters because a wider window of local polynomial regression reduces the variance of the smoothed time series. The experiments with larger diamonds leads to the same relationship of the appropriate span parameters.

5 Calculation of Differences Between Rotations

Differences between rotations can be calculated with the method of dynamic time warping as given by the R package `dtw` of [6], see also [19]. This method can be applied for time series $x = (x_1, \ldots, x_N)$ and $y = (y_1, \ldots, y_L)$ of eventually different lengths N and L as this can be the case for successive rotations. The idea of time warping is to find a warping of the time axis so that the distance between x and y becomes as small as possible. The minimized distance between x and y is then the so-called time warping distance. To define the distance between x and y, consider the matrix $M = (d(x_n, y_l)_{n=1,\ldots,N,l=1,\ldots,L})$ of pointwise distances between x and y where d is a given metric. Usually d is the Euclidean distance. Additionally, define a path through the indices of the matrix M by

$$\Phi : \{1, \ldots, T\} \ni k \to \Phi(k) = (\Phi_x(k), \Phi_y(k)) \in \{1, \ldots, N\} \times \{1, \ldots, L\}$$

with $(\Phi_x(1), \Phi_y(1)) = (1, 1)$, $(\Phi_x(T), \Phi_y(T)) = (N, L)$ and $\Phi_x(k+1) \geq \Phi_x(k)$, $\Phi_y(k+1) \geq \Phi_y(k)$ for $k = 1, \ldots, T$, where $T \in \{\max\{N, L\}, \ldots, N + L - 1\}$. The distance between x and y with respect to Φ is then given by

$$d_\Phi(x, y) := \frac{1}{T} \sum_{k=1}^{T} d(x_{\Phi_x(k)}, y_{\Phi_y(k)})$$

The path Φ^* with

$$d_{\Phi^*}(x, y) := \min_{\Phi} d_\Phi(x, y)$$

provides then the smallest distance between x and y and $d_{\Phi^*}(x, y)$ is called the dynamic time warping (DTW) distance between x and y. Figure 6 demonstrate the principle of time warping using two successive rotations. It also shows the similarity of successive rotations.

Here, dynamic time warping was applied on the F_z time series of successive rotations in the periods of active drilling. In the case of a diamond breakout, it might be reasonably assumed that the drilling performance changes rapidly. Such events should be reflected in high time warping distances. Unfortunately, high peaks in the dynamic time warping distances appear several times in multiple minutes of the B28 and the B29 experiments and do not seem to relate directly with breakouts.

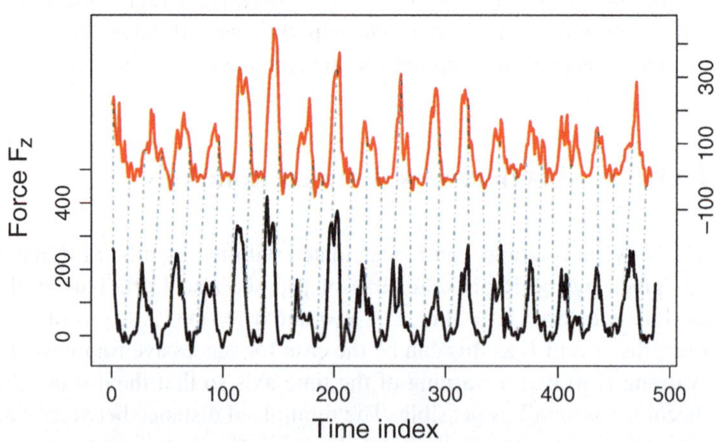

Fig. 6 Example of time warping applied to two successive rotations

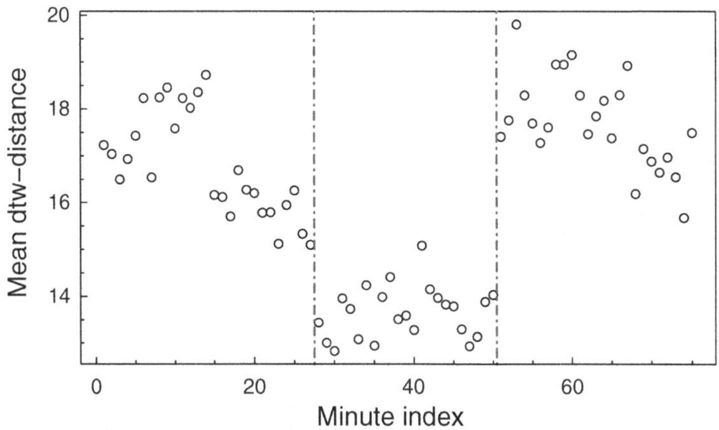

Fig. 7 Means of the dynamic time warping distances in the B29 data over all minutes with suspected changes of the workpiece marked by red lines

However, an interesting observation can be made when investigating the means and standard deviations of the DTW distances for each minute. In both experiments these features show obvious time dependent clusters. The minutes that mark the beginning of a new cluster can be associated with exchanges of the concrete workpiece which occurred at minute 25 and 51 (see Fig. 7). These effects within the DTW distances are more pronounced in the B29 experiments. This might result from the more homogeneous concrete used which leads to more noticeable differences between workpieces.

Since some successive rotations show a different number of peaks (compare Fig. 6), one might assume that the identification of rotations as described in Sect. 4 might not be optimal. Thus, it seems convenient to relax the constraint that beginnings and ends of two rotations have to match allowing a subsequence finding procedure. Applying this modification, which can be computed by setting the parameters `open.end=TRUE` and `open.beginning=TRUE` in the `dtw` function, should eliminate distortions of the DTW distances by suboptimally detected rotation onset- and offset-points. However, the results using this modification are hardly distinguishable from the ones obtained from the regular DTW method so that the classical DTW method is used hereinafter.

6 Feature Generation and Classification

In the following, various features for each minute of the B28 and B29 data were collected which should serve for the task of classifying a diamond breakout. The single rotations of the drilling tool, whose detection has been discussed in Sect. 4, form the basis of all these features. An overview of the 147 generated features can be found in Tables 2 and 3. Note that location and scale parameters are given only by the mean and standard deviation in these tables although their robust counterparts (median and MAD, the median of the absolute deviation from the median) were calculated as well. Several features concern location and scale parameters as well as the number of outliers of parameters calculated from the measurements of single rotations. The parameters calculated for single rotations are again location and scale parameters and additionally the surface under the curve of a single rotation. The surface under the curve was approximated by trapezoids like it is usually done in numerical integration. A rotation has been classified as an outlier in a minute if its feature value is either bigger than the median of the feature plus 3 times its IQR or smaller than the median minus 3 times its IQR. This was done for the acting forces F_y, F_x, and F_z separately, but also two or three of the acting forces were treated simultaneously by calculating the Euclidean distances from the spatial median. Since we, in particular, expected changes in the dynamic time warping distances (DTW distances) of rotations, especially outliers and change points of the DTW distances are considered. All features of change points were computed with the function `cpt.mean` of the `changepoint` package using the PELT method of [11] for the detection of change points. Since the independence assumption of the rotation lengths was often rejected by the runs test and the 3-depth test, their p-values and the differences of their p-values are considered as features as well. Features concerning all three forces F_y, F_x, F_z are given in Table 2. Table 3 contains features which make sense only for one force. These are features based on the time warping distances of rotations and the p-values of the independence tests.

Before the actual classification task, all of the 147 mentioned features in Tables 2 and 3 were put through a feature selection. In the first step, the LASSO method for logistic regression [4], which can be carried out with the function `cv.glmnet` of the `glmnet`-package, was applied for all observations with default settings except the parameter `nfolds` concerning the number of folds used in the cross-validation. Since the number of observations is quite small, namely 75 for the B29-data and 73 for the B28-data, the number of folds was set to the number of observations in both cases which leads to a leave-one-out procedure. The default settings mean here that the l_1-penalty $\lambda\|\beta\|_1$ is used to get the leave-one-out estimates $\widehat{\beta}_\lambda^i$, $i = 1, \ldots, n$, for which the binomial deviance $D(y_i, \pi(\widehat{\beta}_\lambda^i))$ of [16, p.118] for the ith observation y_i is calculated.

For the B29-data only 4 important features were detected by using the penalty factor λ which provides the smallest mean value of $D(y_i, \pi(\widehat{\beta}_\lambda^i))$. These are MaxXZMean, CptsMADZ, MADFlZ, and MADMedianY. It should be noted that, in the case of highly correlated features (which may be expected here due to the fact that all features were also computed with their robust counterparts), the LASSO algorithm only chooses one of those features which may lead to the small number selected. A useful tool to judge whether the choice of only four features is justified is the so-called cross-validation curve which can be plotted for a `cv.glmnet` object. This curve is given in Fig. 8 and shows that the choice of more than 4 features significantly increases the binomial deviance of the model and thus reduces its

Table 2 Designations and descriptions of the generated features based on the single rotations of each minute which are used for all acting forces F_x, F_y, F_z where the terms in brackets (X,Y,Z) are associated with the forces F_x, F_y, F_z

Featurename	Description
MeanAbsMax(X,Y,Z)	Mean of the absolute maximal force values of each rotation
MeanMean(X,Y,Z)	Mean of the means of force value of each rotation
MeanFl(X,Y,Z)	Mean of the surface under the graph of each rotation
MeanSd(X,Y,Z)	Mean of the standard deviation of each rotation
SdAbsMax(X,Y,Z)	Standard deviation of the absolute maximal force values of each rotation
SdMean(X,Y,Z)	Standard deviation the means of force values of each rotation
SdFl(X,Y,Z)	Standard deviation of the surface under the graph of each rotation
SdSd(X,Y,Z)	Standard deviation of standard deviations of force values of each rotation
OutSd(X,Y,Z)	Number of outliers of the standard deviations of rotations
OutMean(X,Y,Z)	Number of outliers of the mean values of rotations
CptsMean(X,Y,Z)	Number of change points in the mean values of rotations
CptsSd(X,Y,Z)	Number of change points in the standard deviation of rotations
Max(XY,XZ,YZ)Mean	Maximal Euclidean distance of the two-dimensional means of rotations from the spatial median
Mean(XY,XZ,YZ)Mean	Mean Euclidean distance of the two-dimensional means of rotations from the spatial median
Out(XY,XZ,YZ)Mean	Number of outliers of the Euclidean distances of the two-dimensional means of rotations from the spatial median
Max(XY,XZ,YZ)Sd	Maximal Euclidean distance of the two-dimensional standard deviations of rotations from the spatial median
Mean(XY,XZ,YZ)Sd	Mean Euclidean distance of the two-dimensional standard deviations of rotations from the spatial median
Out(XY,XZ,YZ)Sd	Number of outliers of the Euclidean distances of the two-dimensional standard deviations of rotations from the spatial median
Max(X,Y,Z)SdMean	Maximal Euclidean distance of the two-dimensional feature of the standard deviation and mean of rotations from the spatial median
Mean(X,Y,Z)SdMean	Mean of the Euclidean distances of the two-dimensional feature of the standard deviation and mean of rotations from the spatial median
Out(X,Y,Z)SdMean	Number of outliers of the Euclidean distances of the Euclidean distances of the two-dimensional feature of the standard deviation and mean of rotations from the spatial median
MaxMeanXYZ	Maximal Euclidean distance of the three-dimensional means of rotations from the spatial median
MeanMeanXYZ	Mean of the Euclidean distances of the three-dimensional means of rotations from the spatial median
OutMeanXYZ	Number of outliers of the Euclidean distances of the three-dimensional means of rotations from the spatial median

(continued)

Table 2 (continued)

Featurename	Description
MaxSdXYZ	Maximal Euclidean distance of the three-dimensional standard deviations of rotations from the spatial median
MeanSdXYZ	Mean of the Euclidean distances of the three-dimensional standard deviations of rotations from the spatial median
OutSdXYZ	Number of outliers of the Euclidean distances of the three-dimensional standard deviations of rotations from the spatial median

Table 3 Designations and descriptions of the generated features based on the single rotations of each minute which are only associated with one acting force

Featurename	Description
DTWMeans	Mean value of the DTW distances based on F_z
DTWSd	Standard deviation of the DTW distances based on F_z
OutDTW	Number of outliers of the DTW distances based on F_z
CptsDTW	Number of change points in the DTW distances between rotations based on F_z
pValueVZT	The p-value of the 3-depth test based on F_x
pValueRuns	The p-value of the runs test based on F_x
DifpValues	The differences of p-values of the correlation tests based on F_x

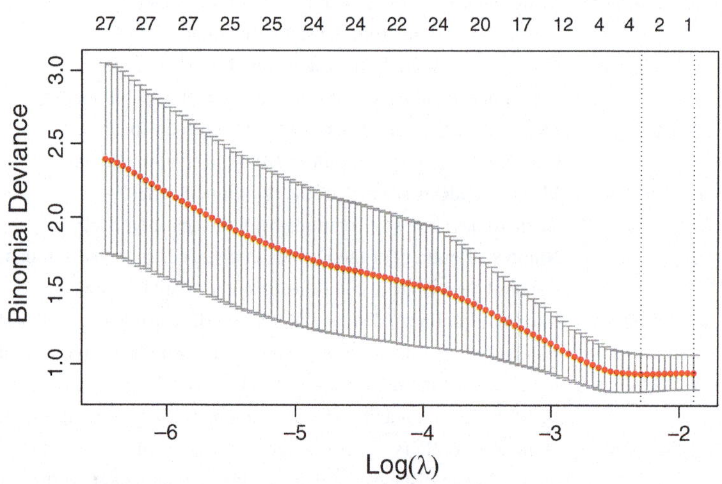

Fig. 8 Leave-one-out cross-validation curve of the B29 data

predictive power which can be seen by an obvious minimum in the curve. Using only those four identified features leads to a misclassification error of 0.186 which is above the "trivial" error of 0.173 that arises, when all observations are classified as "no breakout" and thus is no satisfactory result.

Using the same approach for the B28 data leads to the choice of the features OutMADX, OutSdX, and MADAbsMaxY and a misclassification error of 0.369

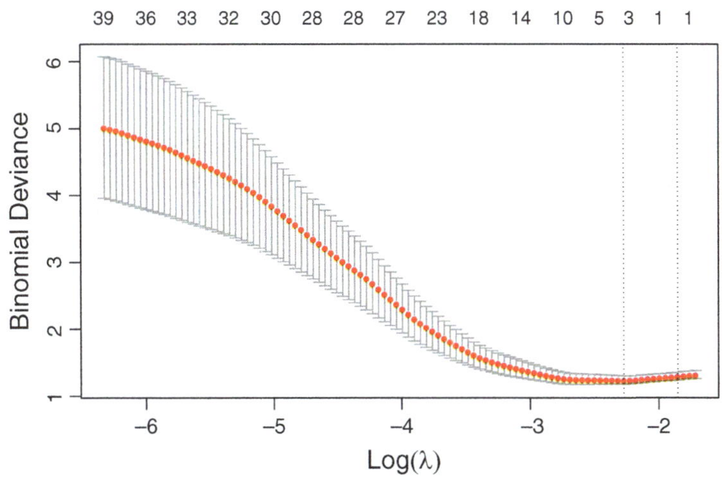

Fig. 9 Leave-one-out cross-validation curve of the B28 data

when the trivial error is 0.342. Studying the shape of the cross-validation curve reveals that for the B28 data the selected number of features is not so clear, see Fig. 9. Here the binomial deviance does not significantly increase up to a number of 14 features. For that reason, it seems to make sense to identify more features which can contribute to the predictive performance of the classification.

For this purpose, the first approach was to use the integrated methods of the random forest. The `randomforest` package provides the possibility to compute the mean decreased Gini index and the mean decreased accuracy for each feature in the random forest based on its bootstrapping approach. This can be carried out by using the `VarImpPlot` function which plots the 30 most important features with corresponding importance values.

These plots show clearly visible gradations in the importances for both datasets. However, when trying to reproduce these results, which means constructing a new random forest, the important variables look quite different. For that reason it is not possible to identify a set of features which has systematically high importance so that the feature selection methods of the random forest do not seem to be very suitable for this classification task. A possible reason for this behavior might be the small sample size which could lead to problems concerning the bootstrapping approach of the random forest, paired with the high amount of features.

Therefore, to identify further important features, the cross-validation-curves of the LASSO were considered once again. This time the maximal amount of variables that does not lead to a significant increase in the binomial deviance was used for the classification task. For B29, there is an obvious minimum in the binomial deviance curve for the mentioned 4 features so that this procedure does not lead to additional important features for the classification task. However, for B28, as already mentioned before, 14 features can be selected this way. These features are

MeanFlY, MeanAbsMaxZ, MADAbsMaxX, SdMeanY, SdMeanZ, MADAbsMaxY, MaxZSdMean, OutMedianXYZ, OutSdX, OutMADX, OutMADY, OutMeanX, OutMedianX, OutMedianZ. Note that these features include the three features identified in the first approach.

Based on the 4 and 14 detected features of the B29 and B28 data, respectively, a random forest and a logistic regression were constructed and the corresponding misclassification errors were computed by performing a leave-one-out cross-validation. Using the 4 features for B29 which are the same as before, the misclassification error for the logistic regression cannot be improved while it is even worse for the random forest with 0.213. Using the 14 features of the B28 data, on the other hand, the misclassification error in the logistic regression managed to outperform the trivial rule by 6 correctly classified observations with a misclassification error of 0.260 while the random forest with a relative error of 0.329 also provides a slight improvement compared to the trivial error of 0.342.

The selected features for the B29 and B28 data do not have much in common besides the result that they are mainly based on one of the three process forces in x, y, and z direction. However, the process forces in drilling direction z as well as the lateral forces in x and y direction all provide selected features. Moreover, both selected feature sets include a feature which is based on the surface under the curve of each rotation, but it is the MAD for the B29-data and the mean for the B28-data. The number of change points is only a selected feature for the B29-data. On the other hand, features based on outliers are only selected for the B28-data. Since only the selected features for the B28-data provide a misclassification error smaller than the trivial classification, these features are more relevant.

7 Discussion

It was tried to classify minutes with and without diamond breakout by generating 147 features based on the ca. 616 rotations per minute. After a feature selection step, random forest and logistic regression were used for the classification. It turned out that the logistic regression combined with LASSO is superior to the random forest in the selection step. Moreover, logistic regression provides better classification results than random forest applied to the selected features. However, only for one of the two considered series of experiments, the misclassification error was smaller than the trivial classification. This is the series of experiments with the higher amount of minutes with diamond breakout. Hence, only the selected features of this series can provide promising features for the classification task. Apart from simple features based on location and scale parameters, mainly outlier based features are selected. Also, a feature based on the surfaces under the graphs of each rotation was considered to be important. The selected features concern all three process forces. However, features based on more than one process force, change points, DTW distances, or p-values do not seem to be very relevant.

One problem of the analysis could be that the diamond breakout happens before the used stationary part of active drilling which was identified in the first step of the analysis. Another problem was the questionable identification of the start and end points of the rotations which could cause features based on differences between rotations which achieve no good classification power. Surprisingly, these features were able to detect differences in the three concrete samples used in the series of experiments which were expected to behave similarly. This showed that in fact, the three concrete samples behaved differently. As a consequence, the three different concrete samples complicated the classification problem. Since only 25 drilling experiment were performed at each of the three concrete samples, the resulting 75 min are not enough to find a good classification rule for both series of experiments so that more experiments are necessary. Moreover, a better identification of the start and end points of the rotations could improve the classification results. However, it may be too difficult to detect diamond breakouts via force measurements because of the inhomogeneity of the concrete.

Acknowledgments

The authors gratefully acknowledge support from the Collaborative Research Center "Statistical Modelling of Nonlinear Dynamic Processes" (SFB 823, B4) of the German Research Foundation (DFG). They also thank the reviewer for valuable comments which improved the paper.

References

1. Carpinteri, A., Dimastrogiovanni, L., Pugno, N.: Fractal coupled theory of drilling and wear. Int. J. Fract. **131**, 131–142 (2005)
2. Cleveland, W. S., Grosse, E.: Computational methods for local regression. Stat. Comput. **1**, 47–62 (1991)
3. Ersoy, A., Buyuksagic, S., Atici, U.: Wear characteristics of circular diamond saws in the cutting of different hard abrasive rocks. Wear **258**, 1422–1436 (2005)
4. Friedman, J., Hastie, T., Tibshirani, R.: Regularization paths for generalized linear models via coordinate descent. J. Stat. Softw. **33**, 1–22 (2010)
5. Gibbons, J., Chakraborti, S.: Nonparametric Statistical Inference. Marcel Dekker Incorporated, New York (2003)
6. Giorgino, T.: Computing and visualizing dynamic time warping alignments in R: The dtw package. J. Stat. Softw. **31**, 1–24 (2009)
7. Holmes, M., Kojadinovic, I., Quessy, J.-F.: Nonparametric tests for change-point detection à la Gombay and Horváth. J. Multivar. Anal. **115**, 16–32 (2013)
8. Hu, Y.N., Wang, C.Y., Ding, H.N., Wang, Z.W.: Wear mechanism of diamond saw blades for dry cutting concrete. Key Eng. Mater. **304–305**, 315–319 (2006)
9. Kansteiner, M., Biermann, D., Dagge, M., Müller, C., Ferreira, M., Tillmann, W.: Statistical evaluation of the wear behaviour of diamond impregnated tools used for the core drilling of concrete. Diamante Applicazioni and Tecnologia **92**, 24–32 (2018)

10. Kansteiner, M., Biermann, D., Malevich, N., Horn, M., Müller, C., Ferreira, M., Tillmann, W.: Analysis of the wear behaviour of diamond impregnated tools used for the core drilling of concrete with statistical lifetime prediction. Diamante Applicazioni and Tecnologia **96**, 17–25 (2018)
11. Killick, R., Fearnhead, P., Eckley, I.A.: Optimal detection of changepoints with a linear computational cost. J. Am. Stat. Assoc. **107**, 1590–1598 (2012)
12. Konstanty, J.S., Tyrala, D.: Wear mechanism of iron-base diamond-impregnated tool composites. Wear **303**, 533–540 (2013)
13. Leckey, K., Malcherczyk, D., Müller, C.H.: Powerful generalized sign tests based on sign depth. SFB Discussion Paper 12/20 (2020) https://eldorado.tu-dortmund.de/handle/2003/39099.
14. Malevich, N., Müller, C.H.: Optimal design of inspection times for interval censoring. Stat. Pap. **60**, 99–114 (2019)
15. Malevich, N., Müller, C.H., Kansteiner, M., Biermann, D., Ferreira, M., Tillmann, W.: Statistical analysis of the lifetime of diamond impregnated tools for core drilling of concrete. In: Ickstadt, K., Trautmann, H., Szepannek, G., Bauer, N., Lübke, K., Vichi, M. (eds.) Applications in Statistical Computing—From Music Data Analysis to Industrial Quality Improvement, pp 233–249. Springer, Berlin (2019)
16. McCullagh, P., Nelder, J.A.: Generalized Linear Models, 2nd edn. Chapman and Hall/CRC, New York (1989)
17. Özçelik, Y.: Multivariate statistical analysis of the wear on diamond beads in the cutting of andesitic rocks. In: Xipeng, X. (ed.) Key Engineering Materials. Machining of Natural Stone Materials, vol. 250, pp. 118–130 (2003)
18. R Development Core Team: R: A language and environment for statistical computing. R Foundation for Statistical Computing. Vienna, Austria (2019). http://cran.r-project.org/
19. Tormene, P., Giorgino, T., Quaglini, S., Stefanelli, M.: Matching incomplete time series with dynamic time warping: An algorithm and an application to post-stroke rehabilitation. Artif. Intell. Med. **45**, 11–34 (2008)
20. Wald, A., Wolfowitz, J.: On a test whether two samples are from the same population. Ann. Math. Stat. **11**, 147–162 (1940)
21. Yu, Y.Q., Xu, X.P.: Improvement of the performance of diamond segments for rock sawing, Part 1: Effects of segment components. In: Xipeng, X. (ed.) Key Engineering Materials. Machining of Natural Stone Materials, vol. 250, pp. 46–53 (2003)
22. Zeileis, A., Grothendieck, G.: zoo: S3 infrastructure for regular and irregular time series. J. Stat. Softw. **14**, 1–27 (2005)

Visualising Complex Data Within a Data Science Loop: A Spatio-Temporal Example from Football

Leo N. Geppert, Katja Ickstadt, Fabian Karl, Jonas Münch, and Michael Steinbrecher

Abstract The cross-sectional research area of data visualisation plays an important role in data science. Graphical presentations provide an accessible way to understand distributions, outliers, processes, trends and patterns in data, and to separate signal from noise. Visualisation tools support the data scientist in representing and analysing Big Data and/or data streams. They are a central tool in all steps of the data science loop. In this contribution we will point out some pitfalls when visualising complex data and will give recommendations on how to avoid them. We will go into more detail about different roles of visualisations, in particular, covering the roles of exploration and presentation and the role of the viewer (data scientist, practitioner, public). For demonstration, we will be using two example data sets from association football.

Keywords Visual analysis loop · Tracking data · Data representation · Derived information · Limitations of data · Static vs. interactive visualisation

1 Introduction

Visualisation plays an important role in all data analyses and is essential for analysing Big Data. Visual representations of data and results complement spreadsheets of mere numbers and summary measures thereof. Our visual perception adds to understanding data and is particularly helpful in finding patterns in data or understanding data structures.

L. N. Geppert (✉) · K. Ickstadt · J. Münch
Department of Statistics, TU Dortmund University, Dortmund, Germany
e-mail: geppert@statistik.tu-dortmund.de; ickstadt@statistik.tu-dortmund.de; jonas.muench@tu-dortmund.de

F. Karl · M. Steinbrecher
Institute of Journalism, TU Dortmund University, Dortmund, Germany
e-mail: fabian.karl@tu-dortmund.de; michael.steinbrecher@tu-dortmund.de

A. Steland, K.-L. Tsui (eds.), *Artificial Intelligence, Big Data and Data Science in Statistics*, https://doi.org/10.1007/978-3-031-07155-3_13

Data visualisation, i.e., the graphical representation of information and data by using visual tools, is a cross-sectional research area that plays an important role in data science. Graphical presentations, like charts, histograms, flow charts, graphs, decision trees, time series or (choropleth) maps, supported by colours and/or shading, provide an accessible way to see and understand distributions, outliers, processes, trends and patterns in data and to separate signal from noise. They enable to highlight useful information and, thus, play a key role in storytelling. When analysing Big Data and/or data streams, visualisation tools are essential to support the data scientist in representing massive amounts of data. For overviews on the topic see, e. g., [4, 6, 10]. Especially the works of William Cleveland inspired the R language [9] and its focus on well-made visualisations.

The main steps in a data science analysis comprise data acquisition, data exploration, data analysis and modelling, model validation and selection, and the representation and deployment of results. From a structural perspective, these are inspired by the famous CRISP-DM (Cross Industry Standard Process for Data Mining) [1], a predecessor of data science. The following six steps are crucial for CRISP-DM:

1. Business understanding
2. Data understanding
3. Data preparation
4. Modelling
5. Evaluation
6. Deployment

Usually, these steps are iterated in a cyclic loop, which leads to the term data science loop or data analysis loop. The number of building blocks for the data science loop may be enlarged, see, e.g., [12] for an enlargement from a statistics point of view, or [2] from a computer science perspective.

While visualisation plays a role in every one of the six steps of the CRISP-DM loop, it is particularly relevant for steps (2) and (3), in which the data are explored and prepared for subsequent analyses. It is also essential for presenting the results of the modelling step (4), and for graphical presentations such as residual diagnostics in support step (5). Visualisation is also a central tool in the deployment step (6). Since visualisation is a key ingredient in all building blocks of the data analysis loop and visual analyses typically exhibit the same adaptive nature as data analyses in general, the loop is also valid and can be transferred to visualisation tasks. In this situation, we can, hence, refer to it as *visual analysis loop*.

Thus, visualising data and results is an adaptive process just as any data analysis is. In this contribution we will point out some pitfalls when visualising complex data, i.e., data that is large in terms of observations or variables or that exhibits intricate structure, and will give recommendations on how to avoid them. We will go into more detail about different roles of visualisations, in particular, covering the roles of exploration and presentation and the role of the viewer (data scientist, practitioner, member of the public). We will be discussing interactive as well as static graphics. Throughout we will be using two example data sets from association football.

2 Data

Our data consists of two matches of the highest German football league, the *Bundesliga*, between SV Darmstadt 98 and Borussia Mönchengladbach as well as FC Augsburg and Hamburger SV. Both matches took place on the 34th and last match day of the 2015/16 season and were kindly provided by Deutsche Fußball-Liga (DFL).

Each data set consists of tracking data recorded by special cameras installed in the stadiums for that purpose. They record the ball's position in three dimensions and every player's position in two dimensions. The cameras record 25 frames per second, resulting in 135,000 observations per player and the ball for a match of 90 min, not including extra time.

Figure 1 shows a processed version of one frame as an example. The frame in question is taken from the match between SV Darmstadt 98 and Borussia Mönchengladbach. We overlay the positions of all players and the ball onto a football pitch. The ball is represented by a black dot, while the players are given in circles with background corresponding to the teams' kit colours, in this case blue for Darmstadt and white for Mönchengladbach. We also add convex hulls around both teams, the filled one corresponding to the Mönchengladbach team indicates that they are in possession of the ball at this moment. The ball is not currently controlled by any player, but the last player who controlled it is a Mönchengladbach player, thus possession has not changed.

In addition, some meta-information are available. For the ball, these are the ball's speed, the minute of the match, whether the ball is in play, and which team has possession of the ball. For the players, the speed as well as the minute of the match are available. Meta-data like identification numbers for players and teams as well as dimensions of the pitch are also given. These make it possible to uniquely identify players and teams over different matches or seasons.

The quality of the data is very good in principle, no observations are missing and documentation is available. However, information about the precision of measurements is not known. Imprecisions may occur, as illustrated by Fig. 2. This frame taken from the match between Darmstadt and Mönchengladbach seemingly shows the ball missing the goal, but it actually is the first goal for Mönchengladbach in the match. This incongruency may be because of measurement inaccuracies or may also be a result of a goal that was not placed completely in the centre of the pitch. For more details, see Sect. 3.1.

3 Pitfalls and Recommendations

In this section, we put the focus on pitfalls a data scientist may encounter when analysing data with the help of visualisations, especially for large or complex data sets. We do not cover how to visualise the data in the first place as this is outside the scope of this contribution. For the purposes of this section, we will assume that

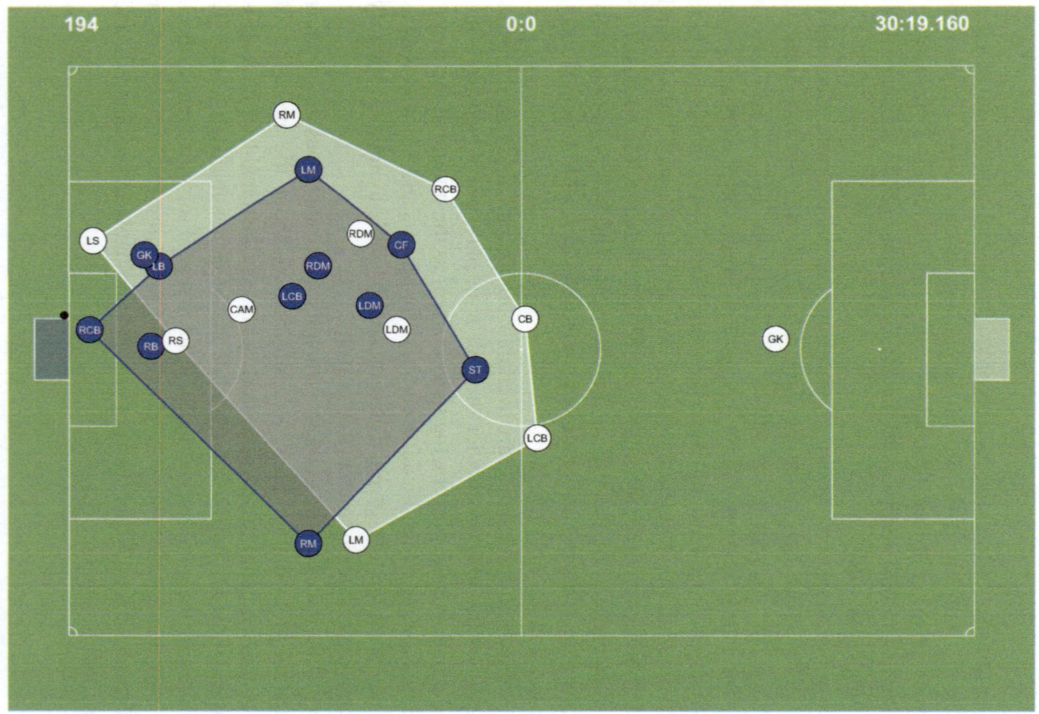

Fig. 1 One exemplary frame taken from the match between SV Darmstadt 98 and Borussia Mönchengladbach. Circles with blue background symbolise Darmstadt's players, circles with white background Mönchengladbach's players. The players' positions are given by their abbreviations. The ball is indicated by a black dot. Convex hulls are drawn around both teams excluding goalkeepers. The white convex hull is filled to indicate that Mönchengladbach is currently in possession of the ball

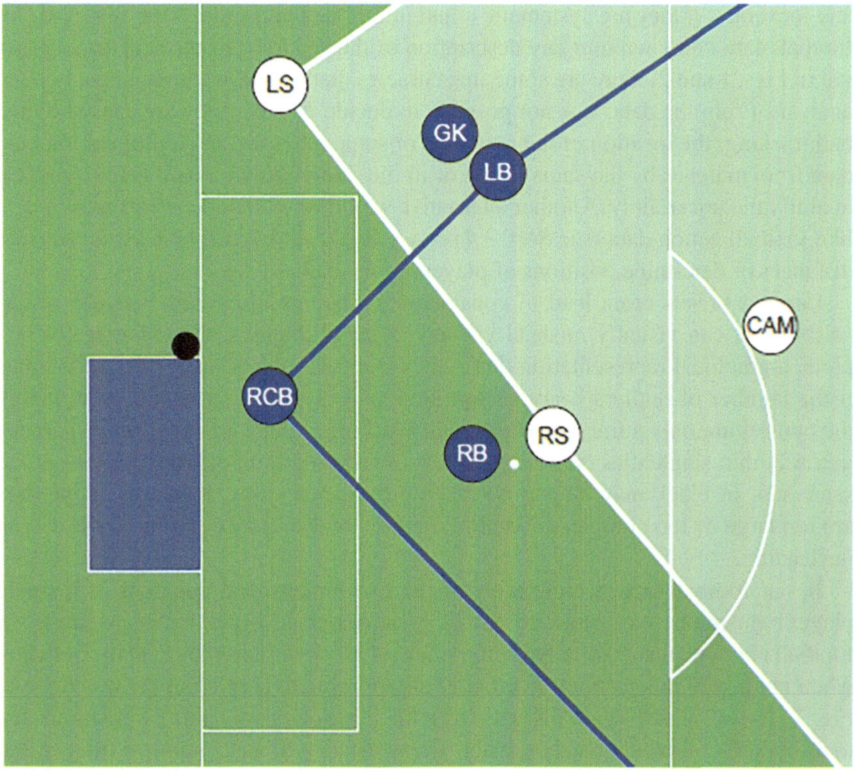

Fig. 2 Same frame as in Fig. 1, zoomed in to focus on ball and goal. Although the ball seems to miss the goal, the frame shows the first goal for Mönchengladbach in the match

visualisations, graphical representations, and sets of colour are chosen in a useful way and graphics in general are well presented. For further reading on these topics, we refer to the excellent books [4, 6, 10].

3.1 Raw Data Might Not Be Suitable for the Underlying Question

In general, it is a good idea to produce simple univariate or bivariate plots of the data that show every observation as one of the first steps of the analysis. This will give some insights about location and dispersion of the data, uni- or multimodality, and seasonal effects, among other things. However, this strategy may reach its limits for large or complex data.

A problem that may arise irrespective of the size of the data is the quantification of uncertainty. Data uncertainties can have an influence on measures of dispersion and location—if they are systematic—and in turn on the results of the analysis. Our football data come without any description of the accuracy of the raw data. As we see in Figs. 1 and 2, there are some inaccuracies that can have a huge impact on the analysis. From the data, it is not possible to decide whether they are caused by the ball tracking, the location of the goal post or some other factor. Additional tracking data from matches by the same teams or in the same stadium could help to (better) quantify the uncertainty. Another alternative would be to include other types of data like football action data (see Sect. 3.4) or a video feed that can be used to measure distances or determine positions of players or the ball.

Large data sets often lead to considerable amounts of overlap between observations that can visually mislead viewers. Such problems can even arise for box plots, a graphical representation that scales well with the number of observations, if the number of outliers becomes too large and visually dominates, even though the outliers are only a fraction of the data. Working with an amount of transparency in a way that single observations show up in grey and multiple observations at the same spot in black may help to some extent in such cases. However, if the data are too large or too complex, just plotting the raw data may be of no avail or even misleading.

In our football data, simple plots of the two-dimensional positions of a single player's positions over time are not of great help, see Fig. 3 for an example. In football matches, teams usually switch sides at half-time. This may lead to confusion when not taken into account. Even after an appropriate correction (as was done in Fig. 3), it is difficult to gain insights from the positions or the path of positions due to the nature of football matches, where every player will make multiple runs up and down the pitch over the course of a match. In Fig. 3 we can see that Lewis Holtby

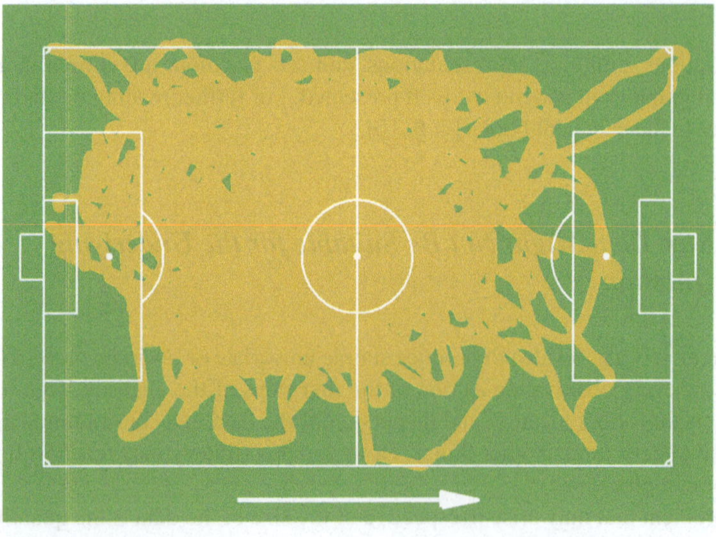

Fig. 3 Two-dimensional scatter plot of the positions of Lewis Holtby (left defensive midfield) over the whole match

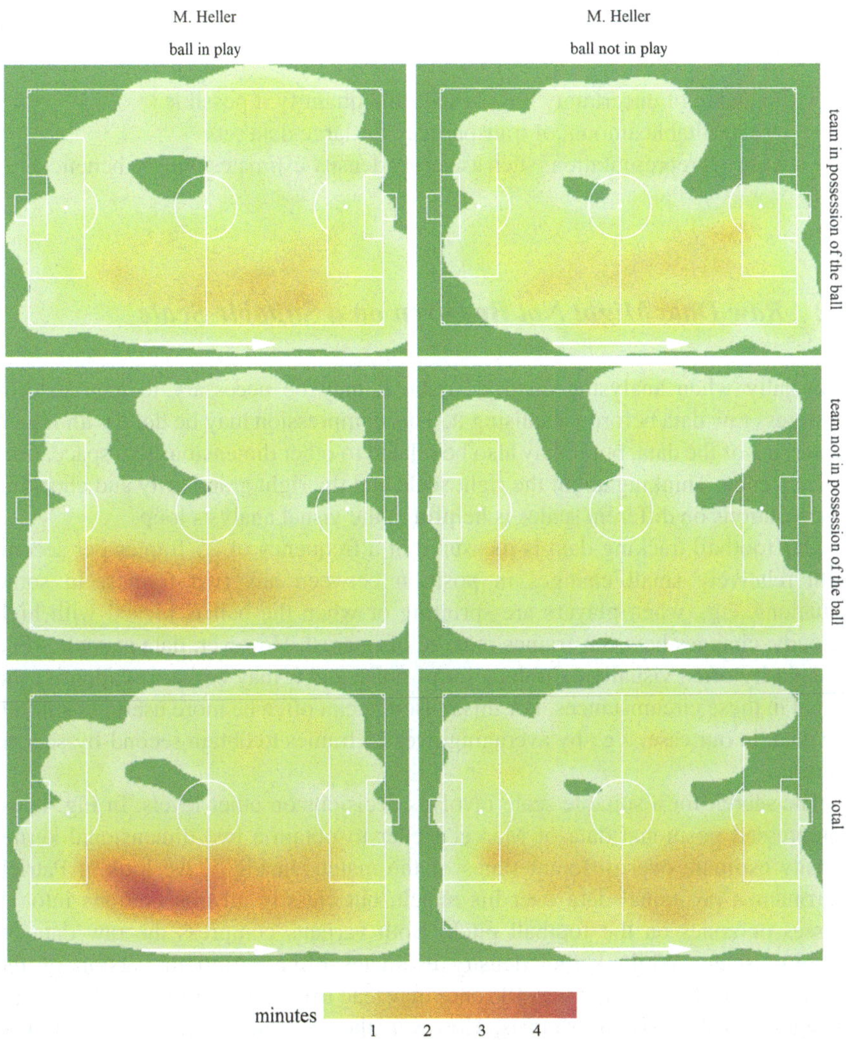

Fig. 4 Two-dimensional kernel density estimate of the positions of Marcel Heller (right midfield), divided into ball in play (left-hand side) and ball not in play (right-hand side) as well as own team in possession of the ball (upper row), opposing team in possession of the ball (middle row), and total (lower row)

covered most of the pitch, but whether he just ran across a position for a split-second or whether he stayed at one place for some time is not clear.

For large or complex data, it may be beneficial to reflect on what the underlying question is. If we are interested in what amount of time a player spends at what position on the pitch, plotting the data as a two-dimensional kernel density estimate will bring a lot more insight compared to the raw positions, see Fig. 4.

Recommendations:

- Be aware of uncertainty in your data and quantify if possible
- Use a suitable amount of transparency for large data sets
- Abstract representations such as kernel density estimates may be beneficial

3.2 Raw Data Might Not Be Given on a Suitable Scale

Especially when analysing complex data, it may be necessary to aggregate or compress raw data before visualising it. This compression may be done with regard to the size of the data, but it may also be related to other dimensions like space, time or frequency. Thinking about the right scale and the right granularity and checking visualisations on different scales is helpful in the visual analysis loop.

Our football tracking data is measured at a frequency of 25 frames per second with relatively small changes in position between any two frames. In some situations, e.g., when players are sprinting or when the ball is kicked with high velocity, changes between frames may be of interest. However, data at such a fine granularity often visually exhibits a lot of noise and it may be hard to identify the signal in these circumstances. For this reason, it can often be more useful to smooth the data, in our case, e.g., by averaging over 25 frames to obtain second-by-second values.

The search for a suitable scale involves decisions on other levels. In Fig. 4, we compressed positional data of Marcel Heller to obtain a two-dimensional kernel density estimate over different states of the match. In Fig. 5, we look at Patrick Herrmann's positional data over his match, but classify all observations into 18 classes or zones on the football pitch. Both variants compress the raw data on a useful scale. While kernel density estimates offer a smooth version of the compression, classification into 18 zones may lead to discontinuities or rather abrupt changes and thus visual artefacts. However, the 18 zones offer a more intuitive explanation in football terms, e.g., time spent in the opposing box, and may thus be more meaningful and relevant to viewers with a strong football background.

Both Figs. 4 and 5 compress the positional data over time. This is useful if we are interested in players' typical positions and the time they spent there. However, such visualisations will not be helpful if we want to analyse players' movements with or without the ball or even a team's attacking patterns. In such cases, we would have compressed the data on the wrong scale and lost the interesting information. Aggregating or compressing data is an important and often necessary tool, but we need to keep its limitations and our aim in mind.

Fig. 5 Time Patrick Herrmann (left midfield) spent in 18 different zones on the pitch, divided by own team in possession of the ball (left-hand side) and opposing team in possession of the ball (right-hand side) as well as first half (upper row), second half (middle row) and whole match (lower row)

Recommendations:

- Compressing data on a suitable scale and granularity enables better identification of signals in the data

(continued)

- Be aware of what you may lose when compressing
- Investigate different granularity levels or subsets of data to obtain a detailed picture and perhaps extended perspectives

3.3 Visualising Derived Information from (Raw) Data

In addition to aggregating and compressing data as described in Sect. 3.2, the visual analysis may benefit from deriving information from the raw data. Our football tracking data sets include information about which team is in possession of the ball for every time frame. From this, we can easily derive wins of possession of the ball, i.e., the moment when possession of the ball switched from one team to the other. We will elaborate on this from a different perspective in Sect. 3.4. Another example are typical positions of all players over the course of a match as illustrated in Fig. 6.

Figure 6 shows the mean position for all players of SV Darmstadt 98 and Borussia Mönchengladbach during the first half with the ball in play. Half-time coincides with the first substitution. The figure is divided into the categories "team is in possession of the ball" and "opposing team is in possession of the ball." This allows for an easy visual comparison of the teams' basic strategies, both with regard to a comparison of both teams and to a comparison of the changes in mean positions on attack and defence for each team.

The lower row of Fig. 6 illustrates the typical defensive positions for both teams. SV Darmstadt 98 chose a 4-4-1-1-formation with four defenders, four defensive midfielders, one attacking midfielder and one forward. In contrast, Borussia Mönchengladbach opted for a 3-5-2-formation. As Borussia Mönchengladbach was in possession of the ball for about two thirds of the time, the subfigures in the upper right corner and the lower left corner are based on more observations than the other two subfigures. The difference between typical positions on offence and defence seem to be greater for Borussia Mönchengladbach than for SV Darmstadt 98. The most illustrious example of this is the Mönchengladbach's left midfielder Patrick Herrmann (number 7) whose position on offence resembles a third attacker more than a midfielder. In contrast, SV Darmstadt 98 keeps more closely to their formation also on attack, with the possible exception of the wide midfielders who both play further up the field.

In addition to Fig. 6, we also calculated the Euclidean distances (in metres) between each player's position on attack and on defence (with the exception of the goalkeepers). Mönchengladbach's player Herrmann indeed shows the highest distance between his positions on attack and defence with 15.39 m. However, when looking at the whole team, the average positional difference is 7.32 m for SV Darmstadt 98 and 6.60 m for Borussia Mönchengladbach and thus higher for Darmstadt. Even though SV Darmstadt kept their formation both on attack and defence, they collectively moved up the pitch substantially. Darmstadt's right

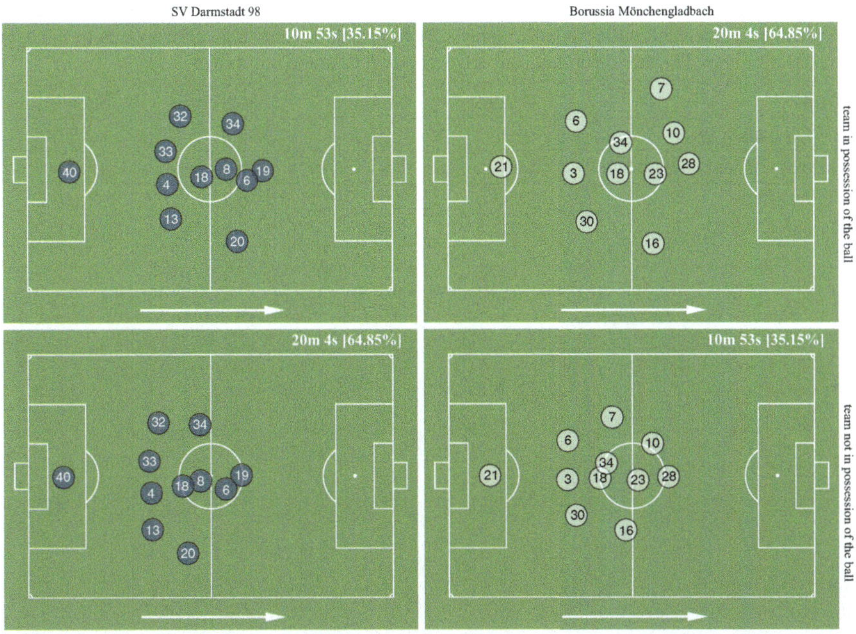

Fig. 6 Mean position of each player of SV Darmstadt 98 (left-hand side) and Borussia Mönchengladbach (right-hand side) while the team is in possession of the ball (upper row) and while the opposing team is in possession of the ball (lower row). This figure is based on times the ball was in play. The numbers in the circles indicate the players' shirt numbers. Both teams are rotated in such in way that the opposing team's goal is on the right-hand side of the pitch as indicated by the arrow. The numbers in the upper right corner of each subfigure indicate how much time the subfigures are based on. The diagonal and the off-diagonal show the same subset of the game from the different teams' perspective, respectively

centre-back Aytaç Sulu (number 4) shows the lowest positional difference of his team, but with still almost 5 m it is substantially higher than the lowest positional difference for Mönchengladbach, 2 m positional difference by centre-back Andreas Christensen (number 3). Indeed, for the other two teams Augsburg and Hamburg, the lowest positional difference is also just over 2 m, see Table 1 that gives an overview of the average positional difference as well as minimum, maximum, and range for the two matches. For both matches, these values cover the first half of the match, during which no substitutions took place in either of the matches.

It is easily possible to include derived data, e.g., when goals were scored, and incorporate them into the visualisation. Figure 7 takes up on the players' mean positions shown in Fig. 6 but distinguishes between the mean positions of Borussia Mönchengladbach's players before and after their first goal after 31 min. Here, it is additionally possible to compare tactical changes that occurred after the lead. On offence, Mönchengladbach's players take more attacking positions after scoring the first goal but also seem to concentrate more on the left wing. The mean positions

Table 1 Summary statistics of the positional differences (in metres) between players' positions on offence and defence with the ball in play. The upper half of the table corresponds to the data underlying Fig. 6. For the lower half, the corresponding figure is not shown

Team	Average	Minimum	Maximum	Range
SV Darmstadt 98	7.32	4.92	14.44	9.52
Borussia Mönchengladbach	6.60	2.05	15.39	13.34
FC Augsburg	5.91	2.06	10.75	8.69
Hamburger SV	5.52	2.08	13.15	11.07

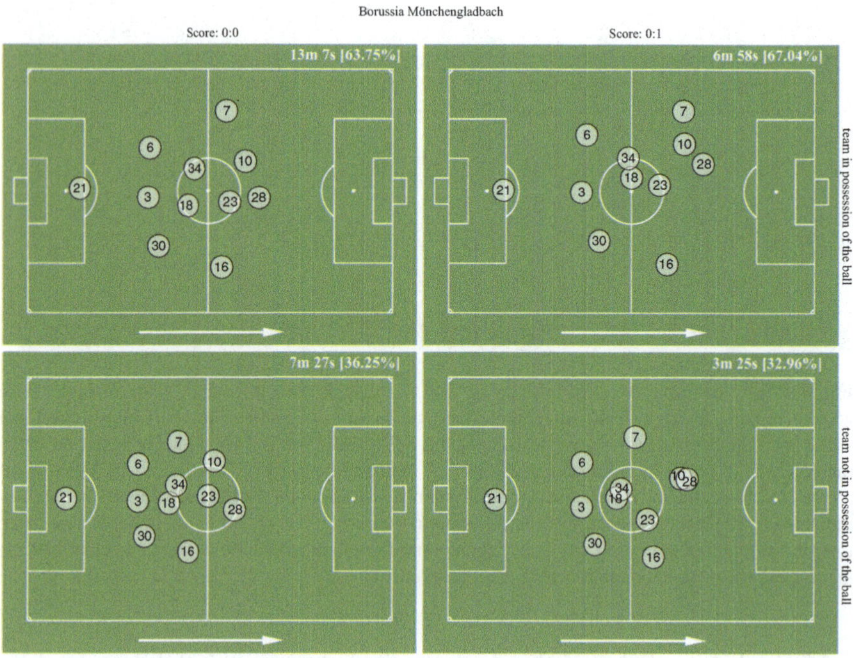

Fig. 7 Mean position of each player of Borussia Mönchengladbach at a score of 0:0 (minutes 1–31, left-hand side) and after Borussia Mönchengladbach's 0:1 (minutes 32–45, right-hand side) while the team is in possession of the ball (upper row) and while the opposing team is in possession of the ball (lower row). This figure is based on times the ball was in play. The numbers in the circles indicate the players' shirt numbers. The opposing team's goal is on the right-hand side of the pitch as indicated by the arrow. The numbers in the upper right corner of each subfigure indicate how much time the subfigures are based on

of SV Darmstadt 98 (not shown) analogously exhibit more defensive positions and a slight change towards the right wing. The numbers in the upper-right corner of each subfigure indicate that Borussia Mönchengladbach spent around two thirds of the time in possession of the ball for both intervals of the first half but became even more dominant after scoring.

Table 2 shows the distances for both teams on offence and defence before and after the goal. The mean positional distance is higher for Mönchengladbach on

Table 2 Summary statistics of the positional differences (in metres) between players' positions with the ball in play up until Borussia Mönchengladbach's first goal (minute 31) and from then on until half-time. The data for Borussia Mönchengladbach correspond to the data underlying Fig. 7, the corresponding figure for SV Darmstadt 98 is not shown

Team	Situation	Average	Minimum	Maximum	Range
SV Darmstadt 98	Offence	8.17	5.80	9.90	4.11
Borussia Mönchengladbach	Offence	6.80	3.51	11.27	7.77
SV Darmstadt 98	Defence	6.57	3.80	10.39	6.59
Borussia Mönchengladbach	Defence	9.10	6.30	13.22	6.91

defence and conversely for Darmstadt on attack. The distances in Fig. 7 may seem more striking for the upper two subfigures, i.e., when Borussia Mönchengladbach is on attack, whereas Table 2 shows a higher mean distance on Mönchengladbach's defence. Both representations of the data capture different aspects; the visual focus lies on changes in the team's tactical formation while Table 2 complements this by stressing that their defensive formation is much more aggressive as they defend in higher positions on average.

Recommendations:

- Data derived from raw data may add visual structure
- Visualisations of derived data might lead to further research questions and answers
- The interpretation of graphics benefits from additional summary statistics

3.4 *Limitations of Data*

Over the course of both the visual analysis loop and the data analysis loop, it is vital to find out what the data can tell you, but also what the data structurally cannot tell you. Our football tracking data offers a rich source of positional data for the ball and the players. It also contains information on the possession of the ball and thus changes in the possession. However, it neither provides means for the quantification of data uncertainty nor information on the actions players took, e.g., whether they shot the ball, passed it to a teammate, tackled an opponent or fought for the ball. Some of these information may be derived from the data (see Sect. 3.3), but uniquely identifying such situations in difficult.

If the information the data does not contain is vital for the analysis, it may be necessary to search for more data to include in the analysis. In football, there also exist game action data, e.g., such data sets note which player took what action out of a catalogue of recognised possible actions. Combining a tracking data set and a

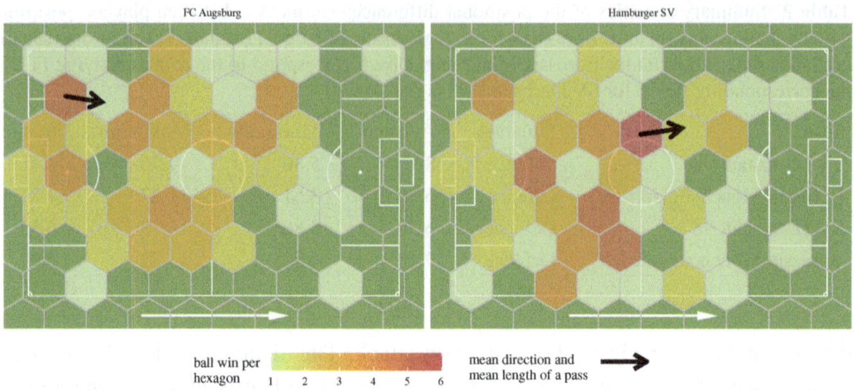

Fig. 8 Pitch overlaid by 57 equally sized hexagons. The colour of each hexagon indicates the number of wins of possession of the ball in open play in this area by FC Augsburg (left subfigure) and Hamburger SV (right subfigure). Green hexagons indicate no wins of possession in this area. The black arrows indicate mean direction and mean length of a pass after a change of possession, centred on the hexagon with the highest number of wins of possession

game action data set may provide an opportunity to gain additional insight. In our analysis, we utilised another technique: using a proxy variable.

Figure 8 shows a characterisation of the teams' transition match, in our case where teams won the ball and where they played the ball next. We visualised the first component, winning the ball, by dividing the pitch into 57 equally sized hexagons and counting the number of wins of possession of the ball within that area. This part of the visual analysis is completely based on the tracking data. The second component, where the ball went next, is not easily extracted from the tracking data, as it is not clear whether the following action was a controlled pass or an attempted clearance kick to touch. We decided to use the difference between the position of the win of possession and the position of the ball a short time afterwards as a proxy. We then calculate the average over all positional differences and indicate them as an arrow overlaid on top of the hexagon connected to the most wins of possession. In the caption of Fig. 8, we relate to the arrow as mean direction and mean length of a pass, but keep in mind that we average over all passes, clearances, lost duels immediately after the win of possession, and other possible actions.

In a reprise of Sect. 3.2, it is possible to look at the typical passes on a different level and, e.g., differentiate between wins of possession in the own half and in the opposing half.

Other limitations of the data may be due to the way the data were collected. Especially convenience samples that are not the result of an experimental design but rather were collected because they were there or easy to obtain, may show biases or otherwise not generalise well from sample to population. In some cases, possible problems may be identified visually, e.g., by looking at the distribution across age-groups in the sample and comparing them to the corresponding distribution in the population or by studying diagnostic plots of a model built on the data like residual plots for regression models. But generally speaking, it is difficult to find out what the data cannot tell you by analysing the data, visually or otherwise. Business understanding, step (1) of the analysis loop, is vital in identifying limitations of the data.

Recommendations:

- Deeply understand your problem and your data set (and how it was collected)
- Identify additional data sources or proxy variables
- Use diagnostic plots
- Compare distribution of a sample to distributions in the population

4 Visualising the Data

Visualising data plays different roles in different stages of the data analysis loop. When understanding the problem or the data is the central task (steps (1) to (3) in Sect. 1), visualisations are more focused towards exploration. Finding typical values, patterns, connections, outliers or otherwise interesting observations are typically central goals in these stages. Visualisations primarily aim at helping the viewer (here especially data scientists) in finding interesting directions to follow with the analysis.

In later steps, the focus shifts towards presentation or sometimes confirmation (steps (4) to (6) in Sect. 1). This is especially true in the evaluation and deployment steps. Here, typical goals are finding out whether a model or its assumptions are suitable as well as conveying findings to other people (practitioners, members of the public) who were typically not involved in the analysis process.

Another, loosely related, dimension of visualisations is their mode of presentation. We classify the mode into two broad categories: interactive visualisations and static visualisations.

Suitable software tools are required to carry out the visual analysis loop. In the R world, a great number of options to produce high-quality visualisations are available. We employ both the built-in graphics system in package `graphics` [9] and `ggplot2` [13] for static visualisations. For interactive visualisations, we combine these packages with `shiny` [5]. To calculate the results we visualise, we use package `MASS` [11] for two-dimensional kernel density estimates and package `hexbin` [3] for hexagonal binning.

4.1 Interactive Visualisations

Interactive visualisations are a very direct and immersive way of presenting data or results. They allow the viewer to set their own course and their own priorities. If the viewer chances upon a finding that sparks his:her interest, s:he can easily follow up on this and delve into a deeper visual analysis. On the other hand, interacting with a data set or model results in such a way may require considerable knowledge on part of the viewer. Recreating the analysis path a person with expert knowledge took may be difficult. Seen from this perspective, it may be more natural for interactive visualisations to be used in the earlier steps of the visual analysis loop or in cases where domain experts and visualisation experts study the results together.

In our analyses of the football tracking data, we employ interactive visualisations mostly in two different situations. One example is to visualise the ball's and the players' positions over time within a relatively short time frame, e.g., thirty seconds leading up to a goal. Our main aims here are to look for typical patterns in players' behaviour and to try out different ways of visualising the positions. Figure 1 is an example taken from such an interactive visualisation.

Interactive visualisations can make it very easy to examine possible connections between different aspects of the data. In football, winning possession of the ball is important, albeit frequent. Generally speaking, winning possession of the ball closer to the opponent's goal is more advantageous as the distance to the goal and the number of defenders in the way is lower. In our analyses, we introduced an overview of all positions where the ball was won in open play on a pitch. This in itself makes an interesting pattern that is also depicted in Fig. 8. In our related interactive version, it is possible to select one of the positions and view the positions of ball and players over the 5 s following the change of possession.

From a practical perspective, interactive visualisations may be a challenge for large and/or complex data sets and complex models. In our football example, calculating the typical position of a player over the duration of a match to visualise their typical positions as in Figs. 4 and 5 can also be done interactively by choosing between the different matches, the different teams, and the different players. However, loading the data for the chosen player or calculating the two-dimensional kernel density estimates takes considerable time. In such cases, where the amount of interactivity is limited to choosing a certain player, it might be more useful to pre-produce a limited number of static visualisations and allow for an interactive and easy way of switching between them.

4.2 Static Visualisations

Static visualisations offer less interaction between viewer and visualisation, but can be optimised for the intended use more easily. If a data scientist wants to share and emphasise an insight s:he gained, it is easier to draw the viewers' attention to it using static visualisations. Static visualisations might also simplify technical choices like which colours to use for which group, as the number of required colours is known beforehand whereas in an interactive case, viewers might select or deselect groups.

In our analysis of the football tracking data, some static visualisations are motivated by the need for a visual summary of insights we gain during the interactive analysis. Some of the static visualisations contain the insights from the interactive visualisations, boiled down or aggregated over time. As a result, these static visualisations typically become more complex compared to each single frame of their interactive predecessors. Figures 6, 7, and 8 are good examples of this. Figure 6 aggregates the information in Figs. 4 and 5 for each player, resulting in a typical location of all players in one team that can be viewed at the same time. Figure 8, on the other hand, combines the relatively straightforward overview of locations where the team won possession of the ball in open play with the more complicated questions of where the ball went next. Not every analysis will show such a link between interactive and static visualisations, but it may be helpful to view as much of the data as possible in an interactive way first, before narrowing down the analysis and introducing more targeted static visualisations.

How well static visualisations can handle very large data sets or streams depends mainly on the graphical representation chosen. A stripchart—also known as one-dimensional scatterplot—shows every single observation and will quickly reach its limits as the number of observations grows. In contrast, a simple box plot depends on the five values minimum, lower quartile, median, upper quartile, and maximum only and thus scales very well with the number of observations.

5 Conclusion

Visualisation plays an important role in all steps within a data analysis loop. It is an important tool for investigating and understanding the data prior to any analysis and is helpful within the process of data preparation. Visualising results supports the modelling task as well as evaluating the models and is an important link towards telling the story of the data analysis to the scientific community, the practitioner, and the public.

Therefore, visualisation is an important task of the data scientist, and graphical tools belong to his or her toolbox. The data scientist needs to decide when and which type of graphical representation to use, which additional information to give, and how to choose symbols and colours in order to present the information in a clear and appropriate way. Keeping the graphical toolbox up-to-date is as important as being aware of new methodology and algorithmic approaches.

This contribution raises the awareness for pitfalls when visualising complex data and gives some recommendations on how to avoid them. It is important to keep the underlying (research) question in mind when visualising the data, and to consider compression of the complex data for graphical representations on a suitable scale or granularity. The data scientist also needs to decide whether to visualise (parts of) the original data or derived information thereof and needs to understand and be aware of the limitations of the data.

Our data only allow conclusions for the two Bundesliga matches at hand, but similar data exist for all other Bundesliga matches in recent years. These data would be suitable for further analyses such as the temporal development of the playing style of a team, how/whether the style of play is affected by the change of a coach, and comparisons of home and away matches. However, these data are not publicly available. In training situations it is common to collect data by player-specific chips; these facilitate obtaining more precise position data and perhaps the strength of body checks.

This contribution is mainly concerned with scientific graphics. However, often there is a smooth transition between visualising complex data and analytic results and infographics that aim at providing an easy-to-understand overview of a topic and usually comprise a collection of imagery, see [7] for a discussion of both mainly from a statistical perspective. A collection of infographics for a variety of sports are presented in [8].

The last step in the data science loop, deployment, is dedicated to translating data into information and telling the story of the analysis to practitioners and the public. Here, visualisation might directly lead to an infographic or may be used as input for further infographics. This can also be seen in our football example. Figure 8 aims at illustrating the transition game. It can be viewed as a scientific graphic that emerged from the research question where the transition game most likely takes place. However, it can also be seen as an infographic that displays a complex data situation in an intuitive way.

References

1. Brown, M.: Data Mining for Dummies. Wiley, London (2014)
2. Cao, L.: Data science: A comprehensive overview. ACM Comput. Surv. **50**(3), 1–42 (2017). https://doi.org/10.1145/3076253
3. Carr, D.: hexbin: Hexagonal Binning Routines (2021). https://CRAN.R-project.org/package=hexbin. Ported by Nicholas Lewin-Koh and Martin Maechler, contains copies of lattice functions written by Deepayan Sarkar
4. Chambers, J.M., Cleveland, W.S., Kleiner, B., Tukey, P.A.: Graphical Methods for Data Analysis, reissued. CRC Press, New York (2018)
5. Chang, W., Cheng, J., Allaire, J., Sievert, C., Schloerke, B., Xie, Y., Allen, J., McPherson, J., Dipert, A., Borges, B.: shiny: Web Application Framework for R (2021). https://CRAN.R-project.org/package=shiny
6. Chen, C., Härdle, W.K., Unwin, A. (eds.): Handbook of Data Visualization. In: Springer Handbooks of Computational Statistics (2008)
7. Gelman, A., Unwin, A.: Infovis and statistical graphics: different goals, different looks. J. Comput. Graph. Stat. **22**(1), 2–28 (2013)
8. Minto, R.: Sports Geek: A Visual Tour of Sporting Myths, Debate and Data. Bloomsbury, London (2016)
9. R Core Team: R: A Language and Environment for Statistical Computing. R Foundation for Statistical Computing, Vienna, Austria (2021). https://www.R-project.org/

10. Unwin, A., Theus, M., Hofmann, H.: Graphics of Large Data Sets—Visualising a Million. Springer, New York (2006)
11. Venables, W.N., Ripley, B.D.: Modern Applied Statistics with S, 4th edn. Springer, New York (2002). https://www.stats.ox.ac.uk/pub/MASS4/. ISBN 0-387-95457-0
12. Weihs, C., Ickstadt, K.: Data science: the impact of statistics. International Journal of Data Science and Analytics **6**(3), 189–194 (2018). https://doi.org/10.1007/s41060-018-0102-5
13. Wickham, H.: ggplot2: Elegant Graphics for Data Analysis. Springer, New York (2016). https://ggplot2.tidyverse.org

Application of the Singular Spectrum Analysis on Electroluminescence Images of Thin-Film Photovoltaic Modules

Evgenii Sovetkin and Bart E. Pieters

Abstract This paper discusses an application of the singular spectrum analysis method (SSA) in the context of electroluminescence (EL) images of thin-film photovoltaic (PV) modules. We propose an EL image decomposition as a sum of three components: global intensity, cell, and aperiodic components. A parametric model of the extracted signal is used to perform several image processing tasks. The cell component is used to identify interconnection lines between PV cells at a sub-pixel accuracy, as well as to correct incorrect stitching of EL images. Furthermore, an explicit expression of the cell component signal is used to estimate the inverse characteristic length, a physical parameter related to the resistances in a PV module.

Keywords Singular spectrum analysis (SSA) · ESPRIT · Electroluminescence images · Photovoltaics · Thin-film modules

1 Introduction

There has been an increasing interest in automated image analysis of spatially resolved characterisation methods for photovoltaic (PV) modules such as electroluminescence (EL) [8–11, 24, 44–46]. Such automated image analysis aims at quality control of modules and is thus of great interest for manufacturers, PV system owners, and insurance companies, as it allows for a systematic inspection of a large number of modules, both prior and after installation.

Electroluminescence is a commonly used imaging technique for PV modules. It relies on the reciprocal operation of the photovoltaic module as a light-emitting diode, so instead of generating an electric current from light, an electric current is driven through the solar cell, which then emits light. As generating electricity and emitting light are reciprocal processes, one process reveals much about the

E. Sovetkin (✉) · B. E. Pieters
IEK5-Photovoltaik, Forschungszentrum Jülich, Germany
e-mail: e.sovetkin@fz-juelich.de

A. Steland, K.-L. Tsui (eds.), *Artificial Intelligence, Big Data and Data Science in Statistics*, https://doi.org/10.1007/978-3-031-07155-3_14

other. Electroluminescence (EL) images provide spatially resolved information on the solar module and is commonly used to locate and identify defects in the device or extract other (local) solar cell properties.

In this paper our focus lies on EL images of thin-film PV modules. For thin-film technology, unlike for a more common crystalline silicon, little is known about shapes and appearances of defects in EL images. For the crystalline silicon PV modules there exists a well-established catalogue of defects visible in EL images (see [25]), whereas such a catalogue does not exist for thin-film modules.

To study defects in EL images, it is important to find a compact way to represent EL image data. This paper proposes such an approach and considers several image processing algorithms for EL images of thin-film PV modules that are based on the singular spectrum analysis (SSA). Our contributions here are manifold.

Firstly, a specific grouping in the SSA algorithm decomposes an EL image into several components: global intensity variation component, local periodic intensity component (or cell component), and a residual image that contains various local aperiodic features. We argue that each of these components has a different physical origin.

Secondly, the extracted components of an EL image can be approximated by a parametric model that represents an EL image as a small dimensional vector, and hence our methods can also be considered as a dimensionality reduction technique. Furthermore, the parametric model of the cell components is used to estimate the position of the interconnection line between individual PV cells. Our algorithm features symbolic differentiation and estimates positions of the interconnection lines at a sub-pixel accuracy. A similar technique is used in the estimation of the inverse characteristic length, a physical characteristic of a PV module that equals the square of the ratio between different resistances in a module.

Lastly, the cell component signal is used to estimate a non-linear transformation of an image to adjust an incorrectly stitched image.

The rest of the paper is organised as follows. Section 2 reviews the methods and the corresponding literature used in this paper. Section 3 describes the data used in the project. The main contribution of this paper is given in Sect. 4, which focuses on various applications of SSA to the EL images of thin-film modules. Lastly, the paper is concluded in Sect. 5.

2 Methods Overview

This section describes the methods used in this paper and overviews related literature. Our main tool is the singular spectrum analysis method (SSA).

The history of SSA can be traced to the works of [3], where an SSA-like method was established and applied in the context of non-linear dynamics for the purpose of reconstructing the attractorof a system from measured time series. Further, in the context non-linear dynamical system, SSA can be also used for phase space reconstruction algorithm, [12].

The so-called Caterpillar methodology is a parallel development of SSA that originated in the former Soviet Union, especially in Saint Petersburg, independently of the mainstream SSA work in the West, [7]. This methodology became known to the rest of the world more recently. "Caterpillar-SSA" [17], emphasises the concept of separability, a concept that leads, for example, to specific recommendations concerning the choice of the SSA parameters.

Originally, the SSA method was applied to the one-dimensional time-series data. In fact, by now, it is not easy to find an applied area related to the analysis of temporal data, where one-dimensional SSA is not being applied. To name a few applications, the method found its way to the analysis of climate and atmospheric data, [13, 49], to meteorological data, [51], as well as to the marine science, [6]. Lima et al. [29] used SSA for gap filling in precipitation data. This method has been also applied in the financial sector to discover hidden economic cycles, [42]. Groth and Ghil [20] used a multivariate extension of SSA and defined a Procrustes test to the analysis of interannual variability in the North Atlantic sea surface temperature. For further references to various applications of SSA in time-series analysis see [19, 55].

More recently, SSA was also used to analyse digital images and other objects that are not necessarily of planar or rectangular form. This particular development is utilised in this paper. Rodriguez-Aragon and Zhigljavsky [39], used SSA to define a distance between images with a possible application in face verification. In [37, 53, 54] the 1D-/2D-SSA variants were used in the context of hyperspectral images for the purpose of denoising, feature extraction, and classification tasks. In [30] SSA was applied in the context of ultrasonic imaging for improving the imaging of brachytherapy seed. In application related to geoscientific data, 2D-SSA was utilised for gap-filling, [56]. 2D-SSA was also applied in texture classification [34], seismology [47], gene expression [23], and medical imaging [43].

In our application of the 2D-SSA, we utilise the ability of the method to separate signal into trend and periodic components. Further, we use the parametric form of the extracted signal to perform various image processing tasks.

In order to explain our algorithms, we give a review of necessary theory in the following subsections. The SSA algorithm is reviewed in Sect. 2.1. SSA itself is non-parametric, however, a parametric model can be given to describe the extracted signal. There are two sets of parameters to be estimated in the model. The ESPRIT method (Sect. 2.2) estimates frequencies and damping factors of a signal, where the least squares provides an estimation of the amplitude and phases (Sect. 2.3). Lastly, remarks on implementation and comparison to similar methods are discussed in Sect. 2.4.

2.1 SSA

Singular Spectrum Analysis (SSA) is a model-free time-series analysis method that belongs to the so-called subspace methods, [48]. In subspace methods, a signal estimation is performed by taking a certain linear subspace. SSA can also be considered as a low-rank approximation method, [31].

The general theory of SSA for one-dimensional time series is elaborated in [17]. Golyandina et al. [19] provide more updated information on extensions with a strong focus on applications.

The output of SSA-like methods is a decomposition of an observed signal x (e.g. a time series, a multivariate time series or an image) into a sum of identifiable components:

$$x = x_1 + \ldots + x_n. \tag{1}$$

Among all SSA-like methods, the following four common algorithm steps can be isolated.

1. **Embedding.** The original signal x (e.g. time series or an image) is mapped into a matrix $\mathbf{X}$, that is called a *trajectory matrix*.
   The embedding is parametrised by a single parameter denoted by L (a number or a vector).
2. **Decomposition.** The second step consists of the decomposition of the trajectory matrix $\mathbf{X}$ into a sum of matrices of rank 1.
   Often the singular value decomposition (SVD) is used for this purpose, which is an optimal rank-one matrix decomposition in the Frobenius norm sense.
3. **Grouping.** The third step is a grouping of the decomposition components. At the grouping step, the elementary rank-one matrices are grouped and summed within groups.
   The grouping algorithm step is often semi-automatic and depends on the type of data and application.
4. **Reconstruction.** The grouped components are not necessarily valid trajectory matrices. Hence a projector operator to trajectory matrix space is applied on the grouped components, resulting in the final signal decomposition (1).

In this paper, we utilise the 2D-SSA variant of the algorithm, [15, 16, 18]. For this variant, the trajectory matrix is a Hankel-block-Hankel matrix and the corresponding embedding is parametrised by a two-dimensional parameter $L = (L_1, L_2) \in \mathbb{N}^2$. The decomposition is performed with SVD, and the reconstruction projection operator is a 2-step diagonal averaging procedure. The precise forms of the embedding, SVD, and reconstruction projection operator are provided in the Appendix.

The grouping step of the SSA algorithm determines the form of the final signal decomposition. A typical SSA decomposition of a signal is the decomposition into a slowly varying trend, regular oscillations, and noise. An important notion in the SSA theory is the notion of *the signal of finite rank*. This notion allows us to classify and group together rank-one matrices into a trend, oscillations, and noise components.

Informally, finite rank signals are those that have a trajectory matrix of a finite rank (see a formal definition in the Appendix). 2D-SSA produces a class of signals of specific objects of finite rank. Those objects have the form described in the following theorem, [15, 18].

Theorem 1 *And infinite 2D-array $\{x(n, m)\}_{n,m\in\mathbb{N}}$ is of finite rank if and only if*

$$x(n, m) = \sum_{k=1}^{K} p_k(n, m)\rho_{1k}^{m}\rho_{2k}^{n} \cos\left(2\pi(\omega_{1k}n + \omega_{2k}m) + \phi_k\right), \quad n, m \in \mathbb{N}, \tag{2}$$

where $p_k(n, m)$ are polynomials in n and m variables, $\rho_{\cdot k}$ are the damping factors, $\omega_{\cdot k}$ the frequency parameters and ϕ_k are the phase parameters.

2.2 ESPRIT

Estimation of Signal Parameters via Rotational Invariance Techniques (ESPRIT) is a method to estimate parameters of a mixture of one-dimensional, [41], and two-dimensional amplitude-modulated sinusoids, [40, 50], in background noise.

For the 2D-ESPRIT method, [40], the observed data y is generated by the following additive model:

$$y(m, n) = x(m, n) + \varepsilon(m, n), \tag{3}$$

where $0 \leq m \leq N_x - 1$ and $0 \leq n \leq N_y - 1$, ε is the zero-mean Gaussian noise with variance σ^2. The model for the signal x is given by the sum of amplitude-modulated two-dimensional sinusoid

$$x(m, n) = \sum_{k=1}^{K} s_k \rho_{1k}^{m}\rho_{2k}^{n} \cos\left(2\pi(\omega_{1k}m + \omega_{2k}n) + \phi_k\right), \tag{4}$$

where ω_{1k}, ω_{2k} are the normalised frequencies in different directions, α_{1k}, α_{2k} are the damping factors, s_k amplitudes, and ϕ_k phases.

By Theorem 1 the signal (4) is of finite rank. The ESPRIT methods utilises the fact of the rank-deficiency of the trajectory matrix of the observed signal y, and a certain transformation matrix between sub-trajectory matrices is obtained. The frequency and damping factor parameters are computed from the argument and absolute values of the complex-valued eigenvalues of the obtained transformation matrix. See more details in [40, 41].

2.3 Amplitude, Phase Estimation

The ESPRIT estimates the frequencies $\omega_{\cdot k}$ and the damping factors $\rho_{\cdot k}$ of the signal model (4). Here only the amplitudes s_k and phases ϕ_k remain to be estimated. This problem can be reformulated as a linear regression model using the formula of the cosine of sums. With this formula, (4) can be rewritten as

$$x(m,n) = \sum_{k=1}^{K} A_k \rho_{1k}^m \rho_{2k}^n \cos(2\pi(\omega_{1k}n + \omega_{2k}m))$$
$$- B_k \rho_{1k}^m \rho_{2k}^n \sin(2\pi(\omega_{1k}n + \omega_{2k}m)), \tag{5}$$

where $\rho_{\cdot k}, \omega_{\cdot k}$ are the parameters estimated from the ESPRIT, and A_k, B_k are the parameters of the linear model to be estimated.

Note that the dependent variable in the linear regression model are the values of $x(n,m)$, the finite rank signal extracted from y. In terms of the SSA algorithm this corresponds to the series reconstructed from the selected components.

The amplitude and phase are given by:

$$s_k = \sqrt{A_k^2 + B_k^2}, \quad \phi_k = \operatorname{atan}(B_k/A_k). \tag{6}$$

2.4 Implementation: Comparison to Alternatives Methods

For our application, we use an R-package "Rssa", [14, 18, 19, 26], where all the required functionality including the ESPRIT method is implemented. It should be noted that the trajectory matrix for 2D-SSA is large and has $O(N^2)$ number of elements, where N is a number of pixels in the signal image. However, the structure of the Hankel-block-Hankel matrix allows implementing of SVD with Lanczos algorithm efficiently in time and memory by computing product of a matrix and a vector with Fast Fourier Transform, [26, 28].

For computing amplitude and phase parameters, we utilise a standard R-function "lm", [4].

The form of the parametric model of the SSA signal suggests that Fourier analysis can be used to obtain similar results. However, in order to obtain a compact representation of a signal a small Fourier coefficient should be discarded. For that purpose a sparse Fourier analysis approach can be used, [21].

However, the parametric model of the SSA signal is more flexible, as every periodic has an amplitude modulation. Furthermore, the Fourier analysis is a low-resolution type of method, as in the context of time series, frequency can be estimated only up to $1/N$, where N is the length of the time series. Whereas ESPRIT is regarded as a high-resolution method.

3 Data

In this section we discuss the data that is used in this paper. The data was acquired within the framework of the PEARL-TF project. The [35] website contains detailed information about the project and the involved partners. In this project, the data from several solar parks with thin-film modules were collected. In addition to EL images, also performance characteristics of the modules were measured.

The EL images are taken at predefined conditions (selected fixed applied current and/or fixed applied voltage). A silicon CCD sensor camera is used to measure subsequently several parts of the module, with the images being stitched afterwards. The applied voltage and the applied current together with the temperature of the module are being recorded. The I/V characteristics are also measured and the solar cell performance parameters are determined.

The database contains over 9500 EL images of thin-film PV modules. The bulk of these EL images (about 6000) are from co-evaporated Copper Indium Gallium di-Selenide (CIGS) modules with a chemical bath Cadmium Sulfide buffer. All EL images shown in this work are from such CIGS modules. Every image is supplied with measured performance data. A typical EL image of a thin-film module from our database is depicted in Fig. 1. The module consists of 150 connected cells in series (in Fig. 1 the cells are recognised as horizontal stripes). The cells are separated by

Fig. 1 Thin-film module EL image. A module consists of 150 cells (positioned horizontally) connected in series. The cells are separated by interconnection lines (horizontal dark lines). The module consists of several submodules separated by vertical isolation lines, which appear dark in the EL image. The EL image is stitched (there are 1 horizontal and 3 vertical stitch lines); overall intensities of different patches of images are different. These intensity differences are attributed to metastable changes during the measurement

interconnection lines (horizontal dark lines in Fig. 1). In addition, the module is separated into 5 parallel submodules by vertical isolation lines (dark vertical lines).

Every EL image consists of several stitched images. Different stitched parts of the image have different overall intensities (see Fig. 1). This is attributed to the metastable behaviour of thin-film solar cells, where the electrical properties of the cell can change during the measurement.

4 Results

In this section we discuss a selection of image processing methods that we build on top of the SSA framework.

Firstly, Sect. 4.1 describes a decomposition of EL images into several components: global intensity, cell, and aperiodic components. This is a direct result of the SSA decomposition, where grouping is performed using prior knowledge of the image size and number of cells in a module.

Secondly, a parametric model of the cell component signal is used to achieve several goals. The symbolic differentiation allows a global search for the minimum points in the cell components that can be used to identify the interconnection lines (Sect. 4.2). Furthermore, we demonstrate the estimation of the inverse characteristic length, a physical parameter that depends on several resistances in a PV module (Sect. 4.3).

Lastly, MSSA and its parametric model is used to obtain a non-linear transformation needed to correct the stitch line in EL images.

4.1 Image Decomposition

The grouping step of the SSA algorithm allows combining of decomposed rank-one components into groups. We define these groups that result in EL image decomposition onto 3 components: global intensity variation, cell, and aperiodic components.

Figure 2 shows a close up image of the module, that clearly identifies a set of 10 parallel cells separated by interconnection lines. The EL intensity varies systematically over the cell width. This is the result of the series resistance of the electrodes. As the transparent zinc-oxide front electrode exhibits a much larger sheet resistance than the molybdenum back contact, the voltage over the diode junction drops from one side to the other [22, 36]. As the EL intensity depends primarily on the cell voltage this leads to a clear intensity gradient over eachcell.

To achieve good separability of the periodic components, the parameter L of the SSA embedding is selected to equal approximately half of the dimensions of the input image, [17, 19]. The ESPRIT estimates frequencies and damping factors of the components, which are used in the decision of the grouping step.

Fig. 2 An enlargement of a 10-cells EL image region. Each cell exhibit periodic behaviour

A thin-film module consists of a fixed number of interconnected cells, therefore, the cell components can be identified as periodic components that have a period smaller than $\frac{\text{image width}}{150}$. The other components are grouped together as the global variation and the residual image captures information of non-low rank signals, such as aperiodic features of an EL image.

The image decomposition algorithm steps are described in the Algorithm 1.

Input: EL image X with dimensions $N_x \times N_y$.
Output: Three images with the same dimensions as X: global intensity component G, cell component S and aperiodic component R.

1. Perform embedding and decomposition steps of the 2D-SSA algorithm for X. Compute first 50 elements in the rank-one decomposition of SVD, denote the sum of this components as $\tilde{\mathbf{X}}$.
2. Apply ESPRIT on the low-rank signal $\tilde{\mathbf{X}}$.
3. • Set the cell components S as in (4), choosing only components with frequency $\omega_{1k} > \frac{150}{N_x}$.
   • The global intensity component G is composed from non-periodic components and components with frequency $\omega_{1k} < \frac{150}{N_x}$.
   • The $R := X - S - G$.

Algorithm 1: EL image decomposition

The choice of the 50 computed components is arbitrary, as 10 components incorporate 99.9% of the Frobenius norm of the trajectory matrix $\mathbf{X}$. The RMSE of

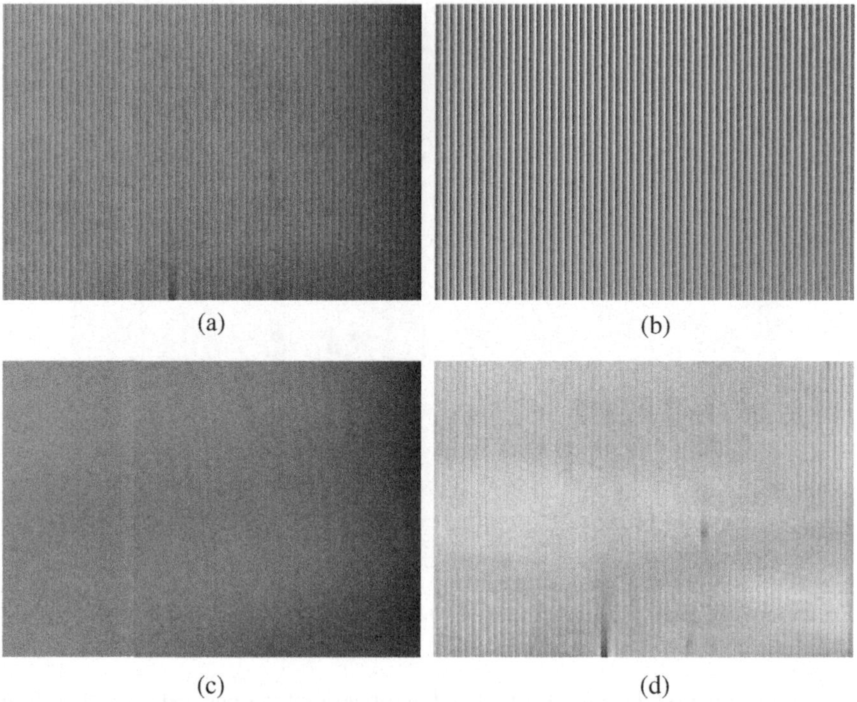

Fig. 3 Decomposition of an EL image onto 3 components. (**a**) Original image. (**b**) Cell component. (**c**) Global variations component. (**d**) Aperiodic component

the linear regression model in the ESPRIT indicates the accuracy of the model (4). For a set of 50 EL images, the mean RMSE equals 0.033.

Figure 3 depicts the non-cell components, the cell components, and the residual image. It can be argued that different components have a different physical origin. By its nature, the global variation component describes changes in a material which results in large losses that spread over large portions of a module. The cell component variation is influenced by the EL measurement conditions. Lastly, the aperiodic component captures effects caused by non-regular changes in material like shunts, or droplets (see [46]).

Shunts are characterised by a more conductive connection between the front and back electrodes than the normal solar cell structure (i.e. the solar cell structure is damaged or missing). There are many causes for shunts. Commonly shunts originate from debris of the copper evaporation source or pinholes in the CIGS absorber [32, 33]. Shunts are generally relevant to the solar module performance, in particular under low light conditions [52].

In addition to shunts we noticed the CIGS modules often exhibit "droplets" in the EL images. The appearance of droplets resembles water stains and thus we speculate these structures originate from the chemical bath deposition. At this point it is unknown what the impact of droplets is on the module performance, however, the bright appearance imply a local change in quantum efficiency according to the reciprocity relations between luminescence and quantum efficiency [38].

We remark that decomposition can be applied to an image (additive model assumption) as well as to the logarithm of an image (multiplicative model assumption). The logarithm of an image corresponds to the internal voltage (see (8), Sect. 4.3).

Furthermore, we remark that it is important to correct any perspective distortion present in images, as a periodic image distorted in such a way is no longer a signal with a finite rank in the settings of the 2D-SSA methodology.

4.2 Interconnection Line Detection

In order to identify interconnection lines, it is sufficient to locate the global minima for every level in the normal direction of the interconnection lines.

The ESPRIT model satisfies Eq. (4), which is a sum of amplitude-modulated cosine functions. A derivative of such function is again a sum of polynomial amplitude-modulated cosine functions, similar to the general form of the signal of finite rank (2). Such derivatives can be computed symbolically. Hence all the global minima in the normal direction of the interconnection lines can be identified by evaluating precise values of the derivatives, and filtering out points that satisfy a minima extreme point requirement.

Algorithm 2 describes the steps of the interconnection line identification. Symbolic derivatives are computed using the "Deriv" R-package, [5].

Input:

- EL image with dimensions $N_x \times N_y$.
- Regular mesh of points $P = \{n, m \in \mathbb{R}, 1 \leq n \leq N_x, 1 \leq m \leq N_y\}$.

Output: Array of coordinates $O \subset P$, corresponding to the estimated locations of interconnection lines.

1. Compute EL image decomposition with Algorithm 1.
2. Use ESPRIT and linear regression to estimate parameters of the model (4), see Sects. 2.2 and 2.3. Denote the resulting signal as $S : P \to \mathbb{R}$.
3. Compute symbolically expression for derivative $dS := \frac{\partial S}{\partial m} : P \to \mathbb{R}$.
4. The output set $O := \{p \in P : dS(p-) < 0 \text{ and } dS(p+) > 0\}$, where $p-, p+$ are the neighbours in P of the point p in the m-variable direction.

Algorithm 2: Interconnection line detection

We remark that the resulting expression for the cell component signal, as well as its derivatives, can be computed on a finer grid P than the original pixel coordinates. Hence the interconnection line identification is performed with sub-pixel accuracy.

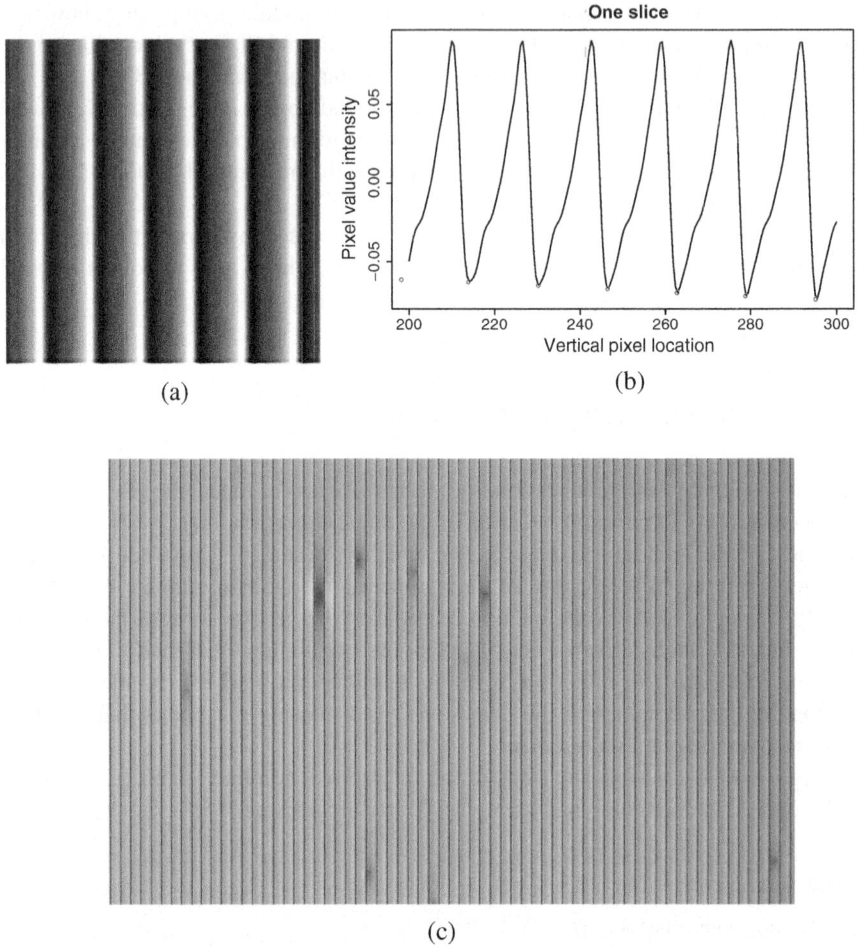

Fig. 4 Detection of the interconnection lines. (**a**) Isolated cell components. (**b**) Pixel value intensity in the normal direction to the interconnection lines. (**c**) A cell with indicated interconnection lines (red)

Figure 4 visually demonstrates the steps of the method. Figure 4a displays the estimated cell component. Figure 4b shows a slice of a pixel value intensities in Fig. 4a, where computed local minima identified with red dots. Lastly, Fig. 4c shows the module with its interconnection lines identified by red lines.

4.3 Inverse Characteristic Length Estimation

Another application of the obtained parametric expression for a module is an estimation of the inverse characteristic length λ [22, 36].

In a defect-free PV module, there are no currents in the vertical direction (along the interconnection line direction), and thus current satisfies the following linearised 1D Poisson equation, [2, 22].

$$\frac{\partial^2 V}{\partial x^2} = \lambda^2 V, \tag{7}$$

where $\lambda = \sqrt{\frac{R_{\text{sheet}}}{r_{\text{j}}}}$ is the inverse characteristic length, where R_{sheet} is the sheet resistance and r_{j} is the local differential junction resistance. Note that (7) is an equation between two fields, where $V(x, y)$ and $\lambda(x, y)$ depend on the position (x, y) in a PV module.

The luminescence intensity relatedto the internal voltage via the following relation

$$I = c \exp(V c_0) \implies V = \frac{1}{c_0} \ln(I/c), \tag{8}$$

where the constant $c_0 = \frac{q}{kT}$ is the thermal voltage (with the elementary charge, q, Boltzmann's constant, k, and the temperature, T), c is a parameter that describes the optical system from the photon generation in the solar cell absorber material to the photon detection within the camera. As such the parameter c depends on the quantum efficiency of both the solar cell and the used camera including all optical components, and the spectral photon density of a black body [38]. Generally, the constant c may vary over the solar cell area due to the camera optics and variations in the solar cell properties. However, in general, the variations in c are small compared to the exponential voltage-dependency [1]. In our analysis, we assume that c is constant over the whole module area and is accessible to a researcher.

The decomposition of a signal onto trend and cell components provides us an EL image signal without small aperiodic defects like shunts (small dark areas within cells boundaries). The global intensity and cell components without small aperiodic defects can be considered as a module without defects. Hence, we consider a module without defects to be an image $G + S$, of the output of Algorithm 1.

Hence combining (7) and (8) allows us to express inverse characteristic length as a function of intensity and its derivative:

$$\frac{\partial^2 V}{\partial x^2} = \frac{\frac{\partial^2 I}{\partial x^2} I - \left(\frac{\partial I}{\partial x}\right)^2}{c_0 I^2}, \tag{9}$$

and, therefore,

$$\lambda^2 = \frac{\frac{\partial^2 I}{\partial x^2} I - \left(\frac{\partial I}{\partial x}\right)^2}{I^2 \ln(I/c)}. \tag{10}$$

All derivatives are computed symbolically from the estimated cell component signal.

4.4 Stitched Image Correction

In order to achieve higher image resolution, several EL images can be stitched together. However, the image aligning can be imperfect, as shown in Fig. 5a. These misaligned image patches and the resulting stitch line can be attributed to incorrect perspective as well as radial distortions. The latter distortion leads to a non-linear transformation that is needed to be applied to an image for the stitch line correction.

Algorithm 3 proposes an approach to correct such distortion. The basic idea of the algorithm is the estimation of phase shifts in neighbouring locations, where each shift is estimated by application of MSSA, [19], in a direction perpendicular to the interconnection lines.

The result of the algorithm is a displacement map that defines shifts in the horizontal direction. Note that shifts have sub-pixel accuracy.

Figure 5 shows the result of the algorithm. Figure 5a shows a patch of the original EL image with a stitched part. Figure 5b depicts the shifted image, resulting from applying the displacement map. As each displacement is estimated locally between neighbours, the resulting transformation is a non-linear transformation.

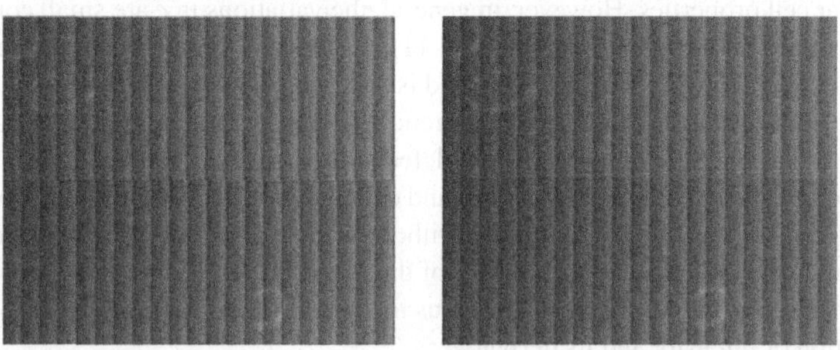

(a) (b)

Fig. 5 An example of the stitched image correction. (**a**) Original stitched image. (**b**) Corrected image

Input: EL image X with dimensions $N_x \times N_y$.
Output: A displacement map M, a vector of dimension $N_x - 1$ indicating size of horizontal shifts in image X (except for the first row).

1. For each image row i starting from the second row:
   a. Compute MSSA of the i and $i - 1$ row of the image I.
   b. Estimate the ESPRIT parameters.
   c. Filter out periods not corresponding to cell components. Let cell component indices be I.
   d. Compute shift for cell each component $s_k := \frac{\phi_k}{2\pi\omega_k}$, $k \in I$.
2. Set $M_i := \max_{k \in I} s_k$.

Algorithm 3: Stitched image correction

To evaluate the accuracy of the phase shift estimation we model the following two time series

$$s_1(x) = \cos(2\pi x/50) + \cos(2\pi x/20) + \cos(2\pi x/30) + \varepsilon_1(x), \tag{11}$$

$$s_2(x) = 2\cos(2\pi x/70) + \underbrace{\cos(2\pi(x+7)/20) + \cos(2\pi(x+7)/30)}_{\text{signal}} + \varepsilon_2(x), \tag{12}$$

where $x \in \{1, 2, \ldots, 1000\}$, components with period 20 and 30 correspond to the signal (models the cell component), components with period 50 and 70 slowly varying trend, and $\varepsilon_1, \varepsilon_2$ are two independent Gaussian iid processes with zero-mean and unit variance.

The signal part of the series s_1 is shifted by 7 units relative to the signal of the series s_2. Figure 6 shows a part of those series on the interval $[1, 100]$.

Table 1 shows the accuracy of the shift estimation of the selected signal components. That simulation was performed using 100 repetitions.

5 Conclusions

In this paper we demonstrated an application of SSA on EL images of thin-film PV modules. This low-rank approach allows capturing several important aspects of those images, namely global and local repetitive variations (or cell components) in an EL image.

Several image processing algorithms based on parametric models of SSA are proposed. The first method identified the interconnection lines between the individual cells at sub-pixel accuracy, and the second method corrects the incorrectly stitched images.

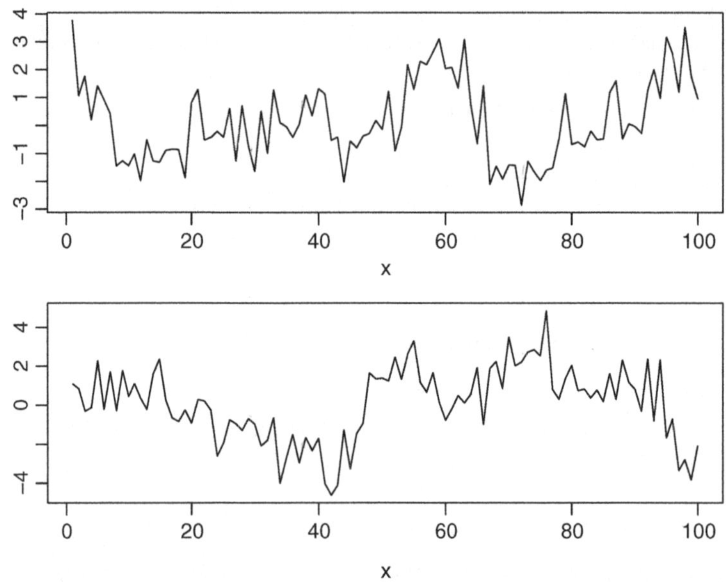

Fig. 6 $s_1(x)$ and $s_2(x)$ time series, for $x \in [1, 100]$. The signal (periods 20 and 30) are "hidden" with a trend and a noise components

Table 1 Accuracy of the shift estimation between signals in time series s_1 and s_2

RMSE	Estimate mean	25% quantile	75% quantile
0.248	7.003	6.836	7.160

Furthermore, we propose an approach based on symbolic differentiation of the SSA signal to estimate the so-called inverse characteristic length, a physical parameter of a module.

We note that in the settings of 2D-SSA it is important to correct perspective distortion in such EL images. The two-dimensional cosine array transformed with perspective distortion is no longer a finite rank signal.

It should be noted that the information captured by SSA is not complete, as local aperiodic features, such as shunts and droplets are not signals of finite rank. Therefore, the full analysis of such PV modules requires other methods. For example, we complement the signal captured by SSA using an encoder-decoder segmentation approach of individual defects (see [45, 46]).

Lastly, we remark that source code and a sample of EL images data are available upon request.

Appendix

2D-SSA Embedding and Projection

Following notations of [16], let x be an 2D-array with dimensions $(N_x, N_y) \in \mathbb{N}^2$:

$$x = \begin{pmatrix} x(0,0) & x(0,1) & \dots & x(0, N_{y-1}) \\ x(1,0) & x(1,1) & \dots & x(1, N_{y-1}) \\ \vdots & \vdots & \ddots & \vdots \\ x(N_{x-1}, 0) & x(N_{x-1}, 1) & \dots & x(N_{x-1}, N_{y-1}) \end{pmatrix}. \tag{13}$$

The 2D-SSA embedding is defined by the window size vector (L_x, L_y), which is restricted by $1 \le L_x \le N_x$, $1 \le L_y \le N_y$, and $1 < L_x L_y < N_x N_y$. Let $K_x := N_x - L_x + 1$ and $K_y := N_y - L_y + 1$, then the trajectory matrix of 2D-SSA is given by the following Hankel-block-Hankel matrix:

$$\mathbf{X} := \begin{pmatrix} \mathbf{X}_0 & \mathbf{X}_1 & \mathbf{X}_2 \dots \mathbf{X}_{K_y-1} \\ \mathbf{X}_1 & \mathbf{X}_2 & \mathbf{X}_3 \dots \mathbf{X}_{K_y} \\ \mathbf{X}_2 & \mathbf{X}_3 & \ddots \ \ddots \ \vdots \\ \vdots & \vdots & \ddots \ \ddots \ \vdots \\ \mathbf{X}_{L_y-1} & \mathbf{X}_{L_y} & \dots\dots \mathbf{X}_{N_y-1} \end{pmatrix}, \tag{14}$$

where each block

$$\mathbf{X}_i := \begin{pmatrix} x(0,i) & x(1,i) & \dots & x(K_x-1, i) \\ x(1,i) & x(2,i) & \dots & x(K_x, i) \\ \vdots & \vdots & \ddots & \vdots \\ x(L_x-1, i) & x(L_x, i) & \dots & x(N_x-1, i) \end{pmatrix}. \tag{15}$$

By construction there is a one-to-one correspondence between 2D-arrays of size $N_x \times N_y$ and the Hankel-block-Hankel matrices.

Let $\mathbf{Z}$ be an arbitrary matrix with a block-structure, where each block $\mathbf{Z}_i$ has the same dimension as the matrix $\mathbf{X}_i$

$$\mathbf{Z} = \begin{pmatrix} \mathbf{Z}_0 & \mathbf{Z}_1 & \mathbf{Z}_2 \dots \mathbf{Z}_{K_y-1} \\ \mathbf{Z}_1 & \mathbf{Z}_2 & \mathbf{Z}_3 \dots \mathbf{Z}_{K_y} \\ \mathbf{Z}_2 & \mathbf{Z}_3 & \ddots \ \ddots \ \vdots \\ \vdots & \vdots & \ddots \ \ddots \ \vdots \\ \mathbf{Z}_{L_y-1} & \mathbf{Z}_{L_y} & \dots\dots \mathbf{Z}_{N_y-1} \end{pmatrix}. \tag{16}$$

Then a projection of $\mathbf{Z}$ to Hankel-block-Hankel matrix can be computed in two steps. Firstly, diagonal averaging is performed within blocks $\mathbf{Z}_i$, $i \in 0, \ldots, N_y - 1$. Secondly, the blocks of the matrix Z are averaged between themselves. The projection can be applied in the reverse order as well.

SVD

Let $\mathbf{S} = \mathbf{X}\mathbf{X}^T$, $\lambda_1 \geq \ldots \geq \lambda_d > 0$ be non-zero eigenvalues of the matrix $\mathbf{S}$, $U_1, \ldots, U_d$ be the corresponding eigenvectors, and $V_i := \mathbf{X}^T U_i / \sqrt{\lambda_i}$, $i = 1, \ldots, d$.

Then SVD of the matrix $\mathbf{X}$ can be written as

$$\mathbf{X} = \mathbf{X}_1 + \ldots + \mathbf{X}_d. \tag{17}$$

The values $\sqrt{\lambda_i}$ are the singular values of $\mathbf{X}$.

Finite Rank Signal

An image x has rank r if the rank of trajectory matrix $\mathbf{X}$ equals $r < \min(L_x, L_y, K_y, K_y)$. In other words, the trajectory matrix is rank-deficient. If rank r does not depend on the choice of L for any sufficiently large dimensions of x, then x is called to have a *finite rank.*

Objects of finite rank are closely related to the linear recurrent sequences, [27]. Linear recurrent formulae are used to build forecast based on the SSA signal.

Computational Complexity

From the computational point of view, the hardest steps of the proposed algorithms are the singular value decomposition (SVD), ESPRIT (Sect. 2.2), and the phase and amplitude least-squares fit (Sect. 2.3).

In the context of the SSA application, the computational complexity of the SVD of a Hankel-block-Hankel matrix is $O(kN \log N + k^2 N)$, where N is the number of pixels in an image and k is the number of computed eigentriples, [26]. The linear equations solved for ESPRIT and the phase and amplitude parameters estimations are solved using QR-decomposition, and require $O(N^3)$ operations. Our numerical experiments show that the SVD decomposition dominates the computational time for the proposed algorithms for image sizes with width and height less than 4000 pixels.

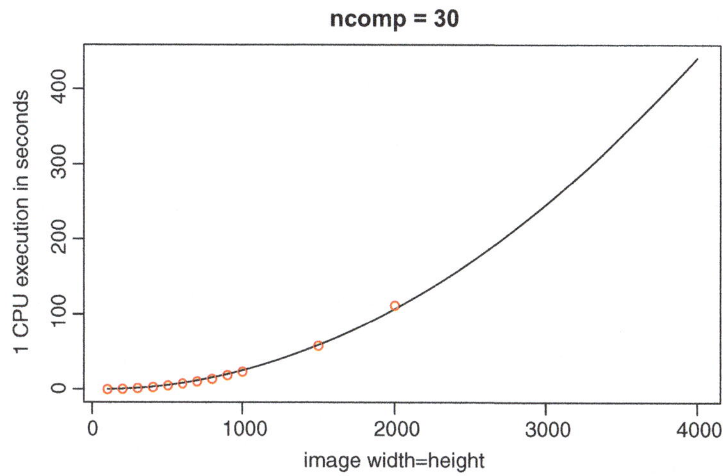

Fig. 7 Computation time required for Algorithm 1 for different input image sizes

Table 2 Maximum memory required for Algorithm 1 for different input image sizes

Image width and height	500	1000	1500	2000
Maximum RAM in GB	0.5	1.6	3.2	5.5

Table 3 Running time of the symbolical differentiation routine used in Algorithm 2 and the inverse characteristic length estimation in Sect. 4.3

Number of components	5	9	13	17
Execution time in seconds	0.15	0.46	1.27	1.79

To measure the required time and memory of the proposed algorithms we utilise a machine with Intel(R) Xeon(R) CPU E5-1620 3.5 GHz processor and 31 GB of RAM. The time measurements are performed on a single CPU.

Figure 7 shows the time required to perform steps of Algorithm 1 for square images (width equals height) for different image widths. The red points correspond to the measured time in seconds, and the black line is a parabola fitted to the points. Table 2 shows the amount of memory required for the steps of Algorithm 1.

Algorithm 2 utilises results of the Algorithm 1 and requires additional symbolical differentiation. Such computation is also required for the inverse characteristic length (Sect. 4.3). The running time of the symbolical differentiation depends on the number of terms of the signal (4). Table 3 demonstrates the relationship between the number of terms in the signal and the running time required for the symbolic differentiation. Algorithm 2 and the inverse characteristic length computation works with the cell component that typically consists of 5–7 terms.

Algorithm 3 utilises a different version of SSA that requires less time and memory, however, the algorithm computes MSSA multiple times for several pairs of neighbour rows. A single iteration of the algorithm loop requires 0.7 s and 0.78

of RAM for an image with width of 4000 pixels. Different iterations of the loop can be run in parallel. Note that most of the used memory is occupied by an image itself, and hence the memory can be shared by multiple processes. Algorithm 3 usually requires running about 100–200 iterations, as an approximate location of the stitching lines is known. Hence, the total running time an 8-core processor can be as little as 20 s.

Lastly, we remark the proposed algorithms run in a deterministic amount of time. Hence, the methods can be run in real-time applications.

Acknowledgments This work is supported by the Solar-era.net framework in the project "PEARL TF-PV" (Förderkennzeichen: 0324193A) and partly funded by the HGF project "Living Lab Energy Campus (LLEC)".

References

1. Abou-Ras, D., Kirchartz, T., Rau, U.: Advanced Characterization Techniques for Thin Film Solar Cells. Wiley, New York (2016)
2. Augarten, Y., Wrigley, A., Rau, U., Pieters, B.: Calculation of the TCO sheet resistance in thin film modules using electroluminescence imaging. In: 2016 IEEE 43rd Photovoltaic Specialists Conference (PVSC), pp. 1527–1531. IEEE, New York (2016)
3. Broomhead, D., King, G.: Extracting qualitative dynamics from experimental data. Physica D: Nonlinear Phenomena **20**(2–3), 217–236 (1986)
4. Chambers, J.M.: Linear models. In: Statistical Models in S (1992)
5. Clausen, A., Sokol, S.: Deriv: R-based Symbolic Differentiation (2019). https://CRAN.R-project.org/package=Deriv, deriv package version 4.0
6. Colebrook, J.: Continuous plankton records-phytoplankton, zooplankton and environment, northeast Atlantic and north-sea, 1958–1980. Oceanol. Acta **5**(4), 473–480 (1982)
7. Danilov, D., Zhigljavsky, A.: Principal Components of Time Series: The "Caterpillar" Method, pp. 1–307. University of St Petersburg, St Petersburg (1997)
8. Deitsch, S., Christlein, V., Berger, S., Buerhop-Lutz, C., Maier, A., Gallwitz, F., Riess, C.: Automatic classification of defective photovoltaic module cells in electroluminescence images. Sol. Energy **185**, 455–468 (2019)
9. Deitsch, S., Buerhop-Lutz, C., Sovetkin, E., Steland, A., Maier, A., Gallwitz, F., Riess, C.: Segmentation of photovoltaic module cells in electroluminescence images. Mach. Vis. Appl. **32**(84), 1432–1769 (2021)
10. Demant, M., Virtue, P., Kovvali, A.S., Yu, S.X., Rein, S.: Deep learning approach to inline quality rating and mapping of multi-crystalline Si-wafers. In: Proceedings of the 35th European Photovoltaic Solar Energy Conference and Exhibition, pp. 814–818 (2018)
11. de Oliveira, A.K.V., Aghaei, M., Rüther, R.: Automatic fault detection of photovoltaic array by convolutional neural networks during aerial infrared thermography. In: Proceedings of the 36th European Photovoltaic Solar Energy Conference and Exhibition (2019)
12. Elsner, J.B., Tsonis, A.A.: Phase Space Reconstruction, pp. 143–155. Springer, Boston (1996). https://doi.org/10.1007/978-1-4757-2514-8_10
13. Fraedrich, K.: Estimating the dimension of weather and climate attraction. J. Atmos. Sci. **43**, 419–432 (1986)
14. Golyandina, N., Korobeynikov, A.: Basic singular spectrum analysis and forecasting with R. Comput. Stat. Data Anal. **71**, 934–954 (2014). r package version 1.0
15. Golyandina, N., Usevich, K.: An algebraic view on finite rank in 2d-ssa. In: Proceedings of the 6th St. Petersburg Workshop on Simulation, pp. 308–313 (2009)

16. Golyandina, N., Usevich, K.: 2D-extension of singular spectrum analysis: algorithm and elements of theory. In: Matrix Methods: Theory, Algorithms and Applications: Dedicated to the Memory of Gene Golub, pp. 449–473. World Scientific, Singapore (2010)
17. Golyandina, N., Nekrutkin, V., Zhigljavsky, A.: Analysis of time series structure: SSA and related techniques. Chapman and Hall/CRC, London (2001)
18. Golyandina, N., Korobeynikov, A., Shlemov, A., Usevich, K.: Multivariate and 2D extensions of singular spectrum analysis with the Rssa package. J. Stat. Softw. **67**(2), 1–78 (2015). https://doi.org/10.18637/jss.v067.i02
19. Golyandina, N., Korobeynikov, A., Zhigljavsky, A.: Singular Spectrum Analysis with R. Springer, Berlin (2018)
20. Groth, A., Ghil, M.: Monte Carlo singular spectrum analysis (SSA) revisited: Detecting oscillator clusters in multivariate datasets. J. Clim. **28**(19), 7873–7893 (2015)
21. Hassanieh, H., Indyk, P., Katabi, D., Price, E.: Simple and practical algorithm for sparse Fourier transform. In: Proceedings of the twenty-third annual ACM-SIAM symposium on Discrete Algorithms, Society for Industrial and Applied Mathematics, pp. 1183–1194 (2012)
22. Helbig, A., Kirchartz, T., Schaeffler, R., Werner, J.H., Rau, U.: Quantitative electroluminescence analysis of resistive losses in Cu (In, Ga) Se2 thin-film modules. Sol. Energy Mater. Sol. Cells **94**(6), 979–984 (2010)
23. Holloway, D.M., Lopes, F.J., da Fontoura Costa, L., Travençolo, B.A., Golyandina, N., Usevich, K., Spirov, A.V.: Gene expression noise in spatial patterning: hunchback promoter structure affects noise amplitude and distribution in drosophila segmentation. PLoS Comput. Biol. **7**(2), e1001069 (2011)
24. Karimi, A.M., Fada, J.S., Hossain, M.A., Yang, S., Peshek, T.J., Braid, J.L., French, R.H.: Automated pipeline for photovoltaic module electroluminescence image processing and degradation feature classification. IEEE J. Photovoltaics **9**(5),1324–1335 (2019)
25. Köntges, M., Kurtz, S., Packard, C., Jahn, U., Berger, K., Kato, K., Friesen, T., Liu, H., Van Iseghem, M.: IEA-PVPS t13-01 2014 review of failures of photovoltaic modules final (2015). Tech. rep., Technical Report IEA-PVPS T13-01: 2014, IEA-PVPS Task 13
26. Korobeynikov, A.: Computation- and space-efficient implementation of SSA. Statistics and Its Interface **3**(3), 357–368 (2010), r package version 1.0
27. Kurakin, V., Kuzmin, A., Mikhalev, A., Nechaev, A.: Linear recurring sequences over rings and modules. J. Math. Sci. **76**(6), 2793–2915 (1995)
28. Lanczos, C.: An Iteration Method for the Solution of the Eigenvalue Problem of Linear Differential and Integral Operators. United States Government Press Office, Los Angeles (1950)
29. Lima, G., et al.: Gap filling of precipitation data by SSA-singular spectrum analysis. In: Journal of Physics: Conference Series, vol. 759, p. 012085. IOP Publishing, New York (2016)
30. Mamou, J., Feleppa, E.J.: Singular spectrum analysis applied to ultrasonic detection and imaging of brachytherapy seeds. J. Acoust. Soc. Am. **121**(3), 1790–1801 (2007)
31. Markovsky, I.: Low-Rank Approximation, vol. 139. Springer, Berlin (2018)
32. Misic, B.: Analysis and Simulation of Macroscopic Defects in Cu(In,Ga)Se2 Photovoltaic thin film modules, Schriften des Forschungszentrums Jülich. Reihe Energie und Umwelt/Energy and Environment, vol. 372, pp. 17–36. Forschungszentrum Jülich GmbH, Germany (2017)
33. Misic, B., Pieters, B.E., Schweitzer, U., Gerber, A., Rau, U.: Defect diagnostics of scribing failures and cu-rich debris in cu(in,ga)se_2 thin-film solar modules with electroluminescence and thermography. IEEE J. Photovoltaics **5**(4), 1179–1187 (2015). https://doi.org/10.1109/JPHOTOV.2015.2422143
34. Monadjemi, A.: Towards efficient texture classification and abnormality detection. PhD thesis, University of Bristol, Bristol (2004)
35. PEARL-TF (2020) PEARL-TF project website. https://pearltf.eu/, accessed: 2020-09-01
36. Pieters, B.E., Rau, U.: A new 2d model for the electrical potential in a cell stripe in thin-film solar modules including local defects. Prog. Photovolt. Res. Appl. **23**(3), 331–339 (2015). https://doi.org/10.1002/pip.2436. https://onlinelibrary.wiley.com/doi/abs/10.1002/pip.2436

37. Qiao, T., Ren, J., Wang, Z., Zabalza, J., Sun, M, Zhao, H., Li, S., Benediktsson, J.A., Dai, Q., Marshall, S.: Effective denoising and classification of hyperspectral images using curvelet transform and singular spectrum analysis. IEEE Trans. Geosci. Remote Sens. **55**(1), 119–133 (2016)
38. Rau, U.: Reciprocity relation between photovoltaic quantum efficiency and electroluminescent emission of solar cells. Phys. Rev. B **76**, 085303 (2007). https://doi.org/10.1103/PhysRevB.76.085303. https://link.aps.org/doi/10.1103/PhysRevB.76.085303
39. Rodriguez-Aragon, L.J., Zhigljavsky, A.: Singular spectrum analysis for image processing. Statistics and Its Interface **3**(3), 419–426 (2010)
40. Rouquette, S., Najim, M.: Estimation of frequencies and damping factors by two-dimensional ESPRIT type methods. IEEE Trans. Signal Process. **49**(1), 237–245 (2001)
41. Roy, R., Kailathm T.: ESPRIT-estimation of signal parameters via rotational invariance techniques. IEEE Trans. Acoust. Speech Signal Process. **37**(7), 984–995 (1989)
42. Sella, L., Vivaldo, G., Groth, A., Ghil, M.: Economic cycles and their synchronization: A comparison of cyclic modes in three European countries. Journal of Business Cycle Research **12**(1), 25–48 (2016)
43. Shin, P.J., Larson, P.E., Ohliger, M.A., Elad, M., Pauly, J.M., Vigneron, D.B., Lustig, M.: Calibrationless parallel imaging reconstruction based on structured low-rank matrix completion. Magn. Reson. Med. **72**(4), 959–970 (2014)
44. Sovetkin, E., Steland, A.: Automatic processing and solar cell detection in photovoltaic electroluminescence images. Integrated Computer-Aided Engineering **26**(2), 123–137 (2019)
45. Sovetkin, E., Pieters, B.E., Weber, T., Achterberg, E.J., Weeber, A., Rau, B., Rennhofer, M., Theelen, M.: PV-AIDED: Photovoltaic artificial intelligence defect identification. multichannel encoder-decoder ensemble models for electroluminescence images of thin-film photovoltaic modules, PEARL TF-PV. In: 37th EU PVSEC (2020)
46. Sovetkin, E., Weber, T., Achterberg, E.J., Pieters, B.E.: Encoder–decoder semantic segmentation models for electroluminescence images of thin-film photovoltaic modules. IEEE J. Photovoltaics **11**(2), 444–452 (2021). https://doi.org/10.1109/JPHOTOV.2020.3041240
47. Trickett, S.: F-xy cadzow noise suppression. In: SEG Technical Program Expanded Abstracts 2008, Society of Exploration Geophysicists, pp. 2586–2590 (2008)
48. Van Der Veen, A.J., Deprettere, E.F., Swindlehurst, A.L.: Subspace-based signal analysis using singular value decomposition. Proc. IEEE **81**(9), 1277–1308 (1993)
49. Vautard, R., Ghil, M.: Singular spectrum analysis in nonlinear dynamics, with applications to paleoclimatic time series. Physica D: Nonlinear Phenomena **35**(3), 395–424 (1989)
50. Wang, Y., Chen, J.W., Liu, Z.: Comments on "estimation of frequencies and damping factors by two-dimensional ESPRIT type methods". IEEE Trans. Signal Process. **53**(8), 3348–3349 (2005)
51. Weare, B.C., Nasstrom, J.S.: Examples of extended empirical orthogonal function analyses. Mon. Weather Rev. **110**(6), 481–485 (1982)
52. Weber, T., Albert, A., Ferretti, N., Roericht, M., Krauter, S., Grunow, P.: Electroluminescence investigation on thin film modules. In: Proceedings of the 26th European Photovoltaic Solar Energy Conference (26th EU PVSEC), pp. 2584–2588 (2011)
53. Zabalza, J., Ren, J., Wang, Z., Marshall, S., Wang, J.: Singular spectrum analysis for effective feature extraction in hyperspectral imaging. IEEE Geosci. Remote Sens. Lett. **11**(11), 1886–1890 (2014)
54. Zabalza, J., Ren, J., Zheng, J., Han, J., Zhao, H., Li, S., Marshall, S.: Novel two-dimensional singular spectrum analysis for effective feature extraction and data classification in hyperspectral imaging. IEEE Trans. Geosci. Remote Sens. **53**(8), 4418–4433 (2015)
55. Zhigljavsky, A.: Singular spectrum analysis for time series: Introduction to this special issue. Statistics and its Interface **3**(3), 255–258 (2010)
56. Zscheischler, J., Mahecha, M., et al.: An extended approach for spatiotemporal gapfilling: Dealing with large and systematic gaps in geoscientific datasets. Nonlinear Process. Geophys. **21**(1), 203–215 (2014)

The Impact of the Lockdown Restrictions on Air Quality During COVID-19 Pandemic in Lombardy, Italy

Paolo Maranzano and Alessandro Fassó

Abstract Environmental agencies and scientists around Europe have reported that COVID-19 lockdown caused an extended environmental clean-up. Considering air quality, we focus on the Lombardy region (Northern Italy), which is at the same time the most populous region and the area most affected by COVID-19 in Italy. Lombardy is also one of the most polluted areas in the European Union. The central research hypothesis concerns if and how the first-wave restrictions imposed during the 2020 spring have improved the air quality in Lombardy and if the improvements are similar throughout the territory. To answer these questions, we use weekly data from January 2015 to mid-June 2020 for 74 ground monitoring stations and provided by the regional environmental protection agency (ARPA Lombardia). We estimate an autoregressive time series model with exogenous covariates (ARX) to assess the combined impact of meteorology, seasonality, trend, and lockdown on the NO_2 concentrations at each monitoring site. We also propose using the LASSO algorithm to select the set of relevant covariates to model the concentrations and then estimate the effect of lockdown restrictions with a maximum likelihood post-LASSO estimator. Statistical modelling confirms a generalised NO_2 reduction due to the lockdown throughout the whole region, despite considerable variability due to the morphological and geographical heterogeneity of Lombardy. Compared to the observed average variations, the estimated lockdown impacts are mitigated by meteorology and natural trends. Expectantly, the most significant and remarkable NO_2 reductions have been estimated near urban and congested areas and in the proximity of industrialised sites.

P. Maranzano (✉)
University of Milano-Bicocca, Dept. of Economics, Management and Statistics (DEMS), Milano, Italy

Fondazione Eni Enrico Mattei (FEEM), Milano, Italy
e-mail: paolo.maranzano@unimib.it

A. Fassó
University of Bergamo, Dept. of Economics (DSE), Bergamo, Italy
e-mail: alessandro.fasso@unibg.it

A. Steland, K.-L. Tsui (eds.), *Artificial Intelligence, Big Data and Data Science in Statistics*, https://doi.org/10.1007/978-3-031-07155-3_15

Keywords COVID-19 lockdown · Lombardy · Air quality · NO_2 · ARX models · LASSO algorithm

1 Introduction

With the global COVID-19 pandemic, the quality of the air we breathe every day has acquired a role of primary importance worldwide, both in scientific research and in daily media storytelling.

Air pollution is seen as having both an active and passive role in managing the epidemic. On the one hand, many research studies have provided evidence of strong correlations between poor air quality and COVID-19 spread, especially in large urban centres with severe air quality conditions. On the other hand, air quality has improved significantly worldwide due to partial and total lockdown measures and to human mobility restrictions imposed on citizens to avoid spreading the virus.

This paper analyses the case of Lombardy, the economic and financial centre of northern Italy, to assess how restrictive measures to control the spread of the COVID-19 virus have influenced the concentrations of oxides, in particular, nitrogen dioxide (NO_2), in the atmosphere. We focus on the first-wave lockdown restrictions imposed on citizens and their productive activities between 9th March and 18th May 2020 all over the country. The choice of NO_2 as the main pollutant of interest is justified by physical-chemical reasons well-known in the literature. Oxides are classified as both primary and secondary pollutants emitted mainly by anthropogenic activities, such as heating systems, motor vehicle traffic, power plants, industrial activities and combustion and, therefore, directly affected by the lockdown restrictions. In particular, the emission inventory for Lombardy [19] estimates that road traffic is responsible for 51% of the annual NO_X emissions in the Lombardy region and the 65% in the metropolitan area of Milan.

Moreover, the choice to analyse NO_2 is also reasonable as it is a pollutant that responds immediately to emission shocks, while other airborne pollutants, such as the atmospheric particulate matter, have more complex and slower reactions. Therefore, nitrogen dioxide is a suitable candidate for a preliminary investigation into the impacts of a severe limitation to anthropogenic activities, especially vehicle traffic, on air quality. We pose some questions: have the restrictions improved the air quality in the region? Are there similar improvements throughout the region? Can we measure the variation of airborne pollutants related explicitly to car traffic?

Although the measures may have generated socio-economic issues in the population, they may have improved air quality in an area that suffers from longstanding environmental problems. Expectations were for substantial reductions in concentrations generalised to all areas of the region. We implemented a time series modelling approach for the available air quality monitoring stations, controlling for the effect of local meteorology, seasonality, and trends.

We propose to use a data-oriented approach based on the combination of time series and statistical learning methods to simultaneously address the statistical variable selection issue and the model estimation using a penalised least squares approach. We estimate the effects of local meteorology, long-term weather trends, and lockdown restrictive measures on NO_2 concentrations in Lombardy implementing an order-one autoregressive with exogenous covariate model, namely ARX(1) model, for each considered station. These models express the response time series as a linear combination of both autoregressive components and exogenous variables and are easily estimated using the maximum likelihood approach. The ARX form allows to estimate the effect of covariates on the response and at the same time allows modelling the temporal dynamics of the dependent variable explicitly. Each time series is modelled as a function of several regressors and one lag of the dependent variable, also known as the ARX(1) model. In our application, the dependent variable is the weekly concentration of NO_2 observed at each ground site, while the covariates include local meteorology, long-run trends and the lockdown. The lockdown's estimated effect will depend both on the estimated coefficient for the lockdown event and the autoregressive dynamics of the dependent variable. The most suitable model to describe the temporal evolution of concentrations for each control unit is determined using the least absolute shrinkage and selection operator (LASSO) algorithm. To validate the correct specification assumptions, we will also provide several diagnostic checks on the residuals. Finally, the estimated effects are discussed and contextualised concerning their geographical location, the area surrounding the station and the type of station they are associated with.

The rest of the paper is organised as follows. In Sect. 2, we report and discuss the most recent contributions to the COVID-19 literature quantifying the impact that the lockdown restrictions on the levels of airborne pollutant concentrations. Section 3 reports the leading facts connected to the pandemic in Italy and Lombardy, and describes the environmental context of the region, focusing on the primary airborne pollution sources and the geographical features. In Sect. 4, we introduce and describe the data set. In Sect. 5, we present the statistical modelling approach by illustrating the main characteristics of the ARX(1) model, how the lockdown effect is computed and the proposed model selection algorithm. In Sect. 6, we discuss the LASSO algorithm's output, the fitting of the models and some diagnostic checks for the regression residuals. In Sect. 7, we discuss the estimated NO_2 variations due to the lockdown restrictions' main findings regarding the estimated lockdown variations, analysing the gaps between the observed and the estimated reductions and contextualising the estimates with respect the geography of Lombardy and its socio-economic structure. Finally, Sect. 8 sums-up the discussion, giving some concluding remarks and outlining some possible extensions to be developed in future works.

2 The Relationship Between the COVID-19 Pandemic and Air Quality in the World

Since the beginning of the pandemic and the early containment measures of the virus, many researchers have focused on the effects of these restrictions on air quality in various parts of the World. In this context, we could refer to the research stream as the analysis of the passive role played by the air quality during the COVID-19 lockdowns. In other words, the focus is on estimating the reduction of airborne pollutant concentrations due to the lockdown restrictions to mobility, industries and, more in general, human activities. Scientists and researchers are compact in stating that, in general, air quality has significantly improved everywhere, reaching minimal levels of pollution. This fact holds particularly true in large urban centres and densely populated areas, often affected by poor air quality. The most considerable improvements were registered in Europe, where oxides and particulate concentrations were reduced by around 40%–70% as in Spain [2, 34], Italy [10], and France [11], and in Brazil, where concentrations were reduced by 50% [25]. Smaller, but still significant, reductions were reported for the USA [4, 36] and other Asian countries, such as Kazakhstan [21] and India [29, 30], in which both the average levels of both oxides and particulates fell from 20% to 30%. These papers measured the impact of lockdown restrictions on air quality controlling for weather conditions, possible counterfactual terms, and long-run trends in the concentrations.

In some cases, including in the models temporal trends as in [36] for New York City, or [7] for the city of Brescia (northern Italy), no significant difference between the years was found, suggesting that the reduction in concentration levels in 2020 was similar to that measured in the previous 5 years.

However, in other studies which analysed the impact both on macro-areas, such as US counties [4], and on micro-areas, such as Sao Paulo in Brazil [25], Los Angeles, New York and Paris [11], or Barcelona [2, 34], the differences proved to be significant while including meteorological factors and time trends. According to all these studies, all the common air pollutants were affected by significant reductions: nitrogen dioxide (NO_2) in Barcelona, Sao Paulo, Los Angeles, New York, Paris, and Almaty (Kazakhstan) fell by 50%, 30%, 38%, 25%, 39%, and 35%, respectively; particulate matter concentrations fell approximately by 31%, 12%, 37%, 36%, and 28% and carbon monoxide (CO) decreased by 40% in São Paulo, 49% in Almaty, 24% in Los Angeles, 19% in New York, and 67% in Paris. By contrast, ozone (O_3) level increased in most parts of the World. It increased by 30% in São Paulo, 7% in New York and 12% in Paris. At the national level, prominent air quality improvements were detected in many countries. As example, in the USA NO_2 and $PM_{2.5}$ reduced at county level of 25.5% and 4.45%, respectively, [4] and in UK [18, 28], which registered reductions in oxide concentrations of -32% (NO_2), -38% (NO_x), and -50% (NO). Or also in India, which registered a general reduction in $PM_{2.5}$, PM_{10}, CO, and NO_2 levels by 43%, 31%, 10%, and 18% during the lockdown period [29].

3 Air Quality in Lombardy During the COVID-19 Lockdown

For many years, the Po Valley in northern Italy has been ranked among the most polluted areas in Europe and the world [12]. According to the report about air quality in 2018 by the European Environment Agency (EEA), in Europe around

3.9 million people live in areas where the limits of the air pollutants are frequently exceeded. Among these, the 95% live in the Po Valley. Moreover, Italy is at the first place in Europe as concerns the number of premature deaths attributable to exposure excess of nitrogen dioxide (around 14,600 victims per year) and ozone (around 3000 victims per year) and at second place after Germany as regards deaths due to fine particulates ($PM_{2.5}$) pollutant. As represented in Fig. 1, which shows the average NO_2 concentrations in Europe in 2018, the Po Valley area is easily identifiable since it reports dark spots (high concentrations) all over its surface.

The main physical and geomorphological features of the Po Valley are depicted in Fig. 2. The area is surrounded by a C-shape mountain range, which acts as a

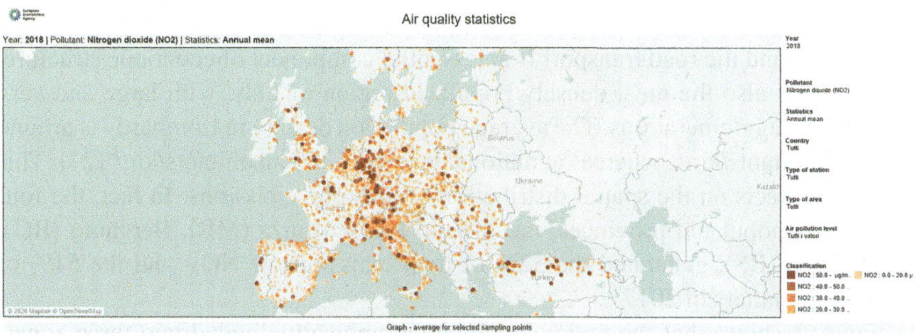

Fig. 1 Average NO_2 concentrations in Europe in 2018. *Source: EEA website - Air quality statistics dashboard*

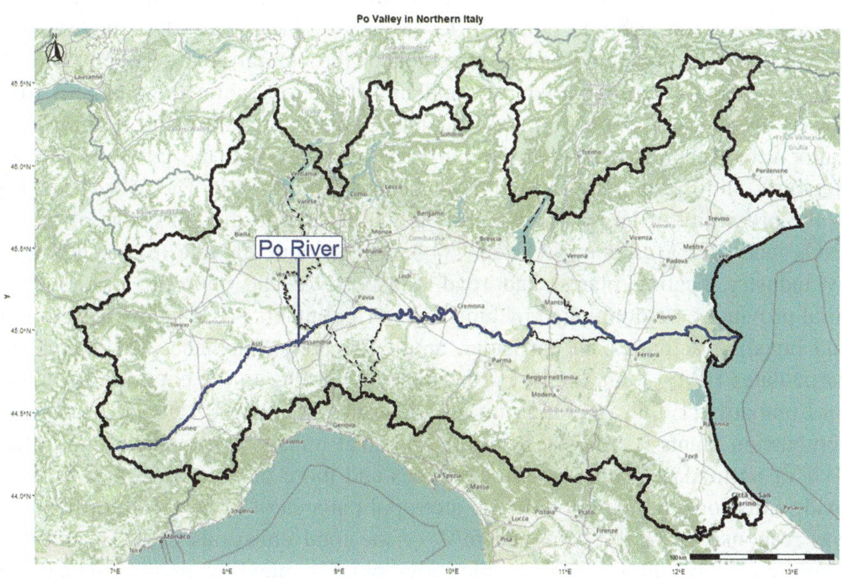

Fig. 2 Physical map of Po Valley in Northern Italy

barrier and prevents wind movement from the West. As a result, the wind speed on the Po Valley is among the lowest in Europe, about 1.5 m/s on average, causing a high accumulation of smog and pollution close to the ground. According to a recent simulation study by Raffaelli et al. [26], if Po Valley had the same meteorological conditions typical of central-northern Europe and kept the same emission levels, average monthly concentrations of PM_{10} and NO_2 would be lowered by 60–80% compared to concentration levels of 2013. Consequently, it is more difficult for the Po Valley region to comply with international air quality standards than other EU and non-EU member states.

The Lombardy region is the economic and financial centre of Po Valley. It is organised in eleven provinces and is home to more than ten million inhabitants. The region holds the highest gross domestic product per inhabitant of the country [27]. In Lombardy are located many industrial facilities, as well as small and medium enterprises, and the road transport is an essential component of economic structure. Lombardy is also the most densely populated region of Italy, with large and very dense urban agglomerations. The average population density in Lombardy is around 419.9 inhabitants/km^2, whereas at national level it is 200 inhabitants/km^2 [27]. This fact also reflects on the spatial distribution of airborne emissions. In fact, the four largest and populated provinces, i.e. Milano (MI), Monza (MB), Bergamo (BG), and Brescia (BS), generated the 52% of total emissions of NO_X and the 51% of particulate matters in 2017.

Figure 3 shows that the region can be geographically divided into three zones. The mountain range of the Alps in the North, the sloping foothills in the mid-north, and the flat southern area. The strong heterogeneous physical conformation of the territory influences the socio-economic organisation of the society and, consequently, the airborne pollution levels in the atmosphere. The mountainous area is sparsely populated and less trafficked, the central area between the hills and the plain is densely urbanised and industrialised, while the southern rural area is less densely populated and oriented towards agriculture and farming. The major urban centres (the metropolitan area of Milan and the cities of Monza, Bergamo, and Brescia) are located in the central foothills belt. The poor air quality in the proximity of large urban centres is mainly due to the emissions of oxides (NO_X and NO_2), carbon (CO and CO_2), and sulphur dioxide (SO_2), produced by industry, heating plants, and road transport. At the opposite, the southern rural provinces are dominated by emissions of pollutants produced by agricultural and breeding activities. In particular, ammonia (NH_3), methane (CH_4), and fine particulates ($PM_{2.5}$) emissions. Furthermore, according to the emission inventory for Lombardy [19], in 2017 the sum of industrial combustion plants, non-industrial combustion plants (i.e. house heating), and road transport represented more than 73% of particulate matter emission sources and more than 76% of total nitrogen oxide emissions in the region. In the metropolitan area of Milan, road traffic alone was responsible for the 65% and 69% of the total emissions of NO_X and CO, respectively. Overall, its unfavourable geographical context, an aggressive land use, climate characteristics, and high pollutant emissions turn in the accumulation of toxic elements in the atmosphere.

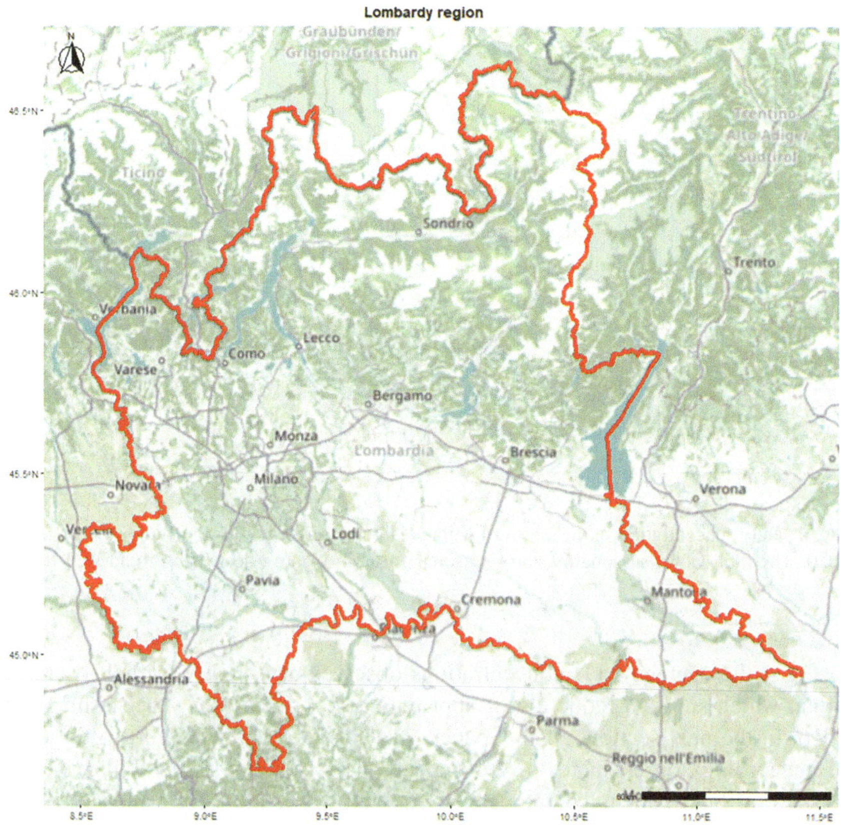

Fig. 3 Physical map of Lombardy

As will be shown in Sect. 7.2, estimates of the effects of the restrictions on nitrogen oxide concentrations will depend strongly on the type of area surrounding the ground station and by the geomorphological structure of the territory and its reliefs. In particular, it will be shown that the proximity of the stations to congested or urban areas leads to substantial reductions in concentrations during the lockdown phase, while in agricultural or rural areas the reductions will be much smaller and, in several cases, zero.

Eventually, seasonal and meteorological variations and human activities have a relevant influence on air quality in Lombardy, as they act as confounders when dealing with the assessment of the lockdown effect on airborne pollutant concentrations. Recent studies of the air quality in Lombardy, such as the case study by Maranzano et al. [24] about the city of Milan or the analysis of [8] concerning the main urban centres of Lombardy, highlighted that the whole region has been experiencing a significant and constant reduction in airborne pollutant concentrations since the early 2000s. To show such decreasing pattern, we reported

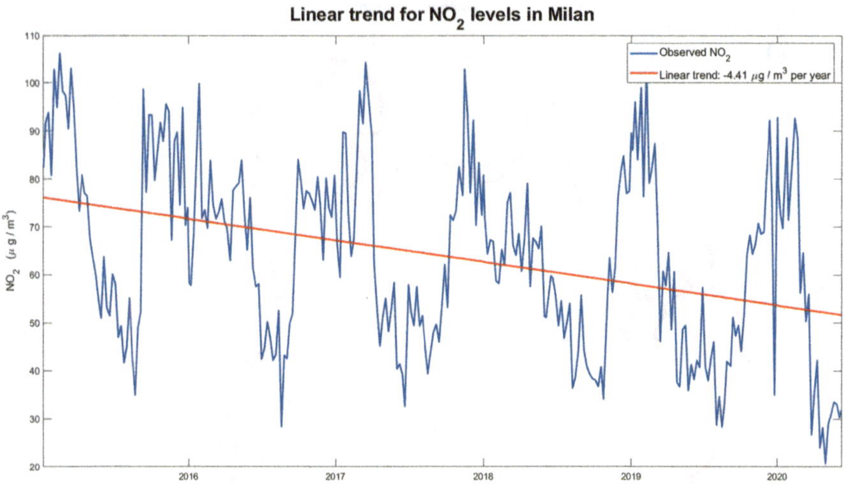

Fig. 4 NO_2 concentration in the North of Milan (Marche traffic monitoring station), years 2015–2020. The plot shows seasonality, intra-seasonal variability and a decreasing trend (red line) of $4\,\mu g/m^3$

in Fig. 4 the average NO_2 concentrations observed at a highly trafficked area of Milan which shows a reduction of approximately $4\,\mu g/m^3$ per year starting from 2016. In the next sections of this paper, we will devote much effort to the effect of meteorology on estimates of the impacts of restrictions. In particular, we will use several environmental parameters, such as air temperature and wind strength, to filter out spurious effects from the regressions and obtain adjusted estimates.

3.1 The COVID-19 Lockdown in Italy

The Italian government imposed a total lockdown period between 9th March and 18th May 2020, totalling 71 days. It was characterised by the closure of all non-essential activities and enterprises and minimised individual mobility and social distancing. It resulted in a generalised reduction of car traffic and personal travel, as discussed by Finazzi and Fassò [14]. According to their estimates, obtained with a sample of 20 thousands Italian users of an earthquake-tracker application, at the peak of the lockdown, the daily mean distance travelled decreased by approximately 50%, and the percentage of users who did not move within 24 h reached 65%.

In principle, the variation due to lockdown can be measured by the difference between the average NO_2 concentration during the event and the corresponding average before. Hence, we could compare air quality during the lockdown and the average of the preceding 71 days. Since this means comparing spring and winter concentrations, it is a biased comparison.

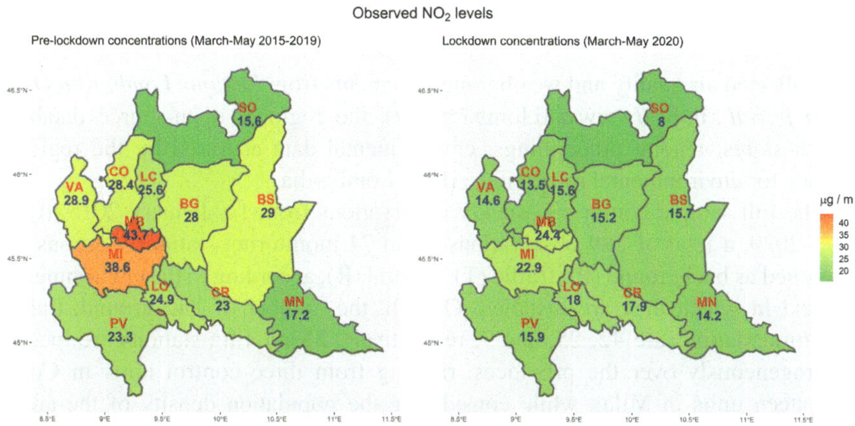

Fig. 5 Observed pre-lockdown and in-lockdown air quality. Average NO_2 levels by province observed between 9th March and 18th May of years 2015–2019 (pre-lockdown) and 2020 (lockdown). Provinces: Varese (VA), Como (CO), Lecco (LC), Sondrio (SO), Bergamo (BG), Brescia (BS), Milano (MI), Monza e Brianza (MB), Pavia (PV), Lodi (LO), Cremona (CR) and Mantova (MN)

To avoid the impact of seasonal variations, we compared the lockdown average concentration with the average of the same period (9th March–18th May) in the preceding years. Figure 5 shows how the average concentrations fell in all provinces[1] during the lockdown, especially in large urban areas, such as Milano, Bergamo, Monza, and Brescia.

At a glance, it is clear that the restrictions generated a positive effect on the whole territory, especially in the urbanised central area. The provinces of Monza-Brianza and Milano experienced reductions close to $20\,\mu g/m^3$, while Bergamo, Brescia, and the northern mountain provinces recorded reductions of around $15\,\mu g/m^3$. In these latter cases, the reductions were around 50% compared to previous years. The southern areas, which are less inhabited and less industrialised, experienced smaller decreases.

Are these reductions due to the lockdown measures, or other factors occurred? In other words, uncertainty and biases may still be present due to variations between years. These may be related to both meteorological variations and anthropogenic variations, such as traffic policies or changes in vehicle emissions [24]. Our purpose is to measure the variation of airborne pollutants such as nitrogen dioxide (NO_2), which is related to car traffic limitations and activity restrictions during the COVID-19 period, controlling for meteorology and trends characterising the phenomenon.

[1] Provinces: Varese (VA), Como (CO), Lecco (LC), Sondrio (SO), Bergamo (BG), Brescia (BS), Milano (MI), Monza e Brianza (MB), Pavia (PV), Lodi (LO), Cremona (CR), and Mantova (MN)

4 Air Quality and Weather Data

We collected air quality and weather measurements from *Regione Lombardia Open Data portal* (https://www.dati.lombardia.it/), the regional open-source database which stores, among other things, environmental data acquired by the regional agency for environmental protection, ARPA Lombardia.

The full sample comprises weekly observations from 1st January 2015 to 8th June 2020, a total of 289 observations, from 74 monitoring stations. Stations are classified as background (B), traffic (T), or rural (R), according to the environmental context in which they are installed. Overall, the number of background, traffic, and rural stations are 42, 25, and 7, respectively. Monitoring stations are located heterogeneously over the provinces, ranging from three control units in Como to sixteen units in Milan while considering the population density of the areas. Moreover, given the geographical specificities of each province, the rural stations are predominantly located in the flat southern strip of land, while traffic stations are placed in the large urban areas in the centres.

The lockdown period (9th March–18th May 2020) is modelled through a dummy variable which assumes value 1 for the lockdown weeks, for a total of 10 observations, and value 0 for the other observations. Thus, the whole period can be divided into three parts: the pre-lockdown period composed of 275 weeks from 1st January 2015 to 4th March 2020, the lockdown period of 10 weeks, and the post-lockdown term from the 20th May 2020 to the latest available observation.

To obtain an unbiased estimate of the effect of lockdown restrictions on nitrogen dioxide concentrations, we considered a set of weather variables to control for possible confounding effects: temperature (measured in degrees Celsius), rainfall (cumulative millimetres), relative humidity (%), and the average wind speed for each quadrant of the Cartesian plane (metres per second).[2] Therefore, overall, we have considered the weekly NO_2 concentrations and eight meteorological variables for each station.

5 Statistical Modelling

The lockdown effect has been estimated for each station using an autoregressive model with covariates, namely ARX model [5]. In particular, we used a steady-state representation of the ARX(1) model, as described by Fassò [13]. The dependent variable is the weekly concentration of NO_2 observed at ground level. The covariates list includes the above seven meteorological variables, a linear trend component, and the dummy for the lockdown period. Given the natural seasonality of the phenomena under analysis, we decided to consider the interactions between the meteorological

[2] We computed four variables measuring the average wind speed blowing from North-East, South-East, South-West, and North-West.

variables and the four climatic seasons' dummies. Instead, the linear trend is treated as a non-seasonal component. In this specific case, we considered spring to be from March to May, summer from June to August, autumn from September to November, and winter from December to February. Hence, the full model considers one lagged term and 30 covariates, that is 28 seasonal weather variables, the linear trend and the lockdown dummy.

Let $y(s_i, t)$ be the weekly observation of NO_2 concentration from the station located in $s_i, i = 1, 2, \ldots, 74$ and time $t = 1, \ldots, 289$ (weeks) and let $\boldsymbol{X_t}$ be the vector of weekly seasonal weather covariates, T_t be the linear trend component, and let L_t be the lockdown dummy variable for station i at time t. The model equation for each location $i = 1, 2, \ldots, 74$ is the following:

$$y_t = \alpha + \beta y_{t-1} + \nu T_t + \gamma L_t + \boldsymbol{\theta X_t} + \varepsilon_t, \tag{1}$$

where ε_t is a Gaussian random noise, γ defines the NO_2 variation at the corresponding station i due to the lockdown, β represents the lag-1 autoregressive coefficient for pollution concentrations, and ν is the linear trend parameter.

In order to consider the best set of covariates able to describe the local meteorology, we performed a variable selection by implementing the Least Absolute Shrinkage and Selection Operator, or LASSO estimator [32], trained with a 20-fold cross-validation setup. For each station, the optimal model has been selected using the *one standard error empirical rule* [16, 17], that is we selected the model with largest penalisation parameter (λ) value such that the MSE is within one standard error of the minimum MSE. This approach ensures that the most parsimonious model whose error is no more than one standard error above the best model's error is selected while considering the randomness generated by the out-of-sample randomisation used to construct LASSO. All the weather covariates, the linear trend, the autoregressive term and the lockdown dummy have been included in the LASSO algorithm. The inclusion of the dummy among the LASSO inputs provides a handy tool to understand the effective impact of restriction measures on oxide concentrations. If the LASSO includes the dummy in the list of relevant variables, this would indicate a great relevance of lockdown restrictions in explaining the concentrations pattern in the weeks of interest; otherwise, its exclusion would be a signal of a modest and negligible variation during the shutdown period.

The model parameters identified by the LASSO algorithm are then re-estimated using a maximum likelihood approach under the hypothesis of Gaussian distribution of the errors. It is well-known that the LASSO introduces a bias in the estimates of the regression parameters to reduce the variance. This bias can be reabsorbed by applying the OLS or ML estimators to estimate the regression parameters of the variables selected by the algorithm. As discussed by Belloni and Chernozhukov [3], this approach, namely OLS post-LASSO estimator, performs at least similarly as the LASSO regression in terms of convergence rate and can achieve a smaller bias for the estimated parameters, even when the algorithm fails in selecting the actual variables. Finally, the post-selection statistical inference has been made by applying the delta method to provide the approximate lockdown impact standard errors.

As stated in [13], the AR(1) dynamics imply that the scalar steady-state impact on weekly NO_2 concentrations is given by

$$\hat{\delta} = \frac{\hat{\gamma}}{1 - \hat{\beta}}. \tag{2}$$

Moreover, ignoring the uncertainty of the pre-intervention estimation of β and applying the delta method to the estimated parameters, it is possible to approximate the variance for δ as follows:

$$VAR(\hat{\delta}) \cong \frac{VAR(\hat{\gamma})}{(1 - \hat{\beta})^2}. \tag{3}$$

The parameter standard errors were simply obtained by calculating the square root of the formula of the above variance.

To correctly assess the estimated coefficients, particularly the one associated with the impact of lockdown, it is necessary to check the whiteness of the regression residuals. In particular, we point out that the variance of the parameters depends on residuals autocorrelations. Indeed, serial correlation generates bias effects on the estimated variances of the coefficients and consequently on the respective p-values and confidence intervals [15]. This necessarily leads to erroneous assessments of the significance of the estimates. Thus, we assessed the goodness of the estimated models by performing several misspecification diagnostics on the regression residuals. In particular, each residuals time series has been checked for serial correlation by analysing the sample ACF function and by using the Ljung-Box test [23]. We also tested for non-normality through the Jarque-Bera test [20].

Several practical and contextual reasons can be argued to justify the choice of the ARX model with a predefined number of lags to model a complex phenomenon such as atmospheric NO_2 concentrations. Firstly, recall that the considered observations have a weekly frequency, allowing the serial correlations to vanish after a few lags. Besides, we want to favour statistical modelling that is as simple but effective as possible, so that the empirical results are unbiased and meaningful. In particular, we recall that in an ARX setting, the effect of each covariate depends both on the estimated coefficient and on the autoregressive dynamics (see Eq. (1)). Thus, fixing a lag order common to all the stations, we make it possible to quantify the local impact of lockdown limitations on NO_2 concentrations, while keeping a standard temporal dependency structure across the region and preserving simplicity and ease interpretation of the estimates. In the end, recall that airborne pollution is a natural phenomenon mainly driven by atmospheric conditions, such as the air temperature and the wind. Thus, by including a suitable set of weather covariates in the regression models, it is possible to reasonably model the seasonality and persistence of the series without requiring more sophisticated and complex models. The residuals diagnostics presented in Sect. 6.3 validate the previous statements. The residuals are in many of the stations non-autocorrelated, and at those stations where some remaining autocorrelation persists, it always shows very small values that do not bias the coefficient estimates and their significance.

6 Model Fitting and Selection

In this section, we discuss in detail the results regarding the model selection performed through the LASSO algorithm and the OLS post-LASSO estimated coefficients obtained by re-estimating the parameters of the optimal model for each station.

6.1 LASSO Performances

To understand how the LASSO algorithm identified the optimal models is primarily necessary to assess which penalty values, i.e. the λ parameter, minimise the cross-validated mean square errors. Recall that we identified the optimal λ using the one standard error empirical rule that is we selected the model with the largest penalisation parameter such that the MSE is within one standard error of the minimum MSE. Figure 6 shows the scatterplot of the one-SE λ and the corresponding cross-validated MSE. For each combination of λ and MSE, it is

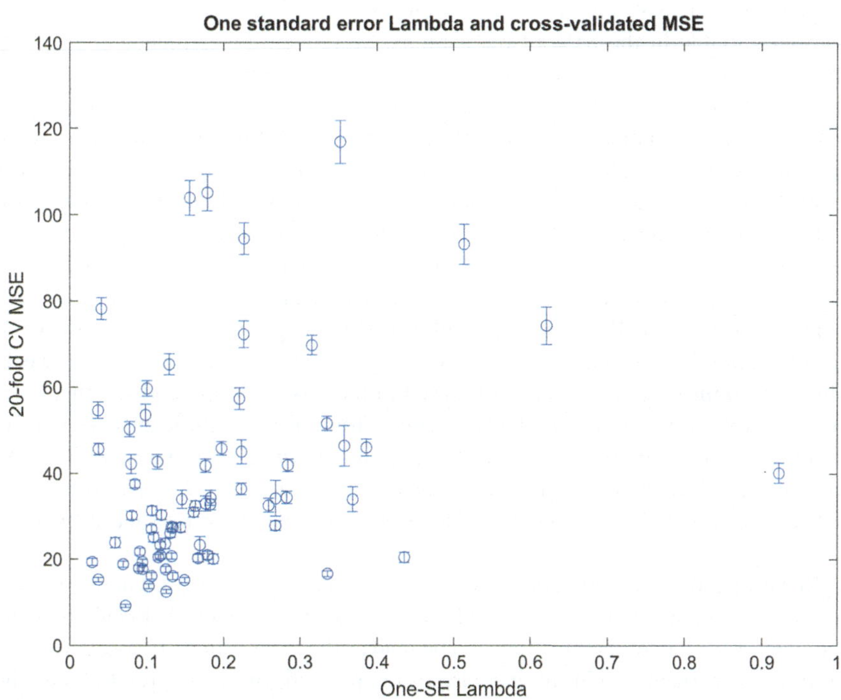

Fig. 6 Selected 1-SE λ and corresponding MSE. Error bars represent the cross-validated standard errors of each penalisation parameter

also reported in an error bar the estimated uncertainty. Recall that the LASSO solution corresponding to a null value of the penalisation parameter is equivalent to the OLS estimator solution. Observing the horizontal axis one can notice that the optimal penalisation values are always lower than one and largely located around 0. Considering the full sample, the largest part reports a λ value below 0.20. It means that the model selection procedure provided estimates that were not very penalised and relatively close to the OLS solution. Hence, the models will contain a large number of relevant covariates. The cross-validated MSE are concentrated between 10 and 30, while the largest mean squared errors are more dispersed. Observing the variability, it can also be noticed that lower λs correspond lower uncertainty, while to greater penalisation value the uncertainty increases.

6.2 Meteorology and Long-Run Trend

In aggregate, the estimated models show that some peculiar features can characterise air quality in the region in terms of long-term trends and meteorology. As reported by Table 1, both the linear trend and the autoregressive term play a key role in explaining the air quality variability.

The autoregressive coefficient has been selected by the LASSO algorithm as a relevant variable in the 85% of the stations. Moreover, the estimated coefficients are statistically significant and with a positive sign, showing how concentrations are particularly persistent even after several days. The estimated variability is low (standard deviation is around 0.10 $\mu g/m^3$), meaning that the same serial dependence structure characterises almost all the stations. Furthermore, such a large majority of significant values, combined with the low serial correlation identified in the residuals, indicate a successful specification of the model regarding NO_2 concentrations' temporal evolution. Also, the trend appears to be a relevant factor. In the 43% of the stations (32 out of 74), the linear trend coefficient has been estimated as statistically significant with a negative sign, whereas it is positive or not significant in some cases. In average, the coefficient is -0.60 $\mu g/m^3$ per week. This fact confirms what we have observed in the city of Milan and in other parts of the region: in the long-run, the NO_2 concentrations are significantly decreasing almost all over the region, with some rare exceptions where the trend is opposite. As we will see below, stations with null or positive trends are often isolated from urban centres and monitor areas where oxides in the atmosphere are low and difficult to reverse.

Regarding the local meteorology, Table 2 summarises the estimated seasonal coefficients for each of the seven climate variables considered. Recall that the coefficients for the temperature have to be read as the change in NO_2 concentrations (in $\mu g/m^3$) associated with an increase in temperature of one degree Celsius; the coefficients for rainfall as the change due to an increase of one mm in rainfall fall; the coefficients for wind as the change due to one m/s increase in wind speed, and those for humidity as the change in NO_2 due to a 1% increase in

Table 1 Summary statistics for the estimated linear trends and autoregressive coefficients

	Negative sign		Positive sign		Null				
	Signif.	Not signif.	Signif.	Not signif.	coefficients	Min	Mean	Max	StdDev
	%	%	%	%	%	$\mu g/m^3$	$\mu g/m^3$	$\mu g/m^3$	$\mu g/m^3$
AR (β)	0	0	85.1	0	14.9	0.30	0.50	0.80	0.10
Trend (ν)	43.2	24.3	2.7	5.4	24.3	−1.9	−0.60	1.50	0.60

Note: we considered as statistically significant the coefficients showing p-values below 5%. Percentages are calculated with respect to the total number of involved stations, that is 74.

The column *Null coefficients* reports the percentage of stations with null estimates, that is those stations in which the LASSO did not selected the covariate and set up to zero by the ML post-LASSO estimator.

Table 2 Summary statistics for the coefficients estimates of the weather covariates

	Negative sign		Positive sign		Null coefficients	Min	Mean	Max	StdDev
	Signif.	Not signif.	Signif.	Not signif.					
	%	%	%	%	%	μg/m^3	μg/m^3	μg/m^3	μg/m^3
Winter									
Temperature	71.6	9.5	1.4	2.7	14.9	−1.8	−0.9	0.6	0.5
Humidity	14.9	25.7	13.5	33.8	12.2	−0.5	−0	0.2	0.1
Wind speed NE	27	12.2	1.4	18.9	40.5	−102.6	−6.7	10.4	17.5
Wind speed SE	43.2	12.2	1.4	8.1	35.1	−39.9	−10.6	19.7	10.9
Wind speed SW	10.8	12.2	5.4	29.7	41.9	−33	−1.2	20.4	9.3
Wind speed NW	41.9	9.5	2.7	9.5	36.5	−19.9	−5	25.9	7.2
Rainfall	47.3	32.4	0	4.1	16.2	−43.7	−16.9	8.9	11.7
Spring									
Temperature	70.3	16.2	0	0	13.5	−1.1	−0.5	−0.1	0.2
Humidity	24.3	14.9	0	0	60.8	−0.5	−0.1	−0	0.1
Wind speed NE	35.1	36.5	0	0	28.4	−41.2	−6.7	−0.8	6.8
Wind speed SE	45.9	33.8	0	0	20.3	−20.2	−7.1	−0.8	5.1
Wind speed SW	32.4	28.4	0	1.4	37.8	−41.7	−8	1.1	8.4
Wind speed NW	47.3	31.1	0	0	21.6	−12.7	−5.7	−1	3
Rainfall	9.5	67.6	0	8.1	14.9	−16.5	−4.7	12.9	4.8

(continued)

Table 2 (continued)

	Negative sign		Positive sign		Null				
	Signif.	Not signif.	Signif.	Not signif.	coefficients	Min	Mean	Max	StdDev
	%	%	%	%	%	μg/m^3	μg/m^3	μg/m^3	μg/m^3
Summer									
Temperature	20.3	27	0	1.4	51.4	−1	−0.4	0	0.3
Humidity	63.5	27	0	0	9.5	−0.6	−0.2	−0	0.1
Wind speed NE	21.6	59.5	0	0	18.9	−24.4	−5.9	−0.9	4.7
Wind speed SE	31.1	50	1.4	0	17.6	−25.2	−5.8	17.2	5.4
Wind speed SW	24.3	40.5	0	0	35.1	−20.4	−6.4	−0.8	4.8
Wind speed NW	21.6	51.4	0	0	27	−16.3	−5.9	−0.7	3.4
Rainfall	1.4	44.6	1.4	10.8	41.9	−17.5	−1.9	8.6	4.1
Autumn									
Temperature	75.7	6.8	0	0	17.6	−1.2	−0.6	−0.1	0.2
Humidity	21.6	0	1.4	5.4	71.6	−0.5	−0.2	0.1	0.2
Wind speed NE	24.3	32.4	1.4	1.4	40.5	−51.7	−7.5	27.7	10.6
Wind speed SE	36.5	32.4	2.7	2.7	25.7	−31.8	−7.4	20.2	7.1
Wind speed SW	9.5	21.6	1.4	5.4	62.2	−33.9	−6	11.6	9.6
Wind speed qNW	14.9	13.5	1.4	12.2	58.1	−17.8	−5.1	7.1	6.9
Rainfall	18.9	62.2	0	5.4	13.5	−13	−5.5	13.8	4.3

Note: we considered as statistically significant the ML post-LASSO coefficients showing p-values below 5%. Percentages are calculated with respect to the 74 involved stations.
The column *Null coefficients* reports the percentage of stations with null estimates, that is those stations in which the LASSO did not selected the covariate and set up to zero by the ML post-LASSO estimator.

relative humidity. The seasonality of the variables of interest is well defined by the alternation of the estimated signs and their respective significance. For example, during spring and summer the coefficients always assume negative and often significant values, indicating climatic factors' strength in reducing concentrations during warm periods. Temperature and wind remain throughout the year the most important factors in reducing the levels of oxide concentrations. Except for summer, the temperature is selected and estimated as negative and significant in 70% of the stations, while the wind is estimated as significant and negative in 30 or 40% of the models. Although the wind in Lombardy is generally very weak, winds blowing from the East and North are of particular importance. Humidity is often discarded, excepting during summer, where it contributes significantly to reducing the concentrations. On the other hand, cumulative rainfall is highlighted as crucial for reducing concentrations especially during the winter.

6.3 Models Fitting and Diagnostic Checks

To assess the goodness of the estimates and model fitting, we performed an analysis of multiple goodness-of-fit indicators. For each station, we evaluated the in-sample adjusted R-squared (R2) index, the residuals root mean squared error (RMSE), and the corrected Akaike Information Criterion (AICc). Considering the 74 stations, the models fitting can be considered satisfactory. Indeed, most of the models report an R^2 above 80% and an RMSE below 6 $\mu g/m^3$. The minimum goodness-of-fit value is above 65%, which can be considered an acceptable share of explained variability. The models that fit better provide also the lowest prediction error. The behaviour of the R^2 indices and prediction errors are consistent with each other, since the models with the best fitting are also associated with the lowest estimated error. The models with greater R^2 index and lower RMSE are also associated with the minimum AICc values. Given these characteristics, from a fitting point of view, it can be stated that the strategy of estimating through ML post-LASSO estimator is an adequate tool for correctly modelling NO_2 concentrations for the chosen sample.

To validate the assumptions concerning the AR(1) structure of the model in Eq. (1), we implemented several diagnostic checks on the residuals. We firstly checked for the residual serial correlations both using graphical and analytical methods. The sample ACF and the estimated autocorrelation tests reported in Fig. 7 provide favourable indications. In particular, looking at 20-lags Ljung-Box test, among the 74 stations, 55 of them had p-values above 5% and 64 report a p-value larger than 1%. Thus, at a 1% significance level, only 10 of 74 present significantly autocorrelated residuals. Moreover, the average sample ACF distribution is always below 10%, while the maximum estimated ACF is around 27%. In Sect. 5, we mentioned that the residual serial correlation induces a bias on the variances of the estimated parameters, and hence on their respective p-values, leading to misjudgments regarding the statistical significance of the coefficients. The above empirical results show that sample autocorrelations are very moderate in absolute

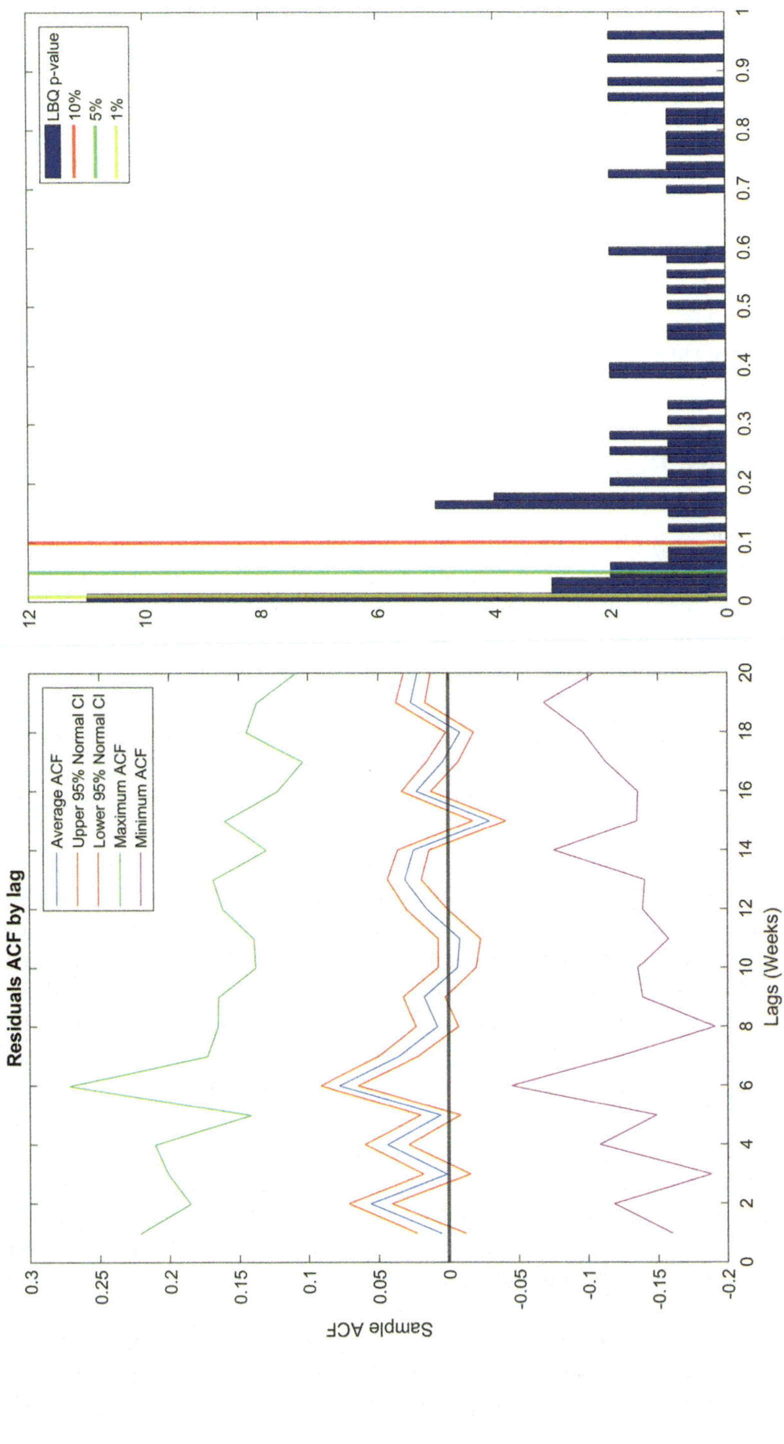

Fig. 7 Residuals ACF by lag (left panel) and empirical distribution of the p-values for the Ljung-Box test with 20 lag applied to the regression residuals (right panel)

value. Therefore, we can consider the bias as negligible and having a minimal impact on the estimated significance.

The Gaussianity assumption of the residuals, verified through the Bera-Jarque test, was confirmed for just five stations out of 74. It means that the stations with independent and normally distributed residuals are only five out of 74. However, the violation of the distributional assumption of the ML estimator can be considered a minor issue. In fact, it can be easily shown that under mild regularity conditions, the QML estimator of the parameters of the autoregressive model considered is consistent and asymptotically normal [6, 15]. Given our large sample size, the distribution of the parameters is then approximately Gaussian and their significance can be assessed with classical hypothesis tests based on likelihood. Moreover, these properties hold for simple autoregressive models [35], double autoregressive models [9], and spatial autoregressive models [22].

7 Lockdown Results

This section discusses the main empirical results related to the estimated variations due to the COVID-19 lockdown. The discussion will focus on the geographical and landscape profile of the considered stations to characterise and justify the estimates. In addition to comments on the estimates by station type and province, we will also comment on some specific stations that have an interesting relevance for an extensive understanding of the lockdown phenomenon. In particular, we will describe the cases of Meda and Schivenoglia as examples of null lockdown effect and other cases in Monza, Bergamo, and Brescia provinces regarding local meteorology and geography. The location of the specific stations is shown in Fig. 8.

The proposed statistical modelling confirms a generalised reduction of NO_2 levels consequently to the lockdown. Compared to the observed average reduction, reported in Fig. 5, the lockdown effect is mitigated by seasonal trends and meteorology. Moreover, a non-null lockdown effect was estimated for a subset of stations among those included in the sample. In fact, the LASSO algorithm selected the lockdown dummy among the covariates 63 times out of 74 stations. For these locations, the lockdown variable can be considered an important factor. For the remaining eleven stations where the dummy was discarded, the lockdown impact on NO_2 concentrations has to be considered null. The reasons for this result are many and will be explored more deeply in the following paragraphs discussing the different types of stations and their geographical distribution. From the statistical modelling perspective, a first intuitive explanation can be ascribed to the presence of external factors other than the lockdown, either embedded in the model or unknown, able to better explain the observed reductions. First of all, the specific local weather conditions. Table 3 reports both the pre-post lockdown observed average reduction and the model-based effects, aggregated by station type.

Regarding the stations where the lockdown dummy was estimated via MLE, the estimates of the coefficients were largely statistically significant. Indeed,

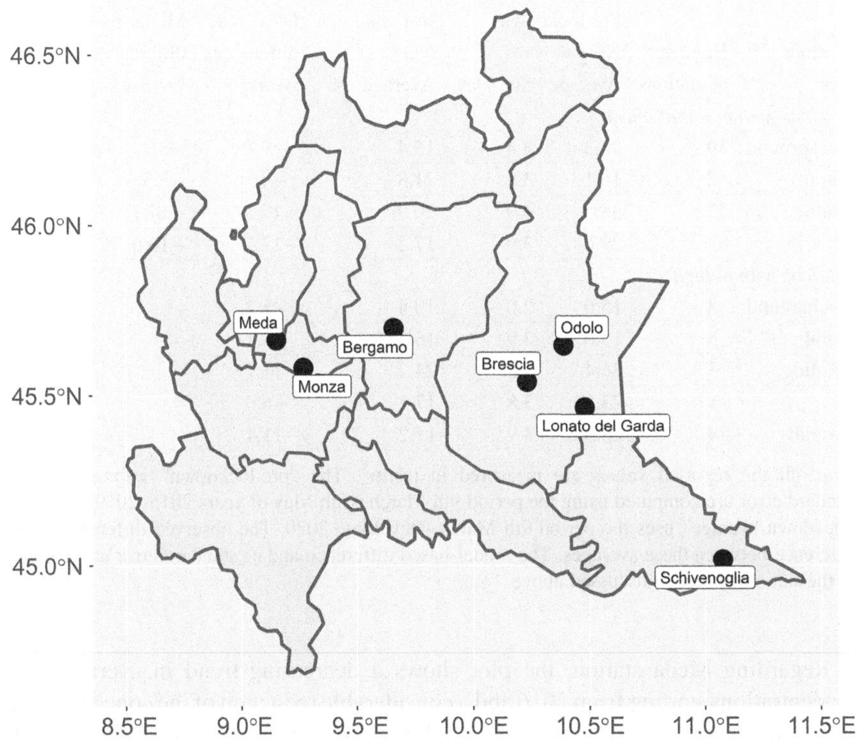

Fig. 8 Geo-location of some sites of interest for the COVID-19 lockdown in Lombardy

considering the alternative hypothesis of a statistically significant negative variation, in 47 over 63 cases, the coefficients were significant at 5%, while 35 of them were statistically significant at a 1% significance level.

To investigate which factors may have led to some null lockdown coefficients, we now provide two illustrative examples. Both examples are represented in Fig. 9, which depicts the time series of the NO_2 concentrations at the Meda station in Monza-Brianza province (upper panel) and Schivenoglia (lower panel), located in Mantova province. In both plots, the blue marks represent the weekly observed concentrations between March and May in 2015–2019, while the red marks are the weekly observed concentrations between March and May 2020.

Although the lockdown impact has been estimated as zero at both locations, the historical evolution of the concentrations and the surrounding anthropogenic context are widely different. Meda control unit is classified as a traffic sensor and is located in a residential area far from the major high traffic roads. Schivenoglia is a rural station surrounded by many kilometres of farmland and livestock. Several secondary rivers and canals flow near its location.

Table 3 Lockdown variations of NO_2 concentration classified by station type

Stations type	Number of stations	Pre-lockdown NO_2 Average	Pre-lockdown NO_2 Std. Dev.	In-Lockdown NO_2 Average	Observed differences Average	Model-based differences Estimate	Model-based differences Std. Error
Stations with non-null effect							
Background	39	25.3	3.4	15.4	−9.8	−8.7	4.2
Rural	2	16.1	3.5	11.8	−4.3	−3.5	2.8
Traffic	22	38.5	4.9	20.9	−17.5	−14.7	5.1
	63	29.6	3.9	17.2	−12.3	−10.6	4.5
Stations with null effect							
Background	3	15.6	2.0	10.4	−5.2	–	–
Rural	5	19.6	2.9	16.5	−3.1	–	–
Traffic	3	36.4	7.0	24.7	−11.7	–	–
	11	23.1	3.8	17.1	−6.0	–	–
Overall	74	28.6	3.9	17.2	−11.4		

Note: all the reported values are measured in $\mu g/m^3$. The "pre-lockdown" average and the standard error are computed using the period 9th March–18th May of years 2015–2019, while the "lockdown average" uses the period 9th March–18th May, 2020. The observed difference is the difference between these averages. The model-based difference and its standard error are obtained by the statistical model discussed above.

Regarding Meda station, the plot shows a decreasing trend in average NO_2 concentrations starting from 2016 and a considerable reduction of the concentrations during the winter season in 2019 and 2020. Between 2015 and 2018, at that location were recorded maximum winter peaks approximately from $80\,\mu/g^3$ to $110\,\mu/g^3$, while in the same period in 2019 and 2020, the observed values hovered around $60\,\mu/g^3$. The concentrations recorded in the 2020 lockdown period are lower and less volatile than those observed in the previous 5-year period. However, the variations do not appear to be large-scale. The abrupt drop in NO_2 concentrations between 2018 and 2019 leads to thinking about possible changes in the viability around the station or at least a drastic change in the surrounding emission sources. The downward trend (common throughout the region) and a possible change in the road network around the station have reduced the concentrations considerably, mitigating the positive effects of the lockdown on air quality. The case of Meda can be considered as a suitable example of air quality improvement due to viability changes.

On the other hand, the second plot depicts a different story about the concentrations in rural areas. NO_2 concentrations in Schivenoglia are always very low but show a marked seasonality. In winter there are always maximum values around $40\,\mu/g^3$, while in summer, the concentrations reach approximately $5\,\mu/g^3$. The chart also shows that the concentrations are always decreasing between March and May and with a slight variability. Despite the lockdown restrictions, 2020 is no exception to the above.

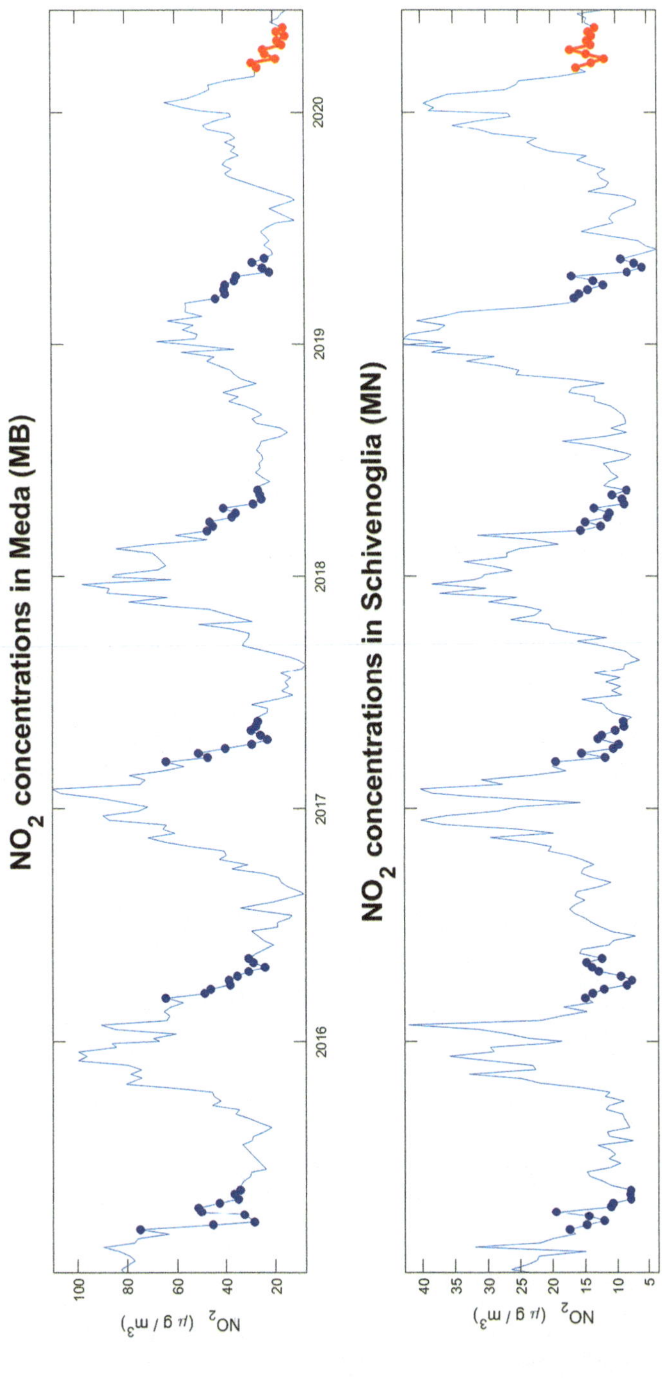

Fig. 9 NO_2 concentrations from 2015 to 2020 at Meda (MB) ground station (upper panel) and NO_2 concentrations from 2015 to 2020 at Schivenoglia (MN) ground station (lower panel). Blue marks are the weekly observed concentrations between March and May in 2015–2019. Red marks are the weekly observed concentrations between March and May in 2020

Moreover, more compact values close to $15\,\mu/g^3$ are observed. The values observed during the lockdown seem even higher than in the past. There is no downward trend, but the scale of values and the variability are much lower than in the previous case. The effect of the lockdown is negligible.

7.1 Evaluation of the Lockdown Effect Based on Area Type

If we simultaneously analyse Table 3 and Fig. 10, which reports the geo-location of the control units dividing the non-null and the null lockdown impact estimates, it is interesting to note that among the stations with null lockdown effect, five are classified as rural, and they are mainly located in the southern plain area. The non-null effects are evenly distributed on the map. This fact is crucial in understanding how the lockdown restrictions acted on pollution concentrations. There was a considerable difference between the variations of traffic, background, and rural stations in both observed and estimated reductions. For traffic control units, the observed and estimated average reductions were $-17.5\,\mu/g^3$ and $-14.7\,\mu/g^3(-38.2\%)$, respectively. At rural sensors, the estimated variations amounted to $-3.5\,\mu/g^3(-21.7\%)$. Estimated variations at background stations stand in the middle, with an average reduction of $-8.7\,\mu g/m^3(-34.4\%)$. Also, the average variability associated with the estimates, i.e. the estimated standard errors, follows the previous order. Indeed, the highest variability has been estimated for traffic sensors ($5.1\,\mu/g^3$), while the lowest is associated with the rural stations ($2.8\,\mu/g^3$). These facts can be primarily attributed to the large differences in the counting of stations for each type (39 background, 22 traffic, and two rural), and secondly to the spatial heterogeneity

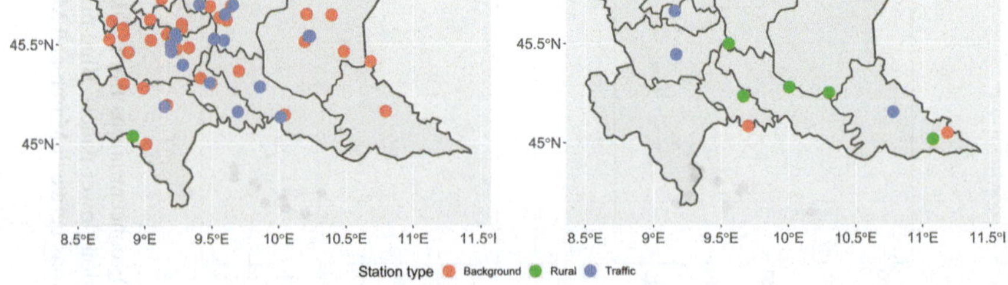

Fig. 10 Geo-location of air quality stations with relevant lockdown effect (left panel) and geo-location of stations with null lockdown effect (right panel)

generated by the geographical location of the stations that could capture different traffic intensities, for traffic stations, for industrial concentrations, and background stations.

As already mentioned in Sect. 3, which describes the air quality conditions in Lombardy, geographical and morphological conformation plays a fundamental role in determining variations due to lockdown. The prevalence of rural stations with null lockdown impact, and the remarkable differences in estimated effects by station types, provided strong evidence of how the limitations to traffic, human mobility, and productive activities worked decisively and sharply in dense urban areas and more lightly in agricultural areas. The former are subject to high levels of oxides (NO_X and NO_2), carbon (CO and CO_2), and sulphur dioxide (SO_2), produced by industry and road transport. At the same time, the latter is characterised by emissions of pollutants produced mainly by agricultural and breeding activities, i.e. ammonia (NH_3), methane (CH_4), and fine particulates ($PM_{2.5}$). See, for example, the annual reports from INEMAR Lombardia [19] and the European Environmental Agency [12] on pollution sources and abatement policies in Lombardy. All these facts are consistent with the analyses on movements and mobility during COVID-19 in Lombardy by Finazzi and Fassò [14], which reported reductions of up to 65% in human movements all over Italy that led to a natural fall in road traffic.

Figure 11 shows the empirical distribution of estimated lockdown effects for the 63 units by the type of station. Estimates in traffic sites show a high variability range, ranging from approximately $-5\ \mu g/m^3$ to $-27\ \mu g/m^3$, while in background sites there were variations between $-1\ \mu g/m^3$ and $-17\ \mu g/m^3$. The figure also highlights that all the stations registered a negative variation, supporting the hypothesis of a generalised improvement of air quality in Lombardy.

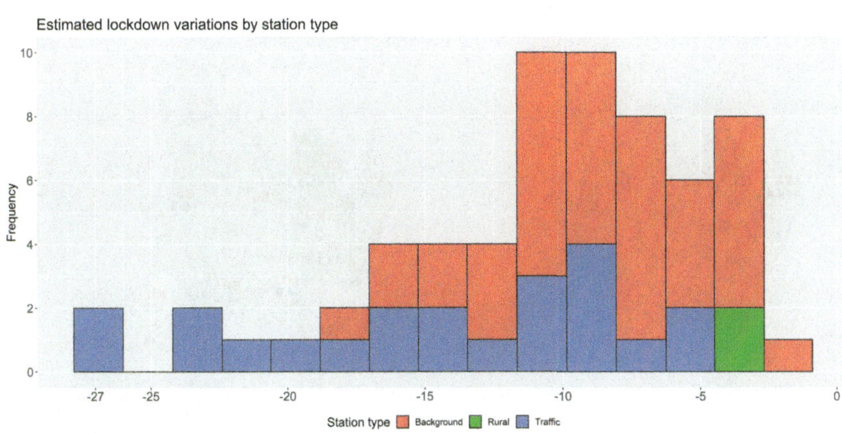

Fig. 11 Frequency distribution of the model-based lockdown variations, in $\mu g/m^3$, by station type

7.2 Geographical Distribution of the Lockdown Effect

To give a synthetic representation of the territorial heterogeneity of the effect of lockdown measures on oxide concentrations, we aggregated the estimated coefficients by province. Observed and estimated lockdown impacts by province are reported in Fig. 12.

The observed average variations are computed by averaging the observed differences of NO_2 previously, and during the 2020 lockdown, whereas to compute the estimated provincial variations, we considered as null the lockdown impact for all the stations in which the LASSO algorithm did not include the dummy variable among the final dataset. See Fig. 10 for the geo-location of these stations. The figure's left panel shows the observed average reductions by province, which are compared with the average provincial reductions estimated by our model, reported in the right panel. The two plots are in line with what has been said above regarding the differences by station types and the gap between rural and urban areas. In almost all the provinces, the estimated impacts are lower than those observed, meaning that, according to our model, the estimated lockdown effect is mitigated by seasonal trends and the local meteorology.

The most remarkable reductions were observed and estimated in the Milan metropolitan area, which is notoriously strongly urbanised, followed by the province of Como. In the latter case, the estimated NO_2 reduction was $-15.8\,\mu g/m^3 (-55.9\%)$. A relevant factor for air quality improvement in the Como area could have been the interruption of cross-border traffic to and from Switzerland. In 2018, about 18 thousands vehicles per month crossed the border at

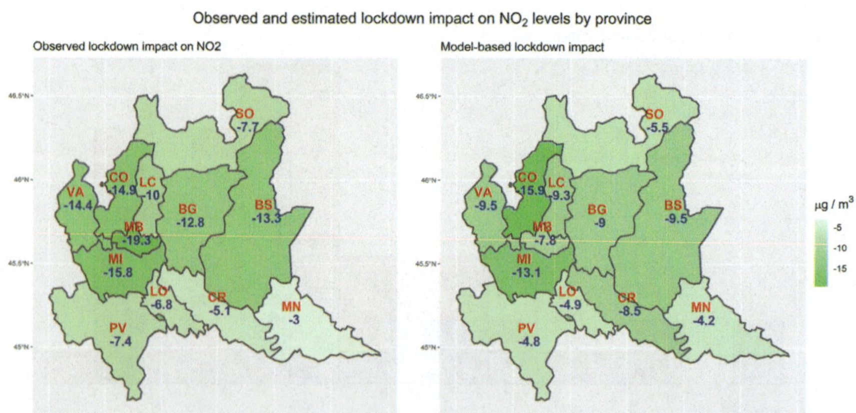

Fig. 12 Variation of NO_2 levels by province. Left panel: Observed average variation computed as the raw difference for the period 9th March–18th May in 2020 and 2015–2019. Right panel: model-based variation for the period 9th March–18th May 2020. Provinces: Varese (VA), Como (CO), Lecco (LC), Sondrio (SO), Bergamo (BG), Brescia (BS), Milano (MI), Monza e Brianza (MB), Pavia (PV), Lodi (LO), Cremona (CR) and Mantova (MN)

Chiasso custom [33], and about 70 thousand Italian cross-border commuters were recorded [31]. According to estimates by the Swiss statistical office, in April 2020, there was an average daily decrease in the number of vehicles entering and leaving the Chiasso border of 79% compared to April 2019, while in May 2020 the decrease was 66% compared to the previous year. The observed reductions are mainly due to the breakdown in the number of transits of commercial or industrial vehicles (−81% in April and −66% in May).

The estimated impact and the observed variation in Milan are quite similar. The observed reduction was $-15.8\,\mu g/m^3$ and the estimated variation amounted to $-13.1\,\mu g/m^3(-36.1\%)$. This result is consistent with the estimates of the provincial agency for the mobility of Milan (AMAT), which reported reductions in road traffic up to 77% (private vehicles) and 66% (commercial vehicles) in the Milan metropolitan area during the lockdown period [1].

Interesting results were obtained for the Monza (MB), Bergamo (BG) and Brescia (BS) provinces. The three provinces have large, densely populated and congested urban centres, especially the provincial capital cities, but they present some peculiarities that can be argued to explain the empirical discrepancies. Regarding Monza, there exists a substantial gap between the observed average variation and the estimated reduction. Indeed, the province passes from an observed reduction of $-19.3\,\mu g/m^3$ to an estimate of $-7.8\,\mu g/m^3(-18.1\%)$. On the Monza territory are located only three monitoring stations, one urban traffic station and two background stations. Both the background stations are located within the Monza city borders, while the urban ground unit is located in Meda. The LASSO model selection algorithm provided a non-null estimate for the background units, as it did not select the dummy at the urban site. The model-based estimate has been fixed at zero despite the observed variation at the traffic station $-17.2\,\mu g/m^3(-46.2\%)$. In Sect. 7, we discussed the case study of Meda, claiming that it can be taken as an excellent example of air quality improvement due to changes in emission sources. Local meteorology is also a contributing factor. The ML post-LASSO estimator provided negative and statistically significant coefficients for the temperature during spring, winter and autumn, and a significant negative trend of $-1.09\,\mu g/m^3$ per week. Rainfall and wind blowing from North-East contribute significantly to reducing concentrations during the winter. Similar results, in particular, regarding the decreasing trend and the cleaning effect of rainfall and wind, hold for both the background stations in Monza city. At the two background sites, the estimated lockdown impact is negative but not statistically significant. In conclusion, regarding the specific case of the province of Monza, it is reasonable to think that the observed reduction can be explained by meteorological factors included in the model, which can have affected the air quality more than the lockdown restrictions.

Concerning the cases of Bergamo and Brescia, some geographical and morphological reasons can be addressed. The estimated impacts are smaller than the observed reductions, although with a smaller gap than in the case of Monza. In Bergamo we estimated a variation of $-9\,\mu g/m^3(-32.1\%)$, while in Brescia $-9.5\,\mu g/m^3(-32.7\%)$. The gaps can be attributed to the strongly heterogeneous physical conformations of both areas. Both provinces are characterised by a northern mountain area, therefore, sparsely inhabited and with high environmental quality, by the flat area in the south with high industrialised districts and medium-large sized lakes. According to the stations' typology, both the observed and estimated reductions may depend on the geomorphological structure of the territory and its reliefs. Considering the cities of Brescia and Bergamo, the urban traffic sites estimated variations of $-27.2\,\mu g/m^3$ and $-22.5\,\mu g/m^3$, respectively, while near less urbanised areas of the provinces the estimated reductions are always lower (in absolute value) than $14\,\mu g/m^3$ and often not statistically significant. To have a better insight into the geographical effect, we consider three stations in the province of Brescia characterised by different territories. The urban station of Brescia city, the urban background station of Lonato del Garda overlooking Lake Garda and the suburban background station of Odolo at 350 meters altitude. At Brescia site, the decisive weather variables are the wind blowing from South-East, which reduces the concentrations significantly during all the seasons ($-12.2\,\mu g/m^3$ in winter and $-11.1\,\mu g/m^3$ in summer), and the summer rainfall ($-17.5\,\mu g/m^3$). The trend is negative ($-1.2\,\mu g/m^3$ per week) and statistically significant. In Odolo, the estimated lockdown reduction is $7.8\,\mu g/m^3$, but it is not statistically significant. The trend is absent, and except for the winter rainfall and the summer temperature, all the weather covariates are not significant. In Lonato, the estimated variation is $-4.8\,\mu g/m^3$. The negative trend effect is weak but significant, while the main reducing drivers are the winds coming from the Garda Lake, i.e. from North and East (around $-8\,\mu g/m^3$ during summer and $-5.5\,\mu g/m^3$ in spring).

The results associated with the southern provinces—that is, Pavia (PV), Lodi (LO), and Cremona (CR)—were in line with expectations: given the local agricultural and rural context, the observed and estimated reductions of NO_2 levels were very modest. In Pavia, the estimated reduction of NO_2 is about $4.8\,\mu g/m^3(-20.6\%)$, in Lodi it is $4.9\,\mu g/m^3(-19.7\%)$, while in Cremona it is $4.2\,\mu g/m^3(-24.4\%)$. The only exception is represented by Mantova (MN), where the estimated NO_2 reduction reached 47.5%.

As shown in Fig. 5, the pre-lockdown NO_2 concentrations in the southern provinces during spring were considerably below the regional average and those of the other provinces. Thus, starting from relatively low levels, the observed reductions are small. In Cremona and Mantova, the estimated lockdown impact appeared even stronger than the observed raw differences. This last fact can be explained by the use of meteorological variables to adjust the estimates. The inclusion of meteorological covariates in the models allows obtaining unbiased estimates of the impact of restrictions on NO_2 concentrations while controlling for possible climatic variations.

It is plausible that the meteorology hid part of the reduction caused by the restrictions during the lockdown period. Indeed, thanks to the available data, it is possible to observe that during the lockdown period (March–May 2020) in both provinces the observed average temperature is about 0.30 to 0.40 Celsius higher than the average in the same period between 2015 and 2019. Moreover, the rainfall occurring in 2020 is on average lower than in the period 2015–2019, as well as the wind speed coming from the North-East and North-West, i.e. from the Alps.

These facts are compatible with a reduction in the atmospheric cleaning capacity in the area. Therefore, the higher temperatures and the failure to recycle the air may have increased the quantities of oxides in the atmosphere, compensating for the considerable reduction generated by the restrictions. In Sect. 7, we introduced the case of the rural ground station in Schivenoglia as an example of null lockdown impact. The associated plot showed that the NO_2 concentrations in the area were already deficient before the lockdown. From the estimated model at that station, we deduce that the meteorology provides a limited effect in reducing concentrations. Indeed, most of the coefficients are shrunk toward low values and are often not statistically significant. In particular, the wind blowing from all the directions are associated with coefficients around $-1\,\mu g/m^3$ to $-5\,\mu g/m^3$. As stated in Sect. 6.2 commenting the estimated coefficients associated with the weather covariates, the wind is often a fundamental factor in explaining the reductions. The only significant effect is due to the winter rainfall, which reduces the concentrations of $-17.3\,\mu g/m^3$.

The previous results on oxide reduction in Lombardy were consistent with the study conducted by Agresti et al. [1], in which the impacts of vehicular traffic on weekly NO_2 concentrations in Lombardy and the metropolitan area of Milan were investigated by implementing physical-chemical models based on emission inventories. Their findings suggest that the lockdown restrictions generated reductions in NO_2 concentrations in March 2020 of 31% in Milan, 35% in Como, 27% in Bergamo, and 23% in Brescia compared to the data observed during previous years.

8 Conclusions and Future Developments

Our findings suggest that during the Italian COVID-19 lockdown, the air quality in Lombardy improved noteworthy due to the restrictions imposed by the Government. In general, all provinces experienced remarkable improvements, especially densely populated, and congested and trade areas.

The adopted time series model provided evidence of statistically significant reductions of NO_2 levels at 47 over 74 ground sites. Overall, the analysis of residuals and diagnostic checks confirmed the goodness of the models used. The estimated models report very high fitting values (often above 80%) and low prediction errors. Moreover, the residuals are weakly or not autocorrelated at many stations, and where some serial correlation persists, it is very poor. The estimates can, therefore, be considered unbiased and meaningful.

Compared to the observed average reduction, the estimated lockdown impact was mitigated by the decreasing temporal trend and local meteorology. The observed raw differences by provinces are almost everywhere larger than the estimated average impact. However, the reductions appear not to be ubiquitous all over the regional territory. The aggregation of the estimated impacts by province showed that all provinces experienced a relevant reduction. However, the effect was stronger in large urban areas located in the central industrialised belt (−36.1% in Milan and −32.7% in Brescia), while the southern territories, characterised by flat rural landscape, showed a significant but weaker reduction of NO_2 concentrations (−20.6% in Pavia and −19.7% in Lodi). Moreover, considering the type of area covered by the stations, the highest reductions in nitrogen dioxide levels were associated with traffic control units (−38.2%), while reductions in rural areas were smaller (−21.7%). These results are consistent with other studies showing that the lockdown restrictions, which affected mobility and production activities, drastically lowered traffic and, therefore, stopped the primary sources of oxide emissions.

In this paper, we considered the spatial dimension indirectly, as it has been used to characterise the results achieved in terms of estimated variations and not directly at the estimation stage. However, we recognise that spatial and spatio-temporal modelling would improve the quality of the estimates considerably. Crucial would be the spatial prediction of variations in sites where there is no direct monitoring and the mapping of the occurred variations over the whole region. This would improve the accuracy of the reduction estimates by province, which currently depends on the number of stations located within the provincial boundaries.

Acknowledgments Map data copyrighted OpenStreetMap contributors and available from https://www.openstreetmap.org

References

1. Agresti, V., Balzarini, A., Bonanno, R., Collino, E., Colzi, F., Lacavalla, Pirovano, M., Riva, G., Toppetti, A., Riva, F., Piccoli, A.: Gli effetti del lockdown sulla qualitá dell'aria a milano e in lombardia (2020). https://dossierse.it/05-2020-gli-effetti-del-lockdown-sulla-qualita-dellaria-a-milano-e-in-lombardia/
2. Baldasano, J.M.: Covid-19 lockdown effects on air quality by no2 in the cities of Barcelona and Madrid (Spain). Sci. Total Environ. **741**, 140353 (2020). ISSN 0048-9697. https://doi.org/10.1016/j.scitotenv.2020.140353. http://www.sciencedirect.com/science/article/pii/S0048969720338754
3. Belloni, A., Chernozhukov, V.: Least squares after model selection in high-dimensional sparse models. Bernoulli **19**(2), 521–547 (2013). ISSN 1350-7265
4. Berman, J.D., Ebisu, K.: Changes in U.S. air pollution during the covid-19 pandemic. Sci. Total Environ. **739**, 139864 (2020). ISSN 0048-9697. https://doi.org/10.1016/j.scitotenv.2020.139864. http://www.sciencedirect.com/science/article/pii/S0048969720333842
5. Bittanti, S.: Model Identification and Data Analysis. Wiley Online Library, New York (2019). ISBN 1119546362
6. Bollerslev, T., Wooldridge, J.M.: Quasi-maximum likelihood estimation and inference in dynamic models with time-varying covariances. Econ. Rev. **11**(2), 143–172 (1992). ISSN 0747-4938. https://doi.org/10.1080/07474939208800229
7. Cameletti, M.: The effect of corona virus lockdown on air pollution: Evidence from the city of Brescia in Lombardia region (Italy). Atmos. Environ. **239**, 117794 (2020). ISSN 1352-2310. https://doi.org/10.1016/j.atmosenv.2020.117794. http://www.sciencedirect.com/science/article/pii/S1352231020305288
8. Carugno, M., Consonni, D., Bertazzi, P.A., Biggeri, A., Baccini, M.: Temporal trends of pm10 and its impact on mortality in Lombardy, Italy. Environ. Pollut. **227**, 280–286 (2017). ISSN 0269-7491

9. Chen, M., Li, D., Ling, S.: Non-stationarity and quasi-maximum likelihood estimation on a double autoregressive model. J. Time Ser. Anal. **35**(3), 189–202 (2014). ISSN 0143-9782. https://doi.org/10.1111/jtsa.12058. https://onlinelibrary.wiley.com/doi/abs/10.1111/jtsa.12058
10. Collivignarelli, M.C., Abbà, A., Bertanza, G., Pedrazzani, R., Ricciardi, P., Carnevale Miino M.: Lockdown for covid-2019 in Milan: what are the effects on air quality? Sci. Total Environ. **732**, 139280 (2020). ISSN 0048-9697. https://doi.org/10.1016/j.scitotenv.2020.139280. http://www.sciencedirect.com/science/article/pii/S0048969720327972
11. Connerton, P., Vicente de Assunção, J., Maura de Miranda, R., Dorothée Slovic, A., José Pérez-Martínez, P., Ribeiro, H.: Air quality during covid-19 in four megacities: lessons and challenges for public health. Int. J. Environ. Res. Public Health **17**(14), 5067 (2020). ISSN 1660-4601. https://www.mdpi.com/1660-4601/17/14/5067
12. European, E., Environmental Agency: Air Quality in Europe—2019 Report 16/10/2019 (2019)
13. Fassò, A.: Statistical assessment of air quality interventions. Stoch. Env. Res. Risk A. **27**(7), 1651–1660 (2013). ISSN 1436-3240
14. Finazzi, F., Fassò, A.: The impact of the covid-19 pandemic on Italian mobility. Significance (Oxford, England) **17**(3), 17 (2020)
15. Hamilton, J.: Time Series Analysis. Princeton University Press, Princeton (1994). ISBN 9780691042893
16. Hastie, T., Tibshirani, R., Wainwright, M.: Statistical Learning with Sparsity: The Lasso and Generalizations. CRC Press, New York (2015). ISBN 1498712177
17. Hastie, T., Tibshirani, R., Friedman, J.: The Elements of Statistical Learning: Data Mining, Inference, and Prediction (Springer Series in Statistics). Springer, New York (2017)
18. Higham, J., Ramírez, C.A., Green, M., Morse, A.: UK covid-19 lockdown: 100 days of air pollution reduction? Air Qual. Atmos. Health **14**(3), 1–8 (2020). ISSN 1873-9326
19. A. INEMAR ARPA Lombardia Settore Aria. Inemar Emission Inventory 2017: Emission in Lombardy Region—Final Data—Public Revision (2020). https://www.inemar.eu/xwiki/bin/view/InemarDatiWeb/Fonti+dei+dati
20. Jarque, C.M., Bera, A.K.: A test for normality of observations and regression residuals. In: International Statistical Review/Revue Internationale de Statistique, pp. 163–172 (1987). ISSN 0306-7734
21. Kerimray, A., Baimatova, N., Ibragimova, O.P., Bukenov, B., Kenessov, B., Plotitsyn, P., Karaca, F.: Assessing air quality changes in large cities during covid-19 lockdowns: The impacts of traffic-free urban conditions in Almaty, Kazakhstan. Sci. Total Environ. **730**, 139179 (2020). ISSN 0048-9697. https://doi.org/10.1016/j.scitotenv.2020.139179. http://www.sciencedirect.com/science/article/pii/S0048969720326966
22. Lee, L.-F.: Asymptotic distributions of quasi-maximum likelihood estimators for spatial autoregressive models. Econometrica **72**(6), 1899–1925 (2004). ISSN 0012-9682. https://doi.org/10.1111/j.1468-0262.2004.00558.x. https://onlinelibrary.wiley.com/doi/abs/10.1111/j.1468-0262.2004.00558.x
23. Ljung, G.M., Box, G.E.: On a measure of lack of fit in time series models. Biometrika **65**(2), 297–303 (1978). ISSN 1464-3510
24. Maranzano, P., Fassò, A., Pelagatti, M., Mudelsee, M.: Statistical modeling of the early-stage impact of a new traffic policy in Milan, Italy. Int. J. Environ. Res. Public Health **17**(3), 1088 (2020). ISSN 1660-4601 (Electronic) 1660-4601 (Linking). https://doi.org/10.3390/ijerph17031088. https://www.mdpi.com/1660-4601/17/3/1088
25. Nakada, L.Y.K., Urban, R.C.: Covid-19 pandemic: Impacts on the air quality during the partial lockdown in são Paulo state, Brazil. Sci. Total Environ. **730**, 139087 (2020). ISSN 0048-9697. https://doi.org/10.1016/j.scitotenv.2020.139087. http://www.sciencedirect.com/science/article/pii/S0048969720326048
26. Raffaelli, K., Deserti, M., Stortini, M., Amorati, R., Vasconi, M., Giovannini, G.: Improving air quality in the Po valley, Italy: Some results by the life-IP-prepair project. Atmosphere **11**(4), 429 (2020)
27. R. Regional Statistical Yearbook: Lombardia Regional Statistical Yearbook 2017/2018 (2017)

28. Ropkins, K., Tate, J.E.: Early observations on the impact of the covid-19 lockdown on air quality trends across the UK. Sci. Total Environ. **754**, 142374 (2020). ISSN 0048-9697
29. Sharma, S., Zhang, M., Anshika, Gao, J., Zhang, H., Kota, S.H.: Effect of restricted emissions during covid-19 on air quality in India. Sci. Total Environ. **728**, 138878 (2020). ISSN 0048-9697. https://doi.org/10.1016/j.scitotenv.2020.138878. http://www.sciencedirect.com/science/article/pii/S0048969720323950
30. Singh, R.P., Chauhan, A.: Impact of lockdown on air quality in India during covid-19 pandemic. Air Qual. Atmos. Health **13**(8), 921–928 (2020). ISSN 1873-9326. https://doi.org/10.1007/s11869-020-00863-1
31. U. Switzerland Federal Statistics Office: Cross-border commuters statistics in 2018 (2019). https://www.bfs.admin.ch/bfs/it/home/statistiche/lavoro-reddito/rilevazioni/staf.assetdetail.7427558.html
32. Tibshirani, R.: Regression shrinkage and selection via the lasso. J. R. Stat. Soc. Ser. B Methodol. **58**(1), 267–288 (1996). ISSN 0035-9246
33. U. Ticino Statistics Office: Traffic in selected stations of Ticino historical data (2020). https://www3.ti.ch/DFE/DR/USTAT/allegati/tabella/T_110302_01C.xls
34. Tobías, A., Carnerero, C., Reche, C., Massagué, J., Via, M., Minguillón, M.C., Alastuey, A., Querol, X.: Changes in air quality during the lockdown in Barcelona (Spain) one month into the SARS-CoV-2 epidemic. Sci. Total Environ. **726**, 138540 (2020). ISSN 0048-9697. https://doi.org/10.1016/j.scitotenv.2020.138540. http://www.sciencedirect.com/science/article/pii/S0048969720320532
35. White, H.: Maximum likelihood estimation of misspecified models. Econometrica **50**(1), 1–25 (1982). ISSN 00129682, 14680262. https://doi.org/10.2307/1912526. http://www.jstor.org/stable/1912526
36. Zangari, S., Hill, D.T., Charette, A.T., Mirowsky, J.E.: Air quality changes in New York city during the covid-19 pandemic. Sci. Total Environ. **742**, 140496 (2020). ISSN 0048-9697. https://doi.org/10.1016/j.scitotenv.2020.140496. http://www.sciencedirect.com/science/article/pii/S0048969720340183.

Author Index

A. Steland, K.-L. Tsui (eds.), *Artificial Intelligence, Big Data and Data Science in Statistics*, https://doi.org/10.1007/978-3-031-07155-3

The manufacturer's authorised representative in the EU is Springer Nature Customer Service Centre GmbH, Europaplatz 3, 69115 Heidelberg, Germany. If you have any concerns regarding our products, please contact ProductSafety@springernature.com

Printed and bound by CPI Group (UK) Ltd, Croydon, CR0 4YY
28/09/2026
02224056-0003